Hyperbolic Problems: Theory, Numerics, Applications

Eighth International Conference in Magdeburg, February/March 2000
Volume I

Edited by

Heinrich Freistühler
Gerald Warnecke

Springer Basel AG

Editors:

Heinrich Freistühler
Max Planck Institute for Mathematics
in the Sciences
Inselstrasse 22–26
04103 Leipzig
Germany
e-mail: hfreist@mis.mpg.de

Gerald Warnecke
Institute of Analysis and Numerical Mathematics
Otto-von-Guericke-University
PSF 4120
39106 Magdeburg
Germany
e-mail: Gerald.Warnecke@Mathematik.Uni-Magdeburg.DE

2000 Mathematics Subject Classification 35LXX; 65MXX, 74B20, 76E25, 76TXX, 83F05

A CIP catalogue record for this book is available from the Library of Congress, Washington D.C., USA

Deutsche Bibliothek Cataloging-in-Publication Data

Hyperbolic problems: theory, numerics, applications : eighth international
conference in Magdeburg, February, March 2000 / ed. by Heinrich Freistühler ;
Gerald Warnecke. - Basel ; Boston ; Berlin : Birkhäuser
ISBN 978-3-0348-9537-8 ISBN 978-3-0348-8370-2 (eBook)
DOI 10.1007/978-3-0348-8370-2

Vol. 1 . – (2001)
(International series of numerical mathematics ; Vol. 140)
ISBN 978-3-0348-9537-8

Originally published by Birkhäuser Verlag in 2001
Softcover reprint of the hardcover 1st edition 2001

Printed on acid-free paper produced of chlorine-free pulp

ISBN 978-3-0348-9537-8

Dedicated to Professor Constantine Dafermos
on the occasion of his 60th birthday

Editors' Preface

The Eighth International Conference on Hyperbolic Problems – Theory, Numerics, Applications, was held in Magdeburg, Germany, from February 27 to March 3, 2000. It was attended by over 220 participants from many European countries as well as Brazil, Canada, China, Georgia, India, Israel, Japan, Taiwan, und the USA. There were 12 plenary lectures, 22 further invited talks, and around 150 contributed talks in parallel sessions as well as posters. The speakers in the parallel sessions were invited to provide a poster in order to enhance the dissemination of information.

Hyperbolic partial differential equations describe phenomena of material or wave transport in physics, biology and engineering, especially in the field of fluid mechanics. Despite considerable progress, the mathematical theory is still struggling with fundamental open problems concerning systems of such equations in multiple space dimensions. For various applications the development of accurate and efficient numerical schemes for computation is of fundamental importance.

Applications touched in these proceedings concern one-phase and multiphase fluid flow, phase transitions, shallow water dynamics, elasticity, extended thermodynamics, electromagnetism, classical and relativistic magnetohydrodynamics, cosmology. Contributions to the abstract theory of hyperbolic systems deal with viscous and relaxation approximations, front tracking and wellposedness, stability of shock profiles and multi-shock patterns, traveling fronts for transport equations. Numerically oriented articles study finite difference, finite volume, and finite element schemes, adaptive, multiresolution, and artificial dissipation methods.

The biannual series of International Conferences on Hyperbolic Problems began in 1986 with a meeting in St. Etienne, France. Further conferences took place in Aachen in Germany, Uppsala in Sweden, Taormina in Italy, Stony Brook, New York, in the USA, Hongkong in China, and Zürich in Switzerland.

The conference in Magdeburg was held under the auspices of the German priority research programme "Analysis and Numerics for Conservation Laws" (ANumE) and the European research network "Hyperbolic Conservation Laws" (HCL). ANumE is a six year (1997-2003) programme of the Deutsche Forschungsgemeinschaft (DFG, Germany's national science foundation), in which over 30 research projects in mathematics, astrophysics, and engineering have received funding. HCL, financed by the Commission of the European Union, gathered pertinent groups all over Europe during the years 1996-2000, prompting a number of novel achievements in the theory of hyperbolic systems.

At the conference, the scientific committee felt that the field is still strongly thriving and offers a multitude of interesting challenges for scientific research to warrant a continuation of the conference series. The next meeting in the series will be organized in 2002 by T. Hou and E. Tadmor at Caltech in Pasadena, USA. The conference in 2004 is planned to be held in Japan.

The scientific committee of the Magdeburg conference consisted of J. Ballmann (Germany), A. Bressan (Italy), C. Dafermos (USA/Greece), B. Engquist (USA/Sweden), M. Feistauer (Czech Republic), J. Glimm (USA), L. Hsiao (China), R. Jeltsch (Switzerland), P. Lax (USA), T.-P. Liu (USA/Taiwan), P. Marcati (Italy), D. Marchesin (Brazil), K.W. Morton (United Kingdom), B. Perthame (France), D. Serre (France), E. Tadmor (USA/Israel), A. Tveito (Norway), and the local organizers. The latter two thank the others for their much appreciated work.

These conference proceedings were refereed, each submitted paper was read by at least two experts. We would like to thank all the referees for this work.

The conference received considerable financial support. The main grants were provided by the DFG and the State of Sachsen-Anhalt, of which Magdeburg is the capital. They provided funds for invited speakers as well as special support for participants from eastern European countries. The Otto-von-Guericke-University hosted the conference, provided rooms and lecture halls as well as staff support, especially for the conference office. A considerable number of participants received support from the programmes ANumE and HCL mentioned above. Some participants were supported through money which was left from the previous Aachen conference. Further support was given by Commerzbank Magdeburg, the computer company Hewlett-Packard, the publisher Birkhäuser and Coca Cola. We are grateful to all sponsors for their generous support.

The performance of the meeting depended very much on many helpers that we had, in particular Nikolai Andrianov, Claudia Bieder, Ines Brückner, Barbara Fischbach, Libor Grigerek, Wolfram Heineken, Christiane Helzel, Ilona Hesse, Maria Lukáčová, Rüdiger Müller, Jitka Saibertova and Yousef Zahaykah. We appreciate their assistance in making conference organization a success.

Finally a great thanks to Nikolai Andrianov, Claudia Bieder, Wolfram Heineken, Stefan Kleeberg, Rüdiger Müller and Yousef Zahaykah for the considerable work they put into producing the final layout of these proceedings.

September 2001 Heinrich Freistühler and Gerald Warnecke

Local conference organizers

Contents

Volume 1

Volume 2

International Series of Numerical Mathematics
Vol. 140, © 2001 Birkhäuser Verlag Basel/Switzerland

Ghost-Fluids for the Poor: A Single Fluid Algorithm for Multifluids

Rémi Abgrall and Smadar Karni

Abstract. Computations of compressible multifluid dynamics often suffer from oscillations near material fronts. Such oscillations do not occur in computations of single fluid dynamics. The paper presents a simple single-fluid framework for the computation of compressible multifluid dynamics. The resulting algorithm produces oscillation-free computations, preserves mass and momentum, and essentially preserves the total energy.

1. Introduction

Understanding the dynamics of fluids consisting of several fluid component is of great interest in a wide range of physical flows. Examples range from studying the dynamics and instabilities of interfaces, to understanding the macroscopic behaviour of liquid suspensions and bubbly flows. This work focuses on multimaterial flows. The fluid components are assumed to be pure fluids separated by propagating material fronts. The multifluid model is solved within the shock capturing framework, which is simple to implement in multidimension, but comes at a price of smearing discontinuities. The numerical difficulties described in this paper are a direct consequence of this choice.

Early attempts to extend successful single-fluid algorithms to multifluids were frustrated by the occurence of oscillations and inaccuracies near the material fronts [9, 1]. It was soon understood that the key to successful computations is good control of the pressure field, namely the method needs to be able to recognize and respect fluid components that are in pressure equilibrium [1, 7, 8]. Within the shock-capturing framework, this cannot be achieved without some degree of sacrificing conservation [2], often only of the total energy [8, 6].

An instructive observation is that such oscillations do not occur in computations of single-fluid flows, even in diffused captured interfaces [1, 7] (see also Section 3.3 below). Indeed, several algorithms that were proposed recently [5, 6] capitalize on this property [3]. In this paper we present an extremely simple single-fluid algorithm to multifluid flow calculations, which amounts to computing *two*

Work supported in part by NSF grant #DMS 99 73291 and ONR grant #N00014-99-1-0449.

intercell flux functions across material fronts: one assuming that the fluids on either side of the interface are both of type A, say, and one assuming that they are both of type B. The intercell flux functions are effectively single-fluid fluxes, hence do not produce oscillations. The resulting scheme conserves mass and momentum, essentially conserves the total energy, and is suitable even for computations involving very strong shocks. Numerical results involving propagating interfaces and shock-interface interactions are presented in one and two space dimensions.

2. The Model

The compressible Euler equations for a two-component multimaterial flow with a single velocity and a single pressure are

$$\begin{pmatrix} \rho \\ \rho u \\ E \\ \rho\psi \end{pmatrix}_t + \begin{pmatrix} \rho u \\ \rho u^2 + p \\ u(E+p) \\ \rho u\psi \end{pmatrix}_x = 0 \tag{1}$$

with ρ, u, p and E denoting the density, velocity, pressure and total energy of the fluid mixture. Following [10], ψ is a smooth indicator function, initialized so that $\psi = 0$ at the material interface, and is positive (negative) in regions occupied fluid A (B). ψ propagates with the fluid velocity hence satisfies the transport equation $\psi_t + u\psi_x = 0$ which, combined with $1a$, gives $1d$. The material front, by definition, is given by the zero level-set of ψ at all later times. The Equation of State (EOS) then becomes

$$p(\rho,\ e,\ \psi) = \begin{cases} f_A(\rho,\ e) & \psi > 0 \\ f_B(\rho,\ e) & \psi < 0 \end{cases} \tag{2}$$

where $f_{A,B}$ are the EOS for materials A and B respectively. The EOS for a stiffened fluid is

$$\rho e = \frac{p + \gamma P_\infty}{\gamma - 1} \tag{3}$$

where γ is the specific heat ratio and setting $P_\infty = 0$ recovers the ideal gas case.

While the discussion here focuses on the level-set formulation [10], we note that several multifluid models that appeared and were extensively discussed in the literature, have the same general form and structure (see [3] for a comprehensive survey).

3. Numerical Preliminaries

Consider a contact surface separating two fluids at possibly two different temperatures, which for simplicity are assumed to be two ideal fluids with specific heat ratios $\gamma_{1,2}$. Solving (1) numerically in time, we have

$$\begin{pmatrix} \rho \\ \rho u \\ E \end{pmatrix}_j^{n+1} = \begin{pmatrix} \rho \\ \rho u \\ E \end{pmatrix}_j^{n} + \begin{pmatrix} \delta\rho \\ \delta\rho u \\ \delta E \end{pmatrix}, \quad \begin{pmatrix} \delta\rho \\ \delta\rho u \\ \delta E \end{pmatrix} = -\nu \begin{pmatrix} \Delta(\rho u) \\ \Delta(\rho u^2 + p) \\ \Delta(u(E+p)) \end{pmatrix} \tag{4}$$

with $\Delta(\,) = (\,)_{j+\frac{1}{2}} - (\,)_{j-\frac{1}{2}}$ denoting spatial variation, and $\nu = \frac{\Delta t}{\Delta x}$. The pressure is recovered using the EOS. Denoting by $\alpha = \frac{1}{\gamma-1}$ we have

$$(p+\delta p)(\alpha+\delta\alpha) = (E+\delta E) - \frac{1}{2}(\rho+\delta\rho)(u+\delta u)^2. \tag{5}$$

Across contact surfaces, $u_j^n = u_{j\pm\frac{1}{2}} = u$, $p_j^n = p_{j\pm\frac{1}{2}} = p$ and (4) reduces to

$$\begin{pmatrix} \delta\rho \\ \delta(\rho u) \\ \delta E \end{pmatrix} = -\nu u \Delta\rho \begin{pmatrix} 1 \\ u \\ \frac{1}{2}u^2 \end{pmatrix} - \nu u p \Delta\alpha \begin{pmatrix} 0 \\ 0 \\ 1 \end{pmatrix} \tag{6}$$

From (6ab) it follows that

$$\delta u = 0 \tag{7}$$

and substituting further (6) and (7) into (5) yields

$$\begin{aligned} (p+\delta p)(\alpha+\delta\alpha) &= (E+\delta E) - \tfrac{1}{2}(\rho+\delta\rho)(u+\delta u)^2 \\ &= E - \tfrac{1}{2}\rho u^2 - \nu u p \Delta(\tfrac{1}{\gamma-1}) \\ &= p\alpha - \nu u p \Delta\alpha. \end{aligned} \tag{8}$$

Several comments are worthy of noting.

(i) In the single fluid case, $\Delta\alpha = \delta\alpha = 0$, and $\delta p = 0$.

(ii) In the multimaterial case one cannot expect in general δp to be zero. Indeed, the pressure often does not remain uniform yielding pressure oscillations at material fronts.

(iii) As is readily seen, and was first noted in [2], in order for δp to vanish, one must have

$$\alpha + \delta\alpha = \alpha - \nu u \Delta\alpha. \tag{9}$$

which imposes a discretization condition on the time evolution of α, consistent with advection of $\alpha = 1/(\gamma-1)$ due to the motion of the inte rface.

(iv) While every state function, $f(\gamma)$, propagate with the fluid flow, hence satisfies a transport equation $f(\gamma)_t + uf(\gamma)_x = 0$, it is $f(\gamma) = 1/(\gamma-1)$ that needs to be transported in order to prevent oscillations.

4. The Algorithm

In the following, U and W are used to denote the vector of primitive and conservative variables respectively. Let $U_j = U_L = (\rho_l,\ u_l,\ p_l,\ \gamma_l)^T$ and $U_{j+1} = U_R =$

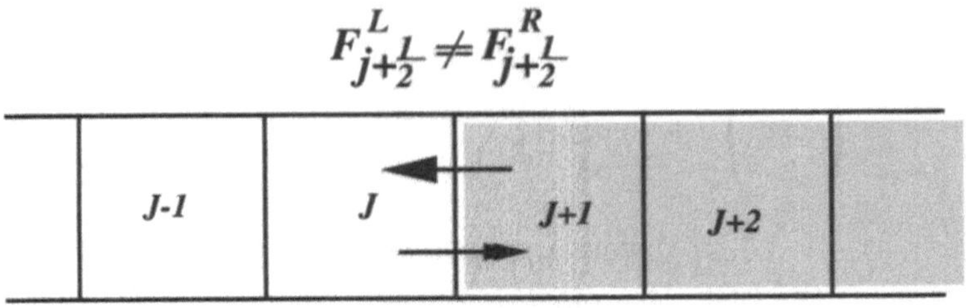

FIGURE 1. A Single Fluid Algorithm using 2 intercell flux calculations

$(\rho_r,\ u_r,\ p_r,\ \gamma_r)^T$ and assume the material interface lies between cell j and $j+1$. The intercell flux, $F_{j+\frac{1}{2}}$, depends on both γ_l and γ_r

$$F_{j+\frac{1}{2}}(W_L,\ W_R\ ;\ \gamma_l,\ \gamma_r).$$

We propose to compute *two* intercell fluxes, one assuming that the fluids in both cells are 'γ_l fluids', and one assuming that they are 'γ_r fluids'. We denote the two fluxes by $F^{L,R}_{j+\frac{1}{2}}$ where

$$F^L_{j+\frac{1}{2}} = F_{j+\frac{1}{2}}((U_L,\ U_R\ ;\ \gamma_l),\ F^R_{j+\frac{1}{2}} = F_{j+\frac{1}{2}}((U_L,\ U_R\ ;\ \gamma_r).$$

We now use $F^L_{j+\frac{1}{2}}$ to update the fluid on the left (cell j) and $F^R_{j+\frac{1}{2}}$ to update the fluid on the right (cell $j+1$).

To implement the algorithm, we assume that the primitive variables $U = (\rho,\ u,\ p,\ \psi)^0_j$ are available, and proceed as follows:

- Use ψ to locate the material interface.
- Compute $W = (\rho,\ \rho u,\ E)^n_j$, using the EOS as determined by (2).
- If $\psi_j \cdot \psi_{j+1} < 0$ (interface, $\gamma_l \neq \gamma_r$) compute *two* intercell fluxes,

$$F^L_{j+\frac{1}{2}}(W_j,\ W_{j+1}\ ;\gamma_l) \neq F^R_{j+\frac{1}{2}}(W_j,\ W_{j+1}\ ;\gamma_r)\ .$$

- Otherwise (no interface, $\gamma_l = \gamma_r \equiv \gamma$) compute *one* intercell flux

$$F^L_{j+\frac{1}{2}} = F_{j+\frac{1}{2}}(W_j,\ W_{j+1}\ ;\gamma) = F^R_{j+\frac{1}{2}}\ .$$

- Update the conserved variables using

$$W^{n+1}_j = W^n_j - \nu(F^L_{j+\frac{1}{2}} - F^R_{j-\frac{1}{2}})$$

- Compute $U = (\rho,\ u,\ p)^{n+1}_j$, using (2) with *frozen* values of $\gamma = \gamma^n_{l,r}$.
- Update the level-set function ψ^{n+1}_j.

We note the following: (i) The method conserves mass and momentum but not total energy (see [3] for more detail). (ii) Conservation error is very small (see numerical results) and tends to zero with mesh refinement ([3]). (iii) Finally, with a very small modification, this single-fluid algorithm may be applied to multifluid models based on mass-fraction [2, 9] (see [3]).

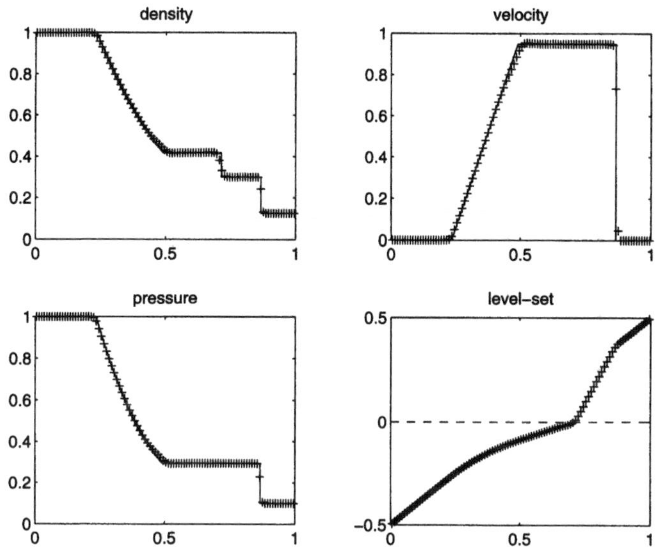

FIGURE 2. Two-fluid shock tube problem.

5. Numerical Results

The algorithm described above does not assume any particular numerical discretization scheme. The numerical results in this section were obtained with a second order Roe's method [13] using the superbee flux limiter.

Shock tube problem: The single fluid algorithm was applied to the two gas shock tube data $U_L = (1.0,\ 0,\ 1.0,\ 1.4)$, $U_R = (0.125,\ 0,\ 0.1,\ 1.2)$, on a 100 point mesh at CFL=0.9. A computation using the level-set model produces strong oscillations near the material front [7, 8]. The present single-fluid algorithm is oscillation free (Figure 2) and has very small (0.2%) conservation error in total energy (Figure 3).

Shock hitting a heavy gas interface: In Figure 4, a right moving shock in air, of shock Mach number $M_S = 2.9520$, is hitting a R22 ($\gamma = 1.249$) interface. The density ratio across the interface is taken to be the ratio of molecular weights $\rho_{R22}/\rho_{air} = 0.287/0.091 = 3.1538$, corresponding to two fluids in thermal equilibrium. For this problem, the incident shock is transmitted, and a shock is reflected. A 200-point mesh was used at CFL=0.9. Numerical and exact solutions are shown.

Shock hitting a light gas interface: In Figure 5, a right moving $M_S = 9.2659$ shock in air is hitting a helium ($\gamma = 1.67$) interface. The density ratio was again set to the ratio of molecular weights $\rho_{helium}/\rho_{air} = 0.287/2.08 = 0.1380$. The incident shock is transmitted, and a rarefaction is reflected. A 200-point mesh was used at CFL=0.9. Numerical and exact solution are shown.

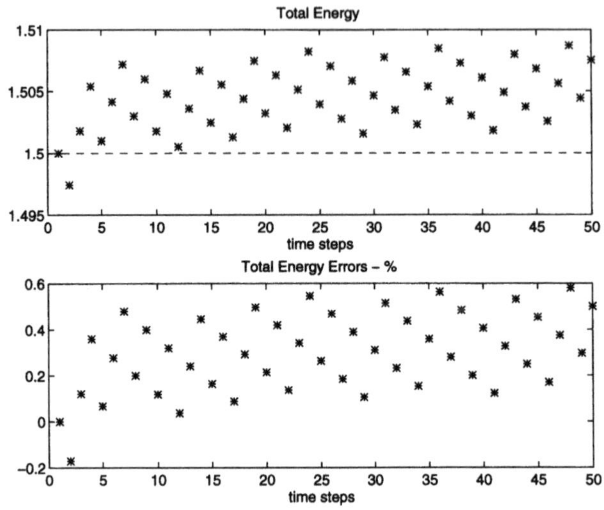

FIGURE 3. Total energy (top) and relative conservation errors (bottom).

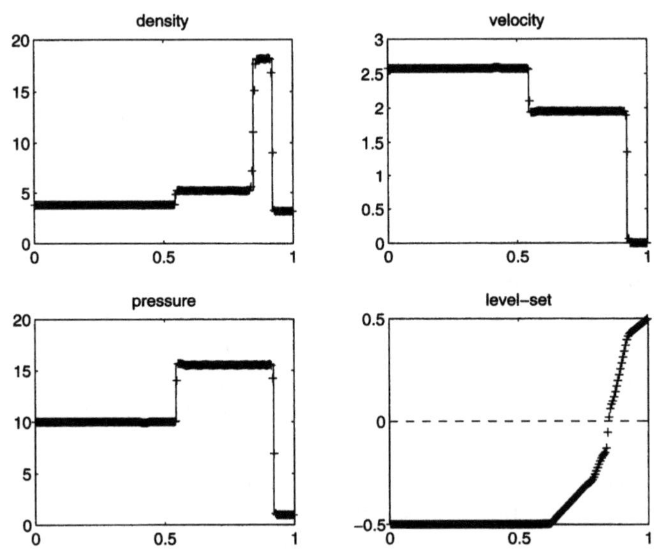

FIGURE 4. A $M_S = 2.9520$ shock wave in air hitting a R22 interface.

Air-Water shock tube problem: For this stiff problem [11], the water phase ($\gamma = 4.4,\ P_\infty = 6\ 10^8$) is initialized with $U_L = (\rho_l,\ u_l,\ p_l) = (1000.0,\ 0,\ 10^9)$ and the air phase ($\gamma = 1.4,\ \infty = 0$) with $U_R = (\rho_r,\ u_r,\ p_r) = (50.0,\ 0,\ 10^5)$. In this problem, the shock wave and material front are propagating with similar speeds.

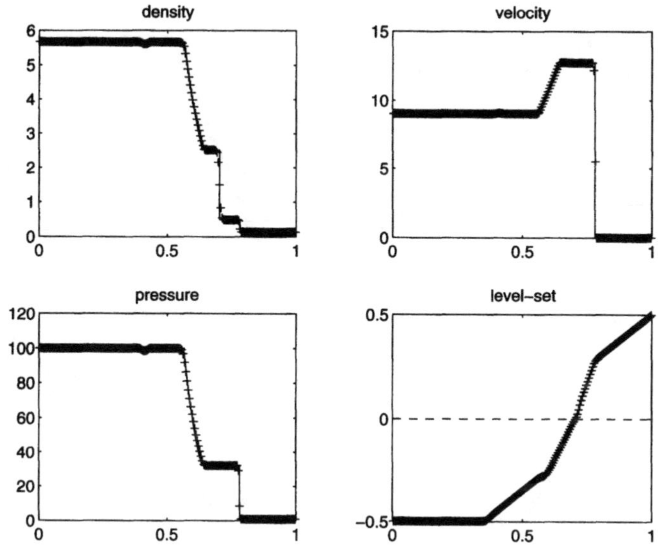

FIGURE 5. A $M_S = 9.2659$ shock wave in air hitting a helium interface.

A finer mesh of 400 points is used in order to allow them to separate clearly in the computation. The CFL number is 0.9. Computed and exact solution are in excellent agreement (Figure 6) and conservation error in energy is still extremely small (0.2%) (Figure 7).

Vortex sheet roll-up: This is a two ideal gas Kelvin-Helmholz shear wave instability. Initial data consists of one fluid with $(\rho,\ \gamma) = (1.0,\ 1.4)$ and another fluid with $(\rho,\ \gamma) = (2/3,\ 1.67)$ at pressure equilibrium, $p = 1$. The velocity field is initialized to correspond to a periodic (incompressible) vortex sheet embedded in a two dimensional flow [14]. The domain is a box of dimensions 2 by 1, the mesh size is 256×128. The initial perturbation to the interface is a sine wave of amplitude 0.05. Oscillations that are produced by the level-set model in [10] are eliminated by using the two-flux single fluid algorithm across the material front (Figure 8).

References

[1] R. Abgrall, *Generalisation of the Roe scheme for the computation of mixture of perfect gases*, Rech. Aérosp., **6**, (1988).

[2] R. Abgrall, *How to prevent pressure oscillations in multicomponent flows : a quasi conservative approach*, J. Comp. Phys., **125**, (1996), 150–160.

[3] R. Abgrall and S. Karni, *Computations of Compressible Multifluids*, J. Comp. Phys., **169**, (2001).

[4] S.F. Davis, *An interface tracking method for hyperbolic systems of conservation laws*, Applied Numerical Mathematics, **10**, (1992) 447–472.

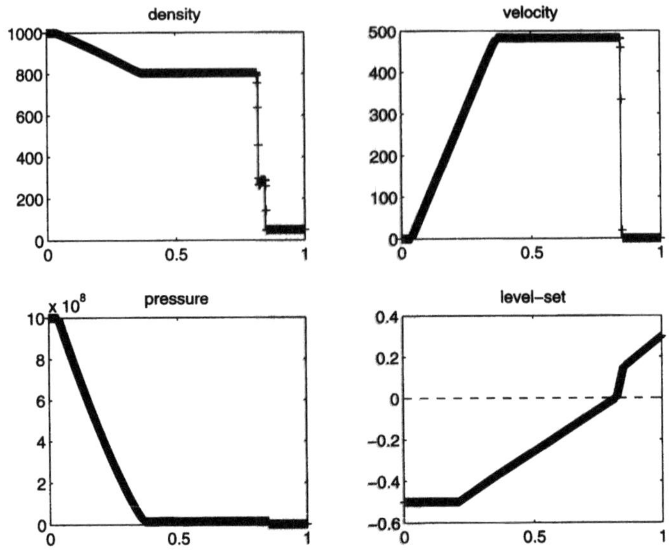

FIGURE 6. Air-Water shock tube problem.

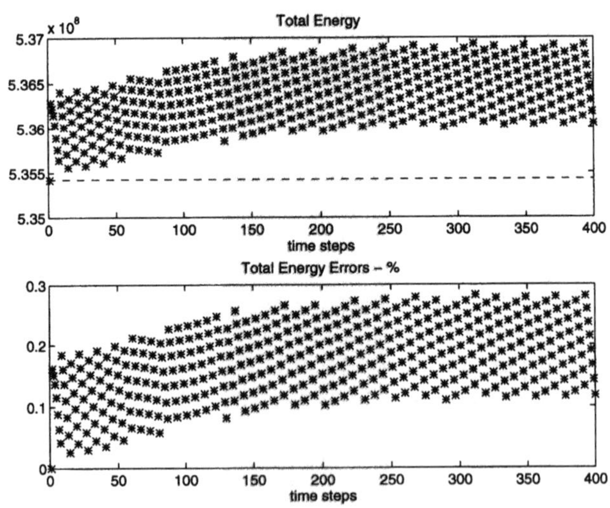

FIGURE 7. Total energy (top) and relative conservation errors (bottom).

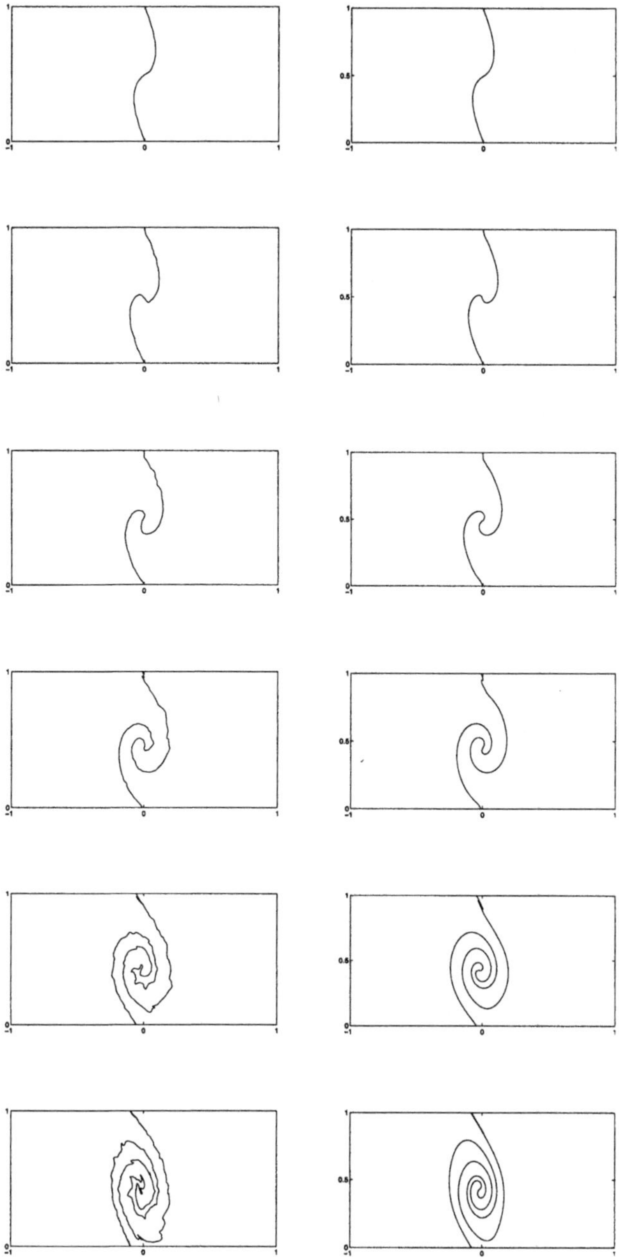

FIGURE 8. Two-fluid vortex sheet roll-up by multi (left) and single (right) fluid algorithms.

[5] R. Fedkiw et al., *A non-oscillatory Eulerian approach to interfaces in multimaterial flows (the Ghost Fluid Method)*, J. Comp. Phys., **152**, (1999), 457–492.

[6] P. Jenny et al., *Correction of conservative Euler solvers for gas mixtures*, J. Comp. Phys., **132**, (1997), 91–107.

[7] S. Karni, *Multi-component flow calculations by a consistent primitive algorithm*, J. Comp. Phys., **112**, (1994), 31–43.

[8] S. Karni, *Hybrid multifluid algorithms*, SIAM J. Sci. Comp., **17**, (1996), 1019–1039.

[9] B. Larouturou, *How to preserve the mass fraction positive when computing compressible multi-component flows*, J. Comp. Phys., **95**, (1991), 59–84.

[10] W. Mulder et al., *Computing interface motion in compressible gas dynamics*, J. Comp. Phys., **100**, (1992) 209–228.

[11] R. Saurel and R. Abgrall, *A simple method for compressible multifluid flows*, Siam J. Sci. Comp., **21**, (1999) 1115–1145.

[12] K.-M. Shyue, *An Efficient Shock-Capturing Algorithm for Compressible Multicomponent Problems*, J. Comp. Phys., **1**, (1998), 208–242.

[13] P.L.Roe, *Fluctuations and signals - A framework for numerical evolution problems*, Numerical methods for fluid dynamics, Editteed by K.W. Morton and M.J. Baines (Academic Press, New York), **1982**, 219–257.

[14] R.Krasny *Desingularization of a periodic vortex sheet roll-up*, J. Comp. Phys., **65**, (1986), 292–313.

Rémi Abgrall
University of Bordeaux, Mathématiques Appliquées,
351 Cours de la Libération,
33 405 Talence Cedex , France
`remi.abgrall@math.u-bordeaux.fr`

Smadar Karni
Department of Mathematics,
University of Michigan,
Ann Arbor, MI 48109, USA
`karni@math.lsa.umich.edu`

International Series of Numerical Mathematics
Vol. 140, © 2001 Birkhäuser Verlag Basel/Switzerland

Propagation of Smoothness for Edge-degenerate Wave Equations

Ruben Airapetyan and Ingo Witt

Abstract. A propagation result for an edge-degenerate wave equation in 1+1 dimensions is proved. The proposed method generalizes to other edge-degenerate wave equations.

1. Introduction

In this paper, we are concerned with hyperbolic differential problems near edges. Writing a differential operator A of order μ locally in cylindrical coordinates (y, r, x), where $y \in \mathbb{R}^q$ is the coordinate along the edge, $r \in \mathbb{R}_+$ is the radial coordinate, and x is the generic point on the cone base X (X a smooth manifold) we consider A as an operator on the wedge $\mathbb{R}^q \times \big((\overline{\mathbb{R}}_+ \times X)/(\{0\} \times X)\big)$. Then A can be represented in the form $A = r^{-\mu} \sum_{j+|\alpha|\le\mu} a_{j\alpha}(y,r)(r\partial_r)^j(r\partial_y)^\alpha$, where $a_{j\alpha} \in C^\infty(\mathbb{R}^q \times \overline{\mathbb{R}}_+; \mathrm{Diff}^{\mu-(j+\alpha)}(X))$. The appearance of derivatives $r\partial_r$, $r\partial_y$, combined with the factor $r^{-\mu}$, is refereed to be *edge-degenerate*.

Here we present a method for investigation of existence, higher regularity, and asymptotics of solutions, and propagation of smoothness as well, for edge-degenerate second-order differential operators which are assumed to be strictly hyperbolic for $r > 0$ in one of the edge directions. This is exemplified on the Cauchy problem for the unknown $u(t, r)$,

$$\partial_t^2 u - \partial_r^2 u - \frac{1}{r}\,\partial_r u + \frac{\nu^2(t)}{r^2}\,u = f(t,r), \quad (t,r) \in \mathbb{R}_+ \times \mathbb{R}_+, \tag{1.1}$$

$$u(0,r) = \phi(r), \;\; \partial_t u(0,r) = \psi(r), \tag{1.2}$$

where $\nu \in C^\infty(\mathbb{R})$ and $\nu(t) > 0$ for all $t \in \mathbb{R}$. For technical details, see the forthcoming paper [1].

Remark 1.1. *In* (1.1), *we have* $q = 1$ *and* $\dim X = 0$. *For most of what follows this is not a serious restriction. The circumstance, however, that no conditions of trace and potential type, respectively, along the edge* $\{r = 0\}$ *are required, is atypical and simplifies things considerably, see* Remark 3.2. *The fact that the parameter* ν *is chosen variably in* t *allows to observe the phenomenon of variable, possibly branching asymptotics.*

Our point of view is:

- Solutions to (1.1), (1.2) exhibit asymptotics of a fixed asymptotic type Q_ν. Since asymptotics of type Q_ν are always present, they will be incorporated into the functional-analytic setting;
- If this is appropriately done, the basic energy inequality (i.e., with the right-hand side f in the analogue of L^2) can be derived by integration by parts. Higher energy estimates (i.e., with the right-hand side f in the analogue of H^s) follow from the basic energy inequality, e.g., by commutator arguments;
- Statements on the propagation of smoothness are microlocalized versions of the energy inequalities. They are available after a corresponding pseudodifferential calculus has been established.

The first item in particular means that all the calculations are performed modulo asymptotics of type Q_ν or, stated it in another way, functions possessing asymptotics of type Q_ν *are considered as* being smooth up to $r = 0$. Since for Eq. (1.1), (1.2) the analogue of the *uniform Lopatinski condition* is fulfilled, we then simply observe reflection of singularities, see Theorem 4.3.

Remark 1.2. *This can be compared to the situation appearing for boundary value problems. Here the corresponding asymptotic type is that resulting from Taylor expansion at $r = 0$; then Boutet de Monvel's calculus for treating boundary value problems is an adequate pseudodifferential calculus.*

The asymptotic type Q_ν mentioned above (notice that ν depends on t) is formally given by

$$u(t,r) \sim \sum_{j=0}^{\infty} \sum_{k=0}^{(2j-\sigma(t))^+} c_{jk}(t)\, r^{\nu(t)+2j} \log^k r \quad \text{as } r \to 0^+, \tag{1.3}$$

where $\sigma(t) \in \mathbb{N} \cup \{\infty\}$ is the order of vanishing of the first derivative ν' at t, with scalar coefficient functions $c_{jk}(t)$, and $a^+ = \max\{a, 0\}$ for $a \in \mathbb{R} \cup \{-\infty\}$.

2. Function Spaces

Instead of giving the definitions of all the function spaces we need, we restrict ourselves to examples, thereby clarifying certain points. The general definitions can be found in [1], [2], [3].

2.1. Asymptotic types

An asymptotic type P for functions $v = v(r)$ on the cone $\mathbb{R}_+$ (recall that $\dim X = 0$) is a sequence $\{(p_j, m_j);\, j = 1, \dots, N(P)\} \subset \mathbb{C} \times \mathbb{N}$, where $N(P) \in \mathbb{N} \cup \{\infty\}$ and $\operatorname{Re} p_j \to -\infty$ as $j \to \infty$ if $N(P) = \infty$. P is associated with the conormal order δ if, in addition, $\operatorname{Re} p_j < 1/2 - \delta$ for all j. An example is $T = \{(-j, 0);\, j \in \mathbb{N}\}$ which is the type for Taylor asymptotics associated with the conormal order 0. Another example is the empty asymptotic type, $\mathcal{O}$, given by $N(\mathcal{O}) = 0$.

An asymptotic type P for functions $u(t,r)$ on the wedge $\mathbb{R}\times\mathbb{R}_+$ is a collection $P(t) = \{(p_j(t), m_j(t));\, j = 1,\ldots,N(P(t))\}$ of asymptotic types for functions on $\mathbb{R}_+$, depending on $t \in \mathbb{R}$, subject to certain further conditions. Here the numeration of the pairs $(m_j(t), p_j(t))$ for various t can be completely independent of each other. If $P(t)$ is actually independent of t, then one speaks about constant asymptotics, otherwise about (genuinely) variable asymptotics. Given P, there is an asymptotic type, $\partial_t P$, obeying the property that if $u(t,r)$ exhibits asymptotics of type P, then $\partial_t u$ exhibits asymptotics of type $\partial_t P$. For example, $Q_\nu = \{(-(\nu(t)+2j), (2j-\sigma(t))^+);\, j \in \mathbb{N}\}$ and $\partial_t Q_\nu = \{(-(\nu(t)+2j), (2j-\sigma(t)+1)^+);\, j \in \mathbb{N}\}$.

2.2. Cone Sobolev spaces

Asymptotics of solutions to edge-degenerate hyperbolic problems are classically known in many cases. The new ingredient which now comes into the analysis is the so-called *conormal order* measuring flatness at $r = 0$. In its consequence, it allows a functional-analytic set-up. In particular, function spaces will depend on two parameters, $s \in \mathbb{R}$, $\delta \in \mathbb{R}$, where s is the Sobolev order and δ is the conormal order.

For $s \in \mathbb{N}$, $\delta \in \mathbb{R}$, $v(r)$ belongs to the space $\mathcal{H}^{s,\delta}(\mathbb{R}_+)$ if $r^{j-\delta}\partial_r^j v \in L^2(\mathbb{R}_+)$ for all $j \in \mathbb{N}$, $j \le s$. For $s \ge 0$, $\delta \in \mathbb{R}$, the space $\mathcal{H}^{s,\delta}(\mathbb{R}_+)$ is then defined by interpolation. Note that $r^\gamma\mathcal{H}^{s,\delta}(\mathbb{R}_+) = \mathcal{H}^{s,\delta+\gamma}(\mathbb{R}_+)$.

For $s \ge 0$, $\delta \in \mathbb{R}$, the space $\mathbb{H}^{s,\delta}(\mathbb{R}_+)$ consists of all functions $v(r)$ such that $\omega(r)v \in \mathcal{H}^{s,s+\delta}(\mathbb{R}_+)$ and $(1-\omega(r))v \in r^\delta H^s(\mathbb{R}_+)$. Here $\omega \in C^\infty(\overline{\mathbb{R}}_+)$ is a fixed cut-off function, i.e., $0 \le \omega \le 1$, $\omega(r) = 1$ for $r \le 1/2$, $\omega(r) = 0$ for $r \ge 1$. For P being an asymptotic type associated with the conormal order δ, we define

$$\mathbb{H}^{s,\delta}_P(\mathbb{R}_+) = \mathbb{H}^{s,\delta}(\mathbb{R}_+) \oplus \operatorname{span}\{\omega(r)r^{-p_j}\log^k r;\; 1 \le j \le N(P),\, 0 \le k \le m_j,\, \operatorname{Re} p_j > 1/2 - s - \delta\} \tag{2.1}$$

provided that $\operatorname{Re} p_j \ne 1/2 - s - \delta$ for all j. This definition can be extended to arbitrary $s \ge 0$. Note that

$$\omega(r)r^{-p_j}\log^k r \in \mathcal{H}^{s,\delta}(\mathbb{R}_+) \iff \operatorname{Re} p_j < \frac{1}{2} - \delta$$

so that the sum on the right-hand side of (2.1) is indeed direct. Moreover, the conormal order is seen to be the parameter with respect to which asymptotics are meant, i.e., by subtracting the first asymptotic terms the conormal order near $r = 0$ is increased. Further it is important to observe that the group $\{\kappa_\lambda^{(\delta)}\}_{\lambda>0}$,

$$\kappa_\lambda^{(\delta)} v(r) = \lambda^{1/2-\delta} v(\lambda r), \tag{2.2}$$

acts on $\mathbb{H}^{s,\delta}_P(\mathbb{R}_+)$. Hereby, $\{\kappa_\lambda^{(\delta)}\}_{\lambda>0}$ is a group of isometries on $\mathbb{H}^{0,\delta}(\mathbb{R}_+)$.

Example 2.1. *For $s \ge 0$, we have $\mathbb{H}^{s,0}_T(\mathbb{R}_+) = H^s(\mathbb{R}_+)$, while $\mathbb{H}^{s,0}(\mathbb{R}_+) = H^s_0(\overline{\mathbb{R}}_+)$ is the space of all functions in $H^s(\mathbb{R})$ supported in $\overline{\mathbb{R}}_+$.*

2.3. Edge Sobolev spaces

For $s \geq 0$, $\delta \in \mathbb{R}$, and P being a type for *constant asymptotics*, we define

$$\mathbb{H}_P^{s,\delta}(\mathbb{R} \times \mathbb{R}_+) = H^0(\mathbb{R}; \mathbb{H}_P^{s,\delta}(\mathbb{R}_+)) \cap H^s(\mathbb{R}; \mathbb{H}^{0,\delta}(\mathbb{R}_+)). \tag{2.3}$$

The key in defining the space $\mathbb{H}_P^{s,\delta}(\mathbb{R} \times \mathbb{R}_+)$ when P is a type for variable asymptotics is the following result:

Lemma 2.2. *An equivalent norm for the space defined in* (2.3) *is given by*

$$\|u\|_{\mathbb{H}_P^{s,\delta}(\mathbb{R}\times\mathbb{R}_+)} = \left\{ \int_{-\infty}^{\infty} \langle\tau\rangle^{2s} \|\kappa^{(\delta)}(\tau)^{-1}\hat{u}(\tau,\cdot)\|^2_{\mathbb{H}_P^{s,\delta}(\mathbb{R}_+)} d\tau \right\}^{1/2}, \tag{2.4}$$

where $Fu(\tau,r) = \hat{u}(\tau,r) = \int_{-\infty}^{\infty} e^{-it\tau}u(t,r)\,dt$ *is the partial Fourier transform of* u *with respect to* t *and* $\kappa^{(\delta)}(\tau) = \kappa_\lambda^{(\delta)}\big|_{\lambda=\langle\tau\rangle}$.

In fact, (2.1), (2.4) together lead to the direct decomposition

$$\begin{aligned} &\mathbb{H}_P^{s,\delta}(\mathbb{R} \times \mathbb{R}_+) \\ &= \mathbb{H}^{s,\delta}(\mathbb{R} \times \mathbb{R}_+) \oplus \operatorname{span}\Big\{ F^{-1}_{\tau\to t}\{\langle\tau\rangle^{1/2-\delta}(r\langle\tau\rangle)^{-p_j}\log^k(r\langle\tau\rangle)\hat{d}_{jk}(\tau)\}; \\ &\qquad d_{jk} \in H^s(\mathbb{R}),\ 1 \leq j \leq N(P),\ 0 \leq k \leq m_j,\ \operatorname{Re} p_j > 1/2 - s - \delta \Big\} \end{aligned}$$

provided that $\operatorname{Re} p_j \neq 1/2 - s - \delta$ for all j. This is the representation which can be further generalized to a type P for variable asymptotics, also giving a precise meaning to formal asymptotic expansions like (1.3).

Lemma 2.3. *If P is a type for variable asymptotics associated with the conormal order δ, then the Hilbert spaces $\mathbb{H}_P^{s,\delta}(\mathbb{R}\times\mathbb{R}_+)$ for $s \geq 0$ form an interpolation scale with respect to the complex interpolation method.*

2.4. Energy inequalities

Theorem 2.4. *Let $s \geq 0$. Let $\phi \in \mathbb{H}^{s+1,-1/2}_{Q_\nu(0),\mathrm{loc}}(\mathbb{R}_+)$, $\psi \in \mathbb{H}^{s,-1/2}_{\partial Q_\nu(0),\mathrm{loc}}(\mathbb{R}_+)$, and $f \in \mathbb{H}^{s,-1/2}_{Q_\nu,\mathrm{loc}}(\mathbb{R}_+ \times \mathbb{R}_+)$. (Here* loc *refers to the behaviour as $t, r \to \infty$ and function spaces on $\mathbb{R}_+\times\mathbb{R}_+$ are defined by restricting from those on $\mathbb{R}\times\mathbb{R}_+$.) Then* Eq. (1.1), (1.2) *possesses a unique solution $u \in \mathbb{H}^{s+1,-1/2}_{Q_\nu,\mathrm{loc}}(\mathbb{R}_+\times\mathbb{R}_+)$ and corresponding energy estimates hold.*

Note that the space $\mathbb{H}^{1,-1/2}_{Q_\nu}(\mathbb{R} \times \mathbb{R}_+) = \mathbb{H}^{1,-1/2}(\mathbb{R} \times \mathbb{R}_+)$ is independent of the function ν; thus allowing integration by parts in deriving the basic energy inquality (with $s = 0$), see also (2.3).

3. Pseudodifferential Calculus

3.1. Edge-degenerate operators

To understand the following construction, one has to know a little bit about the symbolic structure of edge pseudodifferential operators. These operators A, say of order μ, obey an edge-degenerate complete symbol, i.e., a symbol of the form $p(t,r,\tau,\xi) = r^{-\mu}\tilde{p}(t,r,r\tau,r\xi)$, where $\tilde{p}(t,r,\tilde{\tau},\tilde{\xi}) \in S^{\mu}_{\mathrm{cl}}(\mathbb{R}\times\overline{\mathbb{R}}_+\times\mathbb{R}^2)$. They possess a principal pseudodifferential symbol $\sigma^{\mu}_{\psi}(A) \in S^{(\mu)}(T^*_0(\mathbb{R}\times\overline{\mathbb{R}}_+))$ as usual and a principal edge symbol $\sigma^{\mu}_{\wedge}(A)(t,\tau)$, where $(t,\tau) \in T^*_0\mathbb{R}$, that takes values in the cone pseudodifferential operators on $\mathbb{R}_+$ and that is homogeneous in the sense that

$$\sigma^{\mu}_{\wedge}(A)(t,\lambda\tau) = \lambda^{\mu}\kappa^{(\delta)}_{\lambda}\sigma^{\mu}_{\wedge}(A)(t,\tau)\kappa^{(\delta)}_{\lambda^{-1}},$$

see (2.2). Under compositions, we have $\sigma^{\mu+\rho}_{\psi}(AB) = \sigma^{\mu}_{\psi}(A)\sigma^{\rho}_{\psi}(B)$, $\sigma^{\mu+\rho}_{\wedge}(AB) = \sigma^{\mu}_{\wedge}(A)\sigma^{\rho}_{\wedge}(B)$. The invertibility of the pair $\big(\sigma^{\mu}_{\psi}(A),\sigma^{\mu}_{\wedge}(A)\big)$ determines the ellipticity of the operator A.

3.2. The algebra of conormal symbols

In local coordinates, subordinated to the sequence $\{\sigma^{\mu-j}_{\wedge}(A)(t,\tau);\, j\in\mathbb{N}\}$ of edge symbols there is the sequence $\{\sigma^{\mu-j}_{M}(A)(t,z,\tau);\, j\in\mathbb{N}\}$ of conormal symbols. Note that $\sigma^{\mu-j}_{M}(A)(t,z,\tau)$ is a scalar function that is, among others, meromorphic in $z\in\mathbb{C}$, smooth in $t\in\mathbb{R}$ (in a well-defined sense), and polynomial of degree j in $\tau\in\mathbb{R}$. The conormal symbols do not enter in definition of ellipticity, but they are likewise important, since they determine the manner in which asymptotics are mapped.

Our aim is to single out a subcalculus of the general edge pseudodifferential calculus of [3] which leaves asymptotics of type Q_ν invariant. This is achieved by controlling the conormal symbols in a suitable way. The behaviour of conormal symbols under compositions is regulated by the *Leibniz-Mellin translation product*:

$$\sigma^{\mu+\rho-l}_{M}(AB)(t,z,\tau) = \sum_{\substack{j+k=l,\\ \theta\le j}} \frac{1}{\theta!}\left(\partial^{\theta}_{\tau}\sigma^{\mu-j}_{M}(A)\right)(t,z+\rho-k,\tau)\left(D^{\theta}_{t}\sigma^{\rho-k}_{M}(B)\right)(t,z,\tau).$$

Let $g(t,z)$ be a function that is meromorphic in $z\in\mathbb{C}$ and smooth in $t\in\mathbb{R}$ such that, for each $t\in\mathbb{R}$, $g(t,\cdot)$ does not vanish identically. Then all these edge pseudodifferential operators possessing conormal symbols

$$\sigma^{\mu-j}_{M}(A)(t,z,\tau) = g(t,z+\mu-j)\sum_{r=0}^{j}\alpha^{j}_{r}(t)\left(\partial_t+i\tau\right)^{j-r}\left\{\frac{1}{g(t,z)}\right\}, \tag{3.1}$$

where $\alpha^{j}_{r}(t)$ are arbitrary coefficient functions, form a subcalculus of the calculus of all edge pseudodifferential operators. That is, the expression on the right-hand side of (3.1) is invariant under the Leibniz-Mellin translation product, see [4]. In applications, the function $g(t,z)$ is chosen to present a given type P for variable asymptotics. For instance, concerning Eq. (1.1), we choose

$$g(t,z) = 2^{z-1}\,\frac{\Gamma\big((\nu(t)+z)/2\big)}{\Gamma\big((\nu(t)-z)/2+1\big)},$$

where $\Gamma(z)$ is the Gamma function. Note that $g(t,z+2)/g(t,z) = -z^2 + \nu^2(t)$ ($\mu = 2$, $j = 0$ in (3.1)) which is the principal conormal symbol of the operator $\partial_t^2 + r^{-2}(-(r\partial_r)^2 + \nu^2(t))$.

3.3. The pseudodifferential algebra

By further specifying the considerations in the previous section, we obtain:

Theorem 3.1. *There is a subalgebra $\Psi_{Q_\nu}^{0,0}(\mathbb{R}\times\mathbb{R}_+) \subset \mathcal{L}(\mathbb{H}^{0,-1/2}(\mathbb{R}\times\mathbb{R}_+))$ of edge-degenerate pseudodifferential operators A on $\mathbb{R}\times\mathbb{R}_+$ inducing continuous maps*

$$A\colon \mathbb{H}_{Q_\nu}^{s,-1/2}(\mathbb{R}\times\mathbb{R}_+) \to \mathbb{H}_{Q_\nu}^{s,-1/2}(\mathbb{R}\times\mathbb{R}_+) \tag{3.2}$$

for all $s \geq 0$.

Remark 3.2. *The full calculus is $\Psi_{Q_\nu}^{\mu,d}$, where $\mu \in \mathbb{Z}$ is the order of operators involved and $d \in \mathbb{N}$ is their type, like in Boutet de Monvel's pseudodifferential calculus for boundary value problems. The operators in this calculus admit a block structure $\left(\begin{smallmatrix} A & K \\ T & Q \end{smallmatrix}\right)$, where A is as in (3.2), but of the corresponding order, T is a trace operator, K is a potential operator, and Q is pseudodifferential operator acting along the edge. A trace operator essentially assigns to a function $u(t,r)$ its coefficients $c_{jk}(t)$ in the asymptotic expansion* (1.3). *For the purposes below, the case $\mu = d = 0$ suffices.*

Definition 3.3. *$A \in \Psi_{Q_\nu}^{0,0}(\mathbb{R}\times\mathbb{R}_+)$ is called elliptic if $\sigma_\psi^0(A)(t,r,\tau,\xi) \neq 0$ for all $(t,r,\tau,\xi) \in T_0^*(\mathbb{R}\times\overline{\mathbb{R}}_+)$ and, in addition,*

$$\sigma_\wedge^0(A)(t,\tau)\colon \mathbb{H}^{0,-1/2}(\mathbb{R}_+) \to \mathbb{H}^{0,-1/2}(\mathbb{R}_+)$$

is invertible for all $(t,\tau) \in T_0^*\mathbb{R}$.

Note that if A is elliptic then the pair $(\sigma_\psi^0(A)^{-1}, \sigma_\wedge^0(A)^{-1})$, with the inverses taken point-wise, belongs to the principal symbol algebra of $\Psi_{Q_\nu}^{0,0}(\mathbb{R}\times\mathbb{R}_+)$.

Theorem 3.4. *Let $A \in \Psi_{Q_\nu}^{0,0}(\mathbb{R}\times\mathbb{R}_+)$ be elliptic. Then there exists a parametrix $P \in \Psi_{Q_\nu}^{0,0}(\mathbb{R}\times\mathbb{R}_+)$, i.e., an operator P such that*

$$AP - I,\ PA - I \in \Psi_{Q_\nu}^{-\infty,0}(\mathbb{R}\times\mathbb{R}_+).$$

Here $\Psi_{Q_\nu}^{-\infty,0}(\mathbb{R}\times\mathbb{R}_+)$ is the ideal of smoothing operators in $\Psi_{Q_\nu}^{0,0}(\mathbb{R}\times\mathbb{R}_+)$. Furthermore, elliptic regularity holds for A: if $u \in \mathbb{H}^{0,-1/2}(\mathbb{R}\times\mathbb{R}_+)$ and $Au \in \mathbb{H}_{Q_\nu}^{s,-1/2}(\mathbb{R}\times\mathbb{R}_+)$ for some $s > 0$, then $u \in \mathbb{H}_{Q_\nu}^{s,-1/2}(\mathbb{R}\times\mathbb{R}_+)$.

4. Propagation of Smoothness

4.1. Wave front sets

Definition 4.1. *The wave front space* $\mathfrak{M}$ *is the disjoint union* $T_0^*\mathbb{R} \cup T_0^*(\mathbb{R} \times \overline{\mathbb{R}}_+)$. *For* $A \in \Psi_{Q_\nu}^{0,0}(\mathbb{R} \times \mathbb{R}_+)$, *we define*

$$\operatorname{Char}_{Q_\nu}(A) = \{(t,\tau) \in T_0^*\mathbb{R};\ \sigma_\wedge^0(A)(t,\tau) \text{ is not invertible}\} \\ \cup \{(t,r,\tau,\xi) \in T_0^*(\mathbb{R} \times \overline{\mathbb{R}}_+);\ \sigma_\psi^0(A)(t,r,\tau,\xi) = 0\}.$$

There is a natural topology on $\mathfrak{M}$ such that $\operatorname{Char}_{Q_\nu}(A) \subseteq \mathfrak{M}$ is a closed, conic subset for any $A \in \Psi_{Q_\nu}^{0,0}(\mathbb{R} \times \mathbb{R}_+)$.

Definition 4.2. *For* $s \geq 0$ *and* $u \in \mathbb{H}^{0,-1/2}(\mathbb{R} \times \mathbb{R}_+)$, *we define*

$$\mathrm{WF}_{Q_\nu}^{s,-1/2}(u) = \bigcap \operatorname{Char}_{Q_\nu}(A),$$

where the intersection is extended over all operators $A \in \Psi_{Q_\nu}^{0,0}(\mathbb{R} \times \mathbb{R}_+)$ *such that* $Au \in \mathbb{H}_{Q_\nu}^{s,-1/2}(\mathbb{R} \times \mathbb{R}_+)$. $\zeta \in \mathfrak{M} \setminus \mathrm{WF}_{Q_\nu}^{s,-1/2}(u)$ *will also be expressed by writing* $u \in \mathbb{H}_{Q_\nu}^{s,-1/2}(\zeta)$.

The topology induced by $\mathfrak{M}$ on $T_0^*(\mathbb{R}\times\mathbb{R}_+)$ is the standard one. Furthermore, $\mathrm{WF}_{Q_\nu}^{s,-1/2}(u)|_{T_0^*(\mathbb{R}\times\mathbb{R}_+)} = WF^s(u)$ for $s \geq 0$ and $u \in \mathbb{H}^{0,-1/2}(\mathbb{R} \times \mathbb{R}_+)$, due to Hörmander's characterization of wave front sets.

4.2. The propagation result

Theorem 4.3. *Let* $a > 0$, $\zeta^0 = (0, a, \tau^0, \xi^0) \in T_0^*(\mathbb{R} \times \mathbb{R}_+)$ *be a characteristic point for the operator* $\partial_t^2 + L_\nu$, *where* $L_\nu = r^{-2}\big(-(r\partial_r)^2 + \nu^2(t)\big)$, *and* $\gamma_- \cup \gamma_+$ *with* $\gamma_\mp = \{(t, r, \tau^0, \pm\xi^0);\ t \pm r = a\}$ *be the reflected null bicharacteristic through* ζ^0. *Further let* $s > 0$, $\phi \in \mathbb{H}_{\mathrm{loc}}^{1,-1/2}(\mathbb{R}_+) \cap H^{s+1}(\zeta^0)$, $\psi \in \mathbb{H}_{\mathrm{loc}}^{0,-1/2}(\mathbb{R}_+) \cap H^s(\zeta^0)$, *and* $f \in \mathbb{H}_{\mathrm{loc}}^{0,-1/2}(\mathbb{R}_+ \times \mathbb{R}_+) \cap H^s(\gamma_- \cup \gamma_+)|_{\mathbb{R}_+\times\mathbb{R}_+} \cap \mathbb{H}_{Q_\nu}^{s,-1/2}(\zeta_\pm^1)$, *where* $\zeta_\pm^1 = (a, \pm 1) \in T_0^*\mathbb{R}$. *Then, for the unique solution* $u(t,r)$ *to* Eq. (1.1), (1.2), *one has*

$$u \in H^{s+1}(\gamma_- \cup \gamma_+)|_{\mathbb{R}_+\times\mathbb{R}_+} \cap \mathbb{H}_{Q_\nu}^{s+1,-1/2}(\zeta_\pm^1).$$

By standard arguments, we can neglect Cauchy data (ϕ, ψ) and localize near the point $(a, 0)$, i.e., we can assume that $u \in \mathbb{H}^{1,-1/2}(\mathbb{R} \times \mathbb{R}_+) \cap H^{s+1}(\gamma_-)$ is a given solution to (1.1) on $\mathbb{R}\times\mathbb{R}_+$ with bounded support near $(a, 0)$ and right-hand side $f \in \mathbb{H}^{0,-1/2}(\mathbb{R} \times \mathbb{R}_+) \cap H^s(\gamma_- \cup \gamma_+) \cap \mathbb{H}_{Q_\nu}^{s,-1/2}(\zeta_\pm^1)$. Then the key in proving Theorem 4.3 is:

Lemma 4.4. *There exists an operator* $R \in \Psi_{Q_\nu}^{0,0}(\mathbb{R} \times \mathbb{R}_+)$ *that is elliptic along the "curve"* $\{\zeta_+^1\} \cup \{\zeta_-^1\} \cup \gamma_- \cup \gamma_+$ *such that* $Rf \in \mathbb{H}_{Q_\nu}^{s,-1/2}(\mathbb{R} \times \mathbb{R}_+)$,

$$[\partial_t^2 + L_\nu, R] \in \Psi_{Q_\nu}^{-\infty,0}(\mathbb{R} \times \mathbb{R}_+).$$

References

[1] R. Airapetyan and I. Witt, *Diffraction phenomena for edge-degenerate wave equations in* 1+1 *dimensions,* in preparation.

[2] B.-W. Schulze, *The variable discrete asymptotics in pseudo-differential boundary value problems I,* in: M. Demuth, E. Schrohe, and B.-W. Schulze, Eds., Advances in Partial Differential Equations: Pseudo-Differential Calculus and Mathematical Physics, Math. Topics, Vol. 5, Akademie Verlag, Berlin, 1994, 9–96.

[3] B.-W. Schulze, *Boundary Value Problems and Singular Pseudo-differential Operators,* Wiley Ser. Pure Appl. Math., J. Wiley, Chichester, 1998.

[4] I. Witt, *Explicit algebras with the Leibniz–Mellin translation product,* Preprint 99/2, Institute of Mathematics, University of Potsdam, January 1999.

R. Airapetyan
Department of Science & Mathematics,
Kettering University,
1700 West Third Avenue,
Flint, MI 48504-4898, USA
rhayrape@kettering.edu

I. Witt
Institute of Mathematics,
University of Potsdam,
P.O.Box 60 15 53,
D-14415 Potsdam, Germany
ingo@math.uni-potsdam.de

International Series of Numerical Mathematics
Vol. 140, © 2001 Birkhäuser Verlag Basel/Switzerland

Front Tracking for Non Genuinely Nonlinear Conservation Laws

Fabio Ancona and Andrea Marson

1. Preliminaries and Main Result

We are concerned with global weak solutions to the Cauchy problem for an $N \times N$ strictly hyperbolic system of conservation laws

$$u_t + \left[F(u)\right]_x = 0 \tag{1}$$

$$u(0,x) = \overline{u}(x). \tag{2}$$

Here $u = u(t,x) \in \mathbb{R}^N$ is the vector of the conserved quantities, $t \geq 0$ is the time variable and $x \in \mathbb{R}$ is the space one. F is assumed to be a smooth vector field defined on a neighborhood of the origin $\Omega \subseteq \mathbb{R}^N$ and taking values in $\mathbb{R}^N$. Denote with $\lambda_1(u) < \cdots < \lambda_N(u)$ the eigenvalues of the Jacobian matrix $DF(u)$ and with $r_1(u), \ldots, r_N(u)$ the corresponding right eigenvectors. We assume that the characteristic fields are neither genuinely nonlinear nor linearly degenerate in the sense of Lax [9]. Instead, we require that for each characteristic field r_k the directional derivative of λ_k in the direction of r_k

$$D_{r_k}\lambda_k(u) \doteq \lim_{h\to 0} \frac{\lambda_k\big(u + hr_k(u)\big) - \lambda_k(u)}{h},$$

vanishes on a smooth hypersurface transversal to the field r_k. More precisely, we make the following assumption (see Figure 1)

(A): For each characteristic field r_k there exists a connected $(N-1)$-dimensional smooth manifold Ω_k^0, $k = 1, \ldots, N$, such that

$$\Omega_k^0 = \big\{u \in \Omega : \; D_{r_k}\lambda_k(u) = 0\big\}, \tag{3}$$

$$D^2_{r_k}\lambda_k(u) \doteq D_{r_k}\big(D_{r_k}\lambda_k\big)(u) \neq 0, \qquad \forall\, u \in \Omega_k^0. \tag{4}$$

For systems satisfying assumption **(A)** we provide an existence result via a front tracking method. The main idea involved in a front tracking algorithm is to construct an approximate solution to (1)-(2) in a class of piecewise constant functions with small total variation. We start at time $t = 0$ with a piecewise constant approximation $\overline{u}_\varepsilon$ of the initial data $\overline{u}$ (2) and solve the resulting Riemann problems at

This work has been partially supported by TMR project HCL ERBFMRXCT960033

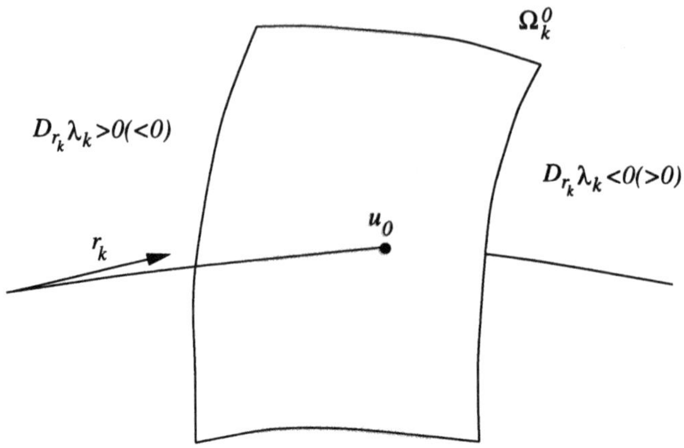

FIGURE 1. Assumption **(A)**

each point of discontinuity of $\overline{u}_\varepsilon$, replacing rarefaction waves by many small jumps of size ε (see Figure 2). Then we track the outgoing fronts until the first time two

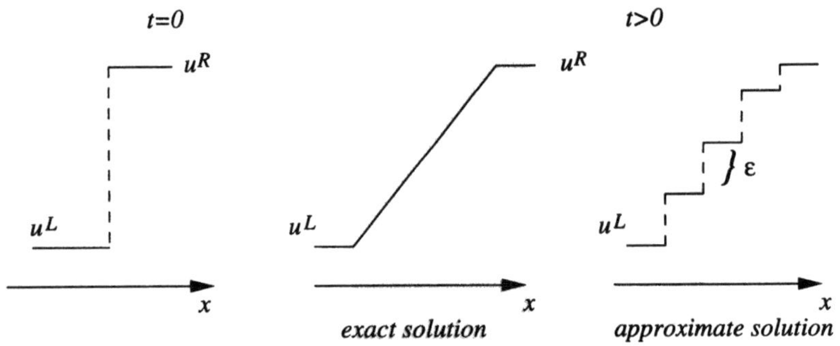

FIGURE 2. Split of rarefaction waves

waves interact. Solve again the corresponding Riemann problem in the same class of functions and go on till the second time interaction, etc. (see Figure 3). The core of the construction consists in showing that

1. the total variation of each approximate solution $u_\varepsilon = u_\varepsilon(t, x)$ remains uniformly small;
2. the number of lines of discontinuity of each u_ε is finite.

Using a compactness argument, it can be proved that the sequence of approximate solutions converges in $\mathbb{L}^1_{loc}((0, +\infty) \times \mathbb{R})$ to a weak solution $u = u(t, x)$ of (1)-(2).

A wave front tracking algorithm for 2×2 genuinely nonlinear systems was introduced in [8]. Next the method was extended in [6, 12] to $N \times N$ systems, where each characteristic field r_k is genuinely nonlinear or linearly degenerate

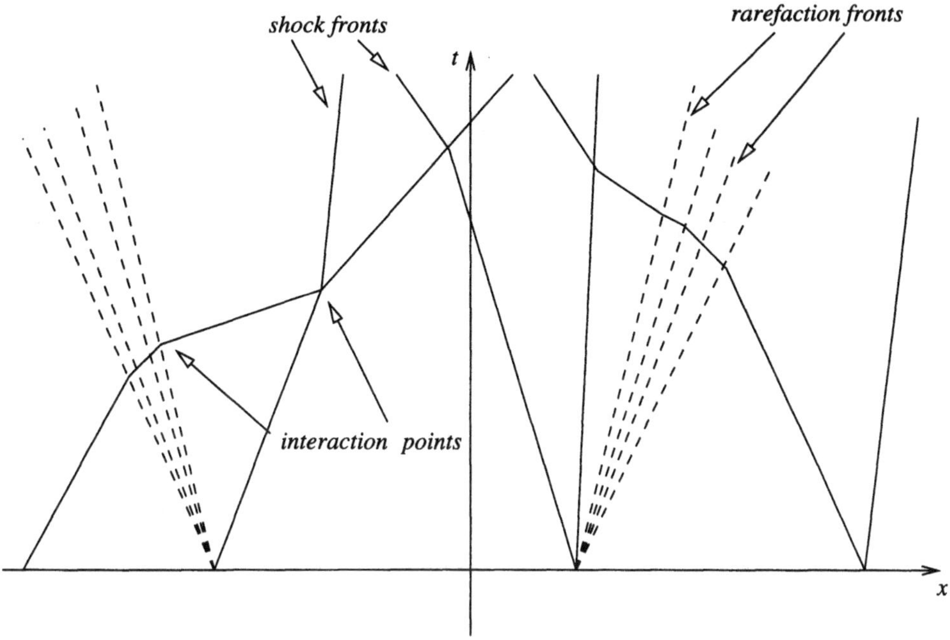

FIGURE 3. A wave front tracking scheme

or satisfies $Dr_k \cdot r_k \equiv 0$, i.e. its integral curves are straight lines. Next, in [2] there was constructed a sequence of approximate front tracking solutions for 2×2 systems satisfying assumption **(A)**. In particular, this algorithm leads to define a sequence of Lipschitz continuous semigroups acting on a class of piecewise constant functions. Such a sequence converges to a Standard Riemann Semigroup [7] of entropy weak solutions of (1) [3].

Throughout the paper, because we will consider only solutions with small total variation, we shall assume that the eigenvalues $\lambda_k(u)$, $k = 1, \dots, N$, $u \in \Omega$, are uniformly bounded. In order to measure the strength of a k-wave, we parameterize the shock and rarefaction curves through a point $u_0 \in \Omega$, $S_k(u_0)[\cdot]$, $R_k(u_0)[\cdot]$, choosing a nonsingular parameter σ_k and letting shock and rarefaction curves of the same characteristic family still have a second order contact at their starting point. For sake of simplicity, we will carry out the analysis under the assumption that

$$D^2_{r_k} \lambda_k(u^0) < 0\,, \qquad u^0 \in \Omega^0_k\,, \tag{5}$$

holds for any $k = 1, \dots, N$. This means that for each $u^0 \in \Omega^0_k$ the function

$$s \mapsto \lambda_k\big(R_k(u^0)[s]\big)\big(u^0\big)$$

attains its maximum value at $s = 0$. The other case is symmetric. Details can be found in [4].

The general entropy admissible solution of a Riemann Problem (u^L, u^R) for a system satisfying assumption **(A)** consists of rarefaction waves, compressive shocks and one side contact-discontinuities adjacent to a rarefaction wave, called *composed waves* (see [1, 10, 11]). In particular, because of (3)-(4), a composed wave of the k-th characteristic family may appear whenever u^L and u^R lie on opposite sides with respect to the manifold Ω_k^0. With the assumption (5), such a wave consists of a left contact-discontinuity on the right of a rarefaction wave (see Figure 4); To take advantage of the symmetry of the system, the general entropy

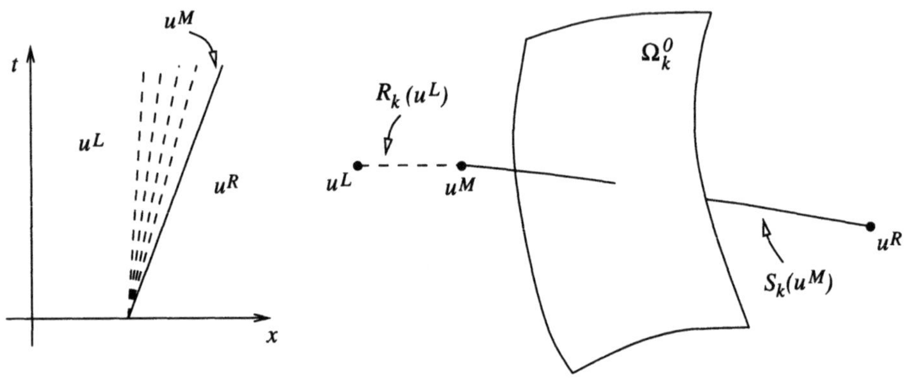

FIGURE 4. Left contact discontinuities

admissible solution of a Riemann problem (u^L, u^R) is expressed in terms of two types of an elementary curve, parametrized on a suitable neighborhood of the origin $\mathcal{J} \subseteq \mathbb{R}$ (see [1, 10, 11]). The *elementary curve through u k-th characteristic family*, consists of all right states connected with the left state u by an elementary wave (rarefaction, shock or composed wave) of the k-th characteristic family and it will be denoted $\Psi_k(u)$.

Given a k-wave (u^L, u^R) of size σ, we will use a superscript ^{r} and a superscript ^{s} to denote respectively its rarefaction and shock components, i.e., with the notations at Figure 4,

$$u^R = S_k(u^M)[\sigma^s], \quad u^M = R_k(u^L)[\sigma^r]. \tag{6}$$

If u^L and u^R are two states connected by two waves of sizes σ' and σ'' respectively, then we define the *quantity of interaction* $\mathcal{I}(\sigma', \sigma'')$ between σ' and σ'' as follows. If σ', σ'' belong to different characteristic families or they belong to the same family and $\sigma'\sigma'' < 0$, we set

$$\mathcal{I}(\sigma', \sigma'') \doteq |\sigma'\sigma''|. \tag{7}$$

If σ', σ'' both belong to the k-th characteristic family, $\sigma'\sigma'' > 0$, we set

$$\mathcal{I}(\sigma', \sigma'') \doteq \Big[\big|(\sigma' + \sigma'')^r - \sigma'^r\big| + \big|\sigma'^s\big|\Big]\,|\sigma''|. \tag{8}$$

The main result of this paper is stated in the following theorem

Theorem 1. *There exists $\delta_0 > 0$ such that for any initial condition $\overline{u} \in \mathbb{L}^1(\mathbb{R}; \mathbb{R}^N)$ satisfying Tot.Var.$(\overline{u}) < \delta_0$ and any $\varepsilon > 0$, the following holds. There exists a uniformly (in ε) Lipschitz continuous map $u_\varepsilon : [0, \infty) \mapsto \mathbb{L}^1_{loc}(\mathbb{R}; \mathbb{R}^N)$ such that $u_\varepsilon(t, \cdot)$ has uniformly (in t and ε) bounded total variation. As a function of two variables, u_ε is piecewise constant with discontinuities occurring along finitely many lines in the (t, x)-plane. Moreover, only finitely many wave fronts interactions occur. A suitable subsequence $u_\nu = u_{\varepsilon_\nu}(t, x)$ converges in $\mathbb{L}^1_{loc}((0, +\infty) \times \mathbb{R})$ to a weak solution $u = u(t, x)$ of the Cauchy problem (1)-(2). Such a solution satisfies the entropy inequality*

$$\left[\eta(u)\right]_t + \left[q(u)\right]_x \leq 0 \tag{9}$$

in the sense of distribution, for any convex entropy $\eta = \eta(u)$ with associated entropy flux $q = q(u)$.

Remark 1. *In [4] it is proved that, in case a Standard Riemann Semigroup [7] exists, then the weak solution obtained at Theorem 1 is admissible in the sense of T.P. Liu [10, 11]. Such an admissibility criterion, introduced for general systems of conservation laws, reduces to the classical Lax condition [9, 7] in case of systems satisfying assumption* **(A)**.

2. Riemann Solvers

Following [5, 6], in order to construct a piecewise constant approximate solution for all positive times, we need to introduce several ways to solve a Riemann problem generated by wave fronts interactions. We will use accurate solvers whenever the quantity of interaction is sufficiently large, while we will use simplified solvers when the quantity of interaction is "small". In this construction, a shock wave or a contact discontinuity of the k-th characteristic family (u^L, u^R) will always travel with its own exact speed $\lambda_k^s(u^L, u^R)$. Instead, we assign to the rarefaction front of the k-th characteristic family (u^L, u^R) the speed of its left state u^L, $\lambda_k(u^L)$. Moreover, sometimes we will force a composed wave (u^L, u^R) (rarefaction plus contact discontinuity) to behave like a unique front. In this case the speed of this front will be defined as

$$\lambda_k^w(u^L, u^R) \doteq \frac{\sigma^r \lambda_k(u^L) + \sigma^s \lambda_k^s(R_k(u^L)[\sigma^r])[\sigma^s]}{\sigma}, \tag{10}$$

where we use the notations at (6). Observe that in any case we have

$$\lambda_k^w(u^L, u^R) = \begin{cases} \lambda_k^s(u^L, u^R) & \text{if } (u^L, u^R) \text{ is a shock front,} \\ \lambda_k(u^L) & \text{if } (u^L, u^R) \text{ is a rarefaction front.} \end{cases} \tag{11}$$

Following [5, 6], in the simplified solution we collect all the "small" waves into a so-called non physical wave front. All non physical waves travel with a fixed speed $\widehat{\lambda}$ satisfying

$$\widehat{\lambda} > \lambda_i(u), \qquad \forall\, u \in \Omega, \quad i = 1, \ldots, N. \tag{12}$$

For notational convenience, we regard a non physical front as belonging to a fictitious linearly degenerate $(N+1)$-th characteristic family.

Now assume that at a positive time $\bar{t}$ an interaction occurs at $\bar{x}$ involving two wave fronts (u^L, u^M), (u^M, u^R), of sizes σ', σ'' and families k', k'' respectively. Here (u^L, u^M) is the left incoming wave. Fix a positive parameter ε controlling the maximum strength of a rarefaction front.

2.0.1. ACCURATE SOLVERS. The general solution of the Riemann Problem (u^L, u^R) is obtained by piecing together all of the approximate solutions of the elementary waves that comes out from the concatenation of the elementary curves Ψ_k. Here, shock fronts and contact discontinuities are solved exactly, while rarefaction waves are split into a rarefaction fan of small jumps, each of size ε (see Figure 2).

2.0.2. SIMPLIFIED SOLVERS. Define

$$x^{\mathcal{NP}}(t) \doteq \bar{x} + \widehat{\lambda}(t - \bar{t}). \tag{13}$$

First assume that σ', σ'' belong to the same characteristic families, i.e. $k' = k'' \doteq k$. Let $\widetilde{u}^R$ be such that

$$\widetilde{u}^R = \Psi_k(u^L)[\sigma' + \sigma'']. \tag{14}$$

Then, the approximate solution of the Riemann problem (u^L, u^R) will consist of two waves, one, $(u^L, \widetilde{u}^R)$, of the k-th characteristic family, and the other, $(\widetilde{u}^R, u^R)$, will be a non-physical front, traveling with speed $\widehat{\lambda}$. The physical wave $(u^L, \widetilde{u}^R)$ will be solved according to the following scheme. Whenever the rarefaction component $(u^L, \widetilde{u}^R)^r$ of $(u^L, \widetilde{u}^R)$ is nonzero, we compare its strength $|(u^L, \widetilde{u}^R)^r|$ with 3ε in order to decide to treat $(u^L, \widetilde{u}^R)$ as a single front or not. Hence we behave as follows (see Figure 5)

1. If $|(u^L, \widetilde{u}^R)^r| < 3\varepsilon$ we consider $(u^L, \widetilde{u}^R)$ as a single wave front traveling with speed $\lambda_k^w(u^L, \widetilde{u}^R)$ defined at (10). This consider also the cases $(u^L, \widetilde{u}^R)$ is a shock front, traveling with speed $\lambda_k^s(u^L, \widetilde{u}^R)$, due to (11).
2. If $|(u^L, \widetilde{u}^R)^r| \geq 3\varepsilon$, we split the rarefaction component $(u^L, \widetilde{u}^R)^r$ in small jumps of size ε (see Figure 2), each traveling with the speed of its left state.

Now assume that σ', σ'' belong to different characteristic families, i.e. $k' > k''$. Let $\widetilde{u}^M$, $\widetilde{u}^R$ be such that

$$\widetilde{u}^M = \Psi_{k''}(u^L)[\sigma''], \qquad \widetilde{u}^R = \Psi_{k''}(\widetilde{u}^M)[\sigma']. \tag{15}$$

Then, the approximate solution of the Riemann problem (u^L, u^R) will consist of three waves, one, $(u^L, \widetilde{u}^M)$, of the k''-th characteristic family, another, $(\widetilde{u}^M, \widetilde{u}^R)$, of the k'-th characteristic family, and the third, $(\widetilde{u}^R, u^R)$, will be a non-physical front, traveling with speed $\widehat{\lambda}$. Each physical wave $(u^L, \widetilde{u}^M)$, $(\widetilde{u}^M, \widetilde{u}^R)$ will be solved according to the scheme described above, splitting the rarefaction components $(u^L, \widetilde{u}^M)^r$, $(\widetilde{u}^M, \widetilde{u}^R)$ whenever they are larger that 3ε.

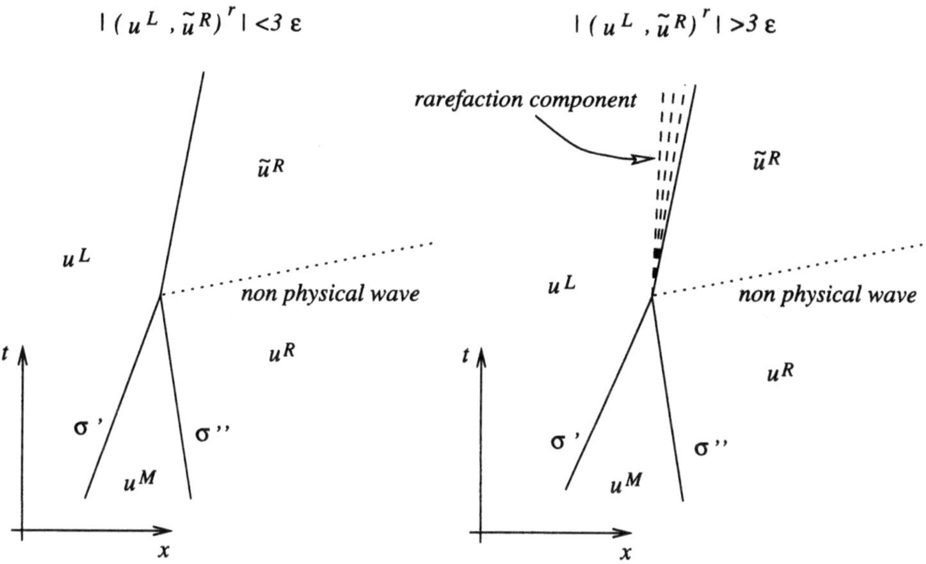

FIGURE 5. Simplified solvers

2.0.3. CRUDE SOLVERS. Assume that (u^M, u^R) is a physical wave-front, i.e. a rarefaction front, a shock front, or a composed wave traveling as a single front. Let (u^L, u^M) be a non-physical wave-front, traveling with speed $\widehat{\lambda} > \lambda_i(u)$, $\forall\, u$, $i = 1, \ldots, N$ and with strength $|\sigma'|$ defined by

$$|\sigma'| \doteq |u^L - u^M|. \tag{16}$$

Let $\widetilde{u}^R$ be defined by

$$\widetilde{u}^R = \Psi_k(u^L)[\sigma'']. \tag{17}$$

Then, the approximate solution is made up of two wave fronts, the physical one $(u^L, \widetilde{u}^R)$ traveling with speed $\lambda_k^w(u^L, \widetilde{u}^R)$, and the non physical one $(\widetilde{u}^R, u^R)$ traveling with speed $\widehat{\lambda}$ satisfying (12).

3. The Approximate Solutions

For any fixed $\varepsilon > 0$, now we construct an approximate solution $u_\varepsilon = u_\varepsilon(t, x)$ of (1)-(2). For a given initial data $\overline{u}$, let $u_\varepsilon(0, \cdot) = \overline{u}_\varepsilon$ be a piecewise constant $\mathbb{L}^1$ function such that

$$\left\|\overline{u} - \overline{u}_\varepsilon\right\|_{\mathbb{L}^1} \leq \varepsilon, \qquad \text{Tot.Var.}(\overline{u}_\varepsilon) \leq \text{Tot.Var.}(\overline{u}).$$

At time $t = 0$ we use Accurate Solvers to construct approximate solutions to the Riemann problems at the points of discontinuity of $\overline{u}_\varepsilon$. By performing a slight perturbation of the speeds of the waves, we can assume that at any positive time any interaction involves only two wave fronts. Now suppose that at time $t > 0$ a collision occurs involving two wave fronts of sizes σ', σ'' respectively, σ' on the left of σ''. The Riemann Problem generated by this interaction is solved as follows. Let $\mathcal{I}(\sigma', \sigma'')$ be defined as at (7)-(8). Let $\rho \ll \varepsilon$ be a small positive parameter to be specified later. We will use

-: the Accurate Solvers if none of the waves is a non physical wave front and $\mathcal{I}(\sigma', \sigma'') > \rho$;

-: the Simplified Solvers if none of the waves is a non-physical wave front and $\mathcal{I}(\sigma', \sigma'') \leq \rho$;

-: the Crude Solvers if one of the wave is a non physical wave front.

Observe that, due to the choice of the speed $\widehat{\lambda}$ at (12), a non-physical front can interact only with physical ones.

The technique involved in the proof of the convergence of the scheme is standard. We introduce two functionals V and Q measuring the total variation and the interaction potential, respectively. At each time $t > 0$ at which no interaction occur, call $x_1(t) < \cdots < x_m(t)$ the position of the jumps in $u_\varepsilon(t, \cdot)$, and $\sigma_1, \ldots, \sigma_m$, $i_1, \ldots, i_m$ their respective sizes and characteristic families. We define the *total strength* $V = V(t)$ of waves in $u_\varepsilon(t, \cdot)$ as

$$V(t) \doteq \sum_{\alpha} |\sigma_\alpha|, \tag{18}$$

and the *interaction potential* $Q = Q(t)$ as

$$Q(t) \doteq \sum_{\substack{i_\alpha = i_\beta \neq n+1 \\ \sigma_\alpha \sigma_\beta > 0}} |\sigma_\alpha \sigma_\beta| + \sum_{\substack{\alpha \\ i_\alpha \neq N+1}} |\sigma^r_\alpha \sigma_\alpha| +$$

$$+ c \Bigg[\sum_{\substack{i_\alpha = i_\beta \neq N+1 \\ \sigma_\alpha \sigma_\beta < 0}} |\sigma_\alpha \sigma_\beta| + \sum_{\substack{i_\alpha > i_\beta \\ x_\alpha(t) < x_\beta(t)}} |\sigma_\alpha \sigma_\beta| \Bigg], \tag{19}$$

where $c > 1$ is a suitable constant. Standard calculations [4] lead to establish that whenever two wave fronts σ' and σ'' interact, there hold

$$\Delta Q(\tau) \leq -\frac{1}{2}\, \mathcal{I}(\sigma', \sigma''), \qquad \Delta V(\tau) + 2C_1 \Delta Q(\tau) \leq 0, \tag{20}$$

for some positive constant C_1. Relying on estimates (20), it can be deduced that the total variation of the approximate solutions u_ε remains uniformly (in ε) bounded.

In order to make the scheme converge, now we must control the total number of wave fronts and the total amount of non physical waves. Regarding the number of wave fronts, observe that, due to (20), the number of interactions involving two wave fronts σ', σ'' such that $\mathcal{I}(\sigma', \sigma'') \geq \rho$ is finite. Then, since new wave fronts can be originated only at interaction points, it suffices to take into account of interactions involving fronts σ', σ'' with $\mathcal{I}(\sigma', \sigma'') < \rho$. Among these interactions, we consider only the ones such that at least one of the outgoing waves has a rarefaction component exceeding 3ε, in the other case no new physical front being originated by construction. Assume that at time $t = \bar{\tau}$ such an interaction occurs. Observe that the rarefaction component of a wave σ' of the k-th family can be increased only by interaction with a wave σ'' of a different characteristic family or belonging to the same family but with $\sigma'\sigma'' < 0$. In any case, because of (7), there holds $\mathcal{I}(\sigma', \sigma'') = |\sigma'\sigma''|$. Relying on estimate $(20)_1$, one can prove that there exist a positive constant C_2 such that the following holds [4]. Let $\bar{\tau}$ be as above. Then it can be found a time $\tilde{\tau} < \bar{\tau}$ such that

$$Q(\bar{\tau}) - Q(\tilde{\tau}) \leq -\frac{1}{2}C_2\varepsilon^2 . \tag{21}$$

From (21) we deduce that in the scheme we use a finite number of times a Simplified Solver generating new physical wave fronts. Hence the total number of physical wave fronts is finite. Since non physical fronts can be originated only at interaction between physical ones, we infer that the number of non physical fronts is finite too.

Regarding the total amount of non physical waves, with arguments similar to the ones used in [5, 7], it can be proved that we can choose $\rho > 0$ so small that

$$\sum_{\text{non physical waves}} |\sigma| = \mathcal{O}(1)\varepsilon .$$

Now, we can pass to the limit in the approximate solutions as $\varepsilon \to 0$. Helly's Theorem implies the existence of a subsequence converging to a weak solution of the Cauchy problem (1)-(2). Such a solution turns out to be entropic in the sense of (9), by means of calculations similar to the ones in [6, 7].

References

[1] F. Ancona, A. Marson, *A note on the Riemann problem for general $n \times n$ systems of conservation laws,* Journal of Math. Anal. Appl., to appear.

[2] F. Ancona, A. Marson, Basic estimates on a front tracking algorithm for general 2×2 conservation laws, Math. Models Methods Appl. Sci., submitted.

[3] F. Ancona, A. Marson, *Well-posedness for general 2×2 systems of conservation laws,* Memoir Amer. Math. Soc., to appear.

[4] F. Ancona, A. Marson, *A wave front tracking algorithm for $N \times N$ non genuinely nonlinear conservation laws,* J. Differential Equations., to appear.

[5] P. Baiti, H.K. Jenssen, *On the front-tracking algorithm,* J. Math. Anal. Appl., **217** (1998), no. 2, 395–404.

[6] A. Bressan, *Global solutions to systems of conservation laws by wave-front tracking,* J. Math. Anal. Appl., **170** (1992), 414–432.

[7] A. Bressan, *Hyperbolic Systems of Conservation Laws - The one-dimensional Cauchy problem,* Oxford Univ. Press.

[8] R. DiPerna, *Global existence of solutions to nonlinear hyperbolic systems of conservation laws,* J. Differential Equations, **20** (1976), 187–212.

[9] P.D. Lax, *Hyperbolic systems of conservation laws II,* Comm. on Pure and Applied Math., **10** (1957), 537–566.

[10] T.P. Liu, *The Riemann problem for general* 2×2 *conservation laws,* Trans. Amer. Math. Soc., **199** (1974), 89–112.

[11] T.P. Liu, *The Riemann problem for general systems of conservation laws,* J. Differential Equations, **18** (1975), 218–234.

[12] N.H. Risebro, *A front-tracking alternative to the random choice method,* Proc. Amer. Math. Soc., **117** (1993), 1125–1139.

F. Ancona
Dipartimento di Matematica and C.I.R.A.M.,
Università di Bologna,
Piazza Porta S. Donato, 5,
40123 - Bologna, Italy
`ancona@ciram3.ing.unibo.it`

A. Marson
Dipartimento di Matematica,
Università di Brescia,
Via Valotti, 9,
25133 - Brescia, Italy
`marson@ing.unibs.it`

International Series of Numerical Mathematics
Vol. 140, © 2001 Birkhäuser Verlag Basel/Switzerland

Well-Posedness for Non Genuinely Nonlinear Conservation Laws

Fabio Ancona and Andrea Marson

1. Introduction

Consider the Cauchy problem for a strictly hyperbolic system of conservation laws in one space dimension

$$u_t + F(u)_x = 0, \tag{1}$$

$$u(0,x) = \overline{u}(x), \tag{2}$$

where $u = u(t,x) \in \mathbf{R}^N$ are the conserved quantities, while $t \geq 0$ and $x \in \mathbf{R}$ are, respectively, the time and space variables. Let the flux function $F : \Omega \mapsto \mathbf{R}^N$ be a smooth vector field defined on a domain $\Omega \subseteq R^N$. Denote with $\lambda_1(u) < \cdots < \lambda_N(u)$ the eigenvalues of the Jacobian matrix $A(u) \doteq DF(u)$ and with $r_1(u), \dots, r_N(u)$, a corresponding basis of eigenvectors. We assume that system (1) admits non genuinely non linear (NGNL) characteristic fields, i.e. characteristic fields that are neither genuinely nonlinear (GNL) nor linearly degenerate (LD) in the sense of Lax. More precisely, we make the following assumption.

(A): If the k-th characteristic family is NGNL, then the set

$$\Omega_k^0 \doteq \{u \in \Omega : \ \nabla\lambda_k \cdot r_k(u) = 0\}$$

is a smooth, $N-1$-dimensional manifold in the u-space, and is transversal to the vector field r_k.

We recall that hyperbolic systems of conservation laws generally do not admit smooth solutions and, therefore, weak solutions in the sense of distributions are considered. Moreover, for the sake of uniqueness, an entropy criterion for admissibility is usually added. In [15] T.P. Liu proposed an admissibility criterion valid for general systems of conservation laws with NGNL characteristic fields. For systems where, for each NGNL characteristic family, the directional derivative $\nabla\lambda_k \cdot r_k$ of the characteristic speed vanishes on a single, connected hypersurface in the u-space, this criterion is equivalent to the classical Lax stability condition.

For general $N \times N$ systems of the type considered in [15], the existence of global entropy weak solutions to (1)-(2), with small total variation, was first obtained in [16], using the Glimm scheme. An alternative method for constructing

This work has been partially supported by TMR project HCL ERBFMRXCT960033 solutions of the Cauchy problem, as limit of piecewise constant approximations defined by a front-tracking algorithm, is developed in [2, 3] for systems of two equations and in [4] for $N \times N$ systems satisfying the assumption **(A)**.

Aim of this note is to provide an account of recent developments concerning the well-posedness theory of the Cauchy problem (1)-(2), for systems with NGNL characteristic fields satisfying the assumption **(A)**. The uniqueness and $\mathbf{L}^1$-stability of entropy admissible weak solutions, when data have small total variation and each characteristic family is GNL or LD, was recently established in a series of papers [6, 8, 9, 10, 11, 12, 18, 19, 14]. The general strategy adopted to tackle the well-posedness problem in all of these works follows the semigroup approach introduced in [6] in which is considered the problem of constructing a whole Lipschitz continuous semigroup of solutions to (1), compatible with the standard solutions of the Riemann problems. Relying on the existence and uniqueness (up to the domain) of such a semigroup, one then obtains the well-posedness theory showing that any entropy weak solution to (1)-(2), that satisfies some mild assumptions either on the oscillation or on the local boundedness of the total variation, must coincide with the trajectory of the semigroup.

In order to prove the existence of a Lipschitz semigroup of solutions generated by (1), two different approaches were proposed in [8, 9], and in [12, 17, 18, 19], for systems with GNL or LD characteristic fields.

(I): Comparison of solutions by homotopy and linearization. In order to estimate the distance between two solutions u and v, one constructs a one parameter path $\gamma_t : \theta \mapsto u^\theta(t)$ of solutions joining u with v, and then study how the $\mathbf{L}^1$-length of γ_t varies in time. As long as all solutions u^θ remain sufficiently regular, the length of γ_t can be computed integrating a suitable defined weighted norm of a generalized tangent vector. By deriving careful a-priori estimates on the weighted norm of these tangent vectors, one provides a bound on the length of γ_t and hence on the distance $\|u(t) - v(t)\|_{\mathbf{L}^1}$.

(II): Construction of a Lyapunov type functional. It is explicited defined a functional $\Gamma = \Gamma(u, v)$ which is equivalent to the $\mathbf{L}^1$-distance:

$$\frac{1}{C} \cdot \|u - v\|_{\mathbf{L}^1} \leq \Gamma(u, v) \leq C \cdot \|u - v\|_{\mathbf{L}^1} \qquad \forall\, u, v\,, \tag{3}$$

for some constant $C > 0$, and which is decreasing in time along any pair of solutions of (1), i.e.

$$\Gamma\big(u(t), v(t)\big) \leq \Gamma\big(u(s), v(s)\big) \qquad \forall\, t > s \geq 0\,. \tag{4}$$

The first approach was pursued in [8], for systems of two equations, and in [9] for $N \times N$ systems. A functional equivalent to the $\mathbf{L}^1$-distance and decreasing along couples of approximate solutions was constructed in [18], in the case of 2×2 systems, and then extended to the general case of $N \times N$ systems in [12, 19].

The main result that we present in this note represents a first step towards the extension of the well-posedness theory to the more general class of NGNL systems, and is achieved following the approach **(I)**. We shall consider systems of two equations having both characteristic families NGNL and satisfying the assumption **(A)**. For such systems, in the same spirit of the construction developed in [8], we construct a front-tracking algorithm which yields a sequence of flow maps S^n whose trajectories are piecewise constant approximate solutions of (1) that depend on the initial conditions in a uniformly Lipschitz continuous way (with a Lipschitz constant independent on n). Such a sequence converges, on a set $\mathcal{D}$ of integrable functions with small total variation, to a globally Lipschitz flow S whose trajectories are weak solutions to (1). More precisely, the following holds.

Theorem 1. *Let F be a smooth map from a domain $\Omega \subset \mathbf{R}^2$ into $\mathbf{R}^2$. Assume that the system (1) is strictly hyperbolic and that each characteristic field $r_k, k = 1, 2,$ is NGNL and satisfies the basic assumption* **(A)**. *Then, there exist a closed domains $\mathcal{D} \subset \mathbf{L}^1(\mathbf{R};\ \mathbf{R}^2)$, constants $L,\ \overline{\delta} > 0$, and a continuous semigroup $S\ :\ [0, \infty[\ \times \mathcal{D} \mapsto \mathcal{D}$, with the properties:*

(i) : *Every function $\overline{u} \in \mathbf{L}^1$ with $Tot.Var.(\overline{u}) \leq \overline{\delta}$ lies in $\mathcal{D}$.*
(ii) : *For all $\overline{u} \in \mathcal{D}$, one has $S_0\overline{u} = \overline{u}$, $\quad S_t S_s \overline{u} = S_{t+s}\overline{u}$, $\quad t, s \geq 0$.*
(iii) : *For all $\overline{u}, \overline{v} \in \mathcal{D}$, $t, s \geq 0$, one has*

$$\left\| S_t\overline{u} - S_s\overline{v} \right\|_{\mathbf{L}^1} \leq L \left(|t - s| + \left\| \overline{u} - \overline{v} \right\|_{\mathbf{L}^1} \right).$$

(iv) : *If $\overline{u} \in \mathcal{D}$ is piecewise constant, then for $t > 0$ sufficiently small the function $u(t, \cdot) \doteq S_t\overline{u}$ coincides with the solution of (1)-(2) obtained by piecing together the entropy-admissible, self-similar solutions of the corresponding Riemann problems.*
(v) : *Each trajectory $t \mapsto u(t, \cdot) \doteq S_t\overline{u}$ yields a weak solution to the Cauchy problem (1)-(2).*

We remark that the rigorous proof of the above theorem is obtained at the price of heavy technicalities and the implementation of the approach **(I)**, in the general case of $N \times N$ NGNL systems, appears to be an almost hopeless task. The authors instead, are now pursuing the construction of a Lyapunov type functional Γ enjoying the properties (3)-(4), for NGNL systems satisfying the assumption **(A)**. The main step towards this goal will be to provide a quadratic functional $\mathcal{E}_\Gamma$ that captures the nonlinearity of nonconvex scalar conservation laws with one inflection point, and plays the same role of the entropy functional introduced in [17]. We conjecture that the basic idea here should be to construct a functional $\mathcal{E}_\Gamma = \mathcal{E}_\Gamma(u, v)$ defined in terms of a suitable decomposition of the $\mathbf{L}^1$-distance between any pair of solutions u, v, that, in a sense, reflects the different types of nonlinearities determined by the nonconvex flux function with a single inflection point. Roughly speaking, for any given pair of states u_0, v_0, the value of the functional $\mathcal{E}_\Gamma(u_0, v_0)$ should depend on three types of parameters measuring the size of the distance between u_0 and v_0 related with the regions of increasing characteristic speed, decreasing characteristic speed and maximum (or minimum) characteristic speed.

In the following sections we will sketch the main arguments envolved in the proof of Theorem 1. For all details we refer to [2, 3].

2. The Algorithm

The standard procedure to construct a front-tracking approximate solution (see [5], [7]) starts at $t = 0$ with a piecewise constant approximation of the initial data. The resulting Riemann problems are then solved within the class of piecewise constant functions by using an approximate Riemann solver. Next, one tracks the outgoing fronts untill the first time two waves interact. The corresponding Riemann problem can be solved applying again the approximate Riemann solver, etc... The main source of technical difficulty in this construction stems from the fact that the number of wave-fronts may approach infinity in finite time. To overcome such a difficulty, a simplified procedure for solving the Riemann problem is usually adopted by introducing "non-physical" fronts that propagate with a constant speed greater than all characteristic speeds. Clearly, the approximate solution generated by a construction of this type will not depend continuously on the initial data.

Instead, the algorithm in [3], as the one developed in [8] for GNL 2×2 systems, allows to keep finite the total number of fronts without introducing any non-physical wave. This is possible thanks to the existence of a coordinate system of Riemann invariants and thanks to a careful approximation of shock profiles.

The main features of our wave-front tracking algorithm are:

1: The approximate solution of every Riemann Problem is solved by the same procedure. In particular, we never introduce any non-physical wave-front.

2: Centered rarefaction waves are cut along a fixed grid of step size ε and approximated by a corresponding piecewise constant rarefaction fan.

3: Shock waves of small amplitude coincide with rarefactions.

A Riemann Solver with the above properties, valid for GNL (or LD) systems, is obtained in [8] by constructing ε-approximate solutions in which

- shocks of strength $|\sigma| \geq 2\sqrt{\varepsilon}$ satisfy the exact Rankine-Hugoniot equations;

- shocks of strength $|\sigma| \leq \sqrt{\varepsilon}$ are rarefactions;

- shocks of strength $|\sigma| \in \left[\sqrt{\varepsilon},\, 2\sqrt{\varepsilon}\right]$ connect a left and right state lying on a smooth interpolation between the shock and the rarefaction curve.

Unfortunately, in the case of NGNL systems, approximate solutions of this type fail to produce a Riemann Solver with the properties **1-3**, depending continuously on the initial data. This is the consequence of the different structure for such systems of the elementary waves contained in the exact solution of the Riemann Problem. In fact, in contrast with the standard GNL systems, the solution of a Riemann problem for NGNL systems may contain *composed waves* made of contact discontinuities adjacent to rarefaction waves. In particular, for systems satisfying the assumption **(A)**, the general self-similar solution of a Riemann problem consists of rarefaction waves, compressive shocks and composed waves made of one side contact-discontinuity adjacent to a rarefaction wave. Then, to guarantee the stability of the composite wave pattern of a piecewise constant function that approximates a wave of this type, we need to modify the Rankine-Hugoniot equations for shock waves of any amplitude. This fact, in turn, forces us to introduce different types of suitable interpolations between the rarefaction and the approximate shock curves defined in connection with the modified Rankine-Hugoniot conditions.

3. Global Existence of Solutions

A basic step in the proof of the global existence of approximate solutions generated by wave-front tracking consists in deriving a-priori bounds on the total variation which, in turn, allow to show that the number of wave-fronts in the approximate solution remains finite. As customary, these a-priori estimates are obtained using a functional Q that measures the *potential interaction* of waves in the solution.

In particular, for NGNL systems, T.P. Liu ([16]) introduced a functional Q in which the amount of potential interaction between waves of the same family is proportional to the product of the strength of the waves and of their angle. This choice is motivated by the fact that no interaction between two waves is expected when their angle is zero. On the other hand, this corresponds to consider an interaction functional that is fourth order w.r.t. the strength of the waves of the same family. In fact, the angle between two waves of strength $|\sigma'|$, $|\sigma''|$, belonging to the the same family, is proportional either to $|\sigma'\sigma''|$ or to $\max\{|\sigma'|^2, |\sigma''|^2\}$.

In our analysis, however, to establish the Lipschitz continuous dependence of the approximate solutions on the initial data, it is essential to have a disposal an interaction-potential that is second order w.r.t. the strength of all waves. For this reason we define here the amount of potential interaction between *any* couple of wave-fronts in the approximate solution as proportional to the product of their strengths, no matter if they belong to the same family or not. More precisely, for any given approximate solution u generated by wave-front tracking, with several wave-fronts, say located at $x_\alpha = x_\alpha(t)$, $\alpha = 1, \ldots, N$, calling $|\sigma_i^\alpha|$ the strength of the wave at x_α belonging to the i-th family, we define the interaction potential $Q(t) \doteq Q(u(t))$ by setting

$$Q(t) = K_1 \Bigg[\sum_{x_\alpha < x_\beta} |\sigma_2^\alpha \cdot \sigma_1^\beta| + \sum_{i=1,2} \sum_{\sigma_i^\alpha \cdot \sigma_i^\beta < 0} |\sigma_i^\alpha \cdot \sigma_i^\beta| \Bigg] + \sum_{i=1,2} \sum_{\sigma_i^\alpha \cdot \sigma_i^\beta > 0} |\sigma_i^\alpha \cdot \sigma_i^\beta| \quad (5)$$

where K_1 is a suitable constant > 1. Notice that, since we are assuming by this definition that the amount of interaction-potential between two wave-fronts of the same family is non zero no matter which angle they make, one would expect that the global interaction-potential may increase after an interaction that produces a composed wave containing a piecewise constant rarefaction fan. However, interactions of this type always involve waves of different families or waves of the same family having an opposite sign. Therefore, observing that the total strength of i-rarefaction waves produced by such an interaction is dominated by the total strength of j-waves (or of i-waves of opposite sign w.r.t. the outgoing waves) involved in the interaction, one can always control the increase of Q due to the new rarefaction waves by assigning a sufficiently large weight K_1 to the amount of interaction-potential of the interacting waves.

4. Stability of Approximate solutions

The proof of the Lipschitz continuous dipendence on the initial data of the approximate solutions constructed by our algorithm, is based on the same technique developed in [8]. The basic idea consists in "differentiating" a family of approximate solutions w.r.t. a parameter which determines the locations of the jumps. More precisely, consider a piecewise constant approximate solution u having

$$u(0,x) = \sum_{\alpha=1}^{N_0} \overline{\omega}_\alpha \cdot \chi_{]\overline{x}_{\alpha-1},\overline{x}_\alpha[}(x),$$

as initial data. Here χ_I denotes the characteristic function of the interval I. Then one can construct a one-parameter family of initial data $u^\theta(0,\cdot)$ obtained from $u(0,\cdot)$ by shifting the positions of the jumps $\overline{x}_0, \overline{x}_1, \ldots, \overline{x}_{N_0}$, at the constant rates $\overline{\xi}_0, \overline{\xi}_1, \ldots, \overline{\xi}_{N_0}$, i.e. letting

$$u^\theta(0,x) = \sum_{\alpha=1}^{N_0} \overline{\omega}_\alpha \cdot \chi_{]x^\theta_{\alpha-1},x^\theta_\alpha[}(x), \qquad x^\theta_\alpha = \overline{x}_\alpha + \theta\,\overline{\xi}_\alpha, \qquad \theta \in [-a,\, a].$$

Here the states $\overline{\omega}_\alpha$ remain fixed for all values of the parameter θ. Let $u^\theta = u^\theta(t,x)$ be the corresponding approximate solution. If a is sufficiently small, at any time t where no interaction occurs, the functions $u^\theta(t,\cdot)$ are still obtained from $u(t,\cdot)$ by shifting the positions of its jumps:

$$u^\theta(t,x) = \sum_{\beta=1}^{N_t} \omega_\beta \cdot \chi_{]x^\theta_{\beta-1}(t),\,x^\theta_\beta[}(x), \qquad x^\theta_\beta = x_\beta + \theta\,\xi_\beta,$$

for some shift rates $\xi_0, \xi_1, \ldots, \xi_{N_t}$. The $\mathbf{L}^1$-length of the path $\gamma_t : \theta \mapsto u^\theta(t,\cdot)$, is then computed by

$$\|\gamma_t\|_{\mathbf{L}^1} = \int_{-a}^{a} \sum_\beta \left|\Delta u^\theta(t,\, x^\theta_\beta)\right| \left|\xi_\beta\right| \, d\theta, \qquad \xi_\beta = \frac{dx^\theta_\beta}{d\theta}, \tag{6}$$

where $\Delta u^\theta = \omega_{\beta+1} - \omega_\beta$ is the jump of u^θ at x^θ_β.

Thus, to estimate the $\mathbf{L}^1$-distance between two nearby approximate solutions u^{-a}, u^a, one has to provide an a-priori bound on the integrand in (6). As soon as an estimate of this type is obtained, a standard argument yields the stability of the approximate solutions. Clearly the integrand in (6) can change (as a function of time) only at times where some interaction takes place between wave-fronts in

u^θ. Therefore the key point is to study how the summation in (6) is changed across each interaction time. As one would expect, the major difference in this analysis, between the standard case of GNL systems and the case of NGNL systems, regards the behaviour of the summation in (6) at any time where two waves of the same family interact. We consider the following basic case.

Example 1. *Let $u = u(t,x)$ be a piecewise constant solution, consisting of two shocks of the first family which interact at some time τ. Let $x'(t) < x''(t)$ be the locations of the shocks for $t < \tau$, and assume that the interaction produces an outgoing shock for each i-th family, at $x_i^+(t)$ (see fig 1). Call σ', σ'' the sizes of the two incoming waves, and σ_i^+ the size of the outgoing i-shock. Assume that σ', σ'' are both positive. Next, consider a family of perturbed solution u^θ, $\theta \in [0, 1]$,*

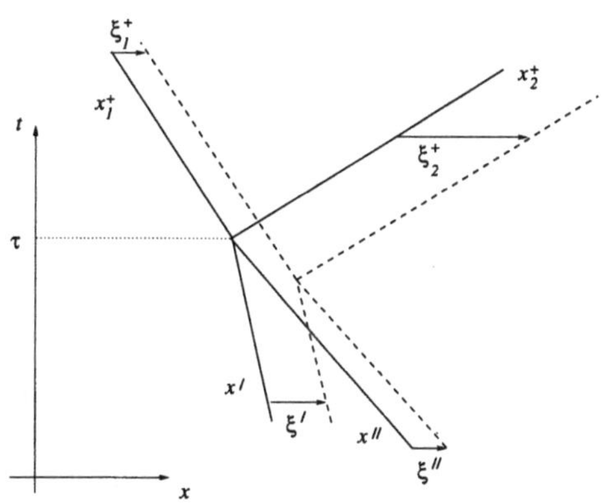

FIGURE 1

obtained from u by shifting the position of x', x'' at the constant rates ξ', ξ''. The incoming shocks of u^θ thus occur on the lines $x = x'_\theta(t)$, $x = x''_\theta(t)$, with

$$x'_\theta(t) = x'(t) + \xi'\,\theta, \qquad x''_\theta(t) = x''(t) + \xi''\,\theta, \qquad \theta \in [0,\,1].$$

For any fixed $\theta \in [0, 1]$, after the interaction is taken place (both in u and in u^θ) the perturbed solution u^θ will still contain wave-fronts of exactly the same strength as u. The locations $x^+_{1,\theta}$, $x^+_{2,\theta}$, of these wave-fronts, however, will be shifted by certain amounts, say ξ_1^+, ξ_2^+. An elementary calculation shows that the shift rates of the outgoing i-th shock is

$$\xi_i^+ = \frac{\partial x^+_{i,\theta}}{\partial \theta} = \frac{\xi''(\dot{x}' - \dot{x}_i^+) - \xi'(\dot{x}'' - \dot{x}_i^+)}{\dot{x}' - \dot{x}''}, \qquad i = 1,2, \tag{7}$$

where $\dot{x}$ denotes the speed of the wave-front at x. By the above arguments, the quantities

$$E^- \doteq |\sigma'\,\xi'| + |\sigma''\,\xi''|, \qquad E^+ \doteq |\sigma_1^+\,\xi_1^+| + |\sigma_2^+\,\xi_2^+|,$$

provide a measure of the $\mathbf{L}^1$*-distance between the two solutions* u, u^1 *before and after the interaction time* τ. *Standard interaction estimates on the size and the speed of the outgoing waves imply that*

$$\left|\sigma_2^+\right| + \left|\sigma_1^+ - (\sigma' + \sigma'')\right| = \mathcal{O}(1) \cdot \left|\sigma' \cdot \sigma''\right| \cdot \max\left\{\left|\sigma'\right|, \left|\sigma''\right|\right\}, \tag{8}$$

$$\left|\dot{x}_1^+ - \frac{\sigma'\dot{x}' + \sigma''\dot{x}''}{\sigma' + \sigma''}\right| = \mathcal{O}(1) \cdot \left|\sigma' \cdot \sigma''\right|, \tag{9}$$

where $\mathcal{O}(1)$ *denotes a quantity uniformly bounded by a constant that depends only on the system (1). Observe that the shifts* ξ_i^+ *in (7) may be very large compared with* ξ', ξ'', *because of the small denominator. However, in the case the system (1) is GNL, a supplementary bound holds*

$$\left|\dot{x}' - \dot{x}''\right| \geq c\left(\left|\sigma'\right| + \left|\sigma''\right|\right) \qquad \textit{for some} \quad c > 0. \tag{10}$$

Thus, from (8)-(10) we obtain the basic estimate

$$E^+ - E^- \leq \left|\sigma_1^+\xi_1^+ - \sigma'\xi' - \sigma''\xi''\right| + \left|\sigma_2^+\xi_2^+\right| = \mathcal{O}(1) \cdot \left|\sigma' \cdot \sigma''\right|\left(\left|\xi'\right| + \left|\xi''\right|\right). \tag{11}$$

The above estimates indicates that, although the quantity in (6) may well increase in time, one can derive a uniform bound on a weighted sum of the form

$$\sum_\beta \left|\Delta u^\theta(t, x_\beta^\theta)\right| \left|\xi_\beta\right| W^\beta \tag{12}$$

provided that the weights W^β are suitably chosen, depending on the amount of waves in u which approach the wave located at x_β. Clearly this arguments fails for NGNL systems since an estimate as in (10) no more holds.

For this reason, it is necessary to provide sharper interaction estimates than the ones in (8)-(9). Indeed, if the system (1) satisfies the assumption **(A)**, the following holds

$$\left|\sigma_2^+\right| + \left|\sigma_1^+ - (\sigma' + \sigma'')\right| = \mathcal{O}(1) \cdot \left|\sigma' \cdot \sigma''\right| \cdot \left(\left|\sigma'\right| + \left|\sigma''\right|\right) \cdot \left(\left|\sigma'\right| + \left|\sigma''\right| + \left|\delta_1\right|\right), \tag{13}$$

$$\left|\dot{x}_1^+ - \frac{\sigma'\dot{x}' + \sigma''\dot{x}''}{\sigma' + \sigma''}\right| = \mathcal{O}(1) \cdot \left|\sigma' \cdot \sigma''\right| \cdot \left(\left|\sigma'\right| + \left|\sigma''\right| + \left|\delta_1\right|\right), \tag{14}$$

$$\left|\dot{x}' - \dot{x}''\right| \geq c' \cdot \left(\left|\sigma'\right| + \left|\sigma''\right|\right) \cdot \left(\left|\sigma'\right| + \left|\sigma''\right| + \left|\delta_1\right|\right) \quad \text{for some} \quad c' > 0, \tag{15}$$

where $\delta_1 \doteq \delta_1(u^L)$ denotes the distance between the state u^L on the left of the wave at x', and the curve $\Gamma_1 = \left\{u \in \Omega : \ \nabla\lambda_1 \cdot r_1(u) = 0\right\}$. Estimates of this type were first established in [16] in the case of interacting waves of the same family and of the same sign, for approximate solutions constructed by the Glimm scheme. A derivation of such estimate in the general case of interacting waves of the same family and of *any* sign, for front-tracking solutions generated by the algorithm in [3], can be found in [2]. Thanks to (13)-(15), one can then recover the basic estimate (11) also in the case of a system satisfying the assumption **(A)**.

Relying on (11), we can now derive a uniform a-priori bound on a weighted sum as in (12), choosing weights of the form

$$W_i^\alpha \doteq \left[1 + K_2 \left[\sum_{\substack{j \neq i \\ (x_\alpha - x_\beta)\cdot(i-j)<0}} |\sigma_j^\beta| + \sum_{\sigma_i^\beta \cdot \sigma_i^\alpha < 0} |\sigma_i^\beta| \right] + K_3 \sum_{\sigma_i^\beta \cdot \sigma_i^\alpha > 0} |\sigma_i^\beta| \right] e^{K_4 Q},$$

where K_2, K_3, K_4, $(K_2 > K_3)$ are suitable positive constants, and Q is the interaction potential at (5). Here, with σ_i^β (σ_j^β) we denote the size of a wave-front of the i-th (j-th) characteristic family, and W_i^α is the weight assigned to an i-th wave located at x_α.

References

[1] F. Ancona and A. Marson, *A note on the Riemann problem for general $n \times n$ systems of conservation laws,* Journal Math. Anal. Appl., **260** (2001), 279–293.

[2] F. Ancona and A. Marson, *Basic estimates for a front tracking algorithm for general 2×2 conservation laws,* Math. Model Meth. & Appl. Sc.-M3AS, to appear.

[3] F. Ancona and A. Marson, *Well-posedness for general 2×2 systems of conservation laws,* Memoir Amer. Math. Soc., to appear.

[4] F. Ancona and A. Marson, *A wave front-tracking algorithm for $N \times N$ non genuinely nonlinear conservation laws,* J. Differential Equations, to appear.

[5] A. Bressan, *Global solutions to systems of conservation laws by wave-front tracking,* J. Math. Anal. Appl., **170** (1992), 414–432.

[6] A. Bressan, *The unique limit of the Glimm scheme,* Arch. Rational Mech. Anal., **130** (1995), 205–230.

[7] A. Bressan, *Hyperbolic Systems of Conservation Laws - The one-dimensional Cauchy problem,* Oxford Univ. Press, 2000.

[8] A. Bressan and R.M. Colombo, *The semigroup generated by 2×2 conservation laws,* Arch. Rational Mech. Anal., **133** (1995), 1–75.

[9] A. Bressan, G. Crasta and B. Piccoli, *Well-posedness of the Cauchy problem for $n \times n$ systems of conservation laws,* Memoir Amer. Math. Soc., **146** (2000), no. 694.

[10] A. Bressan and P. Goatin, *Oleinik type estimates and uniqueness for $n \times n$ conservation laws,* J. Differential Equations, **156** (1999), 26–49.

[11] A. Bressan and M. Lewicka, *A uniqueness condition for hyperbolic systems of conservation laws,* Discr. Cont. Dynam. Syst., **6** no. 3 (2000), 673–682.

[12] A. Bressan, T.P. Liu and T. Yang, $\mathbf{L}^1$ *stability estimates for $n \times n$ conservation laws,* Arch. Rational Mech. Anal., **149** (1999), 1–22.

[13] J. Glimm *A Solutions in the large for nonlinear hyperbolic systems of equations,* Comm. Pure Appl. Math., **18** (1965), 697–715.

[14] J. Hu and P.G. LeFloch, $\mathbf{L}^1$ *continuous dependence property for systems of conservation laws,* Arch. Rational Mech. Anal., **151** (2000), no. 1, 45–93.

[15] T.P. Liu, *The Riemann problem for general systems of conservation laws,* J. Differential Equations, **18** (1975), 218–234.

[16] T.P. Liu, *Admissible solutions of hyperbolic conservation laws,* Memoir Amer. Math. Soc., **30** (1981), no. 240.

[17] T.P. Liu and T. Yang, *A new entropy functional for scalar conservation laws,* Comm. Pure Appl. Math., **52** (1999), no. 11, 1427–1442.

[18] T.P. Liu and T. Yang, $\mathbf{L}^1$*-stability for* 2×2 *systems of hyperbolic conservation laws,* J. Amer. Math. Soc., **12** (1999), no. 3, 729–774.

[19] T.P. Liu and T. Yang, *Well-posedness theory for hyperbolic conservation laws,* Comm. Pure Appl. Math., **52** (1999), no. 12, 1553–1586.

F. Ancona
Dipartimento di Matematica and C.I.R.A.M.,
Università di Bologna,
Piazza P.S. Donato n. 5,
40127 - Bologna, Italy
`ancona@ciram3.ing.unibo.it.`

A. Marson
Dipartimento di Matematica,
Università di Brescia,
Via Valotti n. 9,
25133 - Brescia, Italy
`marson@bsing.ing.unibs.it.`

International Series of Numerical Mathematics
Vol. 140, © 2001 Birkhäuser Verlag Basel/Switzerland

Wave Phenomena at Liquid-solid Interfaces

S. Andreae, J. Ballmann, and U. Specht

1. Introduction

In many problems both in engineering and basic research the effects of interaction of a fluid and a solid phase have to be taken into account. In this paper we present a numerical scheme for the treatment of the fluid-solid interface and its applications. From a mathematical point of view the main problem is the coupling of nonlinear, under certain conditions hyperbolic, partial differential equations. Physically these equations lead to different types of waves and different wave speeds.

The fluid phase is modeled by the time-dependent Navier-Stokes equations for compressible both gases and liquids. Neglecting effects of viscosity and heat conduction, the system is hyperbolic, otherwise parabolic. The system is solved by an implicit Finite-Volume (FV) scheme using structured grids [2]. The solid is assumed to behave elastic-plastically. In the case of a non-decreasing stress-strain relation the governing equations of elastoplasticity are hyperbolic, exhibiting up to four different types of waves: the elastic dilatational- and shear waves and the fast and the slow plastic waves. A priori it is not known, which types of waves will appear. This is part of the solution procedure. The problem is solved by an explicit FV scheme on structured grid using a two step procedure by Lin and Ballmann [4] extended for problems symmetrical with respect to rotation. Applying the jump conditions across the contact discontinuity both nonlinear systems are explicitely coupled using the boundary scheme of the solid algorithm. Taking into account the contact conditions and prescribing for the fluid phase the kinematical quantities from the solution in the solid phase and for the solid phase the dynamical quantities from the solution in the fluid phase, an iterative convergent scheme is obtained.

The effects of interaction discussed here are most important in the case of the fluid being a liquid. The liquid, water in our case, is assumed to be compressible and will obey the Tait equation of state. Since the pressure then depends on approximately the seventh power of the density, even small fluctuations of the density will possibly lead to extreme pressure and loading values of the solid.

As a first application the water shock tube problem will be considered. The initial condition for the solid phase, a static stress distribution, will be obtained

---

Partially supported by the DFG Priority Research Program "Analysis and Numerics for Conservation Laws".

using the scheme for dynamically loaded problems, modified by an artificial viscosity introduced into the momentum conservation equations. In general waves in a solid travel faster than waves in a liquid (the speed of the dilatational wave in aluminium is about 6000 m/s, the speed of sound in water is about 1500 m/s). When the diaphragm of the shock tube bursts a shock and a very steep rarefaction wave will evolve. The shock immediately tends to expand the tube and induces waves in the solid. Since the disturbances in the tube will propagate faster than the shock wave the density and therefore the pressure will decrease in front of the shock wave. In our case even negative pressure in the water is observed. Finally the waves emitted at the fluid solid interface will focus on the axis of symmetry.

Another application is the interaction of a surface wave in a solid induced by a focusing shock in the liquid. We will present different reasons for yielding taking place in the solid. The first reason is yielding directly at the interface due to the extreme pressure in the liquid. The second reason is the so-called Stoneley wave, a surface wave, which causes plastic deformation in the interior of the solid.

2. Governing equations and numerical methods

The fluid is modeled by the time-dependent Navier-Stokes equations for compressible fluid flows. They are solved using an implicit finite volume scheme based on Roe's approximate Riemann solver [5] in a formulation by Grotowsky [2] for two-dimensional flows, extended by Hanke [3] for flows with rotational symmetry. Liquids are modeled using Tait's equation of state:

$$\left(\frac{\rho}{\rho_0}\right)^n = \frac{p+B}{p_0+B}. \tag{1}$$

The properties of different liquids are taken into account by the parameters B and n. The values for pure water are $B = 3.268 \cdot 10^8$ N/m^2 and $n = 7.15$. A reference state ist defined by $p_0 = 101300$ N/m^2 and $\rho_0 = 998.2$ kg/m^3.

The system describing time-dependent elastic-plastic behaviour of solids is solved by an explicit Godunov type two-step scheme according to Lin and Ballmann [4], which corresponds to Zwas' scheme in gas dynamics [1].

3. Coupling fluid-solid

Let the vector of the unknowns of the solid phase denoted by $\mathbf{w}$, its fluxes by $\mathbf{f}$ and $\mathbf{g}$ and the source term due to cylindrical coordinates by $\mathbf{h}$. The respective vectors of the fluid phase are denoted by corresponding capital letters.

Following Lin and Ballmann [4], the scheme for a cylindrical boundary with rotational symmetry of the solid reads (s. Fig. 1):

$$\sum_{i=1}^{2} \int_{\mathbf{w}_i^n}^{\mathbf{w}_{\bar{2}}^{n+1/2}} \mathbf{A} d\mathbf{w} = \frac{\Delta t}{\Delta r}\left(\mathbf{f}_2^n - \mathbf{f}_1^n\right) + \frac{\Delta t}{\Delta z_s}\left(2\mathbf{g}_{\bar{2}}^{n+1/2} - \mathbf{g}_1^n - \mathbf{g}_2^n\right) + \frac{\Delta t}{2}\sum_{i=1}^{2}\frac{\mathbf{h}_i^n}{r_i}. \tag{2}$$

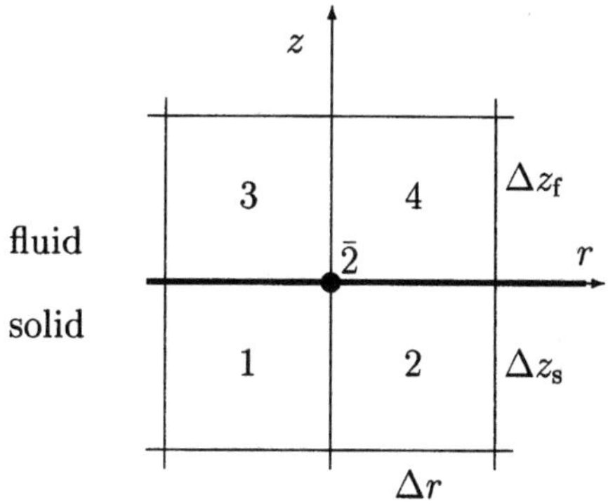

FIGURE 1. Discretization of the fluid-solid interface

The boundary conditions are taken into account by the vector $\mathbf{g}_{\bar{2}}^{n+1/2}$. For the jump conditions at a non-dissipative contact discontinuity the relations $v_{z,\bar{2}} = u_{z,\bar{2}}$ and $\tau_{z,\bar{2}} = -p_{\bar{2}}$ hold, v_z and u_z being the normal velocities in the solid and in the fluid, respectively, and τ_z being the normal stress in the solid and p the pressure in the fluid. For the fluid side the corresponding scheme for the boundary conditions reads:

$$\mathbf{W}_{\bar{2}}^{n+1/2} = \frac{1}{2}(\mathbf{W}_3^n + \mathbf{W}_4^n) + \frac{\Delta t}{2\Delta r}\left(\mathbf{F}_{\mathrm{k},3}^n - \mathbf{F}_{\mathrm{k},4}^n\right) + \frac{\Delta t}{2\Delta z_f}\left(2\mathbf{G}_{\mathrm{k},\bar{2}}^{n+1/2} - \mathbf{G}_{\mathrm{k},3}^n - \mathbf{G}_{\mathrm{k},4}^n\right) - \frac{\Delta t}{4}\sum_{i=3}^{4}\frac{\mathbf{H}_{\mathrm{k},i}^n}{r_i}. \quad (3)$$

The iterative solution (prescribing the dynamic quantities for the solid and the kinematic quantities for the fluid) is obtained by following steps:

1. start with $\mathbf{W}_{\bar{2}}^{i,1/2}$ estimated (iteration index i)
2. solve solid eq. for $\mathbf{w}$ using the dynamic interface condition $\tau_{z,\bar{2}}^{i,1/2} = -p_{\bar{2}}^{i,1/2}$
3. solve fluid eq. for $\mathbf{W}$ using the kinematic interface condition $v_{z,\bar{2}}^{i,1/2} = u_{z,\bar{2}}^{i,1/2}$
4. if $\dfrac{p_{\bar{2}}^{i+1} - p_{\bar{2}}^{i}}{p_{\bar{2}}^{i}} > \epsilon \rightarrow 2$
5. result: $p(\tau_z)$ and v at the interface.

4. Numerical results

4.1. Water shock tube problem

As an important test case for fluid-solid-interaction the shock tube problem with water as medium is treated.

The inner diameter of the cylindrical tube is $d_{\mathrm{i}} = 2$ mm, the outer diameter $d_{\mathrm{o}} = 3.5$ mm. As initial conditions for the liquid at rest (Tait's equation of state

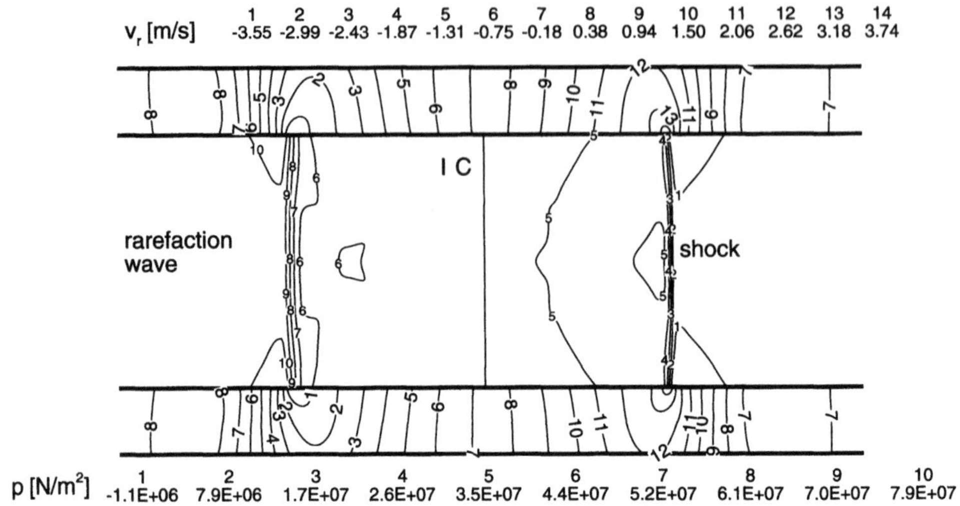

FIGURE 2. Water shock tube problem: radial component of the velocity vector in the solid (upper legend) and pressure in the liquid (lower legend), $t = 0.94\ \mu s$

(eq. 1)) in the high pressure region $p_h = 770$ bar and in the low pressure region $p_l = 1$ bar are prescribed. The position of the discontinuity at the beginning of the numerical calculation is denoted in Fig. 2 by 'IC'. For the solid (aluminium with the material properties: Young modulus $E = 7 \cdot 10^{10}$ N/m^2, shear modulus $G = 2.6 \cdot 10^{10}$ N/m^2, density $\rho = 2700$ kg/m^3) the static stress field ist prescribed.

As soon as the shock propagates into the low pressure region, the tube tends to flare. Since the waves in the solid travel faster than the shock wave in the water a peristaltic disturbance ahead of the shock will develop. Moreover the liquid-solid-interface will accelerate outwards, resulting in a low pressure region in front of the shock. In this example the pressure will reach even negative values[1]. In Fig. 3 the pressure distribution on the axis of symmetry (left diagram) and on the liquid-solid-interface (right diagram) are shown. >From Figs. 2 and 3 one can clearly recognize the focusing waves emitted by the elastic wall on the axis of symmetry. They will lead to a lower pressure behind the shock. The difference in pressure to both the numerical simulations with a rigid wall and the analytically obtained result is about 80 bar which means 20 per cent. In front of the shock on the axis of symmetry the liquid is undisturbed, since the von Schmidt waves (head waves) haven't reached the axis at this time yet, so they have to focus behind the shock.

Contrary to gases rarefaction waves in liquids are very steep, so similar phenomena will evolve there, with reverse sign.

[1]In the time dependent case in liquids for very short times metastable states during the phase change liquid-vapour can lead indeed to negative pressure.

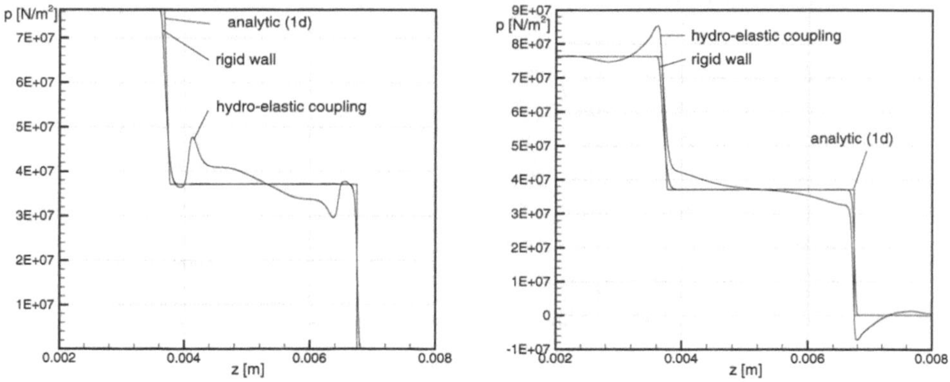

FIGURE 3. Pressure distribution on the axis of symmetry (left) and on the liquid-solid interface (right) for the analytically obtained solution and the numarically obtained results for the coupled and decoupled problems, $t = 0.94\ \mu$s

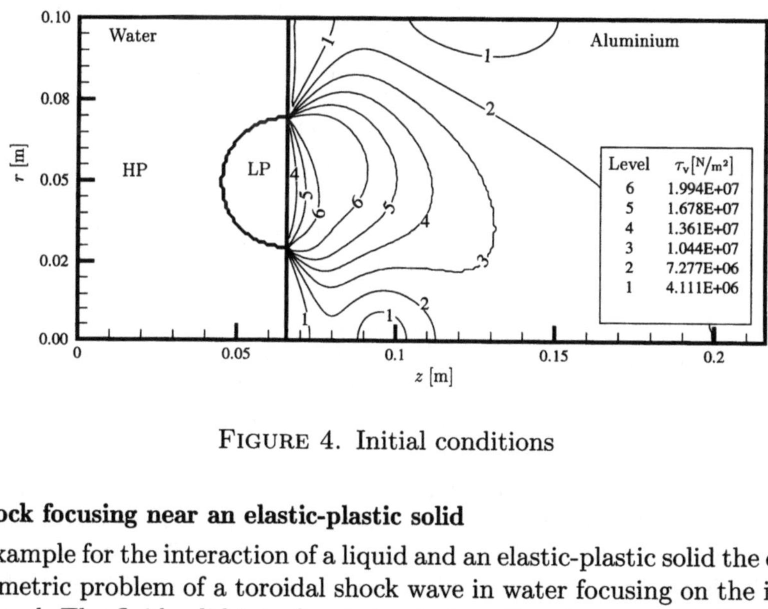

FIGURE 4. Initial conditions

4.2. Shock focusing near an elastic-plastic solid

As an example for the interaction of a liquid and an elastic-plastic solid the cylindrical symmetric problem of a toroidal shock wave in water focusing on the interface is discussed. The fluid-solid interface is located on the plane $z = 0.065$m.

On the liquid side in a region with a higher pressure (HP) a toroidal low pressure (LP) region on the interface is considered (s. Fig. 4 left). On the solid side the von Mises equivalent stress defined by

$$\tau_v = \sqrt{\frac{1}{3}\left((\tau_r - \tau_\Theta)^2 + (\tau_z - \tau_\Theta)^2 - (\tau_r - \tau_\Theta)(\tau_z - \tau_\Theta)\right) + \tau_{rz}^2} \qquad (4)$$

is shown. For the liquid at rest in the low pressure region $p_{LP} = 1$ bar and in the

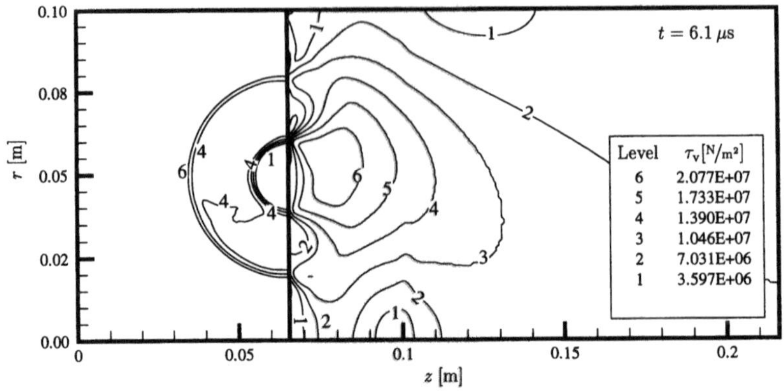

FIGURE 5. Pressure distribution in the liquid ($p_1 = 100$ bar, $p_6 = 600$ bar, $\Delta p = 100$ bar) and von Mises stress distribution in the solid ($t = 6.1$ μs)

high pressure region $p_{\mathrm{HP}} = 700$ bar are prescribed. The corresponding densities are $\rho_{\mathrm{LP}} = 998.2$ kg/m^3 und $\rho_{\mathrm{HP}} = 1025.6$ kg/m^3. Boundaries, which are neither axes of symmetry, nor liquid-solid interfaces, are treated as rigid. First in the solid (aluminium with the material properties: Young modulus $E = 7 \cdot 10^{10}$ N/m^2, shear modulus $G = 2.6 \cdot 10^{10}$ N/m^2, density $\rho = 2700$ kg/m^3) the static stress distribution corresponding to the initial pressure distribution in the water has to be determined. Boundaries, which are neither axes of symmetry, nor liquid-solid interface, are treated as sliding boundaries (vanishing normal velocity and shear stress). For the elastic-plastic constitutive law a plastic shear modulus $G_{\mathrm{p}} = 2.1 \cdot 10^{10}$ N/m^2 and an intial yield stress $\kappa_0 = 6 \cdot 10^7$ N/m^2 are assumed. Then the static stress distribution does not exceed the initial yield sress. For the geometrical situation see Fig. 4. The domains are discretized by 80×120 (liquid) and 180×120 (solid) quadratic cells. For the sake of clearness the labels for the pressure distribution in the liquid are described in the caption by the rule $p_i = p_1 + (i-1)\Delta p$ for the level i.

In Fig. 5 the pressure in the liquid and the von Mises stress distribution in the solid after $t = 6.1$ μs are shown. In the water a shock wave directed towards the solid wall and a steep rarefaction wave have evolved. The stress distribution in the solid is essentially dominated by the static initial conditions. Only at the shock and the rarefaction wave a weak stress concentration can be recognized. In the solid no plastic yielding has taken place yet.

After $t = 12.1$ μs, just before shock focusing (s. Fig. 6), there is still no plastic yielding in the solid. Nevertheless near the focus a strong stress concentration can be recognized. Due to the cylindrical symmetry high stresses also occur near the axis of symmetry.

Shortly after both shock focusing on the interface and rarefaction wave focusing on the axis of symmetry (s. Fig. 7), a small plastic zone near the shock focus

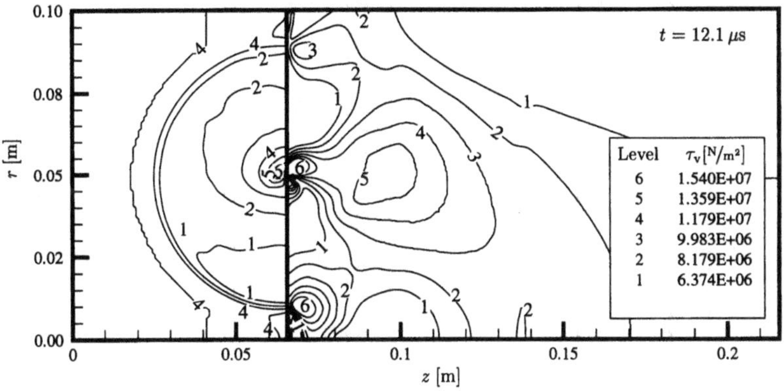

FIGURE 6. Pressure distribution in the liquid ($p_1 = 400$ bar, $p_6 = 900$ bar, $\Delta p = 100$ bar) and von Mises stress distribution in the solid ($t = 12.1$ μs)

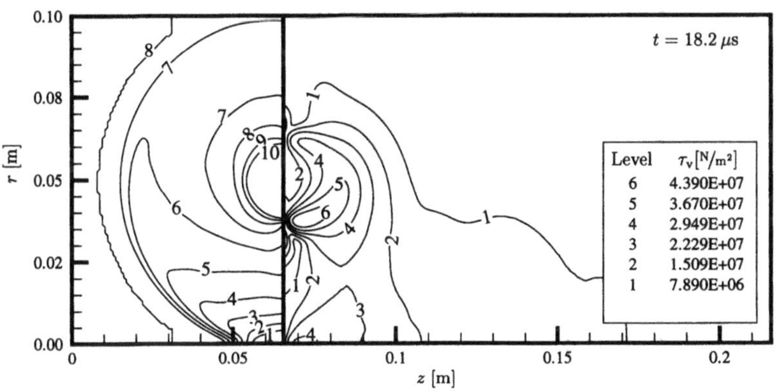

FIGURE 7. Pressure distribution in the liquid ($p_1 = 0$ bar, $p_{10} = 900$ bar, $\Delta p = 100$ bar) and von Mises stress distribution in the solid ($t = 18.2\mu$s)

has developed. Due to this sudden loading both dilatational and shear waves are emitted into the aluminium solid. The strength of these waves can be recognized by the so-called Stoneley wave at the interface (not fully developed yet). The Stoneley wave, a generalized Rayleigh wave, arises due to the interaction of a shear wave with the fluid boundary, which can sustain no or only weak shear stresses. Due to the rarefaction focusing near the axis, the pressure will become very low (in the case of isoline p_1 even negative). This leads to a solid locally loaded by a tension stress. The time history of the plastic zones is shown in Fig. 11.

The Stoneley wave directed towards the axis of symmetry (s. Fig. 8) has lead after $t = 30.3$ μs to a plastified zone inside the solid region (upper right in Fig. 11). Furthermore it is diffracted at the intersection of the axis of symmetry and the

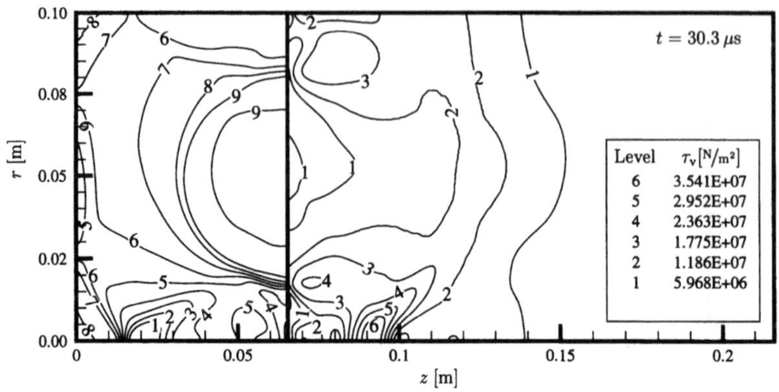

FIGURE 8. Pressure distribution in the liquid ($p_1 = 0$ bar, $p_9 = 800$ bar, $\Delta p = 100$ bar) and von Mises stress distribution in the solid ($t = 30.3$ μs)

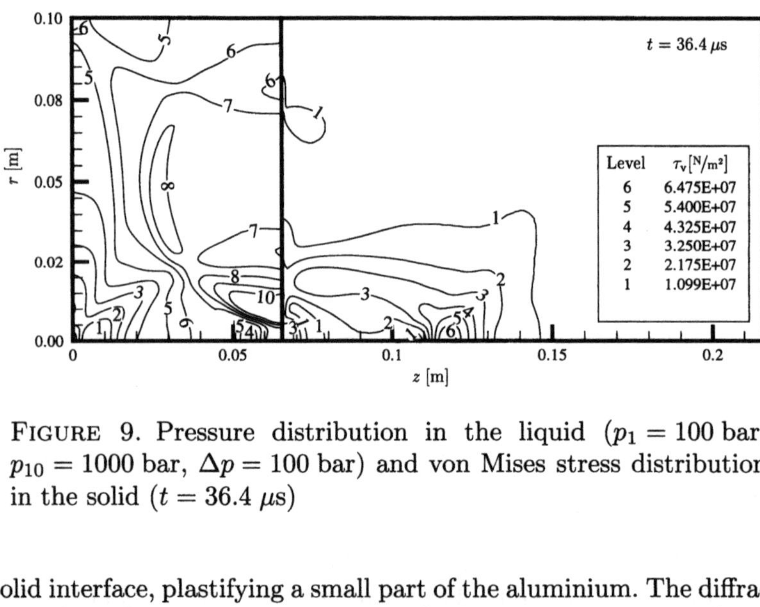

FIGURE 9. Pressure distribution in the liquid ($p_1 = 100$ bar, $p_{10} = 1000$ bar, $\Delta p = 100$ bar) and von Mises stress distribution in the solid ($t = 36.4$ μs)

liquid-solid interface, plastifying a small part of the aluminium. The diffracted part propagates along the axis of symmetry now, leading to an increasing plastified zone after $t = 36.4$ μs (s. Fig. 9 and lower left in Fig. 11).

Focusing of the shock (emitted from the first focus on the interface) on the axis of symmetry leads to a extreme pressures exceeding 2000 bar. At $t = 42.4$ μs this state in shown in Fig. 10. The pressure is loading the solid at the intersection of the axis of symmetry and the interface, leading to an increasing plastic zone (lower right in Fig. 11). Travelling along the axis of symmetry, the diffracted Stoneley wave increases the plastic zone of the solid loosing energy. Therefore in later times no further plastic yielding takes place.

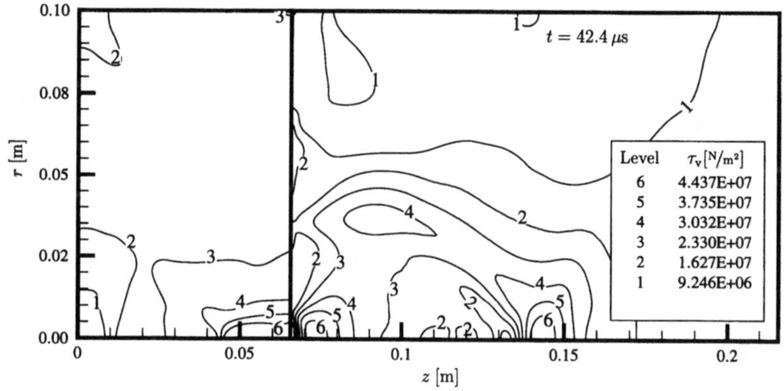

FIGURE 10. Pressure distribution in the liquid ($p_1 = 0$ bar, $p_6 = 2000$ bar, $\Delta p = 400$ bar) and von Mises stress distribution in the solid ($t = 42.4\ \mu s$)

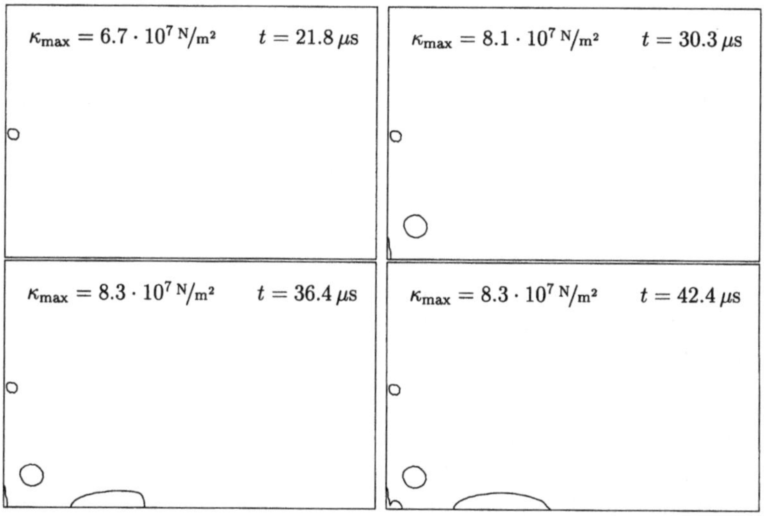

FIGURE 11. Time history of isolines of the yield stress in the solid (left boundaries: liquid-solid interface, lower boundaries: axis of symmetry)

5. Conclusions

Two applications of fluid-solid interaction phenomena have been presented. For the shock tube problem in water the influence of the solid on the liquid could be clearly recognized. The faster waves in the solid cause remarkable pressure fluctuation which would probably lead to cavitating regions using a more sophisticated description of the behaviour of the liquid.

By numerical simulation of a focusing toroidal shock in a liquid on the surface of an elastic-plastic solid different reasons for solid damaging are found. First, plastic yielding has taken place on the surface due to the high pressure caused by the shock focusing. Furthermore the emitted waves in solid lead to the so-called Stoneley wave, which loads the solid in that way, that even inside plastic yielding takes place, which means that not only loads directly applied to solid may lead to its failure but also waves in itself. This might play a role in the cavitation damaging mechanism, the detailed analysis of which will be the aim in our future work.

References

[1] B. Eilon, D. Gottlieb, and G. Zwas. Numerical Stabilizers and Computing Time for Second-Order Accurate Schemes. *J. Comput. Phys.*, 9:387–397, 1972.

[2] I.M.G. Grotowsky. *Ein numerischer Algorithmus zur Lösung der Navier-Stokes-Gleichungen bei Über- und Hyperschallmachzahlen.* PhD thesis, RWTH Aachen, VDI-Verlag, Reihe 20, Nr. 145, 1994.

[3] M. Hanke. Die Navier-Stokes-Gleichungen in krummlinigen Koordinaten; Diskretisierung für Upwind-Schemata. Lehr- und Forschungsgebiet für Mechanik, RWTH Aachen (unpublished), 1995.

[4] X. Lin and J. Ballmann. Numerical Method for Elastic-Plastic Waves in Cracked Solids, Part 2: Plane Strain Problem. *Arch. Appl. Mech.*, 63:283–295, 1993.

[5] P.L. Roe. Approximate Riemann Solvers, Parameter Vectors, and Difference Schemes. *J. Comput. Phys.*, 43:357–372, 1981.

S. Andreae
Lehr- und Forschungsgebiet für Mechanik, RWTH Aachen,
Templergraben 64, 52056 Aachen, Germany
`sigrid.andreae@lufmech.rwth-aachen.de`

J. Ballmann
Lehr- und Forschungsgebiet für Mechanik, RWTH Aachen,
Templergraben 64, 52056 Aachen, Germany
`ballmann@lufmech.rwth-aachen.de`

U. Specht
IABG mbH,
Postfach 12 12, 85503 Ottobrunn, Germany
`specht@iabg.de`

International Series of Numerical Mathematics
Vol. 140, © 2001 Birkhäuser Verlag Basel/Switzerland

Diffusive Discrete BGK Schemes for Nonlinear Hyperbolic-parabolic Systems

Denise Aregba-Driollet, Roberto Natalini, and Shaoqiang Tang

Abstract. We design numerical schemes for nonlinear degenerate parabolic systems. These schemes are based on discrete BGK models where both characteristic velocities and the source-term depend singularly on the relaxation parameter. Although we restrict this paper to cartesian grids, the method may be also adapted to unstructured meshes.

1. Introduction

We design discrete kinetic schemes for possibly degenerate parabolic systems:

$$\partial_t u_k + \sum_{d=1}^{D} \partial_{x_d} \left[A_{kd}\left(u\right)\right] = \Delta_x \left[B_k(u)\right], \quad (x,t) \in \mathbb{R}^D \times]0,\infty[, \ \ 1 \leq k \leq K, \tag{1}$$

$$u(x,0) = u_0(x). \tag{2}$$

We assume that the functions A and B are Lipschitz continuous, $\partial_t + \nabla.A$ is hyperbolic and $B' \geq 0$, *ie* for all u the eigenvalues of $B'(u)$ are real and non negative. If the system is not uniformly parabolic, the solutions can loose their regularity and uniqueness in finite time. In the scalar case $K = 1$, existence and uniqueness hold for entropy solutions of (1)-(2), see [5] and references in this paper. Our aim is to construct discretizations of (1) such that:

i) we have the same framework in both the hyperbolic ($B' \equiv 0$) and the parabolic ($B' > 0$) regime;
ii) convergence holds to the weak, and for $K = 1$ admissible in the sense of Carrillo, solutions to problem (1);
iii) numerical implementation is Riemann solverless and easy, even in the multidimensional case and in more complicate situations (source terms, variable coefficients...);
iv) numerical schemes can be improved easily by using high order algorithms and different methods (finite differences, finite volumes...).

Our approach is inspired, on one part from relaxation schemes [8], on another part from kinetic approximations of hydrodynamic equations, in the hyperbolic setting (see for example [13]) as well as in the diffusive setting (see for example [7]).

The schemes are based on a class of kinetic models which take the form:

$$\begin{cases} \partial_t f_l^\epsilon + \sum_{d=1}^{D} \lambda_{ld} \partial_{x_d} f_l^\epsilon = \frac{1}{\epsilon} \left(M_l(u^\epsilon) - f_l^\epsilon \right), & 1 \le l \le N, \\ \partial_t f_{N+m}^\epsilon + \gamma^\epsilon \sum_{d=1}^{D} \sigma_m^{(d)} \partial_{x_d} f_{N+m}^\epsilon = \frac{1}{\epsilon} \left(\frac{B(u^\epsilon)}{N'\theta^2} - f_{N+m}^\epsilon \right), & 1 \le m \le N', \\ u^\epsilon(x,t) = \sum_{l=1}^{N+N'} f_l^\epsilon(x,t). \end{cases} \tag{3}$$

Here each f_l^ϵ takes values in $\mathbb{R}^K$, the M_l are suitable Lipschitz continuous functions, the coefficients λ_{ld} are real and $\gamma^\epsilon = \mu + \frac{\theta\sqrt{N'}}{\sqrt{\epsilon}}$ with $\theta > 0$, $\mu \ge 0$. The parameter ϵ is positive.

We take $N' \ge D+1$, and $\left\{\sigma^{(1)}, \dots, \sigma^{(D)}\right\}$ is a given orthonormal family in $\left\{X \in \mathbb{R}^{N'},\ \sum_{m=1}^{N'} X_m = 0\right\}$. Consequently, this system is semilinear and weakly coupled.

We fix the initial condition for (3) as

$$f_0(x) = M(u_0(x)) \tag{4}$$

where $M_{N+m} = B/(N'\theta^2)$.

The following compatibility conditions make the link between systems (1) and (3): let u be the solution of (1) we look for, and let I be a fixed rectangle such that $u(x,t) \in I$. For all $w \in I$:

$$\sum_{l=1}^{N} M_l(w) = w - \frac{B(w)}{\theta^2}, \quad \sum_{l=1}^{N} \lambda_{ld} M_l(w) = A_d(w), \ \ 1 \le d \le D. \tag{5}$$

If M satisfies the above conditions we call it a local **Maxwellian Function** for system (1) on I. One can also, modifying slightly these conditions, let the function M depend on ϵ, see [4].

It is easy to see formally that if the sequence u^ϵ converges to some limit function u in a suitable (strong) topology when $\epsilon \to 0$, then the limit function is a weak solution to equation (1). Actually, let us denote

$$u^\epsilon = \sum_{l=1}^{N+N'} f_l^\epsilon, \ \ v_j^\epsilon = \sum_{l=1}^{N} \lambda_{lj} f_l^\epsilon + \gamma^\epsilon \sum_{m=1}^{N'} \sigma_m^{(j)} f_{N+m}^\epsilon \quad 1 \le j \le D. \tag{6}$$

Then a solution of (3) satisfies:

$$\begin{cases} \partial_t u^\epsilon & + \displaystyle\sum_{d=1}^{D} \partial_{x_d} v_d^\epsilon = 0, \\ \partial_t v_j^\epsilon & + \displaystyle\sum_{d=1}^{D} \left[\sum_{l=1}^{N} \lambda_{lj}\lambda_{ld}\partial_{x_d} f_l^\epsilon + \mu^2 \sum_{m=1}^{N'} \sigma_m^{(j)}\sigma_m^{(d)}\partial_{x_d} f_{N+m}^\epsilon \right] \\ & = \dfrac{1}{\epsilon}\left(A_j - v_j^\epsilon - (\sqrt{\epsilon}2\mu\theta\sqrt{N'} + \theta^2 N') \displaystyle\sum_{d=1}^{D}\sum_{m=1}^{N'} \sigma_m^{(j)}\sigma_m^{(d)}\partial_{x_d} f_{N+m}^\epsilon \right). \end{cases} \tag{7}$$

Consequently, if convergence holds, the f_l^ϵ tend to a maxwellian state, we have $\lim_{\epsilon\to 0} v_j^\epsilon = A_j(u) - \partial_{x_j} B(u)$, and u is solution of system (1).

Convergence to the entropy solution of (1)-(2) is proved in [4] for the scalar case $K = 1$. For a class of parabolic systems with nonlinear convection, see [11]. In both results an essential stability requirement is that the eigenvalues of each $M_l'(u)$ are real and positive:

$$\forall u \in I, \;\; M_l'(u) > 0, \;\; 1 \le l \le N, \tag{8}$$

which can be interpreted as a subcharacteristic condition, see the example in section 2. Here we construct models (3) and we use them to design new high order numerical approximations of (1). Stability and convergence properties hold under condition (8). In the purely hyperbolic case our method has already been proved to be efficient, see [2]. The main difficulty in this new framework is that the parabolicity of (1) makes the characteristic velocities depend singularly on ϵ. Two different solutions are proposed to overcome this problem.

2. Models

In the purely hyperbolic case $B = 0$, it is easy to see that for initial conditions in equilibrium (4), the system (3) reduces to its N first equations, and one retrieves the framework introduced in [2]. We gave there some models which can be now extended to the parabolic case.

Example: the one-dimensional diagonal relaxation model. This model is an extension of the relaxation model of Jin and Xin [8], which can be viewed as a kinetic model [2]. We take $N = N' = 2$, $-\lambda_{11} = \lambda_{21} = \lambda > 0$, $-\sigma_1^{(1)} = \sigma_2^{(1)} = 1/\sqrt{2}$ and

$$M(u) = \frac{1}{2}\left(u - \frac{A(u)}{\lambda} - \frac{B(u)}{\theta^2}, u + \frac{A(u)}{\lambda} - \frac{B(u)}{\theta^2}, \frac{B(u)}{\theta^2}, \frac{B(u)}{\theta^2} \right)^T. \tag{9}$$

For a hyperbolic system, we retrieve the model of Jin and Xin. Suppose that for all u, $A'(u)$ and $B'(u)$ have a basis of common eigenvectors and denote $\lambda_k(u)$ and $\theta_k^2(u)$ their respective eigenvalues for $1 \le k \le K$. Condition (8) reads as a generalized subcharacteristic condition:

$$\sup_{u\in I}\sup_{1\le k\le K}\frac{|\lambda_k(u)|}{\lambda}+\frac{\theta_k^2(u)}{\theta^2}<1. \tag{10}$$

The value of μ will be fixed later for numerical convenience. We have also higher dimensional versions of this model. They are different from the one of [8] even for the purely hyperbolic case. Other models, leading to more accurate numerical schemes for (1) are available, see [3].

3. Numerical schemes

In this section we use the system (3) to design numerical schemes for the system (1). The general idea is to construct a discretization of (3) in order to obtain, for sufficiently small ϵ, a scheme for (1). The problem is challenging because we have to deal with characteristic velocities which become infinite when ϵ tends to zero, together with a singular right-hand side. Two methods are presented here. We restrict ourselves to cartesian uniform grids which cells are denoted I_α. For an initial data u_0 we define

$$u_\alpha^0=(vol(\mathrm{I}_\alpha))^{-1}\int_{\mathrm{I}_\alpha}u_0(x)dx,\;\; f_\alpha^{\epsilon,0}=M(u_\alpha^0). \tag{11}$$

Our goal is to obtain a stable and consistent approximation of (1) in the conservation form:

$$\begin{aligned}u_\alpha^{n+1} &= u_\alpha^n-\sum_{d=1}^{D}\frac{\Delta t}{\Delta x_d}\left(\mathbf{A}^n_{\alpha+e_d/2}-\mathbf{A}^n_{\alpha-e_d/2}\right)\\ &+ \sum_{d=1}^{D}\frac{\Delta t}{\Delta x_d^2}\left(\mathbf{B}^n_{\alpha+e_d,d}-2\mathbf{B}^n_{\alpha,d}+\mathbf{B}^n_{\alpha-e_d,d}\right)\end{aligned} \tag{12}$$

with

$$\mathbf{A}^n_{\alpha+e_d/2}=\mathbf{A}_d\left(u^n_{\alpha-a+e_d},\dots,u^n_{\alpha+a}\right),\;\;\mathbf{B}^n_{\alpha,d}=\mathbf{B}_d\left(u^n_{\alpha-b},\dots,u^n_{\alpha+b}\right), \tag{13}$$

where the functions $\mathbf{A}_d$ and $\mathbf{B}_d$ are Lipschitz continuous and satisfy $\mathbf{A}_d(u,\dots,u)=A_d(u)$ and $\mathbf{B}_d(u,\dots,u)=B(u)$.

3.1. A diagonal splitting method

This method works only on cartesian grids. For the sake of simplicity we detail it here for the one-dimensional version of (3):

$$\begin{cases}\partial_t f_l^\epsilon+\lambda_l\partial_x f_l^\epsilon=\dfrac{1}{\epsilon}\left(M_l(u^\epsilon)-f_l^\epsilon\right), & 1\le l\le N,\\ \partial_t f_{N+1}^\epsilon-\left(\dfrac{\mu}{\sqrt{2}}+\dfrac{\theta}{\sqrt{\epsilon}}\right)\partial_x f_{N+1}^\epsilon=\dfrac{1}{\epsilon}\left(\dfrac{B(u^\epsilon)}{2\theta^2}-f_{N+1}^\epsilon\right), & \\ \partial_t f_{N+2}^\epsilon+\left(\dfrac{\mu}{\sqrt{2}}+\dfrac{\theta}{\sqrt{\epsilon}}\right)\partial_x f_{N+2}^\epsilon=\dfrac{1}{\epsilon}\left(\dfrac{B(u^\epsilon)}{2\theta^2}-f_{N+2}^\epsilon\right), & \end{cases} \tag{14}$$

with $u^\epsilon(x,t) = \sum_{i=1}^{N+2} f_i^\epsilon(x,t)$. We split (14) in the following way: in a first step, we solve the $N+2$ transport equations without the source-term. The N first equations are approximated by any high order (for example MUSCL type) scheme, written in conservation form, with numerical fluxes $\Phi^{\epsilon,n}_{\alpha+\frac{1}{2},l}$:

$$\Phi^{\epsilon,n}_{\alpha+\frac{1}{2},l} = \Psi_l\left(f^{\epsilon,n}_{\alpha-k_l+1,l}, \dots, f^{\epsilon,n}_{\alpha+k_l,l}\right), \quad 1 \le l \le N. \tag{15}$$

The last two equations are discretized by the first order upwind scheme. For these last two equations the CFL condition reads:

$$\gamma^\epsilon \frac{\Delta t}{\Delta x} = \left(\frac{\mu}{\sqrt{2}} + \frac{\theta}{\sqrt{\epsilon}}\right)\frac{\Delta t}{\Delta x} \le 1. \tag{16}$$

To take into account the contribution of the singular perturbation term on the right–hand side, we solve on $[t_n, t_{n+1}]$ the associated ordinary differential system:

$$\partial_t f^\epsilon = \frac{1}{\epsilon}(M(u^\epsilon) - f^\epsilon).$$

The solution can be explicitly obtained and letting $\epsilon \to 0$, we obtain:

$$f_\alpha^{\epsilon,n+1} = M(u_\alpha^{\epsilon,n+1/2}).$$

In terms of the macroscopic variable u, we have:

$$u_\alpha^{\epsilon,n+1} = u_\alpha^{\epsilon,n} - \frac{\Delta t_n}{\Delta x}\left(\mathbf{F}^{\epsilon,n}_{\alpha+1/2} - \mathbf{F}^{\epsilon,n}_{\alpha-1/2}\right), \tag{17}$$

where

$$\mathbf{F}^{\epsilon,n}_{\alpha+1/2} = \sum_{l=1}^{N} \lambda_l \Psi_l\left(M_l\left(u^{\epsilon,n}_{\alpha-k_l+1}\right), \dots, M_l\left(u^{\epsilon,n}_{\alpha+k_l}\right)\right) - \frac{\gamma^\epsilon}{2\theta^2}\left(B(u^{\epsilon,n}_{\alpha+1}) - B(u^{\epsilon,n}_\alpha)\right).$$

Therefore we can define the (consistent) numerical flux function associated to A as:

$$\mathbf{A}^n_{\alpha+1/2} = \sum_{l=1}^{N} \lambda_l \Psi_l\left(M_l\left(u^{\epsilon,n}_{\alpha-k_l+1}\right), \dots, M_l\left(u^{\epsilon,n}_{\alpha+k_l}\right)\right), \tag{18}$$

It remains:

$$\mathbf{B}^n_\alpha = B(u_\alpha^{\epsilon,n}), \tag{19}$$

under the condition:

$$\left(\frac{\mu}{\sqrt{2}} + \frac{\theta}{\sqrt{\epsilon}}\right)\frac{1}{2\theta^2} = \frac{1}{\Delta x}, \tag{20}$$

which means that $\epsilon \sim \Delta x^2$. Finally this procedure leads to a numerical scheme with a high order approximation of the hyperbolic part (18) and a second order classical approximation (19) of the diffusion. Remark that we have converted the relaxation parameter in a discretization parameter with a fixed value depending on Δx. This is not a problem here since we are only interested in the design of schemes for (1) and not in what happens in microscopic or mesoscopic scales. The parabolic CFL condition reads:

$$\max\left\{2\theta^2\frac{\Delta t}{\Delta x^2},\ \lambda_l\frac{\Delta t}{\Delta x},\ 1\leq l\leq N\right\}\leq 1. \tag{21}$$

For a scalar equation, we prove in [3] that if the transport discretization is L^∞ stable and TVD (*resp.* monotone) then so is the scheme (17). Also, if the transport discretization owns a cell entropy inequality, then the scheme (17) does and we have a convergence theorem to the entropy solution of (1)-(2) for initial data $u_0 \in L^\infty(\mathbb{R}^D)\cap L^1(\mathbb{R}^D)\cap BV(\mathbb{R}^D)$. This result has to be related to the one of [6].

Example: the parabolic DRM. We apply a first order upwind scheme on each equation of (14) with (9). We obtain:

$$\begin{aligned} u_\alpha^{n+1} = u_\alpha^n - \frac{\Delta t}{2\Delta x}\left(A(u_{\alpha+1}^n) - A(u_{\alpha-1}^n)\right) + \frac{\lambda\Delta t}{2\Delta x}\left(u_{\alpha+1}^n - 2u_\alpha^n + u_{\alpha-1}^n\right)\\ -\lambda\frac{\Delta t}{\Delta x}\frac{B(u_{\alpha+1}^n)-2B(u_\alpha^n)+B(u_{\alpha-1}^n)}{2\theta^2}\\ +\frac{\Delta t}{\Delta x^2}\left(B(u_{\alpha+1}^n)-2B(u_\alpha^n)+B(u_{\alpha-1}^n)\right) \end{aligned} \tag{22}$$

The first line is a Lax-Friedrichs type scheme as for the purely hyperbolic model [2]. The second line is a numerical anti-diffusion coming from the B term of the maxwellian functions M_1 and M_2 (makes part of $\mathbf{A}_{\alpha+1/2}^n$), while the third line is the discretization of the parabolic part under the condition (20). Due to the fact that the schemes are linear, A and B parts are clearly decoupled. If a second order MUSCL correction is added, then one must also add to (22) some correction terms where A and B parts are not decoupled, and therefore our method cannot be reduced to a splitting between hyperbolic and parabolic parts of the equation.

3.2. A non diagonal splitting method

This method works in several space dimensions, regardless the way one meshes the space. Let us consider a multidimensional model under the form (3) with $N' = D+1$. We transform this system in a linear regular way, keeping the N first equations and replacing the $D+1$ last ones by the system (7). Following the ideas of [7], we have put in the RHS all the singular terms. An interesting feature of our model is that we have been able to prove ([3]) that for $\mu > 0$ the LHS is hyperbolic diagonalizable, while for $\mu = 0$ it is not. This allows discretizations by any method for hyperbolic linear systems, including finite volumes on unstructured meshes. We continue in this paper with cartesian grids and applying a stable, high order, conservation form written method on the LHS, we obtain $U^{n+1/2} = (f_1^{n+1/2},\dots,f_N^{n+1/2},u^{n+1/2},v^{n+1/2})$. In particular we have:

$$u_\alpha^{n+1/2} = u_\alpha^n - \sum_{d=1}^{D}\frac{\Delta t_n}{\Delta x_d}\left(v_{d,\alpha+e_d/2}^n - v_{d,\alpha-e_d/2}^n\right). \tag{23}$$

We then solve the RHS and when $\epsilon \to 0$, the solution has a limit which defines U^{n+1} by:

$$u^{n+1} = u^{n+1/2}, \quad f_l^{n+1} = M_l(u^{n+1}), \quad 1 \le l \le N,$$

and

$$v_j^{n+1} = A_j(u^{n+1}) - \partial_{x_j} B(u^{n+1}), \quad 1 \le j \le D.$$

As a consequence:

$$u_\alpha^{n+1} = u_\alpha^n - \Delta t_n \sum_{d=1}^{D} \frac{(A_d(u^n) - \partial_{x_d} B(u^n))_{\alpha+e_d/2} - (A_d(u^n) - \partial_{x_d} B(u^n))_{\alpha-e_d/2}}{\Delta x_d}. \tag{24}$$

This result insures that if converging, the method converges to the right solution. More precisely, a straighforward computation shows that this procedure leads to a consistent scheme where the hyperbolic flux function tends to the one in section 3.1 when $\mu \to 0$ and depends only on the discretization of the N first equations, while the discretization of the parabolic term depends on how one chooses the numerical fluxes on the last $D+1$ equations and how one discretizes $\partial_{x_d} B(u)$. A first choice may be to take a first order upwind scheme for the LHS of (7), together with centered differences for $\partial_{x_d} B(u)$. This leads to :

$$\begin{aligned} u_\alpha^{n+1} = \quad & u_\alpha^n - \sum_{d=1}^{D} \frac{\Delta t_n}{\Delta x_d} \left(\mathbf{A}_{\alpha+e_d/2}^n - \mathbf{A}_{\alpha-e_d/2}^n \right) \\ + \quad & \sum_{d=1}^{D} \frac{\Delta t_n}{4\Delta x_d^2} \left(B(u_{\alpha+2e_d}^n) - 2B(u_\alpha^n) + B(u_{\alpha-2e_d}^n) \right). \end{aligned} \tag{25}$$

Here:

$$\mathbf{B}_{\alpha,d}^n = \frac{1}{4} \left(B(u_{\alpha-e_d}^n) + 2B(u_\alpha^n) + B(u_{\alpha+e_d}^n) \right). \tag{26}$$

Choosing higher order schemes for the LHS of (7) leads to add correction terms in $\mathbf{A}_{\alpha-e_d/2}^n$ and/or $\mathbf{B}_{\alpha,d}^n$. We point out that we can also retrieve (26) by a diagonal splitting, see [3]. Remark that in this method one can keep an ϵ depending, relaxing approximation and then, for fixed Δx make $\epsilon \to 0$, so that the procedure is also valid in mesoscopic scales.

3.3. Different time discretizations

To construct higher order in time approximations we remark that the we have obtained functions $\mathbf{A}_d$ and $\mathbf{B}_d$ such that formula (12) may be viewed as a first order discretization of the differential equation:

$$u'_\alpha = u_\alpha - \sum_{d=1}^{D} \frac{\overline{\mathbf{A}}_{\alpha+e_d/2} - \overline{\mathbf{A}}_{\alpha-e_d/2}}{\Delta x_d} + \sum_{d=1}^{D} \frac{\mathbf{B}_{\alpha+e_d,d} - 2\mathbf{B}_{\alpha,d} + \mathbf{B}_{\alpha-e_d,d}}{\Delta x_d^2}. \tag{27}$$

Here $\overline{\mathbf{A}}_{\alpha+e_d/2}$ is the limit of $\mathbf{A}_d\left(u_{\alpha-a+e_d}, \dots, u_{\alpha+a}\right)$ when $\Delta t \to 0$ and Δx is fixed. Therefore, one may discretize (27) with any high order, for example Runge-Kutta method, or make an implicit treatment of the parabolic part of the equation. The

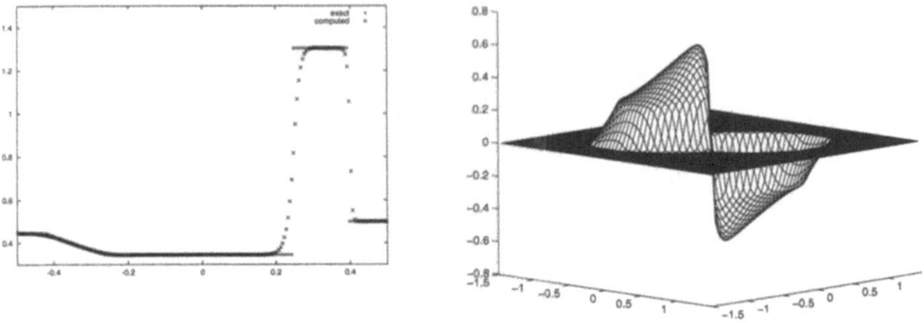

FIGURE 1. a) Lax shock tube, b) 2-D Burgers equation with diffusion

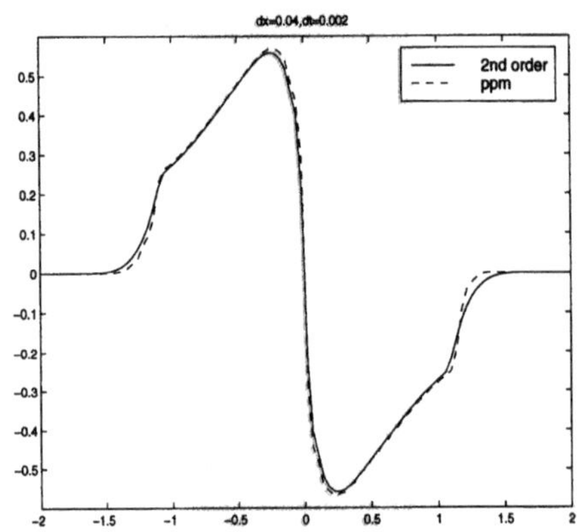

FIGURE 2. A strongly degenerate problem

stability condition on the hyperbolic part remains identical. As usual the order in time has to be related with order in space.

4. Illustration and conclusions

The model (3) and the numerical choices above essentially furnish $\mathbf{A}_d$ and $\mathbf{B}_d$, which can then be mixed with any (stable) discretization in time.

As an illustration we present some numerical results. Figure 1a is the density for a 200 points Lax shock tube where second order in time and space is applied to a flux decomposition model (purely hyperbolic). Figure 1b is the solution of a two-dimensionnal Burgers equation with uniform viscosity : here we use the diagonal relaxation model with second order in time and space and non diagonal splitting. Figure 2 is the solution of 1-D Burgers equation with a strongly degenerate

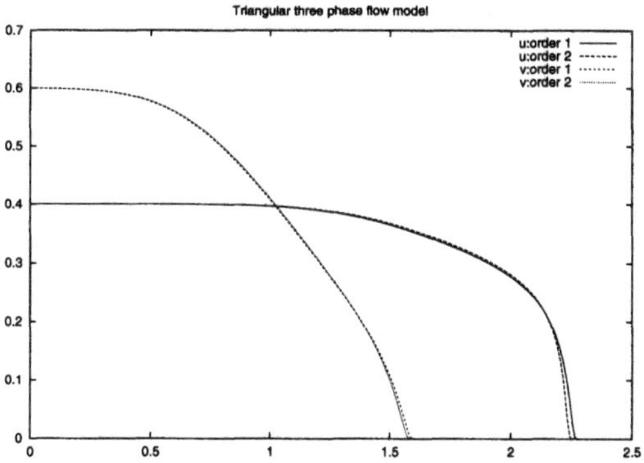

FIGURE 3. A 2x2 system

diffusion proposed in [10]:

$$u_t + u_x^2 = 0.1(\nu(u)u_x)_x, \quad \nu(u) = 0 \text{ if } |u| \leq 0.25, \ 1 \text{ else.}$$

We compare here the results given by a second order MUSCL and by a PPM discretization of the linear transport equations (diagonal splitting). We took the last example, a 2x2 system modelling a three phase flow, in [9]. We present in figure 3 first order and second order results for the nondiagonal splitting method applied to a flux decomposition model. Compared with the results in [9], our computation is satisfying.

When first order spatial discretizations are used our schemes can be retrieved by others Riemann solver free approaches, see for example [10]. With higher order discretizations, we have a really new way of approximating (1), which cannot be reduced to a splitting between hyperbolic and parabolic parts of the equation. The method is also very flexible: it is easy to vary the model, the space discretization, the time discretization.

References

[1] D. Aregba-Driollet and R. Natalini, *Convergence of relaxation schemes for conservation laws,* Appl. Anal. **61** (1996), 163–193.

[2] D. Aregba-Driollet and R. Natalini, *Discrete Kinetic Schemes for Multidimensional Systems of Conservation Laws,* SIAM J. Numer. Anal., **37** (2000), no 6, pp. 1973–2004.

[3] D. Aregba-Driollet, R. Natalini and S. Tang, *Diffusive Kinetic Schemes for Nonlinear Degenerate Parabolic Systems,* 2000, http://www.math.ntnu.no/conservation/2000/

[4] F. Bouchut, F. Guarguaglini and R. Natalini *Diffusive BGK Approximations for Nonlinear Multidimensional Parabolic Equations,* Indiana Univ. Math. J., **49**, (2000), no 2, pp. 723–749.

[5] J. Carrillo, *Entropy solutions for nonlinear degenerate problems,* Arch. Ration. Mech. Anal. **147** (1999) no 4, 269–361.

[6] S. Evje and K.H. Karlsen, *Monotone difference approximation of BV solutions to degenerate convection-diffusion equations,* SIAM J. Numer. Anal., **37** (2000), no 6, pp. 1838–1860.

[7] S. Jin, L. Pareschi, G. Toscani, *Diffusive relaxation schemes for multiscale discrete-velocity kinetic equations,* SIAM J. Numer. Anal. **35** (1998), 2405–2439.

[8] S. Jin and Z. Xin, *The relaxation schemes for systems of conservation laws in arbitrary space dimensions,* Comm. Pure Appl. Math, **48** (1995), 235–277.

[9] K.H. Karlsen, K.A. Lie, J.R. Natvig, H.F. Nordhaug, and H.K. Dahle, *Operator splitting methods for systems of convection-diffusion equations: nonlinear error mechanisms and correction strategies.* Preprint 2000.

[10] A. Kurganov and E. Tadmor, *New high-resolution central schemes for nonlinear conservation laws and convection-diffusion equations,* J. Comput. Phys. **160** (2000), no. 1, 241–282.

[11] C. Lattanzio and R. Natalini, *Convergence of diffusive BGK approximations for non-linear strongly parabolic systems,* to appear in Proceedings Section A: Mathematics-Royal Society of Edinburgh.

[12] R. Natalini, *A discrete kinetic approximation of entropy solutions to multidimensional scalar conservation laws,* J. Diff. Eq. **148** (1998), 292–317.

[13] B. Perthame, *Boltzmann type schemes for gas dynamics and the entropy property,* SIAM J. Numer. Anal. **27** (1990), no 6, 1405–1421.

D. Aregba-Driollet
Mathématiques Appliquées de Bordeaux, Université Bordeaux 1,
351 cours de la Libération, F-33405 Talence cedex, France
aregba@math.u-bordeaux.fr

R. Natalini
Istituto per le Applicazioni del Calcolo "M. Picone" , CNR,
Viale del Policlinico 137, I-00161 Roma, Italy
natalini@iac.rm.cnr.it

S. Tang
Department of Mechanics and Engineering Sciences,
Peking University,
Beijing 100871, P.R. China
maotang@pku.edu.cn

International Series of Numerical Mathematics
Vol. 140, © 2001 Birkhäuser Verlag Basel/Switzerland

Non-oscillatory Lax-Friedrichs Type Central Finite Volume Methods for 3-D Flows on Unstructured Tetrahedral Grids

P. Arminjon, A. Madrane, and A. St-Cyr

Abstract. We present a 3D finite volume generalization of the 1-dimensional Lax-Friedrichs and Nessyahu-Tadmor schemes for hyperbolic equations on unstructured tetrahedral grids. The non-oscillatory central difference scheme of Nessyahu and Tadmor, in which the resolution of the Riemann problem at the cell interfaces is by-passed thanks to the use of the staggered Lax-Friedrichs scheme, is extended here to a two-step, three-dimensional non-oscillatory centered scheme in finite volume formulation. Piecewise linear cell interpolants using Venkatakrishnan's limiter combined with diverse techniques to estimate the gradients lead to a non-oscillatory spatial resolution of order 2. The fact that the expected second order resolution is not fully attained in 3D for nonlinear systems might be caused by the absence of grid adaptation in our calculations. Numerical results for a linear advection problem with continuous initial conditions in 3D show the accuracy and stability of the method. We also include results for the 3D Euler system (shock tube problem)

1. Introduction

In earlier papers [5, 6] we presented a two-dimensional finite volume method generalizing the one-dimensional Lax-Friedrichs [12] and Nessyahu-Tadmor [16] difference schemes for hyperbolic conservation laws to unstructured triangular grids, while in [4, 2] we have also constructed a corresponding extension for 2-dimensional Cartesian grids. Other recent contributions in the case of Cartesian grids can be found in [10, 11, 13, 14].

All these extensions to 2-dimensional problems lead to non-oscillatory, essentially second order accurate finite volume methods.

In [2, 3] we presented an extension to staggered structured three-dimensional Cartesian grids of our two-dimensional finite volume method for Cartesian grids. We proved that the first order three-dimensional finite volume extension of the Lax-Friedrichs scheme is monotone under an appropriate CFL condition. Some difficulties encountered in the numerical experiments led us to consider a special type of limiting function proposed by Venkatakrishnan. The method still displayed too much dissipation and requires some adjustments.

In this paper, we first present a finite volume extension of the first order accurate Lax-Friedrichs scheme to three-dimensional tetrahedral grids. We then construct the extension of our two-dimensional finite volume method to a quasi-second order accurate non-oscillatory finite volume method of "central type", for staggered unstructured tetrahedral grids, which can be viewed as a finite volume generalization of the 1-dimensional Nessyahu-Tadmor difference scheme to 3-dimensional unstructured tetrehedral grids.

2. Mathematical modelling

2.1. Governing equations

Let $\Omega \subset R^3$ be the domain of interest of the flow and Γ be its boundary, we write $\Gamma = \Gamma_B \cup \Gamma_\infty \cup \Gamma_E$, where Γ_B denotes that part of the body boundary which is relevant for the computational domain. Γ_∞ is the (upwind) farfield boundary, and Γ_E is the (downwind) exit part of the boundary. The equations describing 3-dimensional compressible inviscid flows are the Euler equations, written here in their conservation form , given by

$$\frac{\partial U}{\partial t} + \vec{\nabla} \cdot \vec{\mathcal{F}}(U) = 0$$

$$\text{where} \quad U = (\rho, \rho u, \rho v, \rho w, E)^T, \quad \vec{\mathcal{F}}(U) = (F(U), G(U), H(U))^T, \tag{1}$$

$\vec{\mathcal{F}}(U)$ denotes the convective flux , ρ is the density, $\vec{V} = (u, v, w)$ is the velocity vector, $E = \rho e = \rho\epsilon + \frac{1}{2}\rho(u^2 + v^2 + w^2)$ is the total energy per unit volume; p is the pressure of the fluid. Let A,B and C denote the Jacobian matrices $\partial F(U)/\partial U$, $\partial G(U)/\partial U$ and $\partial H(U)/\partial U$, respectively. Eq.(1) can be written in the nonconservative form:

$$\frac{\partial U}{\partial t} + A(U)\frac{\partial U}{\partial x} + B(U)\frac{\partial U}{\partial y} + C(U)\frac{\partial U}{\partial z} = 0 \tag{2}$$

3. Discretization with respect to space and time

3.1. Definitions, description of the first-order method

We assume that Ω is a polyhedral bounded domain of R^3, and start from an arbitrary FEM tetrahedral grid $\mathcal{T}_h$, where h is the maximal length of the edges in $\mathcal{T}_h$. We need the following notations: T_{ij} denotes the set of all tetrahedra which share edge $[i,j]$ as a common edge. Let M denote the set of indices of all sub-tetrahedra of the original tetrahedral grid, defined by the median planes such as: iM_1GG_1,iM_1GG_3,jM_1GG_1 and jM_1GG_3 on fig. 1. Let $m(i,j,\tau,l)$ be one such index contained in M, where $\tau \in T_{ij}$ and l $(1 \le l \le 4)$ is the index of one of the four sub-tetrahedra sharing edge $[i,j]$ and contained in tetrahedron τ. If $l=1,3$ then the index denotes sub-tetrahedra having node a_i in common, and, if $l=2,4$ they share node a_j. Any such sub-tetrahedron will be noted l_k where $k \in M$. $K(i)$ is the set of nodes (vertices) which are neighbours of node a_i. Let nv be the number of vertices of $\tau \in \mathcal{T}_h$. For every vertex a_i, $1 \leq i \leq nv$ (written i for simplicity on fig. 1) of an arbitrary tetrahedron $\tau \in \mathcal{T}_h$, we define the (vertex centered) barycentric cell C_i as follows (this approach was introduced in [1]): every tetrahedron of $\mathcal{T}_h$ is subdivided in 24 sub-tetrahedra l_k by the median planes. The cell C_i is the union of the resulting sub-tetrahedra sharing a_i as a vertex $C_i = \bigcup_{j\in K(i)} \bigcup_{\tau\in T_{ij}} l_{m(i,j,\tau,\{1,3\})}$. The faces of interest associated with each sub-tetrehedron l_k $(k \in M)$ are, first, those which do not share any edge midpoint (such as M_1 in fig. 1) noted ∂l_k^1 (such as jGG_1 or jGG_3 on fig. 1), and secondly, the faces which do not contain vertex a_i in their composition, noted ∂l_k^2. The outward normals to those faces will be noted n_k for ∂l_k^1 and ν_k for ∂l_k^2 (such as, on fig. 1, M_1GG_1 or M_1GG_3). Note that we use normal vectors such that $||n_k|| = Area(\partial l_k^1)$ and $||\nu_k|| = Area(\partial l_k^2)$. The boundary ∂C_i of C_i can be constructed with ∂l_k^2 ($\partial C_i = \bigcup_{j\in K(i)} \bigcup_{\tau\in T_{ij}} \partial l^2_{m(i,j,\tau,\{1,3\})}$). As was the case in our 2-dimensional extensions, our 3-D extension also uses a

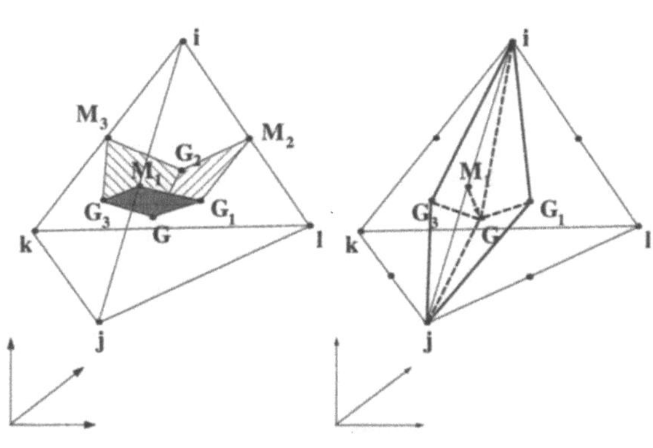

FIGURE 1. **Barycentric cell C_i and diamond cell L_{ij}**

dual grid, with dual cells L_{ij} associated with the edges of $\mathcal{T}_h$. The dual ("diamond") cell L_{ij} is composed by the sub-tetrahedra (defined above) sharing edge $[i,j]$ $(L_{ij} = \bigcup_{\tau\in T_{ij}} l_{m(i,j,\tau,\{1,2,3,4\})})$. Owing to the the fact that ∂l_k^1 denotes one of the facets of ∂L_{ij}, we can formulate the dependence of ∂L_{ij} on the selected faces (i.e. with superscript index 1) of the sub-tetrahedra which compose L_{ij} :

$$\partial L_{ij} = \bigcup_{\tau\in T_{ij}} \partial l^1_{m(i,j,\tau,\{1,2,3,4\})} = \bigcup_{\tau\in T_{ij}} \partial l^1_{m(i,j,\tau,\{1,3\})} \cup \partial l^1_{m(i,j,\tau,\{2,4\})} \tag{3}$$

Let i,j,k,l be the four nodes defining a tetrahedron τ sharing node a_i, then the cell boundary element of cell C_i, associated with τ and edge $[i,j]$ is

$$\partial C_i \cap \tau \cap L_{ij} = Triangle(M_1GG_1) \cup Triangle(M_1GG_3) \quad = \partial l^2_{m(i,j,\tau,\{1,3\})}$$

$$= \partial l^2_{m(i,j,\tau,1)} \cup \partial l^2_{m(i,j,\tau,3)} \tag{4}$$

Let m_{ij} denote the midpoint of edge $[i,j]$, also written as M_1 above, and let $U_i^n \cong U(a_i,t^n)$ and $U_{ij}^{n+1} \cong U(m_{ij},t^{n+1})$ denote the nodal (resp. cell average) values in the first and second grid at time $t = t^n$ and $t = t^{n+1}$, respectively (n even).
The union of all the barycentric cells constitutes a partition of the computational domain Ω_h and the same holds for diamond cells:

$$\Omega_h = \bigcup_{i=1}^{nv} C_i, \quad \Omega_h = \bigcup_{k=1}^{ne} L_k$$

where nv and ne are the number of vertices and number of edges, respectively, of the original finite element triangulation $\mathcal{T}_h$. We now define the two steps of our first order accurate (staggered, Lax-Friedrichs type) finite volume method:
First time step:
In the first (and further odd) time step of our 3-dimensional finite volume scheme, we start from initial (or previously obtained) cell average values $\vec{U}_i$ for the barycentric cells of the first grid, and compute cell average values $\overrightarrow{U}_{ij}$ for the (staggered) dual cells of the second grid. This is done by integrating (1) on an extended control volume $L_{ij} \times [t^n, t^{n+1}]$, where L_{ij} is a (diamond) cell of the dual grid, and by assuming that $\overrightarrow{U}_{L_{ij}}(t^n + \Delta t)$ is piecewise constant on the cells of the second grid. Using (1) and applying the divergence theorem to the flux term, we obtain

$$\int_{t^n}^{t^{n+1}} \int_{L_{ij}} U_t d\vec{x} dt + \int_{t^n}^{t^{n+1}} \int_{\partial L_{ij}} \overrightarrow{\mathcal{F}}(U) \cdot \overrightarrow{n} \, dA dt = 0 \tag{5}$$

Let $m(1,3)$ be equal to $m(i,j,\tau,\{1,3\})$ and $m(2,4) = m(i,j,\tau,\{2,4\})$ for simplicity, (where $m(1,3)$ is in fact dependent on i,j,τ);we obtain

$$\begin{aligned} U_{ij}^{n+1} &= \frac{1}{2}(U_i^n + U_j^n) - \frac{\Delta t}{Vol(L_{ij})} \sum_{\tau \in T_{ij}} \{\overrightarrow{\mathcal{F}}(U_i^n) \cdot \overrightarrow{n}_{m(1,3)} + \overrightarrow{\mathcal{F}}(U_j^n) \cdot \overrightarrow{n}_{m(2,4)}\} \\ &- \frac{\Delta t}{Vol(L_{ij})} \int_{\partial L_{ij} \cap \Gamma_B} \overrightarrow{\mathcal{F}}(U(\vec{x})^n) \cdot \overrightarrow{n} dA - \frac{\Delta t}{Vol(L_{ij})} \int_{\partial L_{ij} \cap \Gamma_\infty} \overrightarrow{\mathcal{F}}(U(\vec{x})^n) \cdot \overrightarrow{n} dA \end{aligned} \tag{6}$$

Second time step: Starting from the known diamond cell value U_{ij}^{n+1}, we compute the barycentric cell value U_i^{n+2} $(1 \le i \le nv)$ by integrating (1) on an extended control volume $C_i \times [t^{n+1}, t^{n+2}]$, applying the divergence theorem to the flux term, and observing that

$$\partial C_i = \bigcup_{j \in K(i)} \{\partial C_i \cap \partial C_j\} \cup \{\partial C_i \cap \Gamma_B\} \cup \{\partial C_i \cap \Gamma_\infty\}.$$

We obtain

$$U_i^{n+2} = \sum_{j\in K(i)} \frac{Vol(C_i\cap L_{ij})}{Vol(C_i)} U_{ij}^{n+1} - \frac{\Delta t}{Vol(C_i)} \sum_{j\in K(i)} \sum_{\tau\in T_{ij}} \vec{\mathcal{F}}(U_i^{n+1})\cdot \vec{\nu}_{m(1,3)}$$
$$- \frac{\Delta t}{Vol(C_i)} \int_{\partial C_i\cap\Gamma_B} \vec{\mathcal{F}}(U^{n+1})\cdot \vec{\nu} dA - \frac{\Delta t}{Vol(C_i)} \int_{\partial C_i\cap\Gamma_\infty} \vec{\mathcal{F}}(U^{n+1})\cdot \vec{\nu} dA \quad (7)$$

3.2. High-order accurate approximations

To obtain second-order accuracy, we introduce cellwise piecewise linear interpolation (MUSCL,[17]), the derivatives of the P_1 (Galerkin) interpolation [8], Least Squares [6], and the Gauss-Green formula [7]. **First step:** We integrate (1) on an extended control volume $L_{ij}\times[t^n, t^{n+1}]$, assuming we have obtained from the cell average values U_i^n piecewise linear reconstructions given by

$$\begin{aligned} U_h(x,y,z,t^n)_{|C_i} =& \mathcal{L}_i(x,y,z,t^n) \\ =& U_i^n + (x-x_i)P_i^n + (y-y_i)Q_i^n + (z-z_i)R_i^n \quad (x,y,z)\in C_i \end{aligned} \quad (8)$$

For the integration with respect to time, in order to ensure "nearly" second-order accuracy, we adopt a " quasi-midpoint formula" time discretization, where the convective flux is computed at the intermediate time $t^{n+\frac{1}{2}}$ (thus requiring the computation of **predicted** values $U_h(x,y,t^{n+\frac{1}{2}})$ along ∂L_{ij}), in the integration with respect to time.

Predictor (First step):

On each face of the cell L_{ij}, using the Euler equations, we define a predicted vector

$$U_{a_i,G_1,G}^{n+1/2} = U_{a_i,G_1,G}^n - \frac{\Delta t}{2}\{F'(U_{a_i,G_1,G}^n)P_i^n + G'(U_{a_i,G_1,G}^n)Q_i^n + H'(U_{a_i,G_1,G}^n)R_i^n\} \quad (9)$$

where, using (8), the value of U_h^n along the face a_i, G_1, G of diamond cell L_{ij} is taken equal to

$$\begin{aligned} U_h(x,y,z,t^n) \cong& U_i^n + \frac{1}{2}(x_{a_i,G_1,G}-x_i)P_i^n + \frac{1}{2}(y_{a_i,G_1,G}-y_i)Q_i^n \\ &+ \frac{1}{2}(z_{a_i,G_1,G}-z_i)R_i^n \equiv U_{a_i,G_1,G}^n \end{aligned} \quad (10)$$

The corrector can now be written, using (9), as

Corrector (First step)

$$Vol(L_{ij})U_{ij}^{n+1} - \left\{\int_{L_{ij}\cap C_i} \mathcal{L}(x,y,z,t^n)d\vec{x} + \int_{L_{ij}\cap C_j} \mathcal{L}(x,y,z,t^n)d\vec{x}\right\}$$
$$+ \Delta t \sum_{\tau\in T_{ij}} \{\int_{\partial l^1_{m(i,j,\tau,\{1,3\})}} + \int_{\partial l^1_{m(i,j,\tau,\{2,4\})}} \}\vec{\mathcal{F}}(U_h^{n+\frac{1}{2}})\cdot \vec{n} dA$$

$$+\Delta t\int_{\partial L_{ij}\cap\Gamma_B}\vec{\mathcal{F}}(U_h^n)\cdot\vec{n_{ij}}dA+\Delta t\int_{\partial L_{ij}\cap\Gamma_\infty}\vec{\mathcal{F}}(U_h^n)\cdot\vec{n_{ij}}dA=0 \quad \text{where} \quad (11)$$

$$\int_{L_{ij}\cap C_i}\mathcal{L}(x,y,z,t^n)d\vec{x}=\sum_{\tau\in T_{ij}}Vol(l_{m(1,3)})\mathcal{L}(x_{\mathcal{G}_{m(1,3)}},y_{\mathcal{G}_{m(1,3)}},z_{\mathcal{G}_{m(1,3)}},t^n) \quad (12)$$

where $\mathcal{G}_{m(1,3)}$ is the centroid of sub-tetrahedron $l_{m(1,3)}$. Note that from the coding viewpoint there is a simplification since the two sub-tetrahedra $(1,3)$ also form a convex polyhedron the barycenter of which can be calculated, and the last integral can be simplified.

Second step

To obtain the second step of the time discretization, we integrate (1) on the cell $C_i\times[t^{n+1},t^{n+2}]$, assuming we have obtained, from the diamond cell average values U_{ij}^{n+1} computed in the first time step, piecewise linear reconstructions, given by

$$\begin{aligned}U_h(x,y,z,t^{n+1})_{|L_{ij}}&=\mathcal{L}_{ij}(x,y,z,t^{n+1})\\&=U_{ij}^{n+1}+(x-x_{ij})P_{ij}^{n+1}+(y-y_{ij})Q_{ij}^{n+1}+(z-z_{ij})R_{ij}^{n+1}\end{aligned} \quad (13)$$

Proceeding as for the first step, we obtain

<u>**Predictor (second step)**</u>

$$\begin{aligned}&U_{G,G_1,M_1,G_3}^{n+\frac{3}{2}}=U_{G,G_1,M_1,G_3}^{n+1}\\&-\frac{\Delta t}{2}\{F'(U_{G,G_1,M_1,G_3}^{n+1})P_{ij}^{n+1}+G'(U_{G,G_1,M_1,G_3}^{n+1})Q_{ij}^{n+1}+H'(U_{G,G_1,M_1,G_3}^{n+1})R_{ij}^{n+1}\}\end{aligned} \quad (14)$$

where

$$\begin{aligned}&U_h(x,y,z,t^{n+1})\cong U_{ij}^{n+1}\\&+\frac{1}{2}(x_G-x_{M_1})P_{ij}^{n+1}+\frac{1}{2}(y_G-y_{M_1})Q_{ij}^{n+1}+\frac{1}{2}(z_G-z_{M_1})R_{ij}^{n+1}\equiv U_{G,G_1,M_1,G_3}^{n+1}\end{aligned} \quad (15)$$

defines an approximation of the value of U on the boundary element $[G,G_1,M_1,G_3]$ of cell C_i.

<u>**Corrector (second step)**</u>

$$\begin{aligned}&Vol(C_i)U_i^{n+2}-\sum_{j\in K(i)}\int_{C_i\cap L_{ij}}\mathcal{L}_{ij}(x,y,z,t^{n+1})d\vec{x}\\&+\Delta t\sum_{j\in K(i)}\sum_{\tau\in T_{ij}}\int_{\partial l^2_{m(i,j,\tau,\{1,3\})}}\vec{\mathcal{F}}(U(x,y,z,t^{n+3/2})\cdot\vec{\nu}dA\\&+\Delta t\int_{\partial C_i\cap\Gamma_B}\vec{\mathcal{F}}(U_h^{n+1})\cdot\vec{\nu}dA+\Delta t\int_{\partial C_i\cap\Gamma_\infty}\vec{\mathcal{F}}(U_h^{n+1})\cdot\vec{\nu}dA=0\end{aligned} \quad (16)$$

where

$$\int_{C_i \cap L_{ij}} \mathcal{L}_{ij}(x,y,z,t^{n+1}) d\vec{x} =$$

$$\sum_{j\in K(i)} \sum_{\tau\in T_{ij}} \{Vol(l_{m(i,j,\tau,1,3)}) \mathcal{L}(x_{\mathcal{G}_m(i,j,\tau,1,3)}, y_{\mathcal{G}_m(i,j,\tau,1,3)}, z_{\mathcal{G}_m(i,j,\tau,1,3)}, t^{n+1})\} \tag{17}$$

is computed according to 12 with the respective barycenters of the sub-tetrahedra (again, a simplification is possible).

3.2.1. APPROXIMATION OF THE SLOPES AND LIMITATION In order to compute the gradient (P_i^n, Q_i^n, R_i^n) of the piecewise linear interpolant $L(x,y,z,t^n)$ for the cell C_i, we use the P_1 (Galerkin) interpolation [8], Least Squares [6] and Gauss-Green [7]. For the limitation we use several procedures, see [17]. Numerical experiments have led us to choose the Gauss-Green formula for the gradients used in the reconstruction for the cells C_i and a least squares weighted procedure for the cells L_{ij}.

3.2.2. STABILITY We refer to [15] for a stability study of linear multidimensional advection models that is also valid in the 3-D case. For the second time step:

$$\Delta t_i = \frac{Vol(C_i)}{\lambda^i_{max} \int_{\partial C_i} dA} \quad \text{where} \quad \lambda^i_{max} = \max(\lambda_i, \max_{j\ neighbour\ of\ i} \lambda_j)$$

$$\text{and} \quad \lambda_i = ||\vec{V}_i|| + c_i$$

$\vec{V}_i$, c_i refer to the values in the cell C_i of the velocity vector and sound speed, respectively. We then choose the minimum Δt_i for all indices i $(1 \le i \le nv)$ $\Delta t = \min_{1\le i\le nv} \{\Delta t_i\}$ with a similar time step definition for the first (odd) step $(t^n \to t^{n+1})$.

4. Numerical experiments

4.1. Linear advection problem

First, to confirm the spatial accuracy of the scheme, a simple advection test is performed on a smooth initial distribution. The distribution is advected near the boundary of the cube $[0,1]^3$ and compared with the exact solution of the following problem:

$$u_t + u_x + u_y + u_z = 0 \tag{18}$$

$$u(x,y,z,t=0) = f(r^2) = \begin{cases} \cos(4\pi r^2) & \text{if } r^2 \le 1/8 \\ 0 & \text{otherwise} \end{cases}$$

where $r^2 = (x-1/2)^2 + (y-1/2)^2 + (z-1/2)^2$. Table 1 shows the L_1-error and ac-

Diameter	Nodes	L^1-error	Order
0.134057	4095	0.0487568	
0.123977	5733	0.0416913	2.0027

TABLE 1. L^1 - error and accuracy order, continuous problem

curacy order; the limiting function used was the one proposed by Venkatakrishnan [18].

4.2. Shock tube problem

To illustrate the accuracy of our scheme, we present numerical results for a 3-D extension of the shock tube problem introduced by Sod . In this problem, an initial discontinuity in the thermodynamical state of the gas breaks into a shock wave followed by a contact discontinuity and a rarefaction wave. Finite element meshes with 909,1809 and 3609 nodes are used and the initial conditions at $t = 0$ are specified by the data

$$\begin{cases} \rho = 1; u = 0; v = 0; w = 0; p = 1.0 & x \in [0, 1/2] \\ \rho = 0.125; u = 0; v = 0; w = 0; p = 0.1 & x \in [1/2, 1] \end{cases} \tag{19}$$

The profiles of density, velocity, pressure and temperature are compared with the analytical solution at $t = 0.16$. The numerical solution of the 3-dimensional shock

h	Nodes	$L^2\ \rho$	Order	$L^2\ V_z$	Order	$L^2\ P$	Order
0.01	909	2.898e-05		9.982e-05		2.778e-05	
0.005	1809	1.555e-05	0.898	6.399e-05	0.641	1.196e-05	1.216

TABLE 2. L^2 - error and accuracy order, Sod's problem

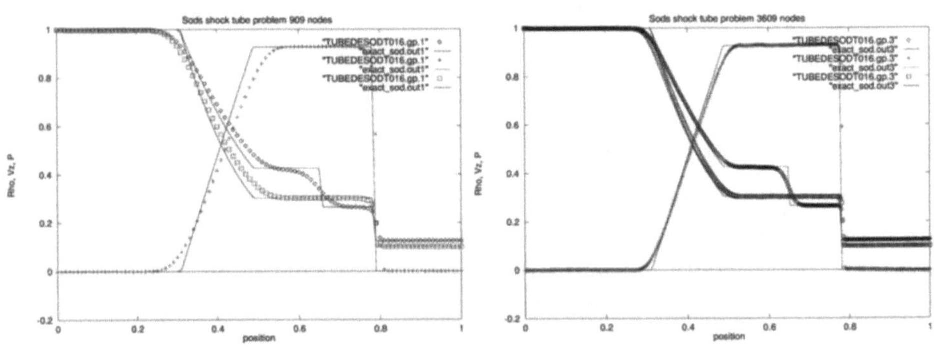

FIGURE 2. Sod's shock tube problem with 909 and 3609 gridpoints

tube problem is clearly much more likely to suffer from excessive numerical dissipation than its 1-dimensional analogue see e. g. [9]. In our case, see table 2 and Fig.

2, this dissipation might be due to the fact that the tetrahedra are fairly stretched (the aspect ratio is more than 1 : 25). This creates an additional difficulty for the capture of the shock and particularly of the contact discontinuity. Mesh adaptation, which was not available to us, would certainly bring a decisive improvement, as was observed in our 2-dimensional experiments [6].

5. Conclusion

In this paper we have presented a new construction of three-dimensional finite volume methods for inviscid flows using staggered unstructured tetrahedral grids. The first and second-order methods can be considered as finite volume generalizations of the Lax-Friedrichs and Nessyahu-Tadmor 1-dimensional difference schemes, respectively. Actual calculations have been restricted to the higher-order method, as the first-order Lax-Friedrichs-type method, in three dimensions, proved to be even more dissipative than in one dimension. Nevertheless the quality of the results does not fully meet original expectations, based on the very high quality obtained with our method for 2-dimensional flows [6], where we were using grid adaptation. We could probably attribute the excessive dissipation in part to the grids, which have a relatively small number of nodes, a high aspect ratio, and have not benefited from the improvements brought about by grid adaptation. We are currently working on this problem.

References

[1] ANGRAND F., BOULARD V., DERVIEUX A., PERIAUX J., VIJAYASUNDARAM G., (1984), *Triangular finite element methods for the Euler equations*, 6th Int. Conf. on Comp. Meth. in Appl. Sciences and Engineering, Glowinski R., Lions J.L., Eds., North Holland .

[2] P. ARMINJON, A. MADRANE AND A. ST-CYR (1999), *New Lax-Friedrichs-type finite volume schemes on 2 and 3D Cartesian staggered grids*, 7th annual conference of the CFD Society of Canada, May 30 to June 1, 1999, Halifax, J. Militzer, editor, pp.(3-3)-(3-10).

[3] P. ARMINJON, A. MADRANE, A. ST-CYR,*New 2 and 3-dimensional non-oscillatory central finite volume methods for staggered Cartesian grids*, to appear in Applied Numer. Math.

[4] P. ARMINJON, D. STANESCU AND M. C. VIALLON, (1995), *A two-dimensional finite volume extension of the Lax-Friedrichs and Nessyahu-Tadmor schemes for compressible flows*, Proc.of the 6th. Int. Symp. on Comp. Fluid Dynamics, Lake Tahoe (Nevada) September 4-8, 1995, M.Hafez and K. Oshima, editors, Vol. IV, pp. 7-14.

[5] P. ARMINJON AND M. C. VIALLON, (1995), *Généralisation du schéma de Nessayahu-Tadmor pour une équation hyperbolique à deux dimensions d'espace*, C.R. Acad. Sci. Paris, t.320, série I, January 1995, pp. 85-88.

[6] P. ARMINJON, M.C. VIALLON AND A. MADRANE, (1994), *A Finite Volume Extension of the Lax-Friedrichs and Nessyahu-Tadmor Schemes for Conservation Laws on*

Unstructured Grids, revised version with numerical applications, Int. J. of Comp. Fluid Dynamics (1997), Vol. 9, No. 1, 1-22.

[7] T. J.Barth and D. C.Jespersen, (1989), *The design and application of upwind schemes on unstructured meshes*,AIAA Paper No. 89-0366, 27th Aerospace sciences meeting, January 9-12, 1989, Reno, Nevada.

[8] Fezoui,L., Lanteri,S. , Larrouturou,B. and Olivier,C., (1989), *Résolution Numérique des Équations de Navier-Stokes pour un Fluide Compressible en Maillage triangulaire*, INRIA Res. Rep. No. 1033, Inst. National de Rech. en Informatique et Automatique, Rocquencourt, 78153 Le Chesnay, France.

[9] Fezoui,L., Steve, H. and Selmin, V. (1988) *Simulation numérique d'écoulements compressibles 3-D par un schéma décentré en maillage non-structuré*, INRIA Res. Rep. No. 0825, Institut National de Recherche en Informatique et Automatique, Roquencourt, 78153 Le Chesnay, France.

[10] G. Jiang and E. Tadmor, (1998), *Non-oscillatory Central Schemes for Multi-dimensional Hyperbolic Conservation Laws*, SIAM J. on Scientific Computing, 19, pp.1892-1917.

[11] T. Katsaounis and D. Levy, (1999), *A modified structured central schme for 2D hyperbolic conservation laws*, Appl. Math. Lett. 12, pp. 89-96.

[12] P.D. Lax, (1954), *Weak solutions of nonlinear hyperbolic equations and their numerical computation*, Comm. Pure and Applied Math.7, pp. 159-193.

[13] D. Levy, G. Puppo and G. Russo, (2000), *A third order central WENO scheme for 2D hyperbolic conservations laws*, Appl. Numer. Math. 33, pp. 415-421.

[14] D. Levy, G. Puppo and G. Russo, (2000), *Compact central WENO scheme for multidimensional conservations laws*, SIAM J. Sci. Comput. 22, pp. 656-672.

[15] A. Madrane, (1998), *Constructions of new finite volume/finite element methods for transonic/supersonic compressible flows*, Ph. D. dissertation, Université de Montréal, Département de Mathématiques et Statistique.

[16] H. Nessyahu and E. Tadmor, (1990), *Non-oscillatory central differencing for hyperbolic conservation laws*, J. Comp. Phys., 87, No. 2, pp. 408–463.

[17] B. van Leer, (1979), *Towards the ultimate conservative difference scheme V. A Second-Order Sequel to Godunov's Method*, J.Comp.Physics, 32, pp. 101-136.

[18] V. Venkatakrishnan, (1995), *Convergence to Steady State Solutions of the Euler Equations on Unstructured Grids with Limiters*, J. Comp. Phys. 118, pp. 120-130.

P. Arminjon, A. St-Cyr
Département de mathématiques et de statistique Université de Montréal,
C.P. 6128, Succ. centre-ville Montréal, Québec, H3C 3J7, Canada
arminjon@crm.umontreal.ca
stcyr@dms.umontreal.ca

A. Madrane
German Aerospace Center DLR Inst. of Design Aerodynamics,
Lilienthalplatz 7, 38108 Braunschweig, Germany
aziz.madrane@dlr.de

International Series of Numerical Mathematics
Vol. 140, © 2001 Birkhäuser Verlag Basel/Switzerland

Stability of Maxwell States in Thermo-Elasticity

Fumioki Asakura

Abstract. We study the Cauchy problem for model equations of 1-D thermo-elasticity that admit a solid-solid phase transition. The Maxwell states are defined to be two constant states such that the entropy is equal in the both states. A small perturbation of these Maxwell states will be our initial data. In the isothermal model, we shall show that: there exists a global in time admissible phase boundary satisfying the Abeyaratne-Knowles kinetic condition; as time goes to infinity, the strain and the velocity outside the phase boundary tend to the Maxwell states and the entropy tends to a certain limit function. In the polytropic case (the internal energy is proportional to the temperature), we shall show that there exist unique Maxwell states for given temperature and there exist at least two admissible solutions to the Riemann problem with certain initial data in a neighborhood of the Maxwell states.

1. Model Equations

We study the model equations of thermo-elasticity in Lagrangian form:

$$u_t - v_x = 0, \quad v_t - \sigma_x = 0 \tag{1}$$

$$\left(e + \tfrac{1}{2}v^2\right)_t - (\sigma v)_x = 0 \tag{2}$$

where u: strain, v: particle velocity, σ: stress and e: internal energy. Thermodynamic quantities e, σ are functions of u and the specific entropy η, and satisfy the thermodynamic identity:

$$de = \theta d\eta + \sigma du \tag{3}$$

where θ is the temperature. The equation (2) is formally equivalent to $\eta_t = 0$. Hence, if η is constant, the system of equations are reduced to the first 2 equations (1) that are called the *isentropic* equations.

In this note we confine ourselves to the *isothermal* model: $\theta = T$ constant and the *polytropic* model: $e = c\theta$. We are interested in thermo-elastic materials in which a solid-solid phase transition occurs. Here, the phase transition is a discontinuous solution to the above system of equations and each side of the discontinuity belongs to different phases: the stress is nearly equal but the strain is different. When the propagation speed of the discontinuity is zero, the stress must be equal and, in the isothermal model, the physically relevant phase transition is composed of the strains satisfying the Maxwell equal-area principle. These strains are admissible in the sense that they minimize the entropy rate for v close to v_m and stationary phase boundaries are not admissible unless $v_- = v_m$ (Hattori's theorem [11]). We expect that: if these Maxwell states are compactly perturbed, then, as time goes on, the states come back to the Maxwell states. This asymptotic stability is true for the isentropic model (Asakura [3]). The isothermal model, as shown below, consists of isentropic equations and the entropy equation. The asymptotic behavior of the entropy function will be studied in this note; this is one aim of this note. The equal-area principle means that the entropy of the both sides is equal, which will be adopted as the definition of the Maxwell states in the polytropic model. We shall suggest in this note that these Maxwell states are not asymptotically stable by showing there exist at least two admissible solutions to the Riemann problem with certain initial data in a neighborhood of the Maxwell states.

In the isothermal model, it follows from the basic identity (3) that the energy function is linear in η and σ depends only on u; the internal energy is expressed as

$$e = T\eta + \psi(u), \quad \psi(u) = \int \sigma(u)\, du. \tag{4}$$

Hence, the first 2 equations (1) involve only u and v and equivalent to the isentropic equations. In general models, the basic identity (3) expressed by the *free energy* : $\psi = e - \theta\eta$ and the *Maxwell relation* imply an important identity

$$\sigma = \left(\frac{\partial e}{\partial u}\right)_\theta + \theta \left(\frac{\partial \sigma}{\partial \theta}\right)_u$$

(see Fermi [8] section 14 for its derivation). Hence, if $e = c\theta$, it follows from the above expression that σ is proportional to θ:

$$\sigma(u, \eta) = \omega(u)\theta. \tag{5}$$

By setting $\theta = T$, we may assume σ has also the above form in isothermal models. Since we are interested in phase transitions, we assume moreover that the stress $\sigma = \sigma(u, \eta)$ is a C^2- function and there exist α, β ($\alpha < \beta$) such that

$$\omega'(u) = \begin{cases} > 0 & \text{for} \quad u < \alpha, \\ < 0 & \text{for} \quad \alpha < u < \beta, \\ > 0 & \text{for} \quad u > \beta. \end{cases} \tag{6}$$

For technical reason, we also assume

$$\omega''(u) \neq 0, \quad \text{for} \quad u < \alpha, \ u > \beta. \tag{7}$$

The region $\Omega_\alpha = \{(u, v, \eta);\ u < \alpha\}$ is called the *α-phase* and $\Omega_\beta = \{(u, v, \eta);\ u > \beta\}$ the *β-phase*. The system of equations is *hyperbolic* in $\Omega_\alpha \cup \Omega_\beta$.

The jump discontinuity of the form:

$$(u, v, \eta) = \begin{cases} (u_-, v_-, \eta_-), & x < st, \\ (u_+, v_+, \eta_+), & x > st, \end{cases} \tag{8}$$

is a weak solution ($u_\pm, v_\pm, \eta_\pm$: constants), if and only if it satisfies the *Rankine-Hugoniot condition*

$$\begin{aligned} s(u_+ - u_-) &= -(v_+ - v_-), \\ s(v_+ - v_-) &= -(\sigma_+ - \sigma_-), \\ s(e_+ + \tfrac{1}{2}v_+^2 - e_- - \tfrac{1}{2}v_-^2) &= -(\sigma_+ v_+ - \sigma_- v_-). \end{aligned} \tag{9}$$

In particular, quantities $e_\pm, \sigma_\pm, u_\pm$ satisfy the Hugoniot equation:

$$e_+ - e_- = \tfrac{1}{2}(\sigma_+ + \sigma_-)(u_+ - u_-) \tag{10}$$

which is a thermodynamic equation. The jump discontinuity (8) is called a *phase boundary*, if those states belong to the different phases and satisfy the Rankine-Hugoniot condition. Clearly the states $(u_\pm, v_\pm, \eta_\pm)$ constitute stationary phase boundary ($s = 0$), if

$$\sigma_+ = \sigma_-, \quad v_+ = v_-. \tag{11}$$

Maxwell states are defined to be a pair of stationary states satisfying

$$\eta_+ = \eta_-. \tag{12}$$

We can see later there exists a unique pair of strains satisfying (12) in the isothermal model, which are called the Maxwell strains. A general phase boundary simply satisfying Rankine-Hugoniot condition is not always physically relevant. We say that the phase boundary is *admissible*, if and only if it satisfies the Abeyaratne-Knowles *kinetic equation*

$$s = -\Phi(\eta_+ - \eta_-) \tag{13}$$

where $\Phi(\xi)$ is a nondecreasing function satisfying $\Phi(0) = 0$. The Maxwell states are admissible stationary phase boundary in this sense. Hereafter, we always assume that Φ is a C^2-function and

$$\Phi(0) = 0, \quad \Phi'(0) > 0. \tag{14}$$

In the next section, we shall review the results in Asakura [4] showing the asymptotic stability of the Maxwell states in the isothermal model. Our previous results on the isentropic equations in Asakura [3] are essential in proving the stability. We shall study the polytropic model in section 3 and show that Maxwell strains constitute a 1-d manifold of strains. Moreover, an admissible solution to the Riemann problem is no longer unique, even if the kinetic condition (13) is imposed, which suggests some instability property of the Maxwell states in the polytropic model.

2. Isothermal Model

In the isothermal model, the Hugoniot equation (10) turns out to be

$$T(\eta_+ - \eta_-) = \tfrac{1}{2}(\sigma_+ + \sigma_-)(u_+ - u_-) - \int_{u_-}^{u_+} \sigma(u)du. \tag{15}$$

Hence the variation of the entropy is expressed by the strains. Denoting $\widehat{s} = su$, we define the *Gibbs function* G to be

$$G = e + \tfrac{1}{2}\widehat{s}^2 - T\eta - \sigma u. \tag{16}$$

Then we have by direct computation

Proposition 2.1.

$$G_+ - G_- = \int_{u_-}^{u_+} \sigma(u)du - \tfrac{1}{2}(\sigma_+ + \sigma_-)(u_+ - u_-). \tag{17}$$

Proposition 2.2.

$$\eta_+ - \eta_- = -\tfrac{1}{T}(G_+ - G_-). \tag{18}$$

There exists a unique pair of strains $u_\pm$ satisfying the equal area principle

$$\sigma(u_+) = \sigma(u_-), \quad \int_{u_-}^{u_+} \sigma(u)du = \sigma(u_\pm)(u_+ - u_-). \tag{19}$$

The last condition is equivalent to the condition $\eta_+ = \eta_-$ which says that these strains, denoted by $u_- = u_m$, $u_+ = u_m^*$, constitute Maxwell strains. By setting $v_+ = v_-$ (denoted by v_m), $U_m = (u_m, v_m)$, $U_m^* = (u_m^*, v_m)$ are the Maxwell states in the isothermal equations. Let us first consider the isentropic equations (Asakura [3], [4]). Suppose that $U_L = (u_L, v_L)$ and $U_R = (u_R, v_R)$ are contained in small neighborhoods of the Maxwell states U_m and U_m^*, respectively. Then we can construct a self-similar solutions which consists of 4 constant regions U_L, $U_- \in \Omega_\alpha$ and U_+, $U_R \in \Omega_\beta$, where U_-, U_+ are connected by a single phase boundary.

These solutions are used to construct approximate solutions with general initial data $U_0(x) = U_L(x) \in \Omega_\alpha$ for $x < 0$, $U_R(x) \in \Omega_\beta$ for $x > 0$. We have the following existence theorem of global solutions by the wave-front tracking alternative of the Glimm method.

Theorem 2.1. *Assume that the initial perturbation is sufficiently small in total variation i.e. the quantity:*

$$T.V.(U_L(x) - U_m)|_{R_-} + T.V.(U_R(x) - U_m^*)|_{R_+} \tag{20}$$

is sufficiently small. Then there exists a weak global solution with a single phase boundary.

Theorem 2.2. *The above phase boundary is Lipschitz continuous curve in xt space. Moreover, the limit:*

$$\lim_{\delta \to \pm 0} U(\chi(t) + \delta, t) = U_\pm(t) \tag{21}$$

exists except for countable t at the phase boundary; the Rankine-Hugoniot condition (9) *and the kinetic equation*

$$\dot{\chi}(t) = -\Phi(\eta_+(t) - \eta_-(t)) \tag{22}$$

holds at these points.

Here $\eta_+(t) - \eta_-(t)$ is taken to be the right side of (18). In order to estimate approximate solutions, we use important ideas of Chern [7]. Existence of the limits $\eta_\pm(t)$ and the relation (22) are guaranteed by the Glimm-Lax theory [10]. The global solutions obtained above are asymptotically stable in the following sense.

Theorem 2.3. *Assume that the initial data satisfy*

$$(u,v)|_{t=0} = \begin{cases} (u_m, v_m) & x < -M \\ (u_m^*, v_m) & x > M \end{cases} \tag{23}$$

$(M > 0)$ and the total variation of the initial perturbation is sufficiently small. Then the speed and the states both side of the phase boundary approach those of the Maxwell phase boundary at the rate $t^{-\frac{3}{2}}$ and the total variation of (u,v) outside of the phase boundary approaches zero at the rate $t^{-\frac{1}{2}}$.

Using these solutions to the isentropic equations, we have weak admissible solutions to the isothermal equations (1), (2) in the following way. By (4), the equation for the specific entropy η is formally expressed as

$$\eta_t = -\tfrac{1}{T}\left[\left(\psi(u) + \tfrac{1}{2}v^2\right)_t - (\sigma v)_x\right]. \tag{24}$$

Obviously the right side depends only on u and v that are already obtained by the isentropic equations. Thus we obtain η simply by integrating above expression (24). The formula (18) is useful for the integration.

For given Riemann data $U_L = (u_L, v_L) \in \Omega_\alpha$ and $U_R = (u_R, v_R) \in \Omega_\beta$ such that $|U_L - U_m|, |U_R - U_m^*|$ are sufficiently small, there exists a unique admissible solution to the Riemann problem, consisting of 4 constant regions U_L, U_R, U_+, U_-, for the isentropic equations (1). For given η_L, η_R, the intermediate states $\eta_\pm$ are defined by the Hugoniot equation (18)

$$\begin{aligned} \eta_- - \eta_L &= -\tfrac{1}{T}(G_- - G_L) \\ \eta_+ - \eta_R &= -\tfrac{1}{T}(G_+ - G_R). \end{aligned} \tag{25}$$

Moreover, we define $\overline{\eta}_\pm$ as the following:

$$\begin{aligned} \overline{\eta}_- - \eta_L &= -\tfrac{1}{T}(G_- - G_L) - \tfrac{1}{T}(G_+ - G_-)_-, \\ \overline{\eta}_+ - \eta_R &= -\tfrac{1}{T}(G_+ - G_R) - \tfrac{1}{T}(G_- - G_+)_-, \end{aligned} \tag{26}$$

where $(G)_- = \min\{G, 0\}$. Then $\overline{\eta}_- - \eta_-$ $(G_+ < G_-)$ and $\overline{\eta}_+ - \eta_+$ $(G_+ > G_-)$ are the changes of entropy across the phase boundary and η defined by

$$\eta = \begin{cases} \overline{\eta}_- & x < 0, \\ \overline{\eta}_+ & x > 0 \end{cases} \tag{27}$$

together with $\overline{\sigma}_\pm = \sigma_-$ (or σ_+) and $\overline{v}_\pm = v_-$ (or v_+) constitutes a contact discontinuity which completes the solution.

Theorem 2.4. *If the Riemann initial data $(u_L,\ v_L,\ \eta_L)$ and $(u_R,\ v_R,\ \eta_L)$ are close to the Maxwell states, then there exists a unique admissible solution which consists of 5 constant regions connected by rarefaction waves, shock waves and* *a phase boundary. Moreover these constant states are Lipschitz continuous with respect to the initial data.*

These Riemann solutions constitute approximate solutions with general initial data

$$(u,v,\eta)=\begin{cases}(u_L(x),v_L(x),\eta_L(x))\in\Omega_\alpha & x<0,\\ (u_R(x),v_R(x),\eta_R(x))\in\Omega_\beta & x>0.\end{cases}\tag{28}$$

We find by (17), (18) that the change of entropy η is expressed by U. Hence $T.V.\eta(*,t)$ is estimated by $T.V.u(*,t)+T.V.v(*,t)$.

Theorem 2.5. *Assume that the initial data is close to the Maxwell states and the perturbation is sufficiently small in total variation. Then there exists a weak global solution with a single phase boundary. This phase boundary is Lipschitz continuous curve in xt-space. Moreover, the limit* (21) *exists except for countable t; the Rankine-Hugoniot conditions* (9) *and the kinetic condition* (22) *holds at these points.*

Using the formula (18) again, we have

Theorem 2.6. *Except for countable x the specific entropy has the form:*

$$\eta(x,t)=\eta_0(x)-\frac{1}{T}\int_0^t d_\tau G(u(x,\tau)).\tag{29}$$

To finish this section, we make two remarks on the above formula.

Remark 2.1. *The expression* (29) *of the entropy function η is also valid for the large amplitude solutions obtained by Nishida* [18] *for model equations of isothermal ideal gas.*

Remark 2.2. *We find by* (29) *that the limit*

$$\eta_\infty(x)=\lim_{t\to\infty}\eta(x,t)$$

exists for almost all x. Moreover, Liu [16] *shows that $\eta_\infty(x)-\eta_m=O(1)|x|^{-\frac{3}{2}}$ and $\|\eta(*,t)-\eta_\infty\|_{L^1}=O(1)t^{-\frac{1}{2}}$ as $t\to\infty$. In this case, we can also show that $\eta(x,t)-\eta_\infty(x)=O(1)t^{-\frac{1}{2}}$.*

3. Polytropic Model

When the material is polytropic: $e=c\theta$, it follows from the basic identity (3) the change of entropy is expressed as

$$\eta_+-\eta_-=c\log\frac{\theta_+}{\theta_-}-\int_{u_-}^{u_+}\omega(u)\,du.\tag{30}$$

Hence the equations for thermodynamic quantities in the Maxwell states are the following.

$$\frac{\theta_+}{\theta_-}=\frac{\omega(u_-)}{\omega(u_+)},\quad \int_{u_-}^{u_+}\omega(u)\,du=c\log\frac{\theta_+}{\theta_-}.\tag{31}$$

We can see from the above identities that Maxwell strains satisfy

$$\int_{u_-}^{u_+} \omega(u)\,du + c\log\frac{\omega(u_+)}{\omega(u_-)} = 0.$$

Hence the set of Maxwell strains (u_+, u_-) forms a 1-dimensional manifold in a neighborhood of given Maxwell strains. Note that $\omega_+ < \omega_-$, if $u_+ > u_-$ and $\omega_\pm > 0$. On the other hand the first equation in (31) indicates that the strains in given distribution of the temperature $\theta_\pm$ also forms a 1-dimensional manifold. Simple computation shows that these two manifolds intersect at given Maxwell states transversally if

$$\frac{\omega'(u_-)}{\omega'(u_+)} \neq \frac{\omega(u_-)^2}{\omega(u_+)^2}.$$

In particular we have

Proposition 3.1. *If the Maxwell strains obtained satisfy*

$$u_+ > u_-,\ \omega(u_\pm) > 0,\ \omega'(u_-) < \omega'(u_+),$$

then they constitute a unique pair of Maxwell states for given distribution of the temperature $\theta_\pm$.

Now we discuss the stability of given Maxwell states $U_m = (u_m, v_m, \eta_m)$, $U_m^* = (u_m^*, v_m, \eta_m)$. In each region of Ω_α and Ω_β, we adopt (σ, v, η) as independent variables. By differentiating the Rankine-Hugoniot condition (9) and (13), we have

Lemma 3.1. *At given Maxwell states, the infinitesimal changes of* σ *and* η *from the Maxwell states are expressed as*

$$d\sigma_+ = d\sigma_- \tag{32}$$

$$d\eta_+ = \frac{[u]_-^+}{\theta_+}d\sigma_- + \frac{\theta_-}{\theta_+}d\eta_- \tag{33}$$

where $[u]_-^+ = u_+ - u_-$.

By using the above expressions, we can define a correspondence of thermodynamic quantities σ and η admitting a phase boundary.

Theorem 3.1. *In a small neighborhood of given Maxwell states, we can define a unique* C^2*-map*

$$\mathcal{P}_0 : (\sigma_-, \eta_-) \to (\sigma_+, \eta_+) \tag{34}$$

such that the Hugoniot equation (9) *and the kinetic condition* (13) *hold.*

It follows from (33) that

$$d\eta_+ - d\eta_- = \frac{1}{\theta_+}\left\{[u]_-^+ d\sigma_- - [\theta]_-^+ d\eta_-\right\}.$$

Hence we have

Proposition 3.2. *In the correspondence $\mathcal{P}_0$, states (σ_-, η_-) satisfying $\eta_+ = \eta_-$ constitute 1-dimensional manifold $\mathcal{N}_0$ whose gradient at the given Maxwell states is*

$$\frac{d\sigma_-}{d\eta_-} = \frac{[\theta]_-^+}{[u]_-^+}. \tag{35}$$

We note that $\frac{d\sigma_-}{d\eta_-} > 0$, if $\theta_+ > \theta_-$. In the isothermal model, $\mathcal{N}_0$ is a vertical line in the $\sigma_-\eta_-$ plane. Next theorem indicates non-uniqueness of solution to the Riemann problem.

Theorem 3.2. *If $\theta_m \neq \theta_m^*$ at given Maxwell states, there are at least two admissible solutions to the Riemann problem, satisfying the kinetic condition* (13), *for Riemann data close to the Maxwell states.*

Outline of Proof: Let $U_L = (\sigma_L, v_L, \eta_L)$ and $U_R = (\sigma_R, v_R, \eta_R)$ be Riemann data close to U_m and U_m^* respectively. We can extend the boundary map $\mathcal{P}_0$ to $\mathcal{P}$ in Ω_α simply by defining $v_+ = v_- + \Phi(\eta_+ - \eta_-)(u_+ - u_-)$. We denote by $\mathcal{W}_F(U_L)$ the forward wave curve issuing from U_L and $\mathcal{W}_B(U_R)$ the backward wave curve issuing from U_R. The image of $\mathcal{W}_F(U_L)$ and the inverse-image of $\mathcal{W}_B(U_R)$ by $\mathcal{P}$ are denoted respectively by $\widetilde{\mathcal{W}}_F(U_F)$ and $\widetilde{\mathcal{W}}_B(U_R)$.

We draw the projection of $\mathcal{W}_F(U_L)$ and $\widetilde{\mathcal{W}}_B(U_R)$ on (σ_-, v_-)-plane. Then we find a unique point of intersection of these both curves. Let $U_- \in \mathcal{W}_F(U_L)$ and $\widetilde{U}_+ \in \widetilde{\mathcal{W}}_B(U_R)$ denote its inverse-images on each curve. Assuming for simplicity $\theta_m^* > \theta_m$, we can adjust the Riemann data so that $\widetilde{U}_+$ is located "over" $\mathcal{N}$ and U_- "under" $\mathcal{N}$, where $\mathcal{N} = \mathcal{N}_0 \times R_v$. Clearly in the same fashion, there exists a point of intersection of the projection of $\widetilde{\mathcal{W}}_F(U_L)$ and $\mathcal{W}_B(U_R)$ on (σ_+, v_+)-plane, and its two inverse-images $U_+, \widetilde{U}_-$ are over and under $\widetilde{\mathcal{N}}$ (the image of $\mathcal{N}$).

Now we can construct two admissible solution as the following. First solution consists of intermediate states $U_-, \widetilde{U}_+, U_+$ where $\widetilde{U}_+$ and U_+ are connected by the admissible phase boundary. Second one consists of $U_-, \widetilde{U}_-, U_+$ where U_- and $\widetilde{U}_-$ are connected by the admissible phase boundary.

References

[1] R. Abeyaratne and J. K. Knowles, Kinetic relations and the propagation of phase boundaries in solids, *Arch. Rational Mech. Anal.*, 114 (1991), 119–154.

[2] R. Abeyaratne and J. K. Knowles, Dynamics of propagating phase boundaries: adiabatic theory for thermoelastic solids, *Physica D*, 79 (1994), 269–288.

[3] F. Asakura, Large Time Stability of Maxwell Phase Boundaries, Methods and Applications of Analysis, Vol. 6, No. 4 (1999) 477–504.

[4] F. Asakura, Kinetic condition and the Gibbs function, Taiwanese J. of Math., Vol.4, No. 1 (2000), 105–117.

[5] F. Bereux-E. Bonnetier-P. LeFloch, Gas dynamics system: two special cases, *SIAM J. Math. Anal.*, 28 (1997), 499–515

[6] G.-Q. Chen-C. Dafermos, The vanishing viscosity method in one-dimensional thermoelasticity, *Trans. AMS*, 147 (1995), 531–541

[7] I.-L. Chern, Stability theorem and truncation error analysis for the Glimm scheme and for a front tracking method for flows with strong discontinuities, Comm. Pure Appl. Math., 42 (1989), 815–844.

[8] E. Fermi, *Thermodynamics*, Dover Pub., New York 1956.

[9] J. Glimm, Solutions in the large for nonlinear hyperbolic systems of equations, Comm. Pure Appl. Math., 18 (1965), 697–715.

[10] J. Glimm–P. D. Lax, *Decay of Solutions of Systems of Nonlinear Hyperbolic Conservation Laws*, Amer. Math. Soc. Memoir, No. 101. A.M.S. Providence, 1970.

[11] H. Hattori, The Riemann problem for a van der Waals fluid with entropy rate admissible criterion – Isothermal Case, *Arch. Rational Mech. Anal.*, 92 (1986), 246–263

[12] H. Hattori, The Riemann problem and the existence of weak solutions to a system of mixed-type in dynamic phase transition, *J. Differential Equations*, 146 (1998), 287-319.

[13] P. D. Lax, Shock waves and entropy, *E. Zarantonello (ed.), Contributions to nonlinear Functional Analysis*, Academic Press, New York, 1971, 603–634.

[14] P. LeFloch, Propagating Phase boundaries: formulation of the problem and existence via the Glimm method, *Arch. Rational Mech. Anal.*, 123 (1993), 153–197

[15] T.-P. Liu, Decay to N-waves of solutions of general systems of nonlinear hyperbolic conservation laws, Comm. Pure Appl. Math., 30 (1977), 585–610.

[16] T.-P. Liu, Linear and nonlinear large time behavior of general systems of hyperbolic conservation laws, Comm. Pure Appl. Math., 30 (1977), 767–796.

[17] R. Menikoff–B. J. Plohr, The Riemann problem for fluid flow of real materials, *Reviews of Modern Physics*, Vol. 61, No.1 (1989), 75–130

[18] T. Nishida, *Nonlinear Hyperbolic Equations and Related Topics in Fluid Dynamics.* Publ. Math. No. 79-02. Univ. Paris-Sud, 1978.

F. Asakura
Faculty of Engineering,
Osaka Electro-Communication University,
18-8 Hatsucho, Neyagawa, Osaka 572-8530, JAPAN
asakura@isc.osakac.ac.jp

International Series of Numerical Mathematics
Vol. 140, © 2001 Birkhäuser Verlag Basel/Switzerland

The Riemann-Problem in Extended Thermodynamics

Jörg D. Au, Daniel Reitebuch, Manuel Torrilhon, and Wolf Weiss

Abstract. In this paper we will approach the shock tube problem, or mathematically speaking the Riemann problem with extended thermodynamics. Extended thermodynamics (ET) provides *dissipative* field equations which are symmetric hyperbolic [1]. Thus, the solution to the shock-tube-problem may be obtained using the well known analytic and numerical methods for hyperbolic systems.

We will observe a large number of acceleration and shock waves for the systems of extended thermodynamics. Due to the dissipation, these waves are strongly damped and the systems of ET lead to *diffusive* solutions of the Riemann problem - although we are dealing with hyperbolic systems.

Furthermore we investigate the start-up phase of the shock tube. For very small times the large systems of ET are needed for a physically valid description. It will be shown numerically that the solution of ET for the start-up phase converges to the solution of the free-flight-equation.

1. Introduction

The physical experiment behind the Riemann problem is the shock tube experiment. It is one of the standard problems in gas dynamics [2]. A shock tube consists of a long tube which is divided into two parts with different pressures separated by a membrane. After the membrane is destroyed, a compression shock runs into the low pressure region, while a rarefaction wave moves into the high pressure part of the tube. Furthermore, usually a contact discontinuity occurs.

In order to describe the shock tube problem theoretically, we need a system of field equations. One of the systems capable of describing the shock tube experiment is the Euler system. This symmetric hyperbolic system consists of the conservation laws of mass, momentum and energy. In this context it is important to note, that the Euler system does not contain dissipative effects such as heat conduction or shear stress. The solution of the Euler system for the Riemann problem is well known (see e.g. [2]). The density along the shock tube coordinate x is shown in Fig.1 (solid) together with the initial condition for the density (thin). The solution exhibits the three typical waves of the shock tube experiment: Starting from the right we see the compression shock, the contact discontinuity and the rarefaction

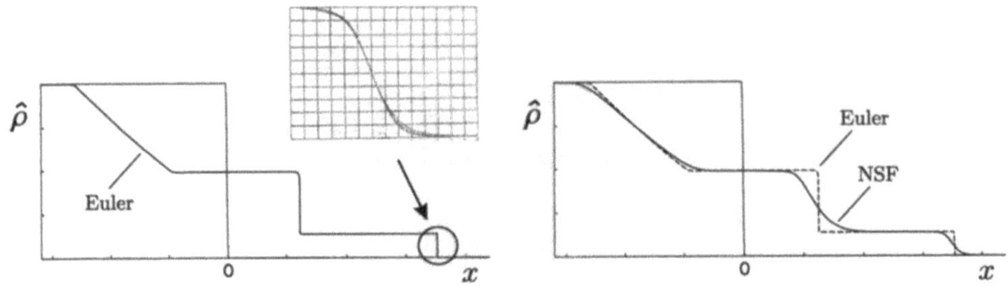

FIGURE 1. Left: Euler solution to the shock tube problem, enlarged we see the measured change in density in a shock wave. Right: NSF-solution together with the Euler solution.

wave. This solution is a similarity-solution in time: It has the above described shape for any time $t > 0$ without any evolutionary behavior. While the solution of the Euler system is not continuous, experimental observations reveal continuous shock structures at a sufficient small scale, see the enlargement in Fig.1. The lack of dissipation in the Euler system is the reason for this behavior.

An alternative system that includes the description of dissipative effects is the parabolic Navier-Stokes-Fourier (NSF) system. The density profile obtained with this system is shown in the right plot of Fig.1. The dissipation leads to a continuous shock profile as well as to a continuous "contact discontinuity" and the kinks of the rarefaction wave are smoothed out. These effects are *not* due to the numerical method. This behavior is closer to reality than the solution of the Euler system. However, careful measurements of the shock structure and shock thickness show that the NSF system does not give quantitatively good results [3].

Several systems have been proposed to remedy the drawbacks of the Euler and the NSF system. One possibility is extended thermodynamics. ET establishes a rational constitutive theory and is able to describe processes with strong gradients and rapid changes, where ordinary theories fail [1]. Thus, the shock tube experiment is an ideal challenge for extended thermodynamics.

In the following we model the initial conditions for the density ρ, the velocity v and the pressure p in the form

$$\rho(x,t=0) = \begin{cases} \rho_1 & x \leq 0 \\ \rho_0 & x > 0 \end{cases} \qquad p(x,t=0) = \begin{cases} p_1 & x \leq 0 \\ p_0 & x > 0 \end{cases} \tag{1}$$
$$v(x,t=0) = 0\,.$$

Furthermore the high and low pressure region is initially in equilibrium. The pressure and density ratios will be equal, that means both regions have the same temperatures in the beginning $T_0 = T_1$. As pressure ratio we choose $p_1/p_0 = 5$. It follows that the resulting shock travels with the Mach number (according to the low pressure region) $M_0 = 1.396$.

2. Equations of Extended Thermodynamics

The equations of extended thermodynamics for monatomic gases may be derived in the context of the kinetic theory [4]. There we have the distribution function f such that $f(\underline{x}, t, \underline{c})\, d\underline{c}$ describes the number density of particles in $\underline{x}$ at the time t with velocities between $\underline{c}$ and $\underline{c}+d\underline{c}$. The basic variables used in extended thermodynamics are the moments

$$F_{i_1 \cdots i_N}(\underline{x}, t) = m \int c_{i_1} \cdots c_{i_N} f(\underline{x}, t, \underline{c})\, d\underline{c} \tag{2}$$

of the distribution function. The first moments are related to the variables of ordinary thermodynamics, for example:

$$\begin{array}{ll} F \equiv \rho & \text{density} \\ F_i \equiv \rho v_i & \text{momentum density} \\ \frac{1}{2}F_{ii} \equiv \frac{1}{2}\rho v_i^2 + \frac{3}{2}p & \text{energy density} \end{array} \tag{3}$$

Throughout this paper we will use the pressure p instead of internal energy or temperature. In extended thermodynamics the ordinary set of variables, density, momentum and energy, is extended by the stress, the heat flux and *higher moments*, which do not have a intuitive interpretation. An increase of the number of variables taken into account may be interpreted as a kind of "series expansion" of the distribution function. The main idea is, that in the case of processes with strong gradients and rapid changes many variables are necessary for an appropriate theoretical description ([1], [5]).

The distribution function obeys the Boltzmann equation

$$\partial_t f + c_k \partial_k f = S(f, f)\ . \tag{4}$$

From that we derive the equations of transfer for the moments, viz.

$$\begin{array}{lcccl} \partial_t F & + & \partial_k F_k & = & 0 \\ \partial_t F_i & + & \partial_k F_{ik} & = & 0 \\ \partial_t F_{ij} & + & \partial_k F_{ijk} & = & P_{\langle ij \rangle} \\ & \vdots & & \vdots & \\ \partial_t F_{i_1 \cdots i_N} & + & \partial_k F_{i_1 \cdots i_N k} & = & P_{i_1 \cdots i_N} \end{array} \tag{5}$$

These equations form a infinite hierarchy of equations with the structure of balance laws. Furthermore the flux of one equation is alway the variable of the next. The first two equations and the trace of the third have zero right hand sides; they represent the conservation laws of mass, momentum and energy. The higher equations, however, have productions, which follow from the collision operator of the Boltzmann equation (4). If we decide on a certain set of variables and cut the hierarchy, we come to a closure problem: The last flux as well as the productions have to be related to the chosen variables in a constitutive manner. These relations are supplied by the constitutive theory of extended thermodynamics. For the left hand side of (5) the main ingredients are the existence of a balance equation for entropy and the definition of entropy from kinetic theory. The productions may be calculated from the collision operator under the assumption of a particle interaction with a Maxwell-potential.

Alltogether extended thermodynamics yields *dissipative, hyperbolic field equations* for an arbitrary choice of variables. Note that these procedure is equivalent to Grad's expansion of the distribution function, if the constitutive equations are linearized with respect to non-equilibrium quantities.

The simplest choice of variables considers F, F_i, F_{ii} or equivalently the mass density ρ, the velocity v_i and the pressure p and the governing equations are the Euler equations of gas dynamics. Thus the Euler equations are a special case in extended thermodynamics. If the set of variables of the Euler case is extended by $F_{\langle ij \rangle}$, F_{kki}, that is the pressure deviator $p_{\langle ij \rangle}$ and the heat flux q_i, the set contains 13 fields. Together with the constitutive equations furnished by ET, the field equations for these 13 variables are equivalent to the 13-moment-case of Grad. In contrast to the Euler equations the equations of the 13-field-case are *dissipative.* In this manner the set of variables may be extended to contain large numbers of moments. It will be seen, that for some processes theses large numbers of variables are indeed necessary for a valid description.

2.1. Example: 13-Field-Case

The most famous example for a hyperbolic dissipative system is Grad's 13 moment system. In the one dimensional case of the shock tube the system reduces to 5 equations for the mass density ρ, the velocity along the axis of the tube v, the pressure p, the $(1,1)$-component of the pressure deviator $\sigma \equiv p_{\langle ij \rangle}$ and the axial heat flux q, viz.

$$\begin{aligned}
\partial_t \rho + \partial_x \rho v &= 0 \\
\partial_t \rho v + \partial_x (\rho v^2 + p + \sigma) &= 0 \\
\partial_t (\rho v^2 + 3p) + \partial_x (\rho v^3 + 5pv + 2\sigma v + 2q) &= 0 \\
\partial_t (\tfrac{2}{3}\rho v^2 + \sigma) + \partial_x (\tfrac{2}{3}\rho v^3 + \tfrac{4}{3}pv + \tfrac{7}{3}\sigma v + \tfrac{8}{15}q) &= -\tfrac{4}{5}B\rho\sigma \\
\partial_t (\rho v^3 + 5pv + 2\sigma v + 2q) + \partial_x (\rho v^4 + 8pv^2 + 5\sigma v^2 + \tfrac{32}{5}qv + \tfrac{p}{\rho}(5p + 7\sigma)) & \\
&= -\tfrac{8}{5}B\rho(\tfrac{2}{3}q + \sigma v)
\end{aligned} \tag{6}$$

The first three equations of the system are conservation laws in contrast to the last two. Setting σ and q to zero and neglecting the last two equations we obtain the well known Euler system. In the case of more moments and hence, more equations the structure of the systems remains the same: The first three equations are conservation laws, the balance equations for the higher moments are not.

The system is written in dimensionless variables. In the productions at the right hand side of (6) a scaling parameter B appears, which corresponds to the inverse of the Knudsen number. It is defined as

$$B = \frac{\text{observation time}}{\text{mean-free-flight-time}} \sim \frac{\text{observation domain}}{\text{mean-free-path}} = \frac{1}{Kn}. \tag{7}$$

The Euler case posesses the three characteristic velocities

$$v,\ v \pm c_0 \tag{8}$$

where $c_0 = \sqrt{5/3\,p/\rho}$ is the speed of sound [2]. These three velocities correspond to three characteristic waves, which may be identified with the phenomena in the experiments. The 13-field-case, however, has five characteristic velocities, which in equilibrium (vanishing σ and q) read (see [1])

$$v,\ v \pm 0.63\,c_0, v \pm 1.65\,c_0\,. \tag{9}$$

Higher moment systems have even more characteristic velocities and the question arises how these systems decribe the experimental phenomena. Because of the greater number of characteristics in these systems, we will observe a greater number of acceleration and discontinuous shock waves. But the dissipation in the higher moment systems of ET results in a strong attenuation of these waves and, in contrast to the Euler equations, leads to diffusive solutions of the shock tube experiment after a sufficient long time - although these systems are hyperbolic.

3. Numerical Methods

In the case of small systems the solutions were calculated by using the package CLAWPACK, which is available in [6]. It implements a upwind method to solve systems of hyperbolic equations of the form

$$\partial_t u + \partial_x f\,(u) = P\,(u)\,. \tag{10}$$

Note that the equations of extended thermodynamics (5)/(6) have exactly this form. Due to the splitting method CLAWPACK requires the solution of the reduced equation

$$\partial_t u = P\,(u) \tag{11}$$

over a certain time step. Furthermore, a Riemann solver has to be supplied.

In extended thermodynamics the equations (11) are easily solved, since the productions are algebraic and thus (11) form a system of ordinary differential equations which decouple spatially. This fact is an advantage of extended thermodynamics compared with the equations of NSF, where the equations (11) couple spatially. As Riemann solver we used the approximate solver of Roe [7]. In the case of the Euler equations it is possible to calculate the required Roe matrix A_R and its eigenvalues and eigenvectors analytically. In extended thermodynamics this is not so easily done and we used the expression

$$A_R = \int_0^1 f'\,(u_l + \theta\,(u_r - u_l))\,d\theta \tag{12}$$

which is a result of the application of the mean value theorem to the condition $A_R\,(u_r - u_l) = f\,(u_r) - f\,(u_l)$ (see [7]). The integral has been calculated numerically by use of the Simpson rule. The eigenvalues and eigenvectors of the resulting Roe matrix were obtained by a routine of the NAG library [8]. An entropy-fix has

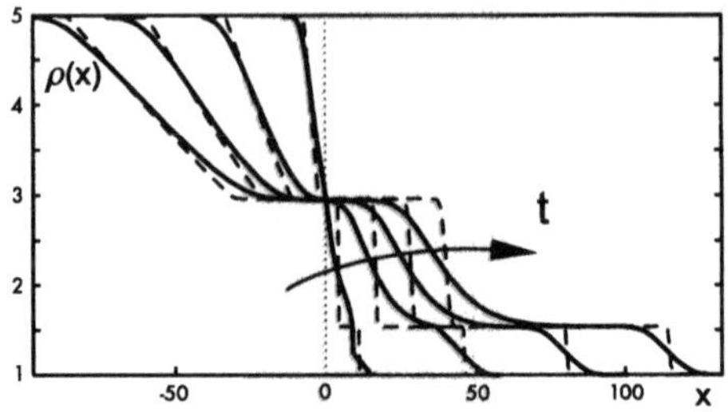

FIGURE 2. Density field in the shock tube for the Euler equations (dashed) and the 13-field-case (solid)

been implemented based on the approach in [9]. This Riemann solver was tested by comparing the results with the solutions obtained from usual Riemann solvers for Euler equations and by using more accurate integration formulas than the Simpson rule. In all considered cases the differences were not significant.

In the case of many moments we preferred the central scheme of Nessayahu & Tadmor described in [10]. This method needs no calculation of local Riemann problems, which becomes a time expensive task for the large systems of extended thermodynamics.

4. Results of the 13-Field-Case

In Fig.2 the solution of the density field for the initial conditions (1) is shown for different times $t = 10, 40, 70, 100$. The dashed line shows the result of the Euler equations and the solid line the result of the 13-field-case, additionally we see the initial condition (thin). Time and space are measured in mean free flight times and mean free paths according to the low pressure region at the right of the membrane.

The result of the Euler equations exhibits the contact discontinuity, the shock and rarefaction wave in the typical discontinous and non-smooth manner. In the result of the 13-field-case we observe fully diffusive profiles for the shock and the contact discontinuity for a sufficient long time. Also the kinks of the rarefaction wave are smoothed out. This solution has an higher physical significance than the Euler solution, since the shock is known to have a structure spread over 10-20 mean free paths. The shock structure of the 13-field-case shows no subshock since its propagation speed is smaller than the biggest characteristic velocity of the 13-field-case [11]. For short times Fig.2 is rather complicated: In the 13-field-case five waves move away from the origin. Some of them are discontinuities (subshocks), they need a close look to be identified in Fig.2. In order to demonstrate the behavior of these waves the results for the density field for different times are rescaled in Fig.3 to the same x-interval. Numerically this procedure is done by choosing appropiate values for the scaling parameter B in (6). Again the dashed line represents the Euler case. Due to its similarity property the solution of the

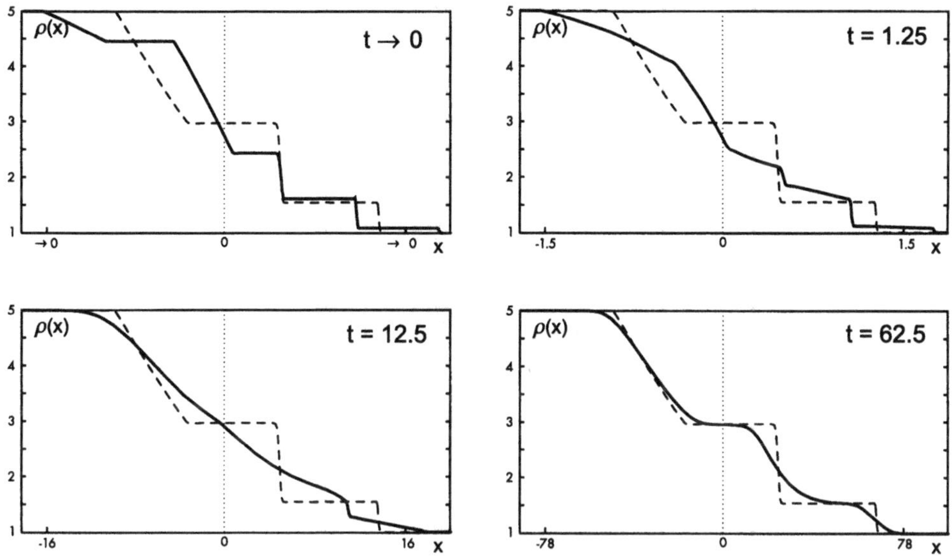

FIGURE 3. Scaled solution of the density field for the Euler equations (dashed) and the 13-field-case (solid)

Euler equations stays the same for all scales. In the first picture the time and length scale are infinitesimally small ($B \to 0$). One may now easily identify the five waves of the 13-field-case. There are three discontinuities, which travel to the right and two rarefaction waves, which travel to the left. There is no connection between the characteristic waves of the Euler equations and the waves occurring in the 13-moment-case. If the time scale is increase the productions in (6) and thus the dissipation comes into play. Due to the dissipation the characteristic waves are damped and for large times the field form the above mentioned diffusive profile. Note that the length scale of the pictures is affected by the scaling process. It is remarkable that the diffusive profiles of the 13-field-case move exactly with the Euler solution, however *not along* the characteristics of the 13-field-case.

For a very small time and length scale, like in the first picture of Fig.3, neither the solution of the Euler case nor the solution of the 13-field-case shows a physically accurate behaviour. We expect a start-up phase which is connected with a very high non-equlibrium. Thus we will need more variables than 13 for a physical description.

5. Start-up Phase

Since the use of more and more variables may be viewed as a series expansion, we rely on the following working-hypothesis: *Given a process, there exists a certain number of variables N_V, such that the results of theories with number of variables*

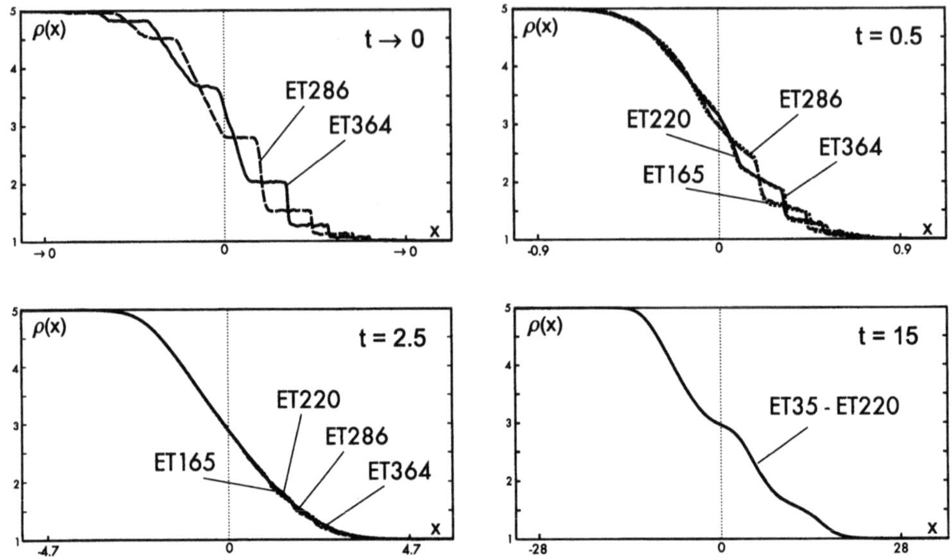

FIGURE 4. Density field in the start-up phase of the shock tube experiment with different theories

$N > N_V$ do not differ (significantly) from the result of theory N_V. We say, the theories *converge*, they become *valid*. Or, to put it in other words: Only if the extension of the variables does not change the result, the solution may be trusted. Experience ([1], [5]) has shown that only valid solutions stand the comparison with experimental data. For an easy derivation of the large systems of extended thermodynamics a software-package named ET_{XX} has been developed [12]. It duplicates the constitutive theory numerically and provides the one-dimensional flux-function and the productions of extended thermodynamics for an arbitrary number of variables. In this paper we present the results of the following theories

ET35	ET56	ET84	ET120	ET165	ET220	ET286	ET364
9	12	16	20	25	30	36	42

The second row shows the number of 1-D equations, i.e. the number of characteristic waves in the system: Here it rises from 9 up to 42.

In Fig.4 we see the density fields of the different theories for the Riemann problem (1) for different times. Time and space are again measured in mean free flight times and mean free paths. In the last picture all theories from ET35 up to ET220 show the same result. That means that for 15 mean free flight times after the membrane has been removed the theory with 35 fields is valid. While the structures of the contact discontinuity and of the shock just begin to develop at $t = 15$, they are not yet visible at a smaller time. Furthermore, at $t = 2.5$, the theories ET165 to ET364 begin to differ in the front. Behind the membrane, however, the theories match and ET165 should be valid in this area. At an even smaller time the theories

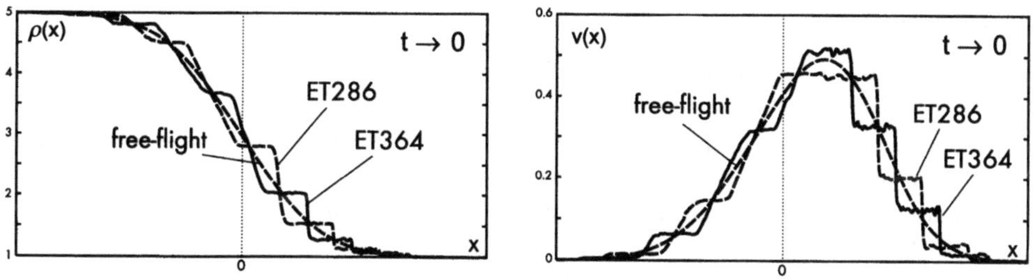

FIGURE 5. Comparison of the result of the free-flight-equation with the result of ET for density and velocity

used differ everywhere; for a valid description one has to increase the number of variables even more. In the first picture the characteristic waves of the theories are recovered. For the theory ET364 there are 42 of them, but only 8 are visible. Most of the characteristic waves carry only a very small amplitude and has already left the computational domain. The convergence behaviour of the theories in Fig.4 depends on the initial pressure ratio. If the pressure ratio is increased, also the initial non-equilibrium is increased and we expect to need more variables for a valid description at a given time.

The first picture in Fig.4 describes the shock tube experiment at an infinitesimally small time and length scale. At that scale no dissipation has any influence. Physically that means that we look at the experiment at a time when collisions of the particles are negligible. This time may also be modeled by the homogenous Boltzmann equation, or free-flight-equation, i.e. equation (4) in the case $S \equiv 0$. This equation is easily solved for the initial conditions (1) by assuming Maxwell distributions for the high pressure and low pressure region of the shock tube. In Fig.5 the results of the free-flight-equations for the density and the velocity are compared with the results of the theories ET286 and ET364. We see an agreement but in a unexpected manner. For the infinitesimally small time and length scale every system of extended thermodynamics gives a result build up of characteristic waves. Thus the smooth curves of the free-flight-equations are approximated in extended thermodynamics by the stepped result of a hyperbolic system. If the number of variables is increased not only large characteristic velocities are added to the system, but also small velocities. Nevertheless in general the spectrum of sucsessive theories does not correspond. Numerical evaluations suggest that in the limit of infinite variables the spectrum becomes dense. That means that an increase of the variables introduces additional "steps" into the result presented in Fig.5 and in the limit the smooth free-flight curve is approximated by more and more and ever finer "steps".

At the right hand side of Fig.5 the solutions of extended thermodynamics exhibit oscillations. They are due to the numerics and indicate that the used methods seem to have problems with large systems containing many waves.

References

[1] Müller, I. and Ruggeri, T.: *Rational Extended Thermodynamics* (2nd edn), Springer Tracts in Natural Philosophy (vol.37), Springer, New York (1998)

[2] Courant, R. and Friedrichs, K.O.: *Supersonic Flow and Shock Waves*, Applied Mathematical Science (vol. 21), Springer, New York (1976)

[3] Schmidt, B.: *Electron Beam Density Measurements in Shock Waves in Argon*, J.Fluid Mech. **39**, (1969) p.361

[4] Chapman, S., Cowling, T.G.: *The Mathematical Theory of Non-Uniform Gases,* 3rd ed., Cambridge (1990)

[5] Weiss, W.: *Zur Hierarchie der Erweiterten Thermodynamik,* dissertation, TU Berlin (1990)

[6] LeVeque, R.J.: *CLAWPACK - A Package for Solving Conservation Laws*, source available in www.amath.washington.edu/~rjl/clawpack.html

[7] Roe, P.L.: *Approximate Riemann Solvers, Parameter Vectors and Difference Schemes,* J.Comp.Phys. **43**, (1981) p.357

[8] Numerical Algorithm Group Ltd: *Fortran 77 Libraries Mark 18,* Wilkinson House, Jordan Hill Road, Oxford, UK.

[9] Harten, A. and Hyman, J.M.: *Self Adjusting Grid Methods for One-Dimensional Hyperbolic Conservation Laws*, J.Comp.Phys. **50**, (1983) p.235

[10] Nessyahu, H. and Tadmor, E.: *Non-oscillatory Central Differencing for Hyperbolic Conservation Laws*, J.Comp.Phys. **87**, (1990) p.408

[11] Weiss, W.: *Continuous Shock Structure in Extended Thermodynamics*, Phys. Rev. E **52**, (1995) p.5760

[12] Au, J.D., Struchtrup, H. and Torrilhon, M.: *ET_{XX} - A Equation Generator for Extended Thermodynamics*, source available on request via M.Torrilhon@vt.tu-berlin.de

J. D. Au
D. Reitebuch
M. Torrilhon
W. Weiss
Institut für Verfahrenstechnik,
Technische Universität Berlin,
Fasanenstr. 90,
D-10623 Berlin
J.Au@vt.tu-berlin.de
D.Reitebuch@vt.tu-berlin.de
M.Torrilhon@vt.tu-berlin.de
W.Weiss@vt.tu-berlin.de

International Series of Numerical Mathematics
Vol. 140, © 2001 Birkhäuser Verlag Basel/Switzerland

Heterogeneous Domain Decomposition Methods for Compressible Magneto-plasma Flows

Monika Auweter–Kurtz, Cristian A. Coclici, Jörg Heiermann, and Wolfgang L. Wendland

Abstract. We present nonoverlapping domain decomposition methods for the numerical treatment of compressible viscous plasma flows inside a self–field magnetoplasmadynamic (MPD) accelerator. The magneto–plasma flow is modelled by a system of conservation laws extended by partial differential equations describing the electromagnetic field. Since the numerical solution of the governing equations is extremely expensive, the flow–field domain is decomposed into two or three model zones, characterized by different physical properties of the flow. The compressible Navier–Stokes equations (extended under the influence of an arc discharge) in the near field of the accelerator are coupled with simplified models of extended conservation laws in the far field via appropriate transmission conditions at the artificial coupling boundaries. Results of our numerical computations are presented.

1. Introduction

MPD thrusters are special devices intensively used within the last decades in the spacecraft industry for the propelling of various spacecrafts. For planetary missions, such as the manned mission to Mars, specific impulses of about 2000 s and high thrust levels are required for the thruster system.

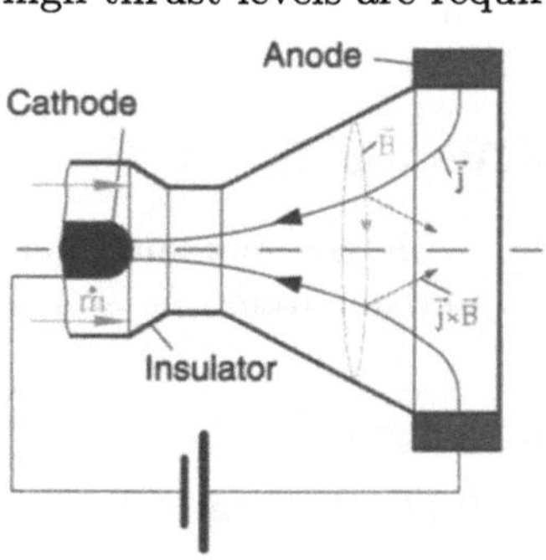

FIGURE 1. Thruster

MPD thrusters are considered for such missions due to their high thrust level and simplicity. Argon is typically chosen as propellant for MPD thrusters because of its low specific ionization energy level.

The principle of a self–field thruster is shown in Figure 1. A cold gas enters the accelerator and is heated up to a hot plasma by an electric discharge. The plasma expands thermally and accelerates into a test tank in the laboratory. In addition, the plasma is accelerated by electromagnetic Lorentz forces.

The flow is described by the conservation equations for mass, momentum and energy for the heavy particles (argon atoms Ar^0 and ions Ar^{1+}, Ar^{2+}), by the conservation equation for electron and internal ionization energy, and by the Maxwell equations of classical electrodynamics. Furthermore, reaction equilibrium, thermal non–equilibrium (two–fluid model), and laminar flow are assumed.

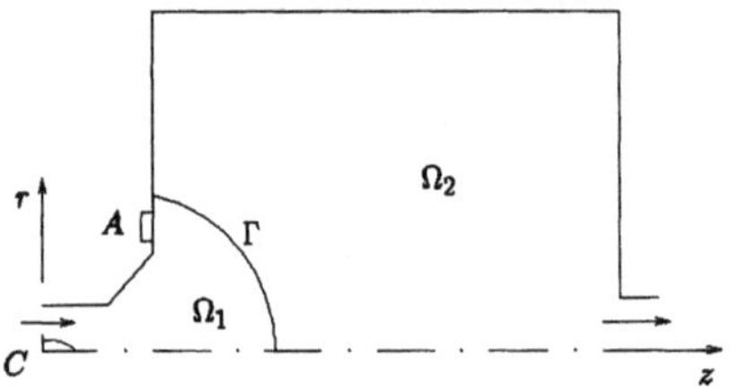

FIGURE 2. First decomposition

Due to the tremendous costs needed for the numerical treatment, the governing system is employed only within an essentially smaller near–field region Ω_1, containing the thruster.

In the complementary far field Ω_2, corresponding to the test tank, we use simplified models of the extended conservation laws. The simplifications should be chosen in such a way, that on one hand the flow in the far–field domain is still modelled accurately enough, and on the other hand, the numerical treatment can be performed efficiently. The axisymmetric plasma flow is described in cylindrical coordinates by the function

$$\mathbf{W} = \mathbf{W}(r, z;\, t) := \left[\mathbf{w}, p_H, T_H;\ \mathbf{w}_e;\ \mathbf{w}_{EB}\right]^\top (r, z;\, t), \quad (r, z) \in \Omega,\ t \in [0, T].$$

Here, the component $\mathbf{w} = (\rho, \rho v_r, \rho v_z, E_H)^\top$ of $\mathbf{W}$ collects the conservative variables with the density ρ, the velocity vector $\mathbf{v} = (v_r, v_z)^\top$, and the energy of the heavy particles E_H. The pressure and the temperature of the heavy particles are denoted by p_H and T_H, respectively. The function $\mathbf{w}_e = (\mathrm{e}_{ei}, p_e, T_e)^\top$ describes the electron component of the plasma, with e_{ei} containing the electron and the ionization energy, and with p_e and T_e representing the pressure and the temperature of the electron component, respectively. Finally, $\mathbf{w}_{EB} = (\mathbf{E}, \mathbf{B}, \mathbf{j})^\top$ contains the electromagnetic field $(\mathbf{E}, \mathbf{B})$ and the electric current density $\mathbf{j}$.

2. Governing equations

We model the heavy–particle flow by the compressible Navier–Stokes equations, extended by the influence of an arc discharge, which in cylindrical coordinates have the form

$$\frac{\partial \mathbf{w}_1}{\partial t} + \mathrm{div}_{(r,z)} \mathbf{F}(\mathbf{W}_1) = \mathrm{div}_{(r,z)} \mathbf{R}(\mathbf{w}_1, \nabla_{(r,z)} \mathbf{w}_1) + \mathbf{G}(\mathbf{W}_1) \ \text{ in } \ \Omega_1 \times [0, T]. \quad (1)$$

The function $\mathbf{F}$ contains the convective part (here, with the pressure field $p = p_H + p_e$), and an electromagnetic pressure term derived from the source terms. We represent $\mathbf{F}$ as

$$\mathbf{F} = (\mathbf{F}_r, \mathbf{F}_z)(\mathbf{W}) = (\mathbf{f}_r, \mathbf{f}_z)(\mathbf{w}, \mathbf{w}_e) + (\mathbf{g}_r, \mathbf{g}_z)(\mathbf{w}_{EB}), \quad (2)$$

where, with the purely azimuthal magnetic field $\mathbf{B} = (0, B_\varphi, 0)^\top$ and with the magnetic permeability of vacuum $\mu_0 > 0$,

$$\begin{aligned}
\mathbf{f}_r(\mathbf{w}, \mathbf{w}_e) &:= \left(\rho v_r,\ \rho v_r^2 + (p_H + p_e),\ \rho v_r v_z,\ [E_H + (p_H + p_e)]\, v_r\right)^\top, \\
\mathbf{f}_z(\mathbf{w}, \mathbf{w}_e) &:= \left(\rho v_z,\ \rho v_z v_r,\ \rho v_z^2 + (p_H + p_e),\ [E_H + (p_H + p_e)]\, v_z\right)^\top, \\
\mathbf{g}_r(\mathbf{w}_{EB}) &:= \left(0,\ B_\varphi^2,\ 0,\ B_\varphi^2 v_r\right)^\top / (2\mu_0),\ \mathbf{g}_z(\mathbf{w}_{EB}) := \left(0,\ 0,\ B_\varphi^2,\ B_\varphi^2 v_z\right)^\top / (2\mu_0).
\end{aligned}$$

The dissipative terms are contained in the function $\mathbf{R}$ represented in the form

$$\mathbf{R} = (\mathbf{R}_r, \mathbf{R}_z)(\mathbf{w}, \nabla_{(r,z)}\mathbf{w}) = (\mathbf{V}_r, \mathbf{V}_z)(\mathbf{w}, \nabla_{(r,z)}\mathbf{w}) + (\mathbf{T}_r, \mathbf{T}_z)(\mathbf{w}). \tag{3}$$

Here, $(\mathbf{V}_r, \mathbf{V}_z)$ collect the components of the viscous part of the stress tensor

$$\mathbf{V}_r(\mathbf{w},\nabla\mathbf{w}) := \left(0, \tau_{rr}, \tau_{rz}, \tau_{rr}v_r + \tau_{rz}v_z\right)^\top, \; \mathbf{V}_z(\mathbf{w},\nabla\mathbf{w}) := \left(0, \tau_{zr}, \tau_{zz}, \tau_{zr}v_r + \tau_{zz}v_z\right)^\top,$$

defined by means of the velocity field $\mathbf{v}$ and of the viscosity coefficient $\mu > 0$ by

$$\tau_{rr} = \mu\Big[2\frac{\partial v_r}{\partial r} - \frac{2}{3}\,\text{div}\,\mathbf{v}\Big],\; \tau_{rz} = \tau_{zr} = \mu\Big[\frac{\partial v_r}{\partial z} + \frac{\partial v_z}{\partial r}\Big],\; \tau_{zz} = \mu\Big[2\frac{\partial v_z}{\partial z} - \frac{2}{3}\,\text{div}\,\mathbf{v}\Big].$$

Furthermore, $(\mathbf{T}_r, \mathbf{T}_z)$ represent the heat conduction terms

$$\mathbf{T}_r(\mathbf{w}) := \left(0,\ 0,\ 0,\ \lambda_H\ \partial T_H/\partial r\right)^\top, \quad \mathbf{T}_z(\mathbf{w}) := \left(0,\ 0,\ 0,\ \lambda_H\ \partial T_H/\partial z\right)^\top,$$

with the heat conductivity $\lambda_H > 0$ of the heavy–particle flow. The function $\mathbf{G}(\mathbf{W})$ contains the electromagnetic force and heat terms as well as quantities describing the heat transfer due to the collisions between the plasma components:

$$\mathbf{G}(\mathbf{W}) := \Big(0,\ \frac{1}{r}\Big[p_H + p_e - \frac{2}{3}\mu\Big(2\frac{v_r}{r} - \frac{\partial v_r}{\partial r} - \frac{\partial v_z}{\partial z}\Big) - \frac{B_\varphi^2}{2\mu_0}\Big],\ 0,$$
$$\Big(p_e + \frac{B_\varphi^2}{2\mu_0}\Big)\,\text{div}\,\mathbf{v} - \frac{v_r}{r}\frac{B_\varphi^2}{\mu_0} + \sum_{\nu=0}^{2} n_\nu n_e \alpha_{e\nu}(T_e - T_H)\Big)^\top.$$

Here, n_ν $(\nu{=}0,1,2)$ and n_e are the densities of the heavy particles and of the electrons, respectively, and $\alpha_{e\nu}$ are heat transfer coefficients. Note that, by including the Lorentz terms $\mathbf{j}\times\mathbf{B}$ as $B_\varphi^2/(2\mu_0)$ in the fluxes, our formulation observes as much conservation as possible.

The conservation of electron and internal energy is in $\Omega_1 \times [0,T]$ given by

$$\frac{\partial e_{ei}}{\partial t} + \text{div}\,(e_{ei}\mathbf{v}) - \text{div}\,(\lambda_{ei}\nabla T_e) = -p_e\,\text{div}\,\mathbf{v} + \frac{5}{2}\frac{k}{e}\mathbf{j}\cdot\nabla T_e - \frac{1}{n_e e}\mathbf{j}\cdot\nabla p_e$$
$$+\sum_{\nu=0}^{2} n_\nu n_e \alpha_{e\nu}(T_H - T_e) + \frac{|\mathbf{j}|^2}{\sigma}. \tag{4}$$

Here, λ_{ei} denotes the heat conductivity for the electron component of the flow, k is the Boltzmann constant, and σ is the electric conductivity. The Maxwell equations and Ohm's law for plasmas read as

$$\text{rot}\,\mathbf{B} = \mu_0\,\mathbf{j},\ \text{rot}\,\mathbf{E} = -\partial\mathbf{B}/\partial t,\ \ \text{div}\,\mathbf{B} = 0; \quad \mathbf{E} = \mathbf{j}/\sigma - \mathbf{v}\times\mathbf{B} + \beta\,\mathbf{j}\times\mathbf{B} - \beta\,\nabla p_e$$

with β – Hall parameter, leading to the discharge equation

$$\frac{\partial^2 B}{\partial r^2} + \frac{\partial^2 B}{\partial z^2} + \Big[\frac{1}{r} - \mu_0\sigma v_r\Big]\frac{\partial B}{\partial r} - \mu_0\sigma v_z\frac{\partial B}{\partial z} - \Big[\frac{1}{r^2} + \mu_0\sigma\Big(\frac{\partial v_r}{\partial r} + \frac{\partial v_z}{\partial z}\Big)\Big]B = F_B, \tag{5}$$

where $F_B = F_B(\sigma, p_e, \mu_0)$ denotes a source term. Additional equilibrium reactions are incorporated into our model but, for brevity, they are here not given explicitly. Due to its high complexity and to the tremendous computational costs, the system has been discretized only in the vicinity Ω_1 of the MPD thruster (see, e.g. [8, 9, 5]), which we identify here as the near field. In these papers, Γ is considered as an outflow (freestream) boundary, and characteristic boundary conditions in terms of data obtained from measurements, are used. In that model one faces the problem that a certain amount of ambient far–field gas *recirculates* into the hot plasma jet in the near field. Consequently, parts of the outflow boundary Γ get "inflow" properties and require additional information about the flow quantities. This makes the numerical treatment of the plasma flow in the far field Ω_2 necessary. We consider there simplified models which take into account the physical properties of the flow, and couple these models to the complete system in Ω_1. Our coupling procedures extend previous work on heterogeneous domain decomposition in aerodynamics (see, e.g. [7, 1, 2]) to the case of compressible magneto–plasma flows.

3. A first domain decomposition method

3.1. Simplified modelling of the far field Ω_2

In a first approximation we assume that far away from the MPD accelerator the shear stresses $\tau_{rr}, .., \tau_{zz}$ and the heat conduction terms $\lambda_H \partial_{r(z)} T_H$, defining the quantity $\mathbf{R}$, are strongly dominated by the convective part:

$$|\tau_{rr}|, ..., |\tau_{zz}| \ll \|\mathbf{F}\| \text{ and } |\lambda_H \partial T_H / \partial r|,\ |\lambda_H \partial T_H / \partial z| \ll \|\mathbf{F}\| \text{ in } \Omega_2 \times [0, T]. \quad (6)$$

Here, $|\cdot|$ and $\|\cdot\|$ denote appropriate norms on $\Omega_2 \times [0, T]$. Hence, we assume the heavy–particles flow to be purely inviscid in Ω_2. We also assume that the magnetic field $\mathbf{B}$ vanishes identically in Ω_2. The governing system takes the reduced form

$$\frac{\partial \mathbf{w}_2}{\partial t} + \operatorname{div}_{(r,z)}(\mathbf{f}_r, \mathbf{f}_z)(\mathbf{w}_2, \mathbf{w}_{e,2}) = \mathbf{H}(\mathbf{w}_2, \mathbf{w}_{e,2}) \quad \text{in} \quad \Omega_2 \times [0, T], \quad (7)$$

with the simplified source term

$$\mathbf{H}(\mathbf{w}_2, \mathbf{w}_{e,2}) := \Big(0,\ \frac{p_H + p_e}{r},\ 0,\ p_e \operatorname{div} \mathbf{v} + \sum_{\nu=0}^{2} n_\nu n_e \alpha_{e\nu} (T_e - T_H)\Big)^{\top}.$$

Furthermore, as a consequence of $\mathbf{j}=\operatorname{rot}\mathbf{B}/\mu_0 \equiv 0$ in Ω_2, the equation of conservation for the electron and the ionization energy (4) is reduced in $\Omega_2 \times [0, T]$ to

$$\frac{\partial e_{ei}}{\partial t} + \operatorname{div}(e_{ei}\mathbf{v}) - \operatorname{div}(\lambda_{ei} \nabla T_e) = -p_e \operatorname{div} \mathbf{v} + \sum_{\nu=0}^{2} n_\nu n_e \alpha_{e\nu} (T_H - T_e). \quad (8)$$

3.2. Transmission conditions

These conditions should be chosen in such a way, that on one hand, the fundamental physical laws are respected, and on the other hand, the resulting coupled problem is well–posed and consistent with the full original problem. The continuity of the characteristic variables could be chosen as transmission condition, but according to the theory of hyperbolic equations, this can be required only across that part of the interface, where the corresponding characteristics enter the hyperbolic region, see e.g. [6, 7, 1, 2]. In accordance with the conservation laws, the continuity of the normal flux yields a transmission condition on the complete interface Γ: the total flux associated with the full model in Ω_1 (containing the inviscid as well as the viscous contributions) is set equal to the normal inviscid flux, that results from the simplified equations in Ω_2:

$$\begin{aligned} -\big[\mathbf{R}_r(\mathbf{w}_1, \nabla \mathbf{w}_1)\, n_r + \mathbf{R}_z(\mathbf{w}_1, \nabla \mathbf{w}_1)\, n_z\big] + \big[\mathbf{f}_r(\mathbf{w}_1, \mathbf{w}_{e,1}) + \mathbf{g}_r(\mathbf{w}_{EB,1})\big]\, n_r \\ + \big[\mathbf{f}_z(\mathbf{w}_1, \mathbf{w}_{e,1}) + \mathbf{g}_z(\mathbf{w}_{EB,1})\big]\, n_z = \mathbf{f}_r(\mathbf{w}_2, \mathbf{w}_{e,2})\, n_r + \mathbf{f}_z(\mathbf{w}_2, \mathbf{w}_{e,2})\, n_z. \quad (9) \end{aligned}$$

The flux condition has successfully been used for pure flow problems (see for example [7, 2]). Note that, imposing this transmission condition, we neglect the viscous fluxes and the heat transfer corresponding to the heavy–particle flow:

$$\lambda_{H,1}\frac{\partial T_{H,1}}{\partial \mathbf{n}} \equiv 0 \quad \text{and} \quad \big[\mathbf{V}_r\, n_r + \mathbf{V}_z\, n_z\big](\mathbf{w}_1, \nabla \mathbf{w}_1) \equiv 0 \quad \text{on} \quad \Gamma. \tag{10}$$

As a consequence, the solutions of the coupled problem may exhibit jumps at the interface, depending on the magnitude of the viscosity and heat transfer terms neglected in Ω_2 (for more details, see [1]). Since the solution of the original problem should satisfy the *natural* transmission conditions at the artificial interface (i.e. continuity of the solution and of the total normal flux), the approximate extended Navier–Stokes / extended Euler solution can only be a first approximation and needs to be corrected by special terms accounting for the loss of continuity and maintaining the continuity of the normal flux. A boundary layer correction for a simplified transmission problem is presented in [3] in the framework of singular perturbation theory. It is extended for the problem under consideration in [4].
In order to assure the conservation of the electron heat flux across the interface, we impose the continuity of the co–normal derivative of the electron temperature:

$$\left[\lambda_{ei,1}\frac{\partial T_{e,1}}{\partial \mathbf{n}}\right](r,z) = \left[\lambda_{ei,2}\frac{\partial T_{e,2}}{\partial \mathbf{n}}\right](r,z) \quad \text{for all} \quad (r,z) \in \Gamma. \tag{11}$$

Finally, we impose $B_\varphi \equiv 0$ on Γ.

3.3. Numerical aspects and results

The extended conservation laws (1) and (7), as well as the electron and the ionization energy equations (4) and (8) are solved on an unstructured, dual mesh by

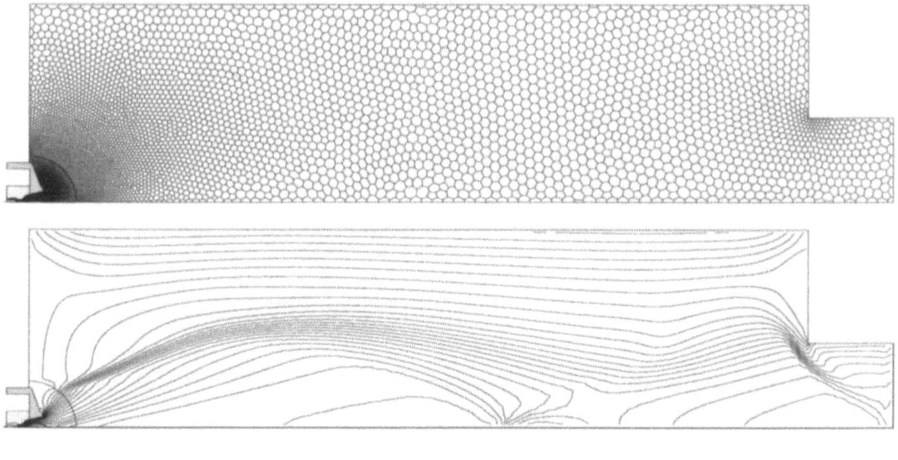

FIGURE 3. Full computational domain and isolines of v_z

using a second–order finite volume upwind scheme based on explicit Euler time–stepping. Equation (5) is solved by linear finite elements on triangular grids, see [5]. The full computational domain including the near field of the MPD accelerator and the far field corresponding to the tank, are shown in Figure 3. The area of the far field is about 80 times larger than that of the near field, emphasizing the necessity of simplifying the mathematical model in the far field. The isolines of the axial velocity component v_z give an overall impression of the plasma flow: The plasma is accelerated in the MPD accelerator, it is then expanded into the tank, and a part of it flows out of the tank at the far right.

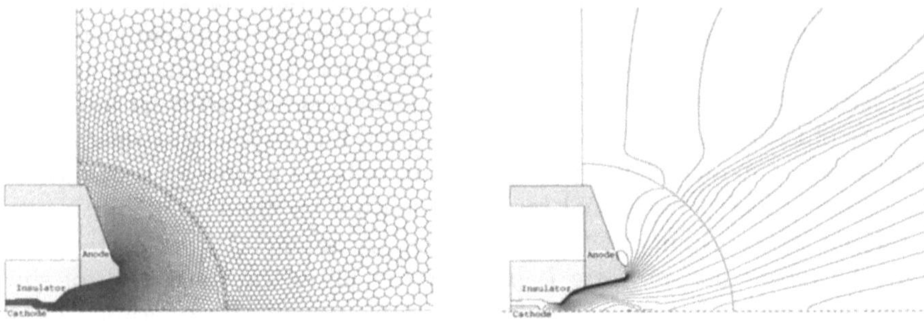

FIGURE 4. Computational domain (left); isolines of v_z (right)

The coupling domain, where dual cells on both sides of the coupling boundary touch each other, is shown in Figure 4 (left). The mesh generator, based on an advancing front algorithm, enforces the global mesh to be conforming at the artificial coupling boundary. The isolines of the longitudinal velocity v_z for the coupled solution are presented on the right. Obviously, the coupling method works very well for the central, hot plasma jet. Up to one nozzle radius above the centerline, v_z passes smoothly the coupling boundary Γ. However, farther away from the centerline, the isolines show small jumps and have kinks across the interface.

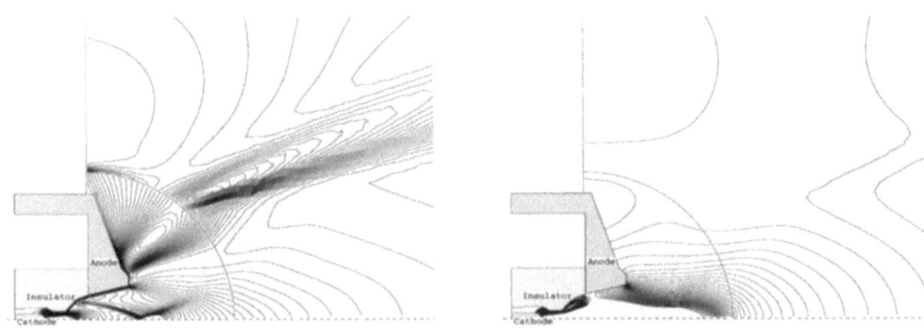

FIGURE 5. Isolines of the heavy–particles temperature T_H (left) and of the electron temperature T_e (right)

Figure 5 (left) shows the effect of neglecting the heat conduction terms corresponding to the heavy–particle flow. While the isolines of the heavy–particle temperature T_H behave smoothly across the part of Γ contained in the central plasma jet, we can see the discontinuities of the solution in the region above very clearly. The (wrong) behavior of T_H at the left (insulator) boundary on top of the anode is also due to the different kind of boundary conditions imposed on T_H: Dirichlet on the near–field part and homogeneous Neumann in the Euler the far–field part at the insulator boundary. On the other hand, the approximate coupled solution also shows that the electron temperature T_e passes the artificial interface smoothly, as can be seen in Figure 5 (right). Thus, the *natural* transmission condition (11), used for the coupling of (4) and (8) is justified also numerically.

4. A second domain decomposition method

The discontinuities in T_H within the region above the central plasma jet show that the heavy–particle heat conduction is still physically relevant with respect to the inviscid Euler energy flux, such that the heavy particle heat conduction cannot be neglected in this geometrical decomposition. Also a shock wave appearing in the immediate vicinity of the coupling boundary Γ is a consequence of employing the purely hyperbolic model (7), (8) too close to the MPD–thruster. Furthermore, the local discontinuity of T_H also causes a slight discontinuity and non–smoothness of other flow variables in the critical region.

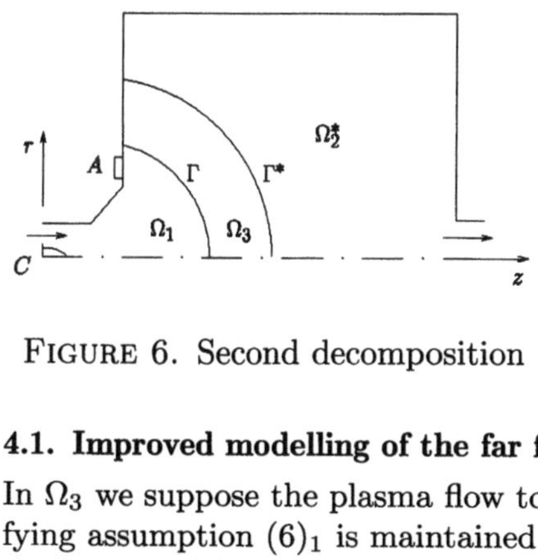

FIGURE 6. Second decomposition

Therefore, we further decompose the far field Ω_2 into a small intermediate domain Ω_3, attached to the near field and to the complementary region Ω_2^*:

$$\overline{\Omega_2} := \overline{\Omega_3} \cup \overline{\Omega_2^*}.$$

Let $\Gamma^* := \partial\Omega_3 \cap \partial\Omega_2^*$ denote the artificial interface separating the two subregions of Ω_2.

4.1. Improved modelling of the far field

In Ω_3 we suppose the plasma flow to be inviscid but *heat conductive*. The simplifying assumption $(6)_1$ is maintained here, whereas the terms $(\mathbf{T}_r, \mathbf{T}_z)$, describing the heat conductivity of the flow, are taken into consideration. We further assume that $\mathbf{B}$ vanishes identically in Ω_3.

The unknown function, denoted here by $\mathbf{W}_3 := \left[\mathbf{w}_3, p_{H,3}, T_{H,3}; \mathbf{w}_{e,3}; \mathbf{0}\right]^{\top}(r, z;\, t)$ is chosen to satisfy in $\Omega_3 \times [0, T]$ the (intermediate) simplified equations:

$$\frac{\partial \mathbf{w}_3}{\partial t} + \operatorname{div}_{(r,z)}(\mathbf{f}_r, \mathbf{f}_z)(\mathbf{w}_3, \mathbf{w}_{e,3}) = \operatorname{div}_{(r,z)}(\mathbf{T}_r, \mathbf{T}_z)(\mathbf{w}_3) + \mathbf{H}(\mathbf{w}_3, \mathbf{w}_{e,3}). \quad (12)$$

The electron and ionization energy equation (8) is employed also in Ω_3. In Ω_2^* we consider the same model as in the far–field region Ω_2 of the previous section.

4.2. Improved transmission conditions

4.2.1. TRANSMISSION CONDITIONS ON Γ For the heavy–particle flow we impose the continuity of the normal flux across Γ. In accordance with the improved model in Ω_3, this condition reads as

$$\begin{aligned}
&-\left[\mathbf{R}_r(\mathbf{w}_1,\nabla\mathbf{w}_1)n_r+\mathbf{R}_z(\mathbf{w}_1,\nabla\mathbf{w}_1)n_z\right]\\
&\qquad+\left[\mathbf{f}_r(\mathbf{w}_1,\mathbf{w}_{e,1})+\mathbf{g}_r(\mathbf{w}_{EB,1})\right]n_r+\left[\mathbf{f}_z(\mathbf{w}_1,\mathbf{w}_{e,1})+\mathbf{g}_z(\mathbf{w}_{EB,1})\right]n_z\\
&=\quad-\left[\mathbf{T}_r(\mathbf{w}_3)\,n_r+\mathbf{T}_z(\mathbf{w}_3)\,n_z\right]+\mathbf{f}_r(\mathbf{w}_3,\mathbf{w}_{e,3})\,n_r+\mathbf{f}_z(\mathbf{w}_3,\mathbf{w}_{e,3})\,n_z. \qquad (13)
\end{aligned}$$

This condition assures the heat transfer corresponding to the heavy–particle flow,

$$\left[\lambda_{H,1}\frac{\partial T_{H,1}}{\partial\mathbf{n}}\right](r,z)=\left[\lambda_{H,3}\frac{\partial T_{H,3}}{\partial\mathbf{n}}\right](r,z),$$

but the viscous fluxes are still neglected, i.e. $\left[\mathbf{V}_r\,n_r+\mathbf{V}_z\,n_z\right](\mathbf{w}_1,\nabla\mathbf{w}_1)\equiv 0$ across Γ. Furthermore, the continuity condition (11) is imposed.

4.2.2. TRANSMISSION CONDITIONS ON $\Gamma^*=\partial\Omega_3\cap\partial\Omega_2^*$ Since Γ^* is situated in a region where the viscosity $(\mathbf{V}_r,\mathbf{V}_z)$ is completely neglected, the continuity of the normal flux takes here the form

$$\begin{aligned}
&-\left[\mathbf{T}_r(\mathbf{w}_3)\,n_r+\mathbf{T}_z(\mathbf{w}_3)\,n_z\right]+\mathbf{f}_r(\mathbf{w}_3,\mathbf{w}_{e,3})\,n_r+\mathbf{f}_z(\mathbf{w}_3,\mathbf{w}_{e,3})\,n_z\\
&\qquad=\mathbf{f}_r(\mathbf{w}_2,\mathbf{w}_{e,2})\,n_r+\mathbf{f}_z(\mathbf{w}_2,\mathbf{w}_{e,2})\,n_z\quad\text{across}\quad\Gamma^*. \qquad (14)
\end{aligned}$$

Note that we neglect both, the heat transfer and the viscous fluxes:

$$\left[\lambda_{H,3}\frac{\partial T_{H,3}}{\partial\mathbf{n}}\right](r,z)\equiv 0\quad\text{and}\quad\left[\mathbf{V}_r\,n_r+\mathbf{V}_z\,n_z\right](\mathbf{w}_3,\nabla\mathbf{w}_3)\equiv 0\quad\text{across}\quad\Gamma^*.$$

Finally, we require the electron heat flux to be continuously transferred

$$\left[\lambda_{ei,3}\frac{\partial T_{e,3}}{\partial\mathbf{n}}\right](r,z)=\left[\lambda_{ei,2}\frac{\partial T_{e,2}}{\partial\mathbf{n}}\right](r,z)\quad\text{across}\quad\Gamma^*. \qquad (15)$$

4.3. Numerical results

Figure 7 shows the dual mesh used for the numerical coupling the the three models, and the isolines of the longitudinal velocity v_z of the coupling solution.

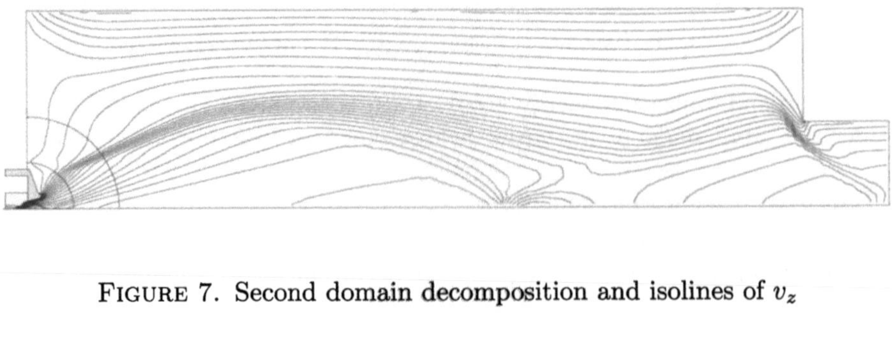

FIGURE 7. Second domain decomposition and isolines of v_z

The latter are plotted in detail in Figure 8 (right) and show a better resolution of the recirculation region than in Figure 3.

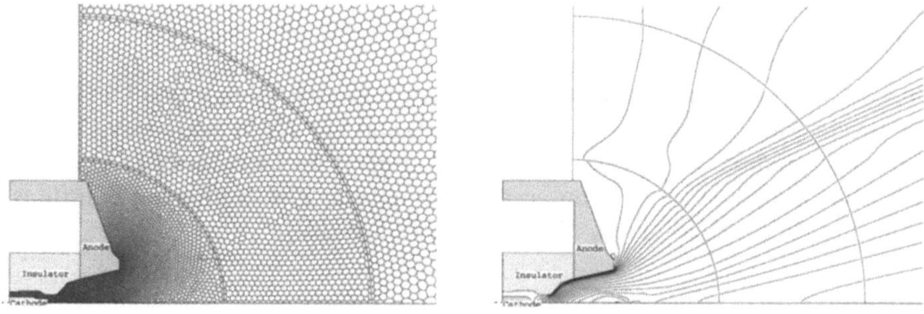

FIGURE 8. Computational domain (left); isolines of v_z (right)

A remarkable improvement can be observed for the heavy–particle flow temperature T_H (Figure 9, left). The "hyperbolic" shock wave created in the two–model coupling (Figure 5, left) is smeared by employing the heat–conducting terms $(\mathbf{T}_r, \mathbf{T}_z)$ in Ω_3; the isolines pass smoothly both interfaces Γ and Γ^*. The boundary layer behavior of T_H can be observed at the left boundary of the thruster. This is due to the change of boundary conditions for T_H on $\partial\Omega_3$, which take the same form as on $\partial\Omega_1$.

Furthermore, the electron temperature (Figure 9, right) shows a similar behaviour as in the two–model coupling; this fact is expectable, since equation (8) is employed in both Ω_3 and Ω_2^*, and the transmission condition (11) is used across both interfaces Γ and Γ^*.

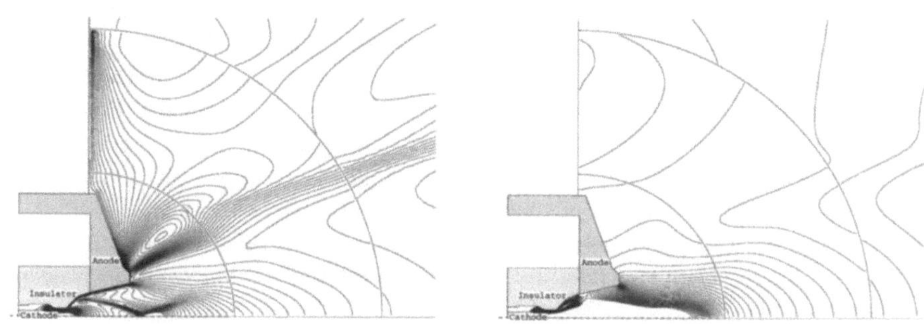

FIGURE 9. Isolines of the heavy–particles temperature T_H (left) and of the electron temperature T_e (right)

Finally, we want to emphasize that, by using the heterogeneous domain decomposition, the rather complex compressible magneto–plasma flow has been computed within the whole MPD accelerator plus tank configuration for the first time, and that the influence of the far field through the recirculating amount of gas has been simulated numerically. Our investigation shows that the heterogeneous domain decomposition method is an efficient tool for the numerical treatment of nonlinear boundary value problems of high complexity.

5. Acknowledgement

This work has been supported by the German Research Foundation DFG through the projects "Domain Decomposition Methods and Far–Field Boundary Conditions for Initial Boundary Value Problems with Conservation Laws and Electromagnetic Fields" (We 659/35–1) and "Conservation Equations and Numerical Solutions for Technically Relevant Magneto–Plasmas" (Au 85/9–1,2).

References

[1] C.A. Coclici, *Domain decomposition methods and far–field boundary conditions for two–dimensional compressible flows around airfoils.* Doctoral Thesis, Mathematical Institute A, University of Stuttgart 1998.

[2] C.A. Coclici and W.L. Wendland, *Analysis of a heterogeneous domain decomposition for compressible viscous flow*, To appear in Math. Models Meth. Appl. Sci., **4** (2001).

[3] C.A. Coclici, G. Moroşanu and W.L. Wendland, *The coupling of hyperbolic and elliptic boundary value problems with variable coefficients*, Math. Meth. Appl. Sci. **23** (2000), 401–440.

[4] C.A. Coclici, G. Moroşanu and W.L. Wendland, *Asymptotic correction of the numerical solution to a heterogeneous domain decomposition for compressible plasma flows*, 2001, in preparation.

[5] J. Heiermann, M. Auweter–Kurtz, A. Eberle, H.J. Kaeppeler, U. Iben, P.C. Sleziona: *Recent Improvements of Numerical Methods for the Simulation of MPD Thruster Flow on Adaptive Meshes*, IEPC-99-169, 26th International Electric Propulsion Conference, Kitakyushu, 1999.

[6] C. Hirsch, *Numerical Computation of Internal and External Flows*, Wiley, 1988.

[7] A. Quarteroni, L. Stolcis, *Homogeneous and heterogeneous domain decomposition for compressible fluid flows at high Reynolds numbers*, Numerical Methods for Fluid Dynamics V, K.W. Morton and M.J. Baines eds., Oxford University Press, Oxford, 113–128, 1995.

[8] P.C. Sleziona, *High Enthalpy Flows for Spaceflight Applications* (in German), Habilitation, Aerospace Faculty, University of Stuttgart, Germany, 1998.

[9] H.P. Wagner, H.J. Kaeppeler, M. Auweter–Kurtz, *Instabilities in MPD thruster flows: 1. Space charge instabilities in unbounded and inhomogeneous plasmas*, J. Phys. D: Appl. Phys. **31** (1998), 519–528.

M. Auweter–Kurtz, C.A. Coclici, J. Heiermann, W.L. Wendland
Mathematical Institute A and Institute for Space Systems,
University of Stuttgart,
D–70569 Stuttgart, Germany
{cristi, wendland}@mathematik.uni-stuttgart.de,
{heierman, kurtz}@irs.uni-stuttgart.de

International Series of Numerical Mathematics
Vol. 140, © 2001 Birkhäuser Verlag Basel/Switzerland

Magnetoplasmadynamic Rocket Thruster Simulation

Monika Auweter–Kurtz, Cristian Coclici, Jörg Heiermann, Claus–Dieter Munz, and Christian Sleziona

Abstract. Magnetoplasmadynamic (MPD) rocket thrusters are investigated numerically and experimentally at the Institut für Raumfahrtsysteme (IRS). The self–magnetic plasma flow in thermal and reaction non–equilibrium is modeled with thirteen conservation equations. A new finite volume solver has been developed to compute the steady–state solution. The upwinding scheme for the species and the time–stepping procedure will be described in detail, and numerical results for the hot anode thruster will be presented.

1. Introduction

Flying to the planets and moons of the solar system in a short time requires rocket thrusters with a thrust of several Newton, high exhaust speed, and low fuel consumption. MPD rocket propulsion may be able to fulfill these requirements at relatively low costs.

MPD thrusters use an arc discharge between two electrodes to heat a gas up to several ten–thousand Kelvin. In these axisymmetric devices, the resulting plasma carries an electric current which induces an azimuthal magnetic field. The plasma is thermally expanded and also accelerated by electromagnetic forces, and thus a very high exhaust velocity for efficiently propelling the spacecraft is achieved.

Theoretical and numerical investigations are performed at the IRS in order to gain a more profound understanding of the complex physical processes involved [1, 2, 3]. In the laboratory, MPD thrusters are investigated experimentally, and the numerical and experimental results are compared in order to develop and verify physical models. This basic research of the complex plasma flow phenomena which influence thrust and efficiency will allow an optimization of the thrusters with respect to performance, reliability and cost.

Differences between experimental and numerical results indicate that the physical and mathematical models used have to be extended. This is done in this paper. To this model an advanced numerical finite volume code is applied to capture the plasma flow physics more accurately.

2. Conservation Equations

The argon plasma flow considered is described by conservation equations for density, momentum and energy for the heavy particles (atoms and ions); by the conservation equation for electron and ionization energy; and by the Maxwell equations of classical electrodynamics for the magnetic field. Furthermore, reaction non–equilibrium, thermal non–equilibrium (two–fluid model), and turbulent flow are assumed. The conservation equations are hyperbolic–parabolic and include strong source terms.

The following physical quantities are used: magnetic field B, electron energy e_e, heavy particle energy e_h, electric current density $\vec{j}$, diffusion flux $\vec{j}_{D,i}$, electron density n_e, heavy particle density n_h, electron pressure p_e, heavy particle pressure p_h, electron temperature T_e, heavy particle temperature T_h, velocity $\vec{v}$, heat transfer coefficient α_{ei}, Hall parameter β, ratio of specific heats γ, heat conductivity of electrons λ_e, atomic and turbulent heat conductivity of heavy particles λ_h and λ_T, atomic and turbulent dynamic viscosity μ and μ_T, magnetic permeability of vacuum μ_0, species densities $n_i(i = 0...6)$, with $i = 0$ neutral and $i = 6$ six–fold ionized, reaction source term ω_i, heavy particle mass density ρ, electric conductivity σ, viscous stress τ and ionization energy $\chi_{i\rightarrow i+1}$.

The plasma flow is assumed to be axisymmetric; the divergence operator in cylindrical coordinates takes the form $div\begin{bmatrix} a_z \\ a_r \end{bmatrix} = \frac{1}{r}\frac{\partial(ra_r)}{\partial r} + \frac{\partial a_z}{\partial z}$. Herewith, the conservation equations for density, momentum and energy of the heavy particles read as

$$\begin{aligned} \frac{\partial n_i}{\partial t} + div\,(n_i\vec{v}) &= -div\,\vec{j}_{D,i} + \omega_i, \\ \frac{\partial(\rho v_z)}{\partial t} + div\,\vec{f}_z &= 0, \\ \frac{\partial(\rho v_r)}{\partial t} + div\,\vec{f}_r &= q_r, \\ \frac{\partial e_h}{\partial t} + div\,\vec{f}_h &= q_h, \end{aligned} \tag{1}$$

with the flux functions

$$\vec{f}_z = \begin{bmatrix} p_h + p_e + \frac{B^2}{2\mu_0} + \rho {v_z}^2 - \tau_{zz} \\ \rho v_z v_r - \tau_{rz} \end{bmatrix}, \vec{f}_r = \begin{bmatrix} \rho v_z v_r - \tau_{rz} \\ p_h + p_e + \frac{B^2}{2\mu_0} + \rho {v_r}^2 - \tau_{rr} \end{bmatrix}, \tag{2}$$

$$\vec{f}_h = \begin{bmatrix} (e_h + p_h + p_e + \frac{B^2}{2\mu_0})v_z - \tau_{zz}v_z - \tau_{zr}v_r - (\lambda_h + \lambda_T)\frac{\partial T_h}{\partial z} \\ (e_h + p_h + p_e + \frac{B^2}{2\mu_0})v_r - \tau_{zr}v_z - \tau_{rr}v_r - (\lambda_h + \lambda_T)\frac{\partial T_h}{\partial r} \end{bmatrix}, \tag{3}$$

and with the source terms

$$q_r = \frac{p_h + p_e - \tau_{\varphi\varphi} - \frac{B^2}{2\mu_0}}{r}, \quad (4)$$

$$q_h = \left(p_e + \frac{B^2}{2\mu_0}\right) div\, \vec{v} - \frac{B^2}{\mu_0\, r}\, v_r + \sum_{\nu=0}^{6} n_\nu n_e \alpha_{e\nu}(T_e - T_h). \quad (5)$$

We also require the equations of state expressing constitutive thermodynamic relations. These equations take the form

$$e_h = \frac{3}{2} n_h k T_h + \frac{1}{2}\rho \vec{v}^{\,2} = \frac{p_h}{\gamma - 1} + \frac{1}{2}\rho\vec{v}^{\,2} \quad \text{and} \quad p_e = n_e k T_e\,. \quad (6)$$

Finally, the dimensionless components of the stress tensor are given by

$$\tau_{zz} = \frac{2}{3}(\mu + \mu_T)\left(2\frac{\partial v_z}{\partial z} - \frac{v_r}{r} - \frac{\partial v_r}{\partial r}\right), \tau_{zr} = (\mu + \mu_T)\left(\frac{\partial v_z}{\partial r} + \frac{\partial v_r}{\partial z}\right), \quad (7)$$

$$\tau_{rr} = \frac{2}{3}(\mu + \mu_T)\left(2\frac{\partial v_r}{\partial r} - \frac{v_r}{r} - \frac{\partial v_z}{\partial z}\right), \tau_{\varphi\varphi} = \frac{2}{3}(\mu + \mu_T)\left(2\frac{v_r}{r} - \frac{\partial v_r}{\partial r} - \frac{\partial v_z}{\partial z}\right). \quad (8)$$

More details of the diffusion fluxes are given in [4, 5]. Data for the ionization reactions ω_i are taken from [6, 7, 8]. With m_h denoting the mass of a single atom or ion, the mass density is computed by $\rho = m_h \sum_i n_i$. Because of the quasineutrality of the plasma, the electron density is given by $n_e = \sum_i i\, n_i$.
The conservation of electron energy is given as

$$\begin{aligned} &\frac{\partial e_e}{\partial t} + div\,(e_e\vec{v}) + p_e div\,\vec{v} - \frac{5}{2}\frac{k}{e}\,\vec{j}\cdot\nabla T_e + \frac{1}{n_e e}\vec{j}\cdot\nabla p_e = \\ &\sum_{\nu=0}^{6} n_\nu n_e \alpha_{e\nu}(T_h - T_e) + div(\lambda_e \nabla T_e) \\ &-div\left(\frac{3}{2}kT_e \sum_{i=1}^{6} Z_i \vec{j}_{D,i}\right) - \sum_{i=0}^{5}\omega_{i+1}\chi_{i\to i+1} + \frac{|\vec{j}|^2}{\sigma}. \end{aligned} \quad (9)$$

In addition, we consider the turbulence conservation equation

$$\begin{aligned} &\frac{\partial(\rho R)}{\partial t} + div\,(\rho R\,\vec{v}) - div\left[\left(\mu + \frac{\mu_T}{\sigma_\varepsilon}\right)\cdot\nabla R\right] \\ &\quad = \rho(C_{\varepsilon 1} - C_{\varepsilon 2} e^{-R_T{}^2})\sqrt{RP} - \nabla\left(\mu + \frac{\mu_T}{\sigma_\varepsilon}\right)\cdot\nabla R - \rho\frac{1}{\sigma_\varepsilon}\nabla\nu_T\cdot\nabla R, \end{aligned} \quad (10)$$

containing the ratio of the square of turbulent kinetic energy and dissipation $R = \nu R_T$, the dynamic atomic and eddy viscosities $\mu = \rho\nu$ and $\mu_T = \rho\nu_T$ and the turbulent production term

$$P = \nu_T\left\{2\left[\left(\frac{\partial v_z}{\partial z}\right)^2 + \left(\frac{\partial v_r}{\partial r}\right)^2 + \left(\frac{v_r}{r}\right)^2\right] + \left(\frac{\partial v_z}{\partial r} + \frac{\partial v_r}{\partial z}\right)^2 - \frac{2}{3}\left(\frac{\partial v_z}{\partial z} + \frac{\partial v_r}{\partial r} + \frac{v_r}{r}\right)^2\right\}.$$

The kinematic eddy viscosity ν_T is given by an empirical function of ρR in [9] as well as the empirical constants $C_{\varepsilon 1}$, $C_{\varepsilon 2}$ and σ_ε.

The Maxwell equations [10] for the discharge are

$$rot\ \vec{B} = \mu_0 \vec{j} \quad \text{and} \quad rot\ \vec{E} = -\frac{\partial \vec{B}}{\partial t}\ . \tag{11}$$

$div\ \vec{B} = 0$ is fulfilled automatically by axisymmetry. Ohm's law for plasmas is given by

$$\vec{E} = \frac{\vec{j}}{\sigma} - \vec{v} \times \vec{B} + \beta\, \vec{j} \times \vec{B} - \beta\, \nabla p_e. \tag{12}$$

The functions $\vec{j}$ and $\vec{E}$ can be eliminated so that one obtains the conservation equation for the azimuthal magnetic field:

$$\frac{\partial B}{\partial t} = -div(B\,\vec{v}) + \frac{B v_r}{r} - rot\left(\frac{rot\vec{B}}{\mu_0 \sigma} + \frac{\beta}{\mu_0}\, rot\vec{B} \times \vec{B} - \beta \nabla p_e\right)_{\varphi}\ . \tag{13}$$

3. Numerical methods

The steady–state solution of the MPD system is obtained by time stabilization from time–stepping unsteady (extended) Navier–Stokes approximations on adaptive, unstructured meshes with dual cells.

In previous versions of the code [3, 12], piece–wise linear finite elements have been used for the electron energy equation and for the discharge equation (both in stationary form). Here, these equations (in non–stationary form) are solved by the finite volume method in order to achieve as much conservation as possible and to easily implement a proper upwinding.

Also, reaction non–equilibrium is considered here. In our previous work, reaction equilibrium was assumed. This led to an equation of sixth order for the electron density that was solved by the computationally relatively expensive Newton–Raphson method. With the introduction of reaction non–equilibrium, we avoid these computational costs and can also obtain frozen flow effects in supersonic flow regions.

3.1. Flux Vector Splitting for the Species

The determination of eigenvalues of the whole system of conservation equations is very difficult. However, it is clear that at least a good approximation is necessary in order to introduce proper upwinding for the hyperbolic part in the seven species conservation equations $\frac{\partial n_i}{\partial t} + div\ (n_i \vec{v}) = -div\, \vec{j}_{D,i} + \omega_i$. A simple and robust flux vector splitting scheme has already been introduced in [12], and here it is also applied to the fluxes of the species conservation equations.

Since the magnetic field is perpendicular to the convective velocity, it is reasonable to use the fast Alfvén or magnetoacoustic wave, which is given by the relation

$$c = \sqrt{\frac{\gamma(p_h + p_e)}{\rho} + \frac{B^2}{\mu_0 \rho}}\ . \tag{14}$$

Now, the central idea is the definition of a variable reference speed s which differs from the sonic speed. This function is introduced by

$$s = \sqrt{\alpha c^2 + q^2\left(1 - 2\alpha + \alpha\frac{q^2}{c^2}\right)} \quad \text{where} \quad q^2 := \min(c^2, {q_n}^2). \tag{15}$$

The free parameter α controls the numerical diffusion of the scheme. This parameter is defined in such a way that in the case of zero velocity the mass flux equals the effusion flux of the particles $\frac{n_h \bar{c}}{4}$, with $\bar{c} = \sqrt{\frac{8kT_h}{\pi m_h}}$. This idea leads to

$$\alpha = \frac{2}{\gamma\,\pi}\ . \tag{16}$$

The speed normal to the cell face is $q_n = v_z n_z + v_r n_r$ and the eigenvalues are defined as $\lambda_0 = q_n$, $\lambda_1 = \lambda_0 + s$ and $\lambda_2 = \lambda_0 - s$.
At this point, the species flux splitting is defined as (l: state left of the cell face, r: state right of the cell face, A: cell face area)

$$h_{1l} = \frac{1}{4}A(\lambda_{1l} + |\lambda_{1l}|), \qquad h_{1r} = \frac{1}{4}A(\lambda_{1r} - |\lambda_{1r}|), \tag{17}$$

$$h_{2l} = \frac{1}{4}A(\lambda_{2l} + |\lambda_{2l}|), \qquad h_{2r} = \frac{1}{4}A(\lambda_{2r} - |\lambda_{2r}|), \tag{18}$$

and, finally, the flux of species i is given by

$$F^{n_i} = n_{il}(h_{1l} + h_{2l}) + n_{ir}(h_{1r} + h_{2r})\ . \tag{19}$$

In order to achieve second–order accuracy, the fluxes are computed with values obtained by linear interpolation of cell mean values. The gradients for the linear interpolation are computed with a WENO scheme [11].

3.2. Random local time steps

In order to accelerate the convergence of the explicit first–order time integration / stabilization, local time steps are used. At any given time level t during the solution process, the time steps for a cell of spatial size Δx are computed as follows:

$$\Delta t_{atoms/ions} = CFL_{atoms/ions} \cdot \min\left\{\frac{\Delta x}{|\vec{v}| + c}, \frac{3n_h k \Delta x^2}{4(\lambda_h + \lambda_T)}\right\}, \tag{20}$$

$$\Delta t_{electrons} = CFL_{electrons} \cdot \frac{3n_e k \Delta x^2}{4\lambda_e}, \tag{21}$$

$$\Delta t_{magnetic} = CFL_{magnetic} \cdot \frac{\mu_0 \sigma \Delta x^2}{2}\ . \tag{22}$$

The CFL numbers further restrict the time step to control non–linear effects and source terms. The time steps computed by using (20), (21) and (22) are multiplied with a random number RND belonging to $[0, 1]$:

$$\Delta t_{atoms/ions,RND} = \Delta t_{atoms/ions} \cdot RND_{atoms/ions}\ , \tag{23}$$

$$\Delta t_{electrons,RND} = \Delta t_{electrons} \cdot RND_{electrons}\ , \tag{24}$$

$$\Delta t_{magnetic,RND} = \Delta t_{magnetic} \cdot RND_{magnetic}\ . \tag{25}$$

Equations (1) and (10) are updated with the time step from (23), equation (9) is updated with the time step from (24), and equation (13) is updated with the time step (25). After the time update, the new state is defined to be the new time level. By using different local time steps for the different equations in each cell, time–accuracy is not achieved. This is not required because only the steady–state solution is of interest. Without random time steps, the CFL–numbers have to be as low as 0.01 during the iteration and about 500 000 time steps are used for a five order of magnitude reduction of the density residual on a given mesh because the source terms cause strong local numerical oscillations for higher CFL–numbers. However, by using random time steps, the CFL–numbers can be as high as 0.9 and the number of time steps is reduced to about 50 000 and lower. Obviously, the random time steps dampen or prevent oscillations caused by the source terms. A mathematical explanation for this convenient behavior is not yet known. In order to check whether the results obtained with the random time steps are meaningful, the results were further iterated with an additional two million global, first–order accurate time steps ($CFL = 0.09$). Because the results did not change, the results obtained with the random time steps are indeed meaningful.

4. Numerical results

The test case chosen was the hot anode thruster HAT with a current of 2000 A and an argon mass flow rate of 0.8 g/s. The computation was started on an initial coarse mesh with 4667 cells. The final mesh after two adaptations using the refinement indicator presented in [13] has 28034 cells, see Fig. 1.
Figure 2 shows the distribution of the electric current and thus, the arc discharge in the plasma. An electric current of 100 A flows between two isolines. The arc is driven out of the laval nozzle mainly by the electron pressure diffusion term $\beta\nabla p_e$ in Ohm's law for plasmas.
An oblique shock is visible in Fig. 3 which shows the heavy particle temperature. The position of the shock correlates nicely with the experiment in the laboratory.

The flux vector splitting scheme for the species densities proved to be very efficient, because of the low number of computational steps required to determine the fluxes and because of its robustness.
In the transonic nozzle throat area as well as in the supersonic flow domain in the diverging nozzle, no oscillations occurred, which also emphasizes that the eigenvalues including the magnetoacoustic speed are very reasonable.

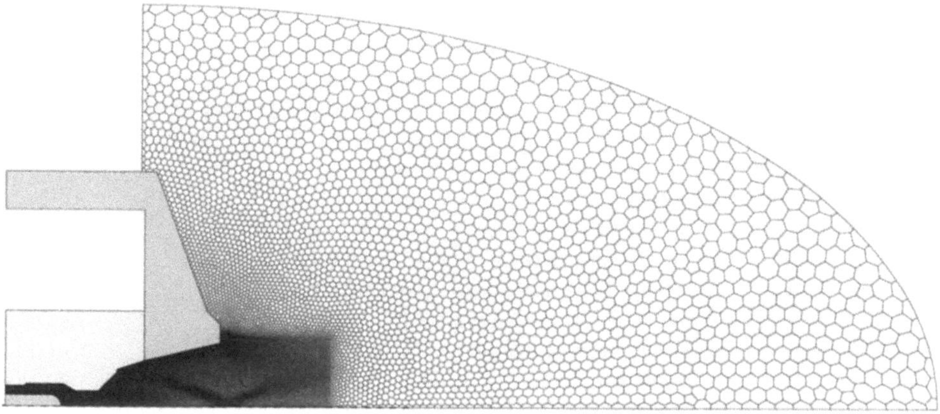

FIGURE 1. Adapted mesh with 28034 cells, HAT, 2000 A, 0.8 g/s

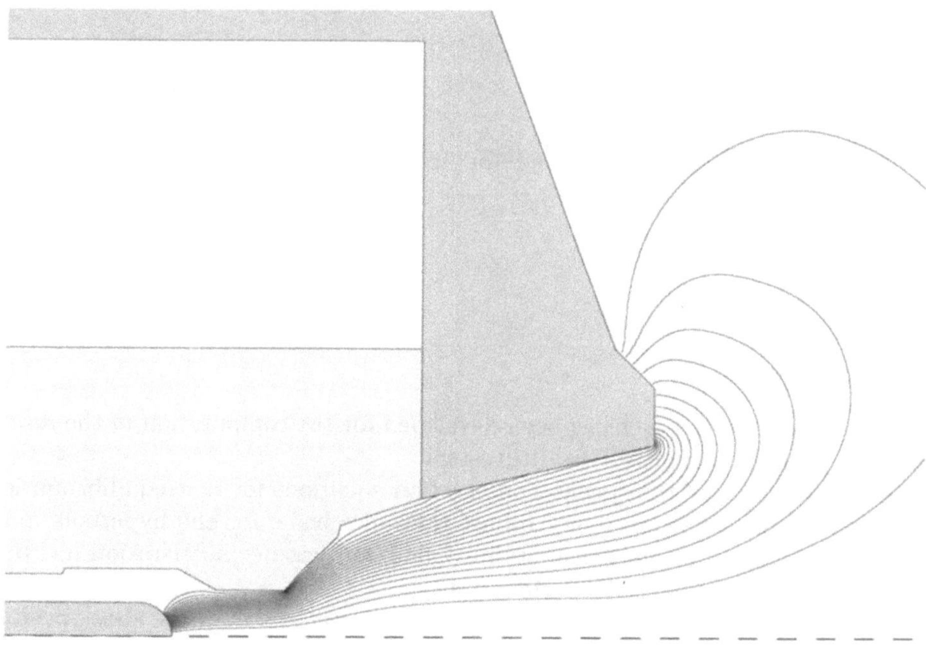

FIGURE 2. Stream function $\Psi = r\,B$, HAT, 2000 A, 0.8 g/s

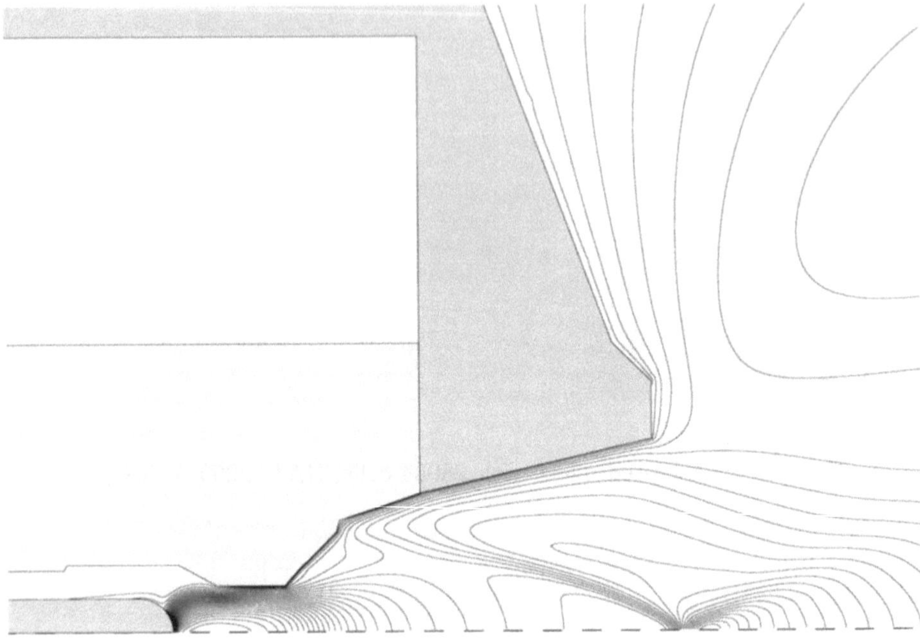

FIGURE 3. Heavy particle temperature T_h, HAT, 2000 A, 0.8 g/s

5. Conclusion

A new finite volume solver has been developed for the computation of the flow in magnetoplasmadynamic self–field thrusters.
The discretization of the species conservation equations for non–equilibrium ionization includes a new and very robust upwind scheme for the hyperbolic part. An important gain of numerical stability and convergence acceleration has been achieved by using random local time steps.
The comparison with the experiment shows good agreement of the shock position in the hot anode thruster.

In the next steps, semi–implicit time integration will be tested to further accelerate convergence. Also, directional adaptive mesh refinement will be implemented to investigate the boundary layers in more detail.

Acknowledgements

This work has been supported by the German Research Foundation DFG (Deutsche Forschungsgemeinschaft) through the projects "Conservation Equations and Numerical Solutions for Technically Relevant Magneto-Plasmas" (Au 85/9-2) and "Domain Decomposition Methods and Far–Field Boundary Conditions for Initial Boundary Value Problems with Conservation Laws and Electromagnetic Fields" (We 659/35–2). The first project belongs to the German Priority Research Program (Schwerpunktprogramm) "Analysis and Numerics for Conservation Laws".
The authors especially thank Prof. T. Sonar and I. Thomas of the University of Braunschweig for their helpful comments on random time steps.

References

[1] H.O. Schrade, P.C. Sleziona, T. Wegmann, H.L. Kurtz: Basic Processes of Plasma Propulsion, Report grant AFOSR-91-0118, Institute of Space Systems, Stuttgart University, 1992.

[2] P.C. Sleziona: High Enthalpy Flows for Spaceflight Applications (in German), Habilitation thesis, Institute of Space Systems, Aerospace Faculty, University of Stuttgart, Germany, 1998.

[3] C. Boie: Numerical simulation of magnetoplasmadynamic self–field thrusters with high resolution adaptive methods (in German), Ph.D. thesis, Institute of Space Systems, University of Stuttgart, Germany, 1999.

[4] J.O. Hirschfelder, C.F. Curtiss, R.B. Bird: Molecular Theory of Gases and Liquids, Wiley, 1967.

[5] A. Sutton, P.A. Gnoffo: Multi–Component Diffusion with Application to Computational Aerothermodynamics, AIAA 98–2575, 7th AIAA/ASME Joint Thermophysics and Heat Transfer Conference, Albuquerque, NM, 1998.

[6] H.W. Drawin, P. Felenbok: Data for plasmas in local thermodynamic equilibrium, Gauthier–Villars, Paris, 1965.

[7] W. Lotz: An Empirical Formula for the Electron–Impact Ionization Cross–Section, Zeitschrift für Physik 206, pp. 205–211, 1967.

[8] W. Lotz: Electron–Impact Ionization Cross–Sections and Ionization Rate Coefficients for Atoms and Ions from Hydrogen to Calcium, Zeitschrift für Physik 216, pp. 241–247, 1968.

[9] U.C. Goldberg, S.V. Ramakrishnan: A pointwise version of Baldwin–Barth turbulence model, Comp. Fluid Dyn., Vol. 1, pp. 321–338, 1993.

[10] J.D. Jackson, Classical Electrodynamics, Wiley, 1962.

[11] O. Friedrich, Weighted Essentially Non–Oscillatory Schemes for the Interpolation of Mean Values on Unstructured Grids, Journal of Computational Physics 144, pp. 194–212, 1998.

[12] J. Heiermann, M. Auweter–Kurtz, A. Eberle, H.J. Kaeppeler, U. Iben, P.C. Sleziona: Recent Improvements of Numerical Methods for the Simulation of MPD Thruster Flow on Adaptive Meshes, IEPC-99-169, 26th International Electric Propulsion Conference, Kitakyushu, Japan, 1999.

[13] U. Iben, J. Heiermann, M. Auweter–Kurtz, G. Warnecke: Adaptive numerics for the simulation of magneto-plasmadynamic rocket thrusters, Proceedings of the Third European Conference on Numerical Mathematics and Advanced Applications (Jyväskyla, Finland) T. Tiihonen and P. Tarvainen (eds.), pp. 121–130, World Scientific, 1999.

M. Auweter-Kurtz
Institut für Raumfahrtsysteme,
Universität Stuttgart,
Pfaffenwaldring 31,
D-70550 Stuttgart, Germany
auweter@irs.uni-stuttgart.de

C. Coclici
Fachbereich Mathematik,
Arbeitsgruppe Technomathematik,
Universität Kaiserslautern,
Erwin-Schrödinger-Str. 48/576,
D-67653 Kaiserslautern, Germany
coclici@mathematik.uni-kl.de

J. Heiermann
Institut für Raumfahrtsysteme,
Universität Stuttgart,
Pfaffenwaldring 31,
D-70550 Stuttgart, Germany
heierman@irs.uni-stuttgart.de

C.-D. Munz
Institut für Aerodynamik und Gasdynamik,
Universität Stuttgart,
Pfaffenwaldring 21,
70550 Stuttgart, Germany
munz@iag.uni-stuttgart.de

P.C. Sleziona
Institut für Raumfahrtsysteme
Universität Stuttgart
Pfaffenwaldring 31
D-70550 Stuttgart, Germany

International Series of Numerical Mathematics
Vol. 140,

The Eikonal Equation on a Manifold. Applications to Grid Generation or Refinement

P. Bagnerini, P. Hoch, and M. Rascle

Abstract. The aim of this work is the generation of anisotropic meshes which are automatically refined in some regions, typically where we want to solve numerically a PDE whose solution is singular. The basic idea is to consider an initial closed curve and to move this curve by the Hamilton-Jacobi equation on a manifold. Similar ideas could also be useful in image processing, in particular for the active contours method.

1. Introduction

Mesh adaption offers the advantages of improving the accuracy of the computed solution in the context of the Finite Element Methods and reducing the number of elements required to capture the behavior of the underlying physical phenomenon. There exists a huge literature on grid generation and refinement algorithms. Among many others, let us mention for instance [5], and references therein.
This paper follows [7, 6]. The purpose is the generation of anisotropic meshes which are automatically refined in regions where a given function $\tilde{z}$, that we want to approximate, has large derivatives (typically z is the numerical solution of some unspecified partial differential equation). Given a previously computed approximation z of this function, the main idea is to generate an adapted mesh by solving the Eikonal equation on the graph of z, equipped with a suitable Riemannian metric. We consider an initial closed curve, typically the boundary of the domain Ω, and we move this curve with a constant Riemannian speed on the graph of z: when we project the curve on Ω in the regions where z is rapidly varying, its horizontal projection moves very slowly, so that the resulting mesh is automatically refined, (see Figure 1). We represent the curve as the zero-level set of a function $\phi : R^2 \times R^+ \longrightarrow R$: the initial position of the curve and its position at further times, constitute the basic structure of the mesh. The method is a variant of the classical level set method [12]. Several Authors have considered similar problems, see e.g. [9], but in general the motivation was different. Although the principle is very natural, the results presented here are not optimal and the version of the

The Authors acknowledge the support from the EC TMR network HCL : n. ERB FMRX CT 96 0033

algorithm described here needs to be improved.
The outline is as follows : in Section 2, we recall a few basic facts in the Riemannian eikonal equation and its numerical approximation, following [7, 6]. In Section 3, we describe the main steps of the algorithm of mesh generation, we show a few numerical results and we present some of difficulties that one has to face in strongly stiff or anisotropic situations, before the conclusion in Section 4.

2. The Riemannian Eikonal Equation

2.1. Geodesics

For convenience, we recall a few basic facts, see [7]. Although the principle is very general, let us consider the typical case of a manifold M in $\mathbb{R}^3$ (equipped with a Riemannian metric G), which is the graph of a smooth function $z : \Omega \to \mathbb{R}$, where $\Omega \in \mathbb{R}^2$ is a bounded domain. In this simple case, it is more convenient to describe M and the metric G via the horizontal projection on Ω. The two natural metrics that we have in mind are based on the positive definite quadratic forms associated either to the matrix:

$$G_1(x) = I + \nabla z(x) \otimes \nabla z(x), \tag{1}$$

which allows to generate a refined mesh in the regions where the gradient of z is very large, or for instance to the matrix

$$G_2(x) = I + |H(z)(x)|, \tag{2}$$

where $H(z)$ is the Hessian matrix, and $|A| := A_+ + A_-$ is defined in the standard way for a symmetric real matrix A. Since the maximum interpolation error E_i in the Finite Element Method depends on the Hessian matrix $H(z)$ (see e.g. [4]), this metric allows to equidistribute E_i on all triangles i of the mesh.

We note $G(x) = (g_{ij}(x))_{1 \leqslant i,j \leqslant 2}$ the positive definite matrix associated to the Riemannian metric, $(u, v)_G = (u^i g_{ij} v^j)$ the scalar product and $|u|_G = ((u, u)_G)^{1/2}$ the corresponding Riemannian norm.
We consider an initial curve C_0 on the plane $\mathbb{R}^2$, see Figure 1 and we want to move this curve C_0 at a constant Riemannian speed.
In the general case, we recall that a curve γ, parametrized by t, on a manifold M, is called *geodesic* if at every point $y \in \gamma$ the covariant derivative of $\dot{\gamma}(t)$ is zero or equivalently the second derivative of $\gamma(t)$ is orthogonal to the tangent space to the manifold. Given two points x, y, if they are close enough, there exists a unique geodesic between x, y which represents the minimal Riemannian distance between x and y, i.e. the curve of minimal length among all curves on M with the same endpoints c.f. e.g. [3, 10, 2].
Given a fixed point $x \in \mathbb{R}^2$ near the curve C_0, we look for a curve $\gamma(t)$ which realizes the Riemannian distance of x to the curve C_0. This Riemannian distance

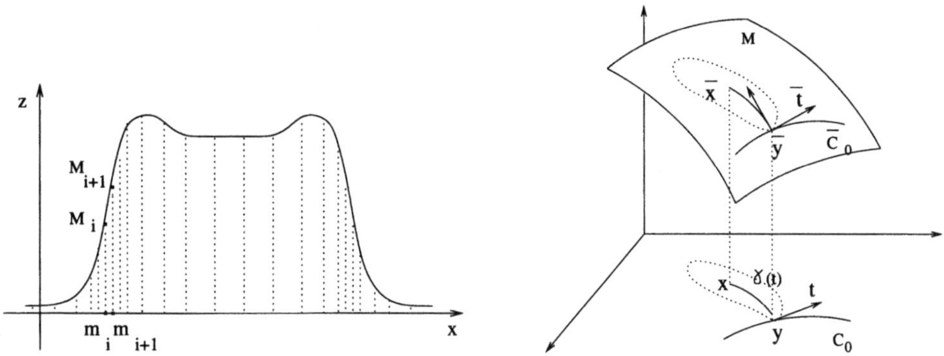

FIGURE 1. Graph of z, in 1D or in 2D.

is given by

$$d^G(x,y) = \inf\left\{\int_0^T L(\gamma(t),\dot\gamma(t))\,dt; y := \gamma(0) \in C_0, \gamma(T) = x\right\}, \tag{3}$$

where the Lagrangian is given by $L(\gamma(t),\dot\gamma(t)) = (G(\gamma(t))\dot\gamma(t),\dot\gamma(t))^{(1/2)}$. If the point x is close enough to the curve C_0, such $\gamma(t)$ is a geodesic. Using basic tools of Calculus of Variations (see again [10, 2, 7]), it is easy to prove that this (minimal) geodesic $\gamma(t)$ is characterized by the Euler-Lagrange equation

$$-\frac{d}{dt}\left(\frac{\partial L}{\partial\dot\gamma}(\gamma(t),\dot\gamma(t))\right) + \frac{\partial L}{\partial\gamma}(\gamma(t),\dot\gamma(t)) = 0, \tag{4}$$

and by

$$\frac{\partial L}{\partial\dot\gamma}(\gamma(0),\dot\gamma(0))\cdot\tau = 0, \tag{5}$$

where τ is the tangent vector to C_0 at point $y = \gamma(0)$. Then $\frac{\partial L}{\partial\dot\gamma}(\gamma(0),\dot\gamma(0))$ or equivalently $G(\gamma(0))\,\dot\gamma(0)$ must be parallel to the normal vector n to C_0. Finally the unit tangent vector to the minimal geodesic at point $y = \gamma(0)$ is

$$\dot\gamma(t) = \frac{G^{-1}(\gamma(t))\,n}{|G^{-1}(\gamma(t))\,n|_G}. \tag{6}$$

2.2. The Eikonal equation on a manifold

We consider an initial curve C_0. In order to move each point of C_0 in the direction of the above-mentioned Riemannian geodesic, we use the *level set method* of Osher-Sethian, see e.g. [12, 14], i.e. we describe the front C_t at each time $t \geqslant 0$ as the zero-level curve (see Figure 2):

$$C_t = \{x \in \bar\Omega; \phi(x,t) = 0\} \tag{7}$$

with $\phi : \mathbb{R}^2 \times \mathbb{R}^+ \to \mathbb{R}$. Now, differentiating (7) with respect to t, we get :

$$\partial_t\phi + \nabla\phi(x(t),t)\cdot\dot x(t) = 0, \tag{8}$$

and using (6), we finally obtain the Riemannian Eikonal equation :

$$\partial_t \phi + |G^{-1/2}\nabla\phi| = 0, \tag{9}$$

where $|G^{-1/2}\nabla\phi|$ denotes the Euclidean norm of $G^{-1/2}\nabla\phi$. We also note that the Hamiltonian $H(x,p) := |G^{-1/2}(x)p|$ is convex with respect to p. The principle is to use the evolution equation (9) to move the initial closed curve C_0 (in fact the boundary $\partial\Omega$), inside Ω, as a frame to construct the basic structure of the mesh.

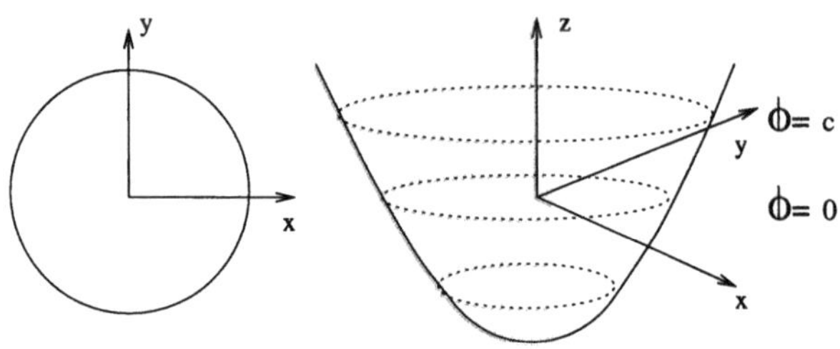

FIGURE 2. Zero Level set of ϕ_t

Now, the initial curve C_0 is the boundary of the domain Ω on which we want to construct our mesh. The corresponding initial datum ϕ_0 has to satisfy the assumptions ([14])

$$\phi_0|_{\partial\Omega} = 0, \quad \phi_0 > 0 \text{ inside } \Omega, \quad |\nabla\phi_0| \neq 0 \text{ on } \partial\Omega. \tag{10}$$

For reasons which are explained in details in [7, 6], a good choice of boundary conditions for (9) is:

$$\phi(x,t) = -t \text{ on } \partial\Omega \times (0,T), \tag{11}$$

and with this choice, adapting a result in [11], one can prove:

Proposition 2.1. *For any Lipschitz function ϕ_0 satisfying the conditions* (10), *the initial boundary value problem*

$$\begin{cases} \partial_t\phi + |G^{-1/2}\nabla\phi| = 0 & \text{in } Q = \Omega \times (0,T), \\ \phi(x,t) = -t & \text{on } \partial\Omega \times (0,T), \\ \phi(x,0) = \phi_0(x) & \text{in } \Omega, \end{cases} \tag{12}$$

has a unique viscosity solution ϕ, given by Oleinik-Lax-Krushkov type of formula

$$\phi(x,t) = \begin{cases} d^G(x,\partial\Omega) - t = u(x) - t, & d^G(x,\partial\Omega) \leqslant t, \\ \min\{\phi_0(y),\ d^G(x,y) \leqslant t,\ y \in \Omega\}, & d^G(x,\partial\Omega) \geqslant t. \end{cases} \tag{13}$$

Indeed, (11) is compatible with the particular solution $\phi(x,t) = u(x) - t$, where u is the solution to the stationary problem.

Proposition 2.2. *Let us consider the stationary problem:*

$$\begin{cases} |G^{-1/2}\nabla u(x)| = 1 & \text{in } \Omega \\ u(x) = 0 & \text{on } \partial\Omega \end{cases} \tag{14}$$

Then $u(x) := d^G(x, \partial\Omega) := \inf_{y\in\partial\Omega} d^G(x,y)$ *is the unique viscosity solution to Problem* (14).

Sketch of proof:
The uniqueness is guaranteed by standard assumptions for a general stationary problem, which are satisfied here with $H(x, \nabla u) = |G^{-1/2}\nabla u(x)| - 1 = 0$, see [1], theorem 2.7, page 37.
As to the existence, we have to show that $u(x) := d^G(x, \partial\Omega)$ is a viscosity *sub-solution* (the proof of the other part is similar). We consider a C^1 test-function v, and we suppose that x is a local maximum of $(u - v)$: for all y in a neighbourhood $V = \{y; d^G(x,y) \leqslant h\}$ of x, we have : $u(x) - v(x) \geqslant u(y) - v(y)$. Since $u(x) = d^G(x, \partial\Omega) \leqslant d^G(x,y) + d^G(y, \partial\Omega) \leqslant h + u(y)$, we have

$$0 \leqslant h + (\nabla v(x), y - x) + o(h) \leqslant \inf_V \{h + (\nabla v(x), y - x) + o(h)\} =$$

$$= h \inf_V \left\{1 + \left(\nabla v(x), \frac{y-x}{h}\right) + o(1)\right\} \leqslant h\left(1 + \inf_V \left\{\left(\nabla v(x), \frac{y-x}{h}\right) + o(1)\right\}\right)$$

We set $z = G^{1/2}(x)\frac{y-x}{h}$ and we obtain

$$0 \leqslant 1 + \inf_{z\leqslant 1} \left\{\left(G^{1/2}(x)\nabla v(x), z\right) + o(1)\right\} = 1 - |G^{-1/2}\nabla v(x)| + o(1).$$

Finally, both for theoretical [8] and numerical [7, 6], reasons of compatibility with the stationary solution, it is important to impose that the initial data satisfy an approximation of the condition :

$$|G^{-1/2}(x)\nabla\phi_0(x)| = 1 \quad \text{on } \partial\Omega. \tag{15}$$

2.3. The numerical approximation

Numerically, the Hamilton-Jacobi equation (9) is hyperbolic in $\nabla\phi = U = (u, v)$ and we use a finite differences scheme. We again refer to [7, 6], for the details. Classically, there are three steps in the algorithm:

(i) We use a second order TVD interpolation of $\nabla\phi = (u, v)$ with a minmod limiter [12].
(ii) We semi-discretize in space and we approximate the hamiltonian H by a two dimensional monotone Riemann Solver, typically by a Lax-Friedrichs type numerical hamiltonian.
(iii) We discretize in time the ordinary differential equation $\partial_t\phi = L(\phi) = -|G^{-1/2}\nabla\phi|$ by a TVD second order Runge Kutta method [13].

3. Construction of the mesh

We have seen that the direction of the Riemannian geodesic at point $x(t)$ is given by (6). In order to construct the mesh, we choose the initial curve C_0, for instance the boundary of the domain to be meshed and M points M_i belong to C_0. We set $n = 0$ and we repeat for every $n\Delta t$ the following algorithm :
For every edge $M_i^n M_{i+1}^n$ we compute the mid-point $M_{i+1/2}^n$ and we move this point according to the ordinary differential equation (6)

$$\dot{x}(t) = \frac{G^{-1}(x(t))\nabla\phi(x,t)}{|G^{-1}(x(t))\nabla\phi(x,t)|_G} \tag{16}$$

where $\nabla\phi(x,t)$ is computed by the PDE (9). We obtain, in this way, the point $M_{i+1/2}^{n+1}$, which is located on the level curve of the viscosity solution $\phi_n(x)$ to the problem (12). Pratically, there are several elementary steps between the points M^n and M^{n+1}. The curves $\dot{x}(t)$ are the horizontal projections of the bicharacteristics (see [8]).
To construct the triangles, we connect the new points $M_{i+1/2}^{n+1}$ with M_i^n and M_{i+1}^n and we repeat the same process, until we have constructed the whole mesh.

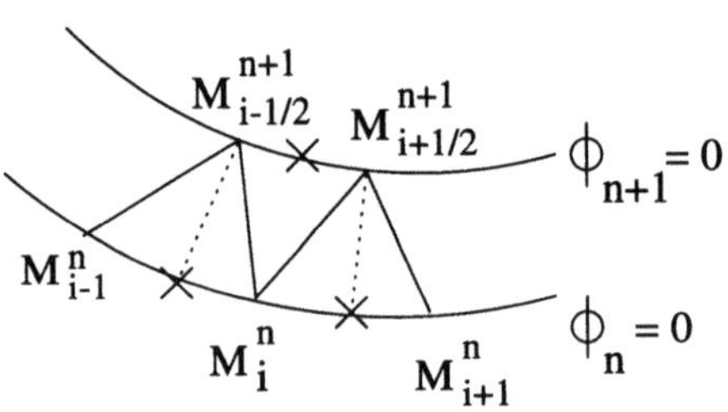

We try to obtain triangles as equilateral as possible with respect to the Riemannian metric G, and the edges of the same given Riemannian length. Naturally we have to face a number of difficulties : the Riemannian distance between two points can be too large or too small, we can have missed the new level curve, or there may be changes of topologies in the level curves. To try to solve these difficulties, we have a certain number of tests.

Quality test: Let d_{ref} a given Riemannian length. For every triangle based upon the level curves at times n and $n+1$, we test its quality: different tests are possible, and we have chosen to compute the three ratios between the Riemannian lengths of the edges of the triangle: $r_1 = \frac{l1}{l2}$, $r_2 = \frac{l1}{l3}$, $r_3 = \frac{l2}{l3}$ with li, $i = 1,3$ the Riemannian lengths of the edges.
If $\frac{1}{\sqrt{2}} \leqslant r_i \leqslant \sqrt{2}$, $i = 1,3$ is verified, the quality of the element is considered good. Otherwise,

Test of the length: We compute the Riemannian length of $M_{i-1/2}^{n+1}\, M_{i+1/2}^{n+1}$:
If $d^G(M_{i-1/2}^{n+1}, M_{i+1/2}^{n+1}) < c1$ the edge is suppressed,
If $d^G(M_{i-1/2}^{n+1}, M_{i+1/2}^{n+1}) > c2$ we add a suitable number of points.

Here c_1 and c_2 positive constants, for instance $c_1 = 0.5\, d_{ref}$ and $c_2 = 1.5\, d_{ref}$.

Edge swapping: A simple tool to improve considerably the quality of the final mesh is the edge swapping, keeping the same number of vertices, edges and elements. For every edge a, we consider the two adjacent triangles having a as common edge. We test the quality of the two possible configurations, and we choose the configuration with the best quality of the elements.

Changes of topology: Given a point M_i^n at time n and M_i^{n+1} at time $n+1$, the point M_i^{n+1} may have missed the corresponding level curve. In this case, we use a Newton method to find the intersection of the level curve at time $(n+1)$ with the straight line $M^n M^{n+1}$: we look for a point (see Figure 3)

$$M = M^{n+1} + \epsilon M^n M^{n+1};$$

in regions where there is a change of topology, it is not possible to find such a point M with $\phi_{n+1}(M) = \phi(M, t_{n+1}) = 0$. Then, if after a certain number of iterations, the Newton method has not converged, we know that there is a change of topology.

In this case, we explore the sign of ϕ_{n+1} between the neighbour points, in order to detect the pieces of the new level curve, and we use a local Delaunay type algorithm to generate the mesh in the (small) regions where this change of topology has occured, see Figure 3.

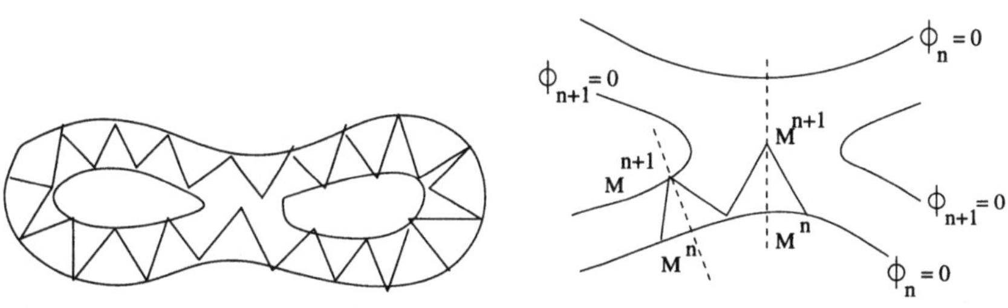

FIGURE 3. change of topology

We have tested the algorithm on several examples, associated either to the Euclidean metric (for the piston) or to the Riemannian metric $G_1(x) = I + \nabla z(x) \otimes \nabla z(x)$, where z is given by $z(x,y) = c_1 \arctan\,(c_2\, f(x,y))$, with suitable constants c_1 and c_2, so that the curves move very slowly near $f(x,y) = 0$.

In Figures 4, 5, 6, the initial curve is the contour of a piston with a valve, and the metric is Euclidean: no previous solution z was available. Note the changes of topology in the level curves. In Figure 7, the curve $f(x,y) = 0$ is a spiral: $f(x,y) = r^2 - (0.25 - \theta/8/\pi/3)$ where $r^2 = (x-.5)^2 + (y-.5)^2$, and $\tan\theta = \frac{y-.5}{x-.5}$ with $\theta \in (0, 6\pi)$. In this Figure, z is rapidly varying near a spiral. Therefore, the level curves of the solution ϕ to the eikonal equation can not climb this spiral shaped mountain, and can only progress in the "plane" between these spirals. In Figure 8, $f(x,y) = 0$ is

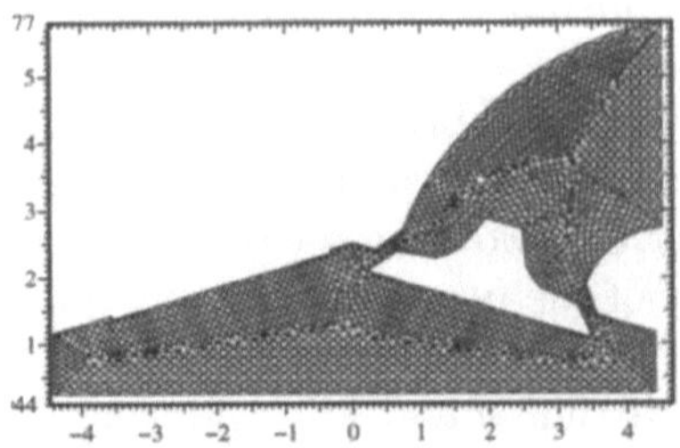

FIGURE 4. A mesh of the piston.

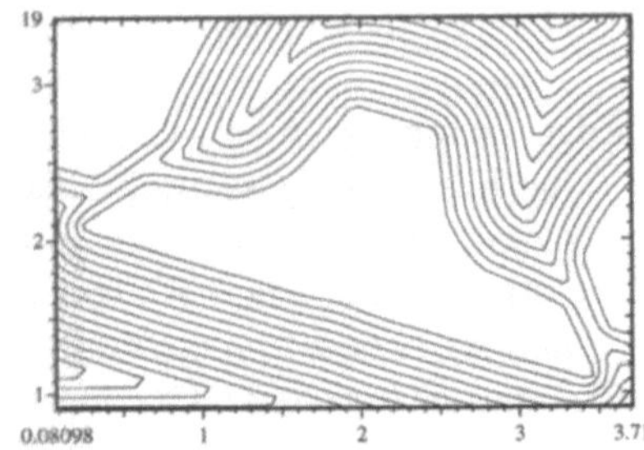

FIGURE 5. Zoom of the level curves.

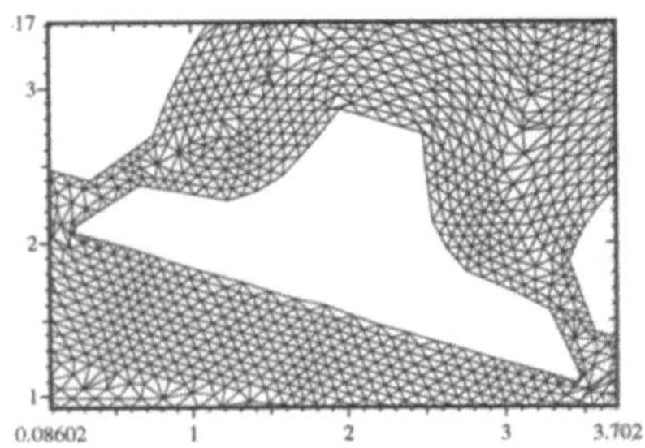

FIGURE 6. Zoom of the mesh.

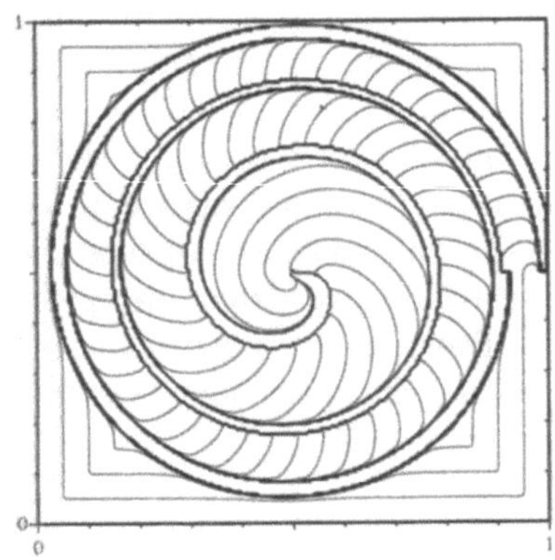

FIGURE 7. The spiral : level curves.

the alpha-shaped, i.e. $f(x,y) = x^3 - 5*x^2/3 + 8*x/9 - 4/27 - 8*(y-.5)^2/27$ and in Pictures 9, $f(x,y) = y - a*x^3 - 1.5*a*x^2 + (a/2-1)*x + 1$ where $a = -4$.

4. Conclusion

As we said in the Introduction, this work is an extension of [7]. On the theoretical side, clearly, the concept is very neat and general. On the numerical side, there are two parts in the algorithm :

(i) the generation of the level curves, i.e. the Eulerian part, which works quite well

(ii) the Lagrangian part, i.e. the motion of grid points.

In moderately anisotropic situations, this second part of the algorithm works well, but more severe difficulties can arise when the metric G is both stiff and strongly anisotropic. In such cases, the changes of topology are still well resolved, but sometimes, the characteristics can be too oblique with respect to the level curves,

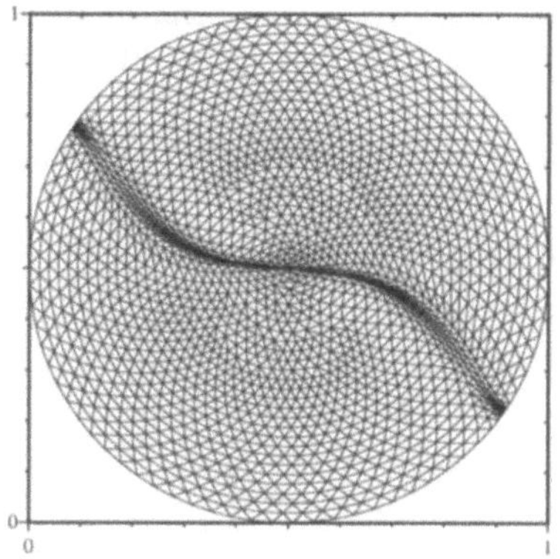

FIGURE 8. α-curve : example of mesh.

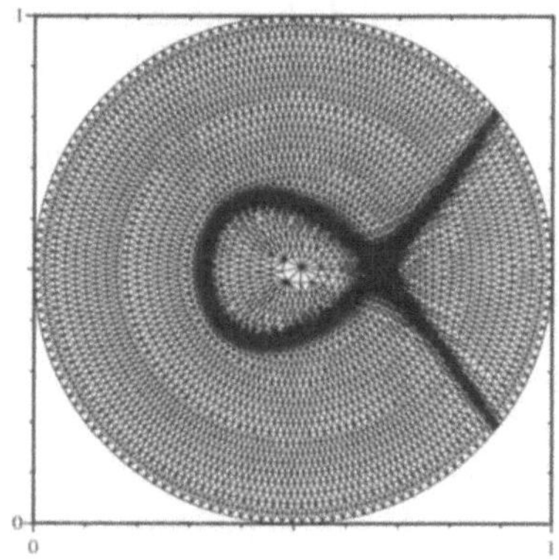

FIGURE 9. S-curve : example of mesh.

so that the Newton method can fail or even converge to another intersection point, which can produce the overlapping of some triangles. We are currently testing another algorithm of local exploration, to try to solve this difficulty.
Finally, another strategy could be, at each time step, a combination of step (i) with an algorithm of exploration of the new level curve and a Riemannian Delaunay algorithm, to construct a new set of triangles, and move forward.

References

[1] G. Barles, *Solutions de viscosité des équations de Hamilton-Jacobi,* Springer Verlag, 1994.

[2] M. Berger and B. Gostiaux , *Géométrie différentielle : variétés, courbes et surfaces,* Presses Universitaires de France, 1987.

[3] M-P.do. Carmo, *Differential geometry of curves and surfaces,* Prentice Hall, 1976.

[4] M.J. Castro Diaz and F. Hecht and B. Mohammadi, *New progress in anisotropic grid adaptation for inviscid and viscous flows simulations,* Technical Report 2671 INRIA, 1995.

[5] P-L. George, *Génération automatique de maillages. Applications aux méthodes d'éléments finis,* Masson, 1991.

[6] Ph. Hoch, *Approximation de problèmes hyperboliques non linéaires, équations de Hamilton-Jacobi et applications,* PhD Thesis, 2000.

[7] Ph. Hoch and M. Rascle, *Hamilton-Jacobi Equations and Applications to mesh generation and refinement,* submitted SIAM J. Sci. Compu., (1999).

[8] F. John, *Partial differential equations,* Springer Verlag, 1971.

[9] R. Kimmel, *Curve evolution on surfaces,* PhD Thesis, 1995.

[10] J. Lelong-Ferrand, *Géométrie différentielle,* Masson, 1963.

[11] P-L. Lions, *Generalized solutions of Hamilton-Jacobi equations,* Pitman, 1982.

[12] S. Osher and J. A. Sethian, *Fronts propagating with curvature-dependent speed: Algorithms based on Hamilton-Jacobi formulations,* J. Comput. Phys., 79(1) (1988), 12-49.

[13] S. Osher and C.-W. Shu, *High-order essentially nonoscillatory schemes for Hamilton-Jacobi equations,* SIAM J. Numer. Anal., 28(4) (1991), 907-922.

[14] J. A. Sethian, *Level set methods and fast marching methods,* Cambridge University Press, 1999.

P. Bagnerini
P. Hoch
M. Rascle
Laboratoire de Mathématiques CNRS n. 6621,
Université de Nice Sophia-Antipolis,
Parc Valrose B.P. 71,
06108 Nice Cedex 2, France
pbagneri@math.unice.fr
philippe.hoch@inria.fr
rascle@math.unice.fr

International Series of Numerical Mathematics
Vol. 140, © 2001 Birkhäuser Verlag Basel/Switzerland

Crossflow Instabilities in the Approximation of Detonation Waves

Derek S. Bale and Christiane Helzel

Abstract. We describe a mechanism which can cause crossflow instabilities in the approximation of multidimensional detonation waves. Crossflow instabilities can grow if a discontinuity is nearly aligned with the mesh and if the velocity transverse to the discontinuity is close to zero. When these conditions are met a continuous perturbation of the constant state transverse to the discontinuity can cause the crossflow instability to grow. In both the reactive and the nonreactive case such a continuous perturbation can be caused by oscillations which arise behind the discontinuity. In the nonreactive case these oscillations are a numerical artifact of the scheme, whereas in the reactive case the oscillations reflect a physical instability inherent in the problem. We suggest a fix which avoids the crossflow instability without limiting the oscillatory wave structure, which can arise behind detonation waves.

1. Introduction

We consider hyperbolic systems of conservation laws with source terms to model flow in the presence of detonation waves. The theory describes detonation waves as reactive gasdynamic interactions consisting of a strong leading shock followed by a reaction zone. The chemical reactions are usually very complex and many chemical species are involved. As a simplified model the reactive Euler equations with only two chemical species, namely unburnt gas and burnt gas, can be used. In conservative variables, the reactive Euler equations can be obtained as a combination of the Euler equations with a reaction model. In addition to the Euler equations we then have in the 2D case an equation of the form:

$$(\rho Z)_t + (\rho u Z)_x + (\rho v Z)_y \quad = \quad -\rho K(T) Z.$$

The variable Z is the mass fraction of unburnt gas, where $Z = 1$ describes the unburnt state and $Z = 0$ describes the completely burnt state. Furthermore, ρ is the density, T the temperature and u and v are the velocity components of the gas in x and y direction, respectively. The equation of state has the form

$$E = \frac{p}{\gamma - 1} + \frac{1}{2}\rho(u^2 + v^2) + q_0 \rho Z,$$

with the ratio of specific heats γ and the heat release q_0. The term $q_0 \rho Z$ is the chemical energy, which is released as heat during the burning process. The reaction rate depends on the temperature via an *Arrhenius law* of the form

$$K(T) = K_0 \exp(-E^+/T).$$

Here K_0 is the rate constant and E^+ is the activation energy, see for instance [1].

We restrict our considerations to numerical solutions which are computed by using a fractional step method. The numerical approximation of the reactive Euler equations can lead to various numerical difficulties. One problem can arise if the time scale of the chemical reaction is orders of magnitude smaller than the time scale of the gas transport. In this case the problem is said to be stiff and an underresolved approximation can lead to an unphysical numerical solution. A modification of the fractional step scheme, was recently developed in [3]. This scheme allows the correct approximation of the global solution structure of detonation waves without resolving the very small reaction zone. However, in order to approximate the correct stability behavior of a detonation wave as it will be our goal here, the reaction zone has to be resolved. In this contribution we restrict our considerations to a numerical problem, namely *crossflow instabilities*, which can arise in the resolved approximation of multidimensional detonation waves.

It is a known problem that the approximation of strong shocks modeled by the 2D Euler equations using finite volume schemes on a Cartesian grid may become unstable if the shock is strong enough and nearly aligned with the grid. An initially very small perturbation can grow without limit due to the solution of Riemann problems transverse to the strong shock. This numerical problem can arise for various finite volume schemes. It was first documented and analyzed by Quirk [8]. For blunt body flow calculations similar problems can arise and are known as the *carbuncle effect*. During the last years, this numerical problem found quite a lot of attention in the literature, see also [10]. Whereas the instability is quite well understood for the nonreactive Euler equations, this is not the case for the reactive Euler equations. In Quirk [9], crossflow instabilities were documented for the approximation of detonation waves. The mechanism, which leads to the instability for detonation waves was not further investigated. This is the main object of our considerations. We give a brief description of the instability for the nonreactive Euler equations in Section 2 and describe an easy fix of the problem. In Section 3, we consider the mechanism which leads to the crossflow instability in the approximation of detonation waves, modeled by the reactive Euler equations. For our numerical calculations, we used the high-resolution wave propagation algorithm, which was developed by LeVeque [5] and which is implemented in CLAWPACK [4].

2. Crossflow instabilities in the nonreactive case

In order to characterize the crossflow instability, we first consider the approximation of strong shocks modeled by the multidimensional nonreactive Euler equations. Assume that a strong shock is aligned with the mesh and moving in the x

direction. A small perturbation in pressure or density along the shock can lead to an instability, due to the solution of Riemann problems transverse to the shock. For the calculation shown in Quirk [8], he slightly perturbed one grid line (therefore he chose the name odd even decoupling). However, in his analytical considerations of the crossflow instability he studied the performance of numerical schemes if a sawtooth perturbation of very low amplitude is added to the pressure and density along the line of a strong shock. In LeVeque [6], an example is given where the crossflow instability arises if a small perturbation is only added in one mesh cell. Here we reproduce this calculation but consider in addition also the 3D case. The problem is described in Example 2.1. The exact solution of this problem consists of two symmetric shock waves which are propagating to the left and right.

Example 2.1. *We consider the 2D and 3D Euler equations with the Riemann initial values $\rho = 1$, $p = 1$ and $v = 0$ at both sides of a discontinuity whereas the velocity in x-direction u is equal to* 20 *in the left half plane and equal to* -20 *in the right half plane. For the 3D calculation we consider the cases $w = 0$ (Figure 1 (a)) and $w = 1$ (Figure 1 (b)). The discontinuity is initially located at $x = 30$. In one mesh cell at the center of the computational domain $[0, 60] \times [0, 30]$ the pressure was perturbed to $p = 1 + 10^{-6}$. The ratio of specific heat is set to the value $\gamma = 1.05$.*

In Figures 1 (a) and (b) we show numerical results obtained with the 3D CLAWPACK algorithm of Langseth and LeVeque. They indicate that the crossflow instability grows most strongly if both transverse velocities are close to zero. Nevertheless, if only one transverse velocity is zero the crossflow instability can clearly be observed. The numerical solution of Example 2.1 calculated with the 2D high-resolution wave propagation algorithm and a Roe Riemann solver is shown in Figure 1 (c), the correct solution structure is completely destroyed. In (d) we show the numerical results obtained with the modified solver described below. This modification can also be applied in the 3D case.

In order to understand the numerical problem in the 2D case one can consider 1D solutions across the shock structure, i.e. in x-direction, as well as transverse to the shock structure, i.e. in y-direction, separately. The two shock waves, which are the correct solution of Example 2.1, are slowly moving shocks. Several numerical schemes, e.g. the Roe solver, produce an oscillatory wave structure behind such shock waves. In a one-dimensional calculation these oscillations do not grow without bound. Nevertheless, in a two-dimensional calculation the oscillations can introduce a small jump in the pressure and/or density during every time step in the region behind the shock. Such a perturbation will then increase due to the solution of the Riemann problems in transverse direction. In order to show, how a small perturbation can grow in the solution of Riemann problems transverse to the strong shock, we now consider the exact solution of a Riemann problem consisting of a constant state in density ($\rho = 1$) and velocity ($v = 0$). The pressure is set to $p_l = 1.01$ and $p_r = 1$. Figure 2 shows the solution of this Riemann problem in pressure and density. The solution consists of a left going rarefaction wave, which is here spread-out over a very thin region and a right going shock wave. In the

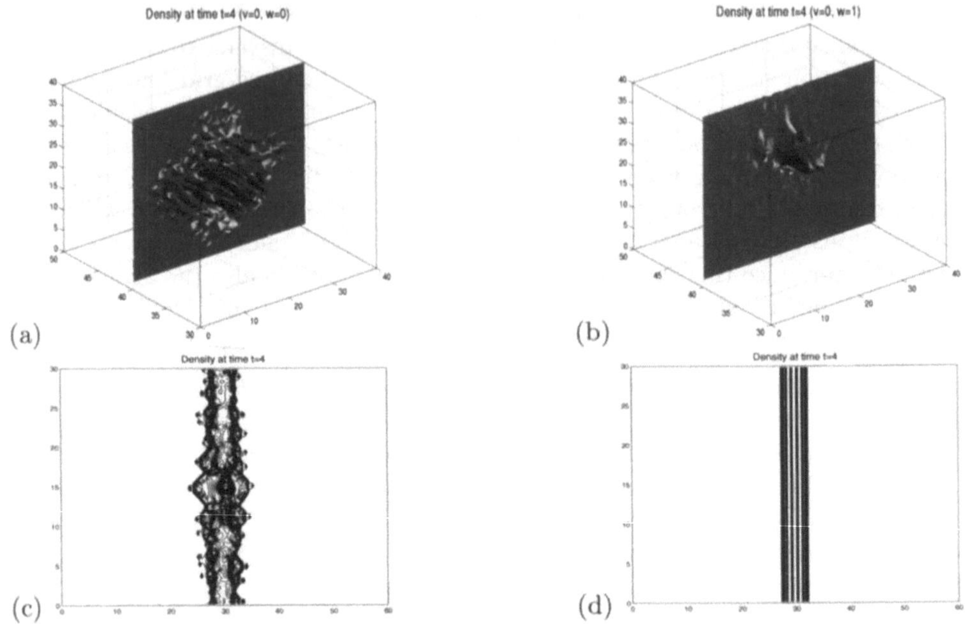

FIGURE 1. Numerical solutions for Example 2.1. Isosurface plot of density from 3D calculation with (a) $v = w = 0$ and (b) $v = 0$, $w = 1$. Contour plots of density from 2D simulation (c) using the Roe Riemann solver (d) with a modified solver that avoids the crossflow instability.

solution of this Riemann problem, the perturbation in pressure is averaged out but causes a perturbation in the density. The largest jump in density between the

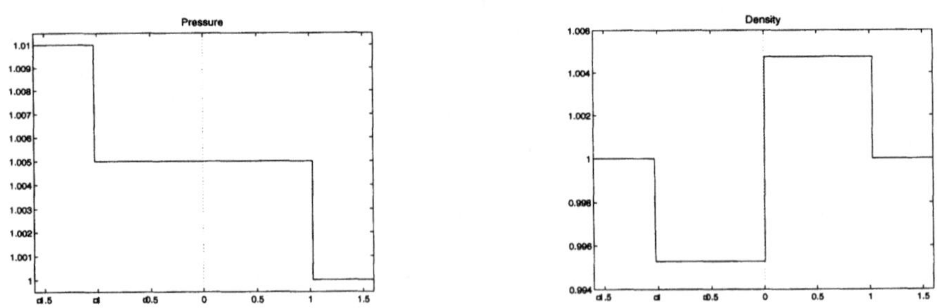

FIGURE 2. Exact solution of a Riemann problem consisting of constant states in density and velocity and a small perturbation in the pressure.

cell average of two neighboring mesh cells will be obtained if the contact discontinuity has a speed close to zero. During further time steps in a 1D calculation the initially introduced perturbation would be propagated through the domain, but can not grow further. If in each time step a perturbation in pressure is introduced, then the jump in density grows and leads to the crossflow instability as shown in Figure 1. In a 2D calculation, such a continuous perturbation can be caused by the oscillations which arise in numerical simulations of slowly moving shocks. Across the viscous shock profile of a numerically approximated slowly moving shock the eigenvalue corresponding to the shock wave changes sign. Additionally, the flow speed transverse to the shock is close to zero. Therefore, the occurrence of crossflow instabilities for our example is predicted by the analytical results of [10].

In order to avoid the crossflow instability, one could use a Riemann solver which performs better for the approximation of slowly moving shock waves, e.g. the Osher scheme or the Marquina solver. On the other hand, one can modify the Riemann solver in the direction transverse to the strong shock if the transverse velocity component is close to zero. Such an approach is used here, were we averaged out the jump in the conservative variables across the contact discontinuity over the region bounded by the 1-wave and the 3-wave. It can be used in an exact Riemann solver as well as in an approximative Riemann solver. Furthermore, the same modification can be applied for calculations on curvilinear grids, using the CLAWPACK extension on curved manifolds, in order to avoid the carbuncle problem. This is similar to an entropy-fix applied to the contact discontinuity, as mentioned by Quirk [8]. Using the wave propagation scheme this fix can be implemented by replacing the 2-wave (contact discontinuity), which is moving with the speed $\lambda_2 = u^*$ by two separate waves, moving with speeds $u^* - \varepsilon$ and $u^* + \varepsilon$. The jump in density and energy across the contact discontinuity has to be split up into the two new waves such that conservativity remains preserved. In the case $\varepsilon = c$ (the speed of sound) this scheme would agree with the HLLE solver in the sense that only the 1-wave and the 3-wave are approximated. For our examples less numerical viscosity ($\varepsilon \approx 0.2$) was sufficient in order to avoid the crossflow instability. Note that the introduction of further waves in the wave propagation scheme is also useful in other applications as recently described by LeVeque and Pelanti [7]. Sanders et. al [10] introduce numerical viscosity based on all three waves arising in the Riemann problem.

Alternatively, one could switch to any Riemann solver which does not resolve the contact discontinuity of the Riemann solution, e.g. the HLLE solver. By using the modified Riemann solver in the wave propagation scheme we can still apply the second order correction terms used in CLAWPACK. Figure 1 (d) shows a numerical simulation, where the modified Riemann solver was used in the direction transverse to the strong shock. The contour lines indicate the numerical oscillations behind the slowly moving shock, which were not avoided by our scheme.

3. Crossflow instabilities in the approximation of detonation waves

The approximation of detonation waves modeled by the 2D reactive Euler equations can also suffer from the described crossflow instability. We will show that again post-shock oscillations lead to the instability. In contrast to the nonreactive case, these oscillations are not caused by an inappropriate approximation of slowly moving shocks. Instead, they display an instability inherent in the equations. The conditions which can lead to crossflow instabilities, i.e. a transverse velocity close to zero and a continuous perturbation for instance in pressure during every time step, can also arise in the approximation of detonation waves. This will now be demonstrated on the example of the approximation of 2D detonation waves which are moving through a narrow channel.

The ZND theory predicts that a detonation wave consists of a fluid dynamic shock followed by a reaction zone, see Fickett and Davis [2]. Although this reflects the main features of a detonation wave in two space dimensions, such a quasi one-dimensional structure is not stable for most parameter values. Two dimensional detonation waves are characterized by transverse waves which are moving across the front inside the reaction zone, see [2]. The detonation front is not a straight line. It consists of a series of intersecting shocks, with an alternation of *Mach stems*, i.e. strong shocks with large pressure ratio, and *incident shocks*, i.e. weak shocks with low pressure ratio. The *transverse waves* are the reflected shocks in such a triple point configuration. Transverse waves are moving inside the reaction zone as the triple points are moving along the detonation front. Furthermore, there is a slip line starting at each triple point, which divides the material which passes through the Mach stem from those, which passes through the incident and reflected shock. The interaction of incident shocks, Mach stems and transverse waves can form very regular patterns, so called cellular structures, inside the reaction zone. These patterns can be simulated by considering a detonation wave which is traveling down a narrow channel, as it was done by Bourlioux and Majda [1]. Their numerical scheme consists of a front-tracking, which correctly approximates the leading detonation front, combined with an adaptive mesh refinement and the PPM method of Colella and Woodward. We now consider the results obtained with CLAWPACK for two different cases considered in [1]. We used an exact Riemann solver and a fractional step approach in order to incorporate the source term.

Following [1] the initial values for the unburnt state in Example 3.1 and 3.2 are set to $\rho_u = 1$, $u_u = 0$, $v_u = 0$, $p_u = 1$ and $Z_u = 1$. For the ratio of specific heats the value $\gamma = 1.2$ is used. The unburnt state is connected to the burnt state by a ZND detonation wave with a given degree of overdrive f ($f = (s/s_{CJ})^2$ describes the ratio between the speed of the detonation wave and the speed of the Chapman-Jouguet detonation wave for the same unburnt state), an activation energy E^+ and the heat release q_0. These parameter values vary in the two examples which will be considered. The interface between the unburnt and the burnt state is a sine curve along the channel width with very small amplitude and a period which is specified for each Example.

Example 3.1. *Here we consider the case of* small heat release *using the parameter values* $E^+ = 20$, $q_0 = 2$ *for the activation energy and the heat release. The degree of overdrive is equal to* $f = 1.1$. *The channel is* 50 *units long and* 5.7 *units wide.*

The initial perturbation of the detonation front is a sine wave with period equal to the channel width. Periodic boundary conditions are used on the top and the bottom of the computational domain, outflow boundary conditions (implemented by first order extrapolation) are used on the left and the right end of the channel.

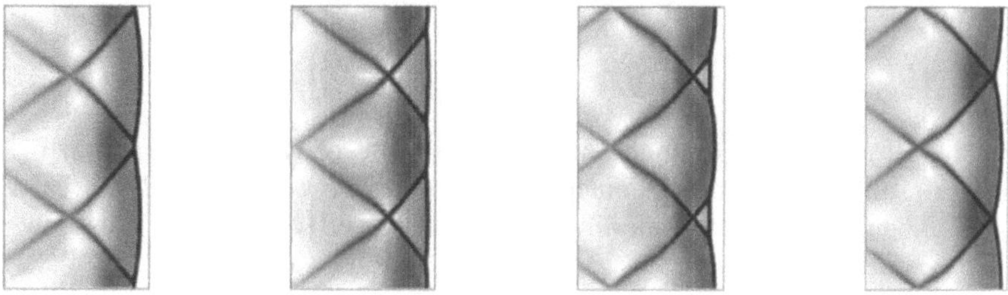

FIGURE 3. Schlieren plots of the pressure at different time steps between 16 and 19 (plotted twice).

Figure 3 shows schlieren plots of the pressure for Example 3.1 at different time steps. The combustion front contains incident shocks and Mach stems. The first plot shows the combustion front right after the collision of two triple points. The reflection produces a new Mach stem which moves over the incident shocks as it can be seen in the second plot. The third plot is taken shortly before the next triple point interaction, which is shown in the fourth plot. After this interaction a new cycle starts. The obtained cellular structure is in agreement to those shown in [1].

The structure of the detonation wave depends very strongly on the ignition temperature and the heat release as can be seen from the next example, where we get a more complicated structure.

Example 3.2. *Now we consider the case of* large heat release and low activation energy. *The parameter values are set to* $E^+ = 10$, $q_0 = 50$ *and* $f = 1.2$. *In this example the channel is* 80 *units long and* 10 *units wide. The initial perturbation has a period which is equal to the half channel width. Again periodic boundary conditions are used on the top and the bottom of the computational domain, outflow boundary conditions are used on the left and the right end of the channel.*

The higher heat release in Example 3.2 produces a more complex structure of the transverse waves. The collision of the triple points causes a creation of vortex sheets which can be seen very well in the plots of the mass fraction of unburnt gas, see Figure 4. The crossflow instability did arise in Example 3.2 but not in Example 3.1. The modified Riemann solver was successfully used to avoid the instability in the numerical solution of Example 3.2. In Figure 5, schlieren plots of the numerical results for Example 3.2 without this extended 'entropy-fix' at

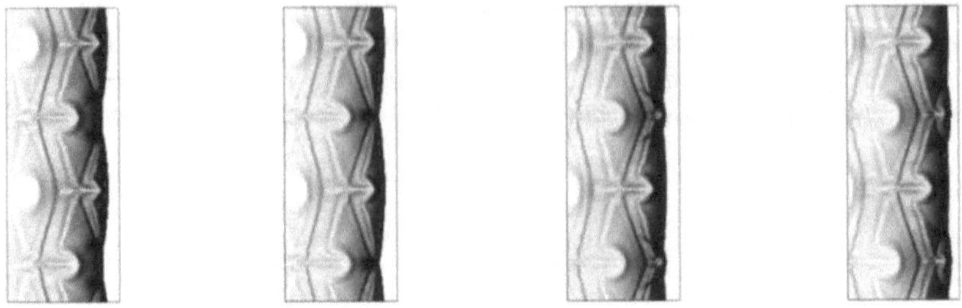

FIGURE 4. Sequence of Schlieren plots of the mass fraction of unburnt gas times density at time steps between 8.4 and 9.0 (calculated with the modified Riemann solver to avoid the crossflow instability).

the contact discontinuity are plotted. The first snapshot shows the effects of the crossflow instability before the complex cell structure could develop. The influence of the crossflow instability is seen at the later time steps, where the solution is more disturbed.

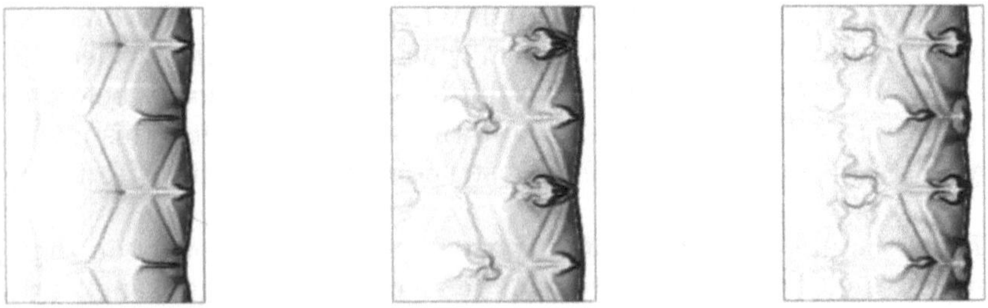

FIGURE 5. Schlieren plots of the mass fraction of unburnt gas times density at three different time steps ($t = 3, 6, 9$), where the crossflow instability arise in the numerical solution.

The detonation front in Examples 3.1 and 3.2 is by far not a slowly moving shock. Indeed the crossflow instability did arise behind the faster detonation front. The reason for the growing instability transverse to the detonation front in Example 3.2 is again a post-shock oscillation but now caused by the stability behavior of the reactive Euler equations.

In Figure 6 we show the pressure peak history of the pressure at the so-called von Neumann state for the 1D ZND detonation waves corresponding to Examples 3.1 and 3.2. These plots show the pressure right behind the leading shock of the detonation wave as it evolves in time. Here we added a perturbation with very low amplitude and with a frequency analogous to the perturbation of the detonation front in the 2D calculations to the density in the unburnt state. The dotted line is the pressure at the von Neumann state for an unperturbed ZND detonation wave. The pressure peak history corresponding to Example 3.1 predicts a stable behavior, whereas the pressure peak history for the 1D detonation wave corresponding to Example 3.2 predicts an unstable behavior.

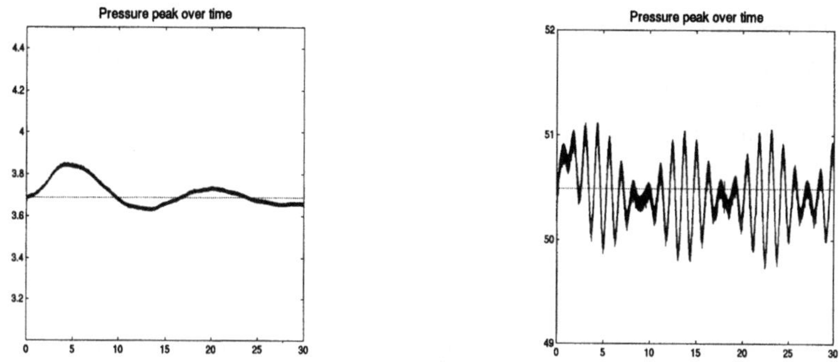

(a) (b)

FIGURE 6. Pressure peak history over time $t \leq 30$ of the 1D detonation waves corresponding to (a) Example 3.1 and (b) Example 3.2. The dotted line is the pressure at the von Neumann state for the unperturbed ZND detonation wave.

In most situations the ZND structure is not the stable long time solution and we therefore have to expect post-shock oscillations. As indicated by our examples, crossflow instabilities arise in situations where the ZND structure of a detonation wave becomes more unstable. Here the post-shock pressure oscillations can provide a continuous perturbation of the constant state behind the detonation front in crossflow direction. This can cause the numerical crossflow instability to grow.

For more unstable ZND detonation waves also the structure of the 2D detonation wave becomes more complex and eventually irregular, as it can already be seen by comparing Figure 3 and 4. As the structure becomes more complex it may be difficult to distinguish between crossflow instabilities introduced by the numerical scheme and physically relevant instabilities.

4. Conclusions

We have shown that in the approximation of 2D detonation waves the occurrence of the crossflow instability is strongly coupled with the post-shock oscillations that arise for unstable 1D detonation waves. These post-shock oscillations arise due to an instability inherent in the equations. In a 2D calculation these physically relevant oscillations can cause the crossflow instability to grow. This is in contrast to the nonreactive case, where a continuous perturbation, which is necessary to trigger the crossflow instability, is caused by a numerical artifact, namely oscillations which arise behind slowly moving shocks. A modification of the Riemann solver transverse to the discontinuity can avoid the crossflow instability without limiting the post-shock oscillations. This is important for calculations of detonation waves, since the physically relevant oscillations influence the complex 2D detonation wave structure.

Acknowledgments
This work was supported in part by DOE grants DE-FG03-96ER25292, DE-FG02-88ER25053 and NSF grant DMS-9803442 as well as DFG grants Wa 633/7-1, 2 and GIF grant I-318-195.06/93.

References

[1] A. Bourlioux and A.J. Majda *Theoretical and numerical structure for unstable two-dimensional detonations.* Combust. Flame, **90** (1992) 211-229.

[2] W. Fickett and W. C. Davis *Detonation.*, University of California Press, 1979

[3] C. Helzel, R. J. LeVeque and G. Warnecke *A modified fractional step method for the accurate approximation of detonation waves.* SIAM J. Sci. Comput., **22** (2000) 1489-1510.

[4] R. J. LeVeque CLAWPACK software. available from `http://www.amath.washington.edu/~rjl/clawpack.html`.

[5] R. J. LeVeque *Wave propagation algorithms for multidimensional hyperbolic systems.* J. Comput. Phys., **131** (1997), 327-353.

[6] R. J. LeVeque *Nonlinear conservation laws and finite volume methods for astrophysical fluid flow.* In O. Steiner and A. Gautschy, editors, Computational Methods for Astrophysical Fluid Flow, 27th Saas-Fee advanced course lecture note. Springer-Verlag, 1998

[7] R. J. LeVeque and M. Pelanti *A class of approximate Riemann solvers and their relation to relaxation schemes.* J. Comput. Phys. (2001)

[8] J. J. Quirk *A contribution of the great Riemann solver debate.* Int. J. Numer. Methods Fluids, **18** (1994), 555-574.

[9] J. J. Quirk *Godunov-type schemes applied to detonation flows.* Combustion in High-Speed Flows, 575-596, 1994.

[10] R. Sanders, E. Morano and M.-C. Druguet *Multidimensional dissipation for upwind schemes: stability and applications to gas dynamics.* J. Comput. Phys., **145** (1998), 511-537.

D. S. Bale
Dept. of Applied Mathematics,
University of Washington,
Box 352420,
Seattle, WA 98195-2420, USA
dbale@amath.washington.edu

C. Helzel
Courant Institute,
251 Mercer Street,
New York, NY 10012, USA
helzel@cims.nyu.edu

International Series of Numerical Mathematics
Vol. 140, © 2001 Birkhäuser Verlag Basel/Switzerland

Wave Propagation Algorithms for Hyperbolic Systems on Curved Manifolds

Derek S. Bale and Randall J. LeVeque

Abstract. We develop a modified version of the wave propagation algorithm based on the solution of Riemann problems to solve hyperbolic systems of partial differential equations on curved manifolds specified by a stationary metric. The algorithm is implemented in the software package CLAWPACK. The implementation of our modified algorithm maintains both the ease of use, and versatility of CLAWPACK, making it useful to a broad class of users. We demonstrate that the method works well on both flat and curved manifolds alike by presenting examples of the algorithm for the Euler equations in curvilinear coordinates, and the propagation of linear waves on a curved surface.

1. Introduction

We develop a modified version of the wave propagation algorithm [5] based on the solution of Riemann problems to solve hyperbolic systems of partial differential equations on curved manifolds. We will present an implementation of this high resolution algorithm in the software package CLAWPACK (Conservation LAWs PACKage), a collection of freely available Fortran routines [1]. The standard CLAWPACK software requires the user to provide, in addition to initial and boundary conditions, only the Riemann solver for the system of equations at hand. Consequently, CLAWPACK is easily applied to a variety of systems on logically rectangular grids in flat space. Many applications of interest, however, involve the solution of hyperbolic systems on curved manifolds that cannot be covered by flat grids. It is well known that the representation of such systems on manifolds introduces a number of necessary modifications to any finite volume scheme. Successful algorithms have long been developed for hyperbolic systems in curvilinear coordinates [3, 2], and for specific systems of equations in more general manifolds [6, 7, 8], but codes that are able to handle different systems on a variety of manifolds are rare. The implementation of our modified algorithm maintains both the ease of use, and versatility of CLAWPACK, making it useful to a broad class of users.

Our algorithm is based on solving Riemann problems in sets of locally orthonormal basis vectors as was done in [7]. Tangent vector components are locally transformed from the computational basis at the center of a finite volume cell to an orthonormal basis located at each cell edge. This has advantages over directly solving Riemann problems in the non-constant basis located at the cell center. Typically, Riemann solvers become more complicated in a curvilinear basis. One benefit of solving Riemann problems in sets of locally orthonormal basis vectors is that no modification to Cartesian Riemann solvers is necessary. The implementation presented here uses standard flat-space Riemann solvers. Another benefit of this algorithm is that it avoids the introduction of geometrically induced source terms for scalar conservation laws. Typically, non-constant basis vectors that cover a patch of a manifold introduce geometric source terms that must be incorporated into the numerical integration by a splitting technique. It is well known that split integration schemes can give rise to large errors if source terms nearly balance flux gradients [4].

Our implementation in CLAWPACK requires only three routines to define the geometry in which a system of equations will be solved. These routines set the metric tensor components, Christoffel symbols (which are a combination of metric components and their derivatives), and the appropriate derivatives of Christoffel symbols. These can be easily computed with a simple Maple routine.

We demonstrate that the method works well on both flat and curved manifolds via two examples : the flow of an inviscid gas in a curvilinear set of coordinates, and the propagation of linear waves on a curved surface. Examples of this algorithm used for flow on the surface of the sphere can be found elsewhere in these proceedings [9].

2. Standard Wave Propagation

The wave propagation algorithm as implemented in CLAWPACK which can be applied to hyperbolic partial differential equations of the form

$$\frac{\partial}{\partial t}\boldsymbol{q}(t,\vec{x}) + \sum_{k=1}^{n_d} A^k(\boldsymbol{q},\vec{x})\,\frac{\partial}{\partial x^k}\boldsymbol{q} = \boldsymbol{\psi}(\boldsymbol{q},t,\vec{x}). \tag{1}$$

The summation in this equation is over the number of spatial dimensions n_d and typically the grid volume is covered with a standard Cartesian basis $\{\boldsymbol{e}_{(x)}, \boldsymbol{e}_{(y)}, \dots\}$. Throughout this work vectors will be denoted by either the standard arrow over the symbol (i.e. $\vec{v}$), or braces around subscripts and superscripts (i.e. $\boldsymbol{e}_{(x)} = (1,0,0)^T$). This avoids confusing the index of a vector with the index of its components. The state vector $\boldsymbol{q}(t,\vec{x})$ in equation (1) is a function $\boldsymbol{q} : \mathbb{R}^{n_d} \times \mathbb{R}^+ \to \mathbb{R}^{n_{eq}}$ where its bold type indicates that it is a vector with n_{eq} components $\left(q^1, \cdots, q^{n_{eq}}\right)^T$. The A^k are $n_{eq} \times n_{eq}$ matrices for each $k \in \{1 \dots n_d\}$ and $\boldsymbol{\psi}$ is the source. We assume each A^k has a complete set of eigenvectors $\{\boldsymbol{r}^{(1)}, \cdots, \boldsymbol{r}^{(n_{eq})}\}$ associated with which is an ordered set of eigenvalues $\{\lambda^1, \cdots, \lambda^{n_{eq}}\}$.

The class of quasilinear equations (1) includes conservation laws of the form

$$\boldsymbol{q}_{,t} + \boldsymbol{f}^{(k)}_{,k} = 0, \tag{2}$$

where both subscripts and superscripts preceeded by a comma will denote partial differentiation (i.e. $q_{,k} \equiv \partial q/\partial x^k$), and a sum is implied on repeated indices if one is raised and the other is lowered. The conservative properties of equation (2) can be realized by integrating over a volume $\mathcal{V}$ to get

$$\frac{d}{dt}\int_{\mathcal{V}} \boldsymbol{q}\, d\mathcal{V} + \int_{\partial\mathcal{V}} \boldsymbol{f}\cdot \boldsymbol{d\sigma} = 0, \tag{3}$$

where $\boldsymbol{d\sigma}$ is the outward pointing normal to the volume boundary $\partial\mathcal{V}$. The time rate of change of the integral of $\boldsymbol{q}$ over the volume $\mathcal{V}$ is due only to flux through the volume boundary, independent of a coordinate basis. High-resolution finite volume methods are able to evolve systems of the form (3) in Cartesian coordinates accurately and stably. We describe the algorithm from the point of view of the more general quasi-linear system (1) since the conservation law (2) is just a special case of that system.

Data on the nth time slice in each cell $\mathcal{C} = \{i, j, \cdots\}$ is represented by piecewise constant functions $\boldsymbol{Q}^n_{\mathcal{C}}$ representing the finite volume average at cell center. These averages define one dimensional planar Riemann problems at each cell interface. Each Riemann problem generates a set of n_{eq} waves $\{\boldsymbol{\mathcal{W}}^1_{i-1/2}, \cdots \boldsymbol{\mathcal{W}}^{n_{eq}}_{i-1/2}\}$ with associated wave speeds $\{s^1_{i-1/2}, \cdots, s^{n_{eq}}_{i-1/2}\}$. These waves carry jumps in the state vector $\boldsymbol{Q}^n_{\mathcal{C}}$

$$\boldsymbol{Q}_i - \boldsymbol{Q}_{i-1} = \sum_{p=1}^{n_{eq}} \boldsymbol{\mathcal{W}}^p_{i-1/2}, \tag{4}$$

and induce fluctuations across cell boundaries, denoted by

$$\boldsymbol{\mathcal{A}}^{\pm}_k \boldsymbol{\Delta Q}_{i-1/2} = \sum_p \left[s^p_{i-1/2}\right]^{\pm} \boldsymbol{\mathcal{W}}^p_{i-1/2}, \tag{5}$$

where we have closely followed the notation of [5].

We motivate the standard algorithm in Cartesian coordinates by considering the time advancement of the multidimensional constant coefficient advection equation

$$q_{,t} + u^k q_{,k} = 0. \tag{6}$$

Starting with information at time t, the Taylor series approximation for the solution a time Δt later is

$$q(t+\Delta t) = q(t) - \underbrace{\Delta t u^k q_{,k}(t)}_{I} + \frac{1}{2}\Delta t^2 \underbrace{u^k u^j q_{,kj}(t)}_{II} + O(\Delta t^3), \tag{7}$$

where once again we sum on repeated indices. A high-resolution finite volume scheme must model both type I and II terms shown in (7) while avoiding oscillations. Type I terms are one dimensional upwind terms and we will focus our discussion on them. Type II terms are modeled by high-resolution terms that are not discussed here.

If we are sweeping through the grid in the x direction, then type I terms in the Taylor series (7) are of the form $u^x q_{,x}$ (no sum on x). These terms neglect

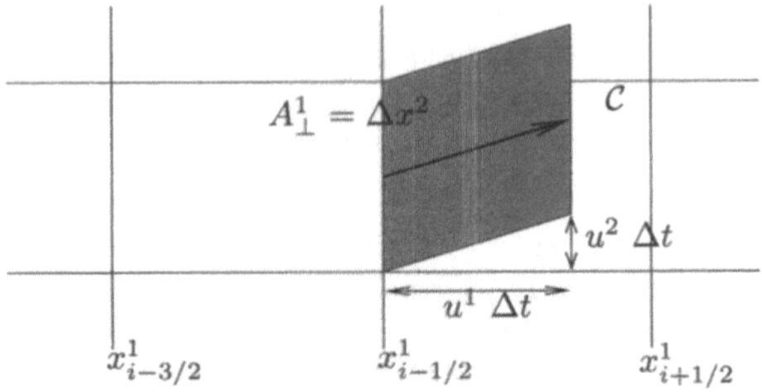

FIGURE 1. A wave propagating through a Cartesian grid

the multidimensional propagation of waves. In this direction, a wave travels a distance $u^x_{i-1/2}\Delta t$ into the upwind cell over one timestep as depicted in Fig. 1. The influence of this wave is $A^x_\perp \Delta t u^x_{i-1/2}\mathcal{W}^x_{i-1/2}$ which gives the first order cell update (for $u^x > 0$)

$$Q^{n+1}_{\mathcal{C}} = Q^n_{\mathcal{C}} - \sum_{k=1}^{n_d} \frac{A^x_\perp}{V(\mathcal{C})}\Delta t \mathcal{A}^+_x \Delta Q_{i-1/2}, \tag{8}$$

where $A^x_\perp$ is the area of the cell edge through which the wave is travelling, and we have used (5) to write $u^x_{i-1/2}\mathcal{W}^x_{i-1/2} = \mathcal{A}^+_x \Delta Q_{i-1/2}$.

By incorporating the influence of waves from each direction, and noting that $A^k_\perp / V(\mathcal{C}) = 1/\Delta x^k$ we write the standard first order update as

$$Q^{n+1}_{\mathcal{C}} = Q^n_{\mathcal{C}} - \sum_{k=1}^{n_d} \frac{\Delta t}{\Delta x^k}\Delta^k_{\mathcal{C}}, \tag{9-a}$$

where

$$\Delta^k_{\mathcal{C}} = \mathcal{A}^+_k \Delta Q_{i-1/2} + \mathcal{A}^-_k \Delta Q_{i+1/2}. \tag{9-b}$$

In addition to these first order updates, as mentioned above, high resolution corrections are included taking care of type II terms in the expansion (7). The details of these limited fluxes can be found in [5].

3. Wave Propagation on Curved Manifolds

We work with a general manifold $(\mathcal{M}, \boldsymbol{h})$ described by the coordinate basis $\{\boldsymbol{e}_{(1)}, \boldsymbol{e}_{(2)}, \cdots, \boldsymbol{e}_{(n_d)}\}$. The metric tensor $\boldsymbol{h}$ has components $h_{ij} = \boldsymbol{h}(\boldsymbol{e}_{(i)}, \boldsymbol{e}_{(j)})$ while its inverse has components $h^{ij} = \boldsymbol{h}(\boldsymbol{e}^{(i)}, \boldsymbol{e}^{(j)})$ so that $\boldsymbol{e}_{(i)} \cdot \boldsymbol{e}^{(j)} = \delta^j_i$. The algorithm presented here is based on the solution of Riemann problems in sets of locally orthonormal bases at cell interfaces as in [7]. One such orthonormal basis is shown in Fig. 2 (a). Solving Riemann problems in such bases allow us to use existing Cartesian Riemann solvers, avoiding complications due to the changing coordinate basis.

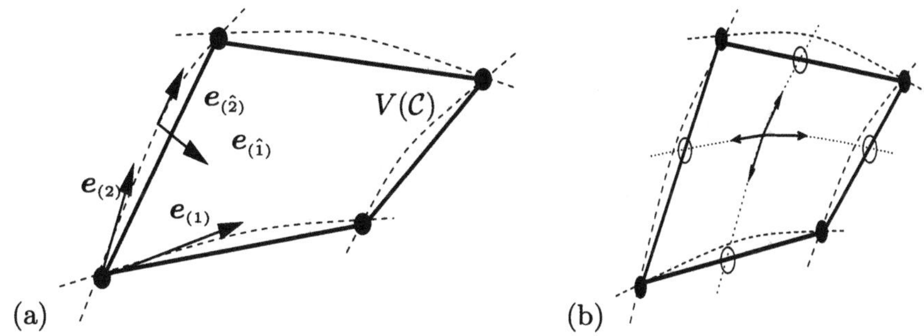

FIGURE 2. (a) The orthonormal coordinate system at cell edge where Riemann problems will be solved for the left interface. (b) Vectors must undergo parallel transport to set up cell edge Riemann problems.

Applying equation (3) over a finite volume cell in a manifold one gets

$$\frac{d}{dt}\int_C \sqrt{h}\boldsymbol{q}\;dx^{n_d} + \sum_{j=1}^{n_d}\left[\int_{\Sigma^j_{i+1/2}} \sqrt{h}\boldsymbol{f}\cdot\boldsymbol{e}^{(j)} + \int_{\Sigma^j_{i-1/2}} \sqrt{h}\boldsymbol{f}\cdot(-\boldsymbol{e}^{(j)})\right] = 0, \qquad (10)$$

where Σ^j is the cell edge with x^j = constant. Note that in this coordinate basis the flux is scaled by the determinant of the metric. This scale factor can be computed by considering waves propagating in a manifold.

3.1. Scalar conservation laws

A scalar conservation law has the form

$$q_{,t} + \boldsymbol{\nabla}\cdot\vec{f}(q) = 0 \qquad (11)$$

independent of any coordinate basis. However, in order to numerically evolve equations of this form on general manifolds, we represent this equation on a coordinate basis

$$q_{,t} + f^k(q)_{,k} = -\Gamma^k_{jk} f^j(q) \quad\Leftrightarrow\quad q_{,t} + \frac{1}{\sqrt{h}}\left(\sqrt{h} f^k(q)\right)_{,k} = 0. \qquad (12)$$

where $\boldsymbol{\nabla}_m \boldsymbol{e}_{(n)} = \Gamma^j_{nm}\boldsymbol{e}_{(j)}$. Both equations in (12) are equivalent. However it is desirable to integrate the equation in the form that does not have a source. That is, our algorithm will model scalar conservation laws in the form on the right hand side of (12). In order to do that, one can see that there will need to be scaled wave speeds as also indicated in equation (10).

Waves carry scalar densities into adjacent cells for equations of the form (11). The velocity of the wave emanating from the cell edge has edge-normal component

$$s^{\hat{k}} \equiv \vec{s}\cdot\boldsymbol{e}^{(\hat{k})} = \vec{s}\cdot\frac{\boldsymbol{e}^{(k)}}{\sqrt{h^{kk}}} = \frac{s^k}{\sqrt{h^{kk}}}, \qquad (13)$$

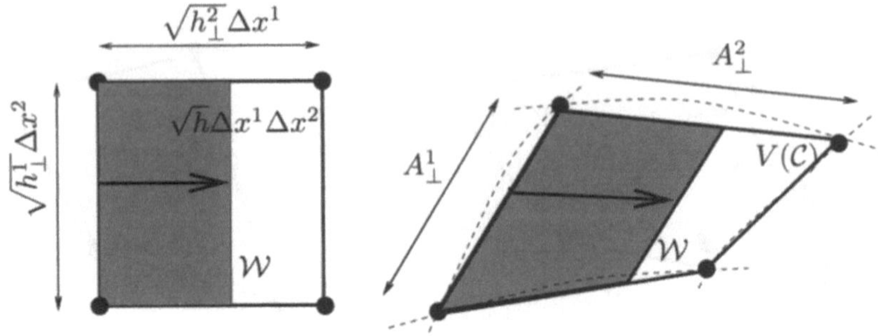

FIGURE 3. The left figure shows a wave propagating through a uniform grid in coordinate space. The right figure shows the same wave propagating in physical space.

where a hat over a superscript (or subscript) denotes a component in the orthonormal basis $\{\boldsymbol{e}_{(\hat{1})}, \boldsymbol{e}_{(\hat{2})}, \cdots, \boldsymbol{e}_{(\hat{n_d})}\}$. This allows us to write this wave speed in terms of the coordinate basis in which we are updating the solution. Fig. 3 depicts the propagation of a wave from the left interface along the x^1 direction. On the left, the propagation is depicted in the coordinate basis with uniform spacing. One the right, the same wave is shown propagating in physical space. The wave propagates a distance

$$s^{\hat{1}}\Delta t = \frac{s^k}{\sqrt{h^{11}}}\Delta t \tag{14}$$

normal to the cell edge while it covers a physical area

$$A_\perp^1 = \sqrt{h_\perp^1}\Delta x^2, \tag{15}$$

where $\boldsymbol{h}_\perp^k$ is the metric tensor projected onto the x^k = constant coordinate plane, and $h_\perp^k = \det(\boldsymbol{h}_\perp^k)$. The influence of this wave in the adjacent coordinate cell is then

$$s^{\hat{1}}\Delta t A_\perp^1 \mathcal{W}_{i-1/2}^1 = \Delta x^2 \Delta t \sqrt{h_\perp^1}\, s^{\hat{1}} \mathcal{W}_{i-1/2}^1 = \Delta x^2 \Delta t \sqrt{h} s^1 \mathcal{W}_{i-1/2}^1, \tag{16}$$

where we have used the fact that $h_\perp^k \equiv hh^{kk}$ (no sum on k) in conjunction with equation (14). Notice that the right most expression in (16) has the coordinate flux scaled by $\sqrt{h}$, exactly as was to be expected from equation (10).

3.2. Vector conservation laws

It is often the case that we must deal not only with scalar conservation laws alone, but systems of conservation laws with vectors such as momentum. The flux is then no longer a simple vector, but a second rank (and usually) symmetric tensor. These equations have the coordinate-independent representation

$$\partial_t \vec{s} + \boldsymbol{\nabla} \cdot \boldsymbol{T}(q) = 0. \tag{17}$$

The coordinate representation has the form

$$s^j_{,t} + T^{kj}_{,k} = -\Gamma^k_{mk} T^{jm} - \Gamma^j_{km} T^{km} \Leftrightarrow s^j_{,t} + \frac{1}{\sqrt{h}} \left(\sqrt{h} T^{kj} \right)_{,k} = -\Gamma^j_{mk} T^{mk}. \quad (18)$$

Notice that the equation on the right side of (18) is the form involving the least number of source terms. It is this form of the conservation law that we integrate. The flux term has the same form as for the scalar equation, but we must now transport vector components from cell center to cell edge along coordinate lines. That is, we must parallel transport vectors from cell centers to cell edges, then write them in terms of the orthonormal components at the edge. The right hand side of the equation is taken care of by this process. Parallel propagation of vector components is depicted in Fig. 2(b).

We represent parallel propagation as a matrix multiplication

$$v^i_{\text{edge}} = P^i_m v^m_{\text{center}}, \quad (19)$$

where the components of the matrix $\boldsymbol{P}$ that take vectors from cell center to cell edge along the x^1 coordinate line are defined by

$$P^i_m \approx \delta^i_m - \Delta x^1 \Gamma^i_{m1} + \frac{1}{2} (\Delta x^1)^2 \left(\Gamma^i_{n1} \Gamma^n_{1m} - \mathcal{G}^i_{m11} \right), \quad (20)$$

with $\mathcal{G}^i_{mkk} \equiv \Gamma^i_{mk,k}$.

Once the vector components from both left and right states are at the cell edge, we must project them onto the orthonormal basis shown in Fig. 2(a). This is done with another matrix multiplication

$$v^{\hat{i}}_{\text{edge}} = O^{\hat{i}}_j v^j_{\text{edge}}, \quad (21)$$

with the matrix $\boldsymbol{O}$ of the form

$$\left[O^{\hat{i}}_j \right] = \begin{pmatrix} 1/\sqrt{h^{11}} & 0 \\ h_{21}/\sqrt{h_{22}} & \sqrt{h_{22}} \end{pmatrix} \quad (22)$$

in the x^1 direction and $n_d = 2$, for example. The resulting Riemann problem will be between left and right components that undergo transformations of the form

$$v^{\hat{i}}_{\text{edge}} = O^{\hat{i}}_j P^j_k v^k_{\text{center}}, \quad (23)$$

parallel transport followed by orthonormalization.

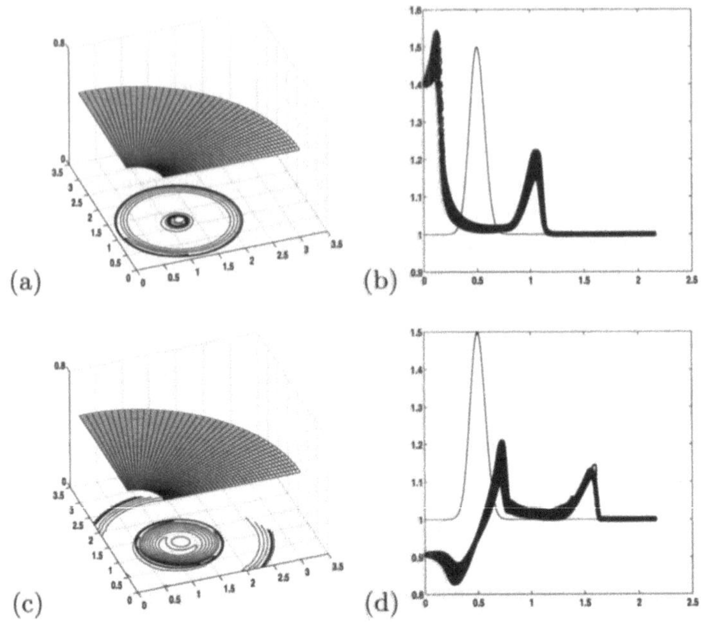

FIGURE 4. Evolution of a perfect gas on a flat plane in polar coordinates.

4. Examples

4.1. The Euler Equations

The Euler equations describing the time evolution on a general manifold of an inviscid, non-heat conducting, perfect fluid are

$$\rho_{,t} + \frac{1}{\sqrt{h}}\left(\sqrt{h}\rho u^j\right)_{,j} = 0 \tag{24-a}$$

$$\left(\rho u^k\right)_{,t} + \frac{1}{\sqrt{h}}\left(\sqrt{h}T^{kj}\right)_{,j} = -\Gamma^k_{ij}T^{ij} \tag{24-b}$$

$$e_{,t} + \frac{1}{\sqrt{h}}\left[\sqrt{h}(e+p)u^j\right]_{,j} = 0, \tag{24-c}$$

where ρ is the density, p the pressure, $\vec{u}$ the velosity field, e the fluid energy, and $\mathbf{T}$ is the momentum tensor with components $T^{ij} = \rho u^i u^j + ph^{ij}$. We apply the algorithm presented here to these equations on a flat plane covered by a polar grid. The metric takes the form $\mathbf{h} = \mathrm{diag}(1, r^2)$. In the first example presented here, we solve these equations with a radially symmetric pressure disturbance that is placed off center on the polar grid. This will test the method's ability to maintain radial symmetry while the waves propagate at arbitrary angles to the grid cell interfaces.

Fig. 4 shows two snap shots of this evolution. The top row of plots is at an early time when the disturbance is still entirely within the grid. In the top left

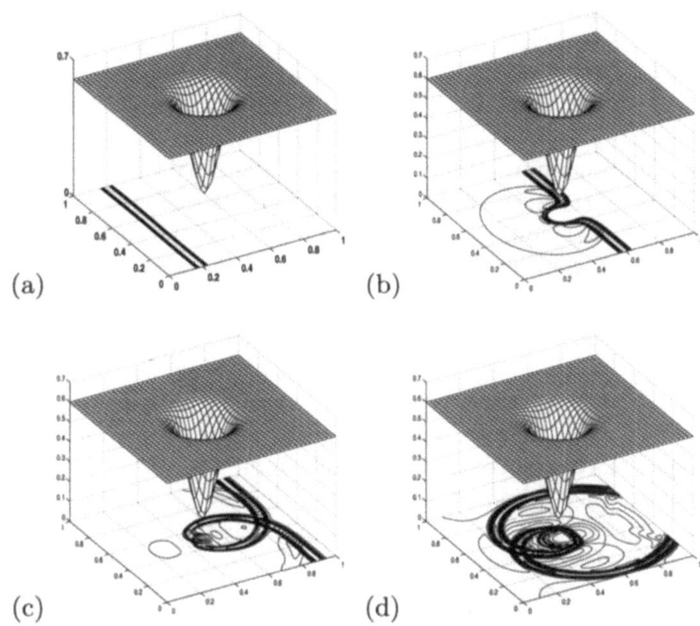

FIGURE 5. Scalar wave propagating on a surface

plot we display the flat surface covered by a polar grid on which we are solving the equations, and project a contour plot of the solution for easy viewing. Notice the radial evolution. In the top right plot we show a scatter plot (circles) about the symmetry axis of the disturbance (which is not the grid axis). We also show a one-dimensional, very refined calculation (solid line under circles) for comparison. The solid line away from the circles is the initial disturbance. The bottom row of plots is the same set of plots, but at a later time when the waves are leaving the grid.

4.2. The wave equation

The second order wave equation on a curved manifold can be written

$$-\varphi_{,tt} + \boldsymbol{\nabla}_k(\varphi^{,k}) = 0, \tag{25}$$

where $\boldsymbol{\nabla}_k$ is the metric-induced covariant derivative operator. This second order equation can be reduced to the standard first order system by using the definitions

$$\phi = \varphi_{,t} \tag{26}$$

$$\psi^k = \varphi^{,k} = h^{km}\varphi_{,m}. \tag{27}$$

In terms of the new variables ϕ and ψ^k, this equation takes the first order form

$$\phi_{,t} - \frac{1}{\sqrt{h}}\left(\sqrt{h}\psi^j\right)_{,j} = 0 \tag{28-a}$$

$$\psi^k_{,t} - h^{kj}\phi_{,j} = 0. \tag{28-b}$$

We solve this equation on an asymptotically flat 2-surface that has a Gaussian dip in the center. The surface can be seen in all the plots of Fig. 5. We use the same plotting form as the previous example whereby we plot the surface covered by the coordinates in which we are working. Contours of the solution are projected for easy viewing, and the sequence shows a right-going wave moving through the dip, which generates a reflected wave.

References

[1] R.J. LeVeque. CLAWPACK software. Available from `http://www.amath.washington.edu/~claw`

[2] M. Vinokur. Conservation Equations of Gasdynamics in Curvilinear Coordinate Systems. J. Comput. Phys., 81:1-52,1989.

[3] D.L. Brown. An Unsplit Godunov Method for Systems of Conservation Laws on Curvilinear Overlapping Grids. Mathl. Comput. Modelling, 20:29-48,1994.

[4] R.J. LeVeque. Balancing Source Terms and Flux Gradients in High-Resolution Godunov Methods: The Quasi-Steady Wave-Propagation Algorithm. J. Comput. Phys., 146:346-365,1998.

[5] R.J. LeVeque. Wave Propagation Algorithms for Multidimensional Hyperbolic Systems. J. Comput. Phys., 131:327-353,1997.

[6] R. Eulderink and G. Mellema. General relativistic hydrodynamics with a Roe solver. Astron. Astrophys. Suppl. Ser., 110:587-623,1995.

[7] J. A. Pons and J. A. Font and J. M. Ibanez and J. A. Miralles. General relativistic hydrodynamics with special relativistic Riemann solvers. Astron. Astrophys., 339:638-642,1998.

[8] F. Banyuls and J. A. Font and J. M. Ibanez and J. M. Marti and J. A. Miralles. General Relativistic Hydrodynamics: A Local Characteristic Approach. The Astrophysical Journal, 476:221-231,1997.

[9] J. A. Rossmanith and R.J. LeVeque. A Wave Propagation Algorithm for the Solution of PDEs on the Surface of a Sphere. These Proceedings, 2000.

D. S. Bale
Randall J. LeVeque
Department of Applied Mathematics,
University of Washington,
Box 352420, Seattle, WA 98195-2420
`dbale@amath.washington.edu`
`rjl@amath.washington.edu`

International Series of Numerical Mathematics
Vol. 140, © 2001 Birkhäuser Verlag Basel/Switzerland

The Random Projection Method for Stiff Multi-species Detonation Computation

Weizhu Bao and Shi Jin

Abstract. In this note we review the random projection method, recently introduced by the authors, for numerical simulations of the hyperbolic conservation laws with stiff reaction terms:

$$U_t + F(U)_x = \frac{1}{\varepsilon}\Psi(U).$$

In this problem, the reaction time ε is small, making the problem numerically stiff. A classic spurious numerical phenomenon – the incorrect shock speed – occurs when the reaction time scale is not properly resolved numerically. The random projection method, a fractional step method that solves the homogeneous convection by any shock capturing method, followed by a random projection for the reaction term, was introduced in [1] to handle this numerical difficulty. For a scalar model problem, one can prove that the random projection methods capture the correct shock speed with a first order accuracy, if a monotonicity-preserving method is used in the convection step. This method can be extended to compute stiff multi-species detonation waves by randomizing the ignition temperatures. Numerical results of a problem with five species and two reactions will be presented.

1. Introduction

In this note we review the random projection method, proposed recently by the authors [1, 2, 3], for numerical simulations of the hyperbolic conservation laws with stiff reaction terms:

$$U_t + F(U)_x = \frac{1}{\varepsilon}\Psi(U), \tag{1}$$

where U is the vector of dependent variables with components: mass, momentum, total energy, and mass fractions of species in the reacting mixture. The flux function $F(U)$ describes the fluid dynamical convection while the source term $\Psi(U)$ arises from the chemistry of the reacting species. ε is the reaction time. In a reacting flow, the reacting time ε is much smaller than the characteristic time scale of the fluid, making the source term in (1) stiff.

Research supported in part by NSF grant No. DMS-9704957

Numerical methods for the stiff reaction term has attracted lots of attention in the last decade. It was first observed by Colella, Majda and Roytburd [8] that underresolved numerical methods, where ε is not resolved by suitable small time steps and grid sizes, lead to spurious shock that travels at an incorrect speed. Since then, lots of attention have been paid to study this peculiar numerical phenomenon (see [4], [15], [16], [6] and references therein) and or find a robust way to fix it (see [9, 10, 5, 13]). It is known that the numerical shock profile, an essential mechanism in all shock capturing methods, leads to too early chemical reaction once the smeared temperature value in the numerical shock layer is above the ignition temperature.

Recently we proposed the random projection method for the hyperbolic system with stiff reaction terms (1) [1, 2, 3]. Unlike the classical random choice method for reacting flow [7], originated from Glimm's scheme [11], which requires solving a generalized Riemann problem for hyperbolic systems with source terms, our method is a fraction step method, which combines a standard –no Riemann solver is needed –shock capturing method for the homogeneous convection with a strikingly simple random projection step for the reaction term. In the random projection step, the ignition temperature is chosen to be uniformly distributed random variable between the initial left and right states. Although at each time step, this random projection will move the shock with an incorrect speed, the statistical average yields the correct speed, even though the small time scale ε is not numerically resolved. In particular, when the random sequence is chosen to be the equidistributed van der Corput's sampling sequence [12], we can prove, for a scalar model problem, a first order accuracy if a monotonicity-preserving method is used in the convection step. This method can be extended straightforwardly to the calculation of detonation waves with two- or multi-species, where in the projection step for the reaction terms, the ignition temperatures are made random. A large amount of numerical experiments for one and two dimensional detonation waves with two- or multi-species demonstrate the robustness of this novel approach [1, 2, 3].

The note is organized as follows: In section 2 we review the random projection method introduced in [1] and its error estimate for a scalar model problem. In section 3 we present the local random projection method for one-dimensional multi-species detonation waves. In section 4 we report numerical results for stiff multi-species detonation capturing.

2. A scalar model problem

In this section we review the random projection method introduced in [1] for the following hyperbolic conservation law with stiff reaction term,

$$u_t + f(u)_x = -\frac{1}{\varepsilon}(u-\alpha)(u^2-1), \qquad -1 < \alpha < 1, \tag{1}$$

with piecewise constant initial data

$$u(x,0) = u_0(x) = \begin{cases} 1, & x \le x_0, \\ -1, & x > x_0. \end{cases} \tag{2}$$

Here ε is the reaction time, f is a convex function of u, i.e. $f''(u) > 0$, and x_0 is a given point.

The source term in (1) admits three local equilibria, i.e., the unstable one $u = \alpha$, and the stable ones $u = \pm 1$. When the solution is at equilibrium, the reaction term has no effect. Thus the exact solution is a shock discontinuity connecting $u = 1$ with $u = -1$ and propagating to the right with the speed determined by the Rankine-Hugoniot jump condition [15]:

$$s = \frac{1}{2}[f(1) - f(-1)]\,. \tag{3}$$

Namely,

$$u(x,t) = \begin{cases} 1, & \text{if } x \le x_0 + st, \\ -1, & \text{if } x > x_0 + st. \end{cases} \tag{4}$$

Let h be the spatial increment and k be the time step, such that x_0 is a grid point, i.e. $x_0 = l(0)h$ with $l(0)$ an integer. The numerical solution is evaluated at the points (jh, nk), $j = 0, \pm 1, \pm 2, \cdots$, $n = 0, 1, 2, \cdots$. Let u_j^n approximate $u(jh, nk)$ and u^n be the solution vector of $u(\cdot, nk)$ at time $t = t_n = nk$. When the reaction term is resolved, i.e. $k = O(h) < \varepsilon$, any method which works well for the homogeneous hyperbolic conservation law still works well here. Here we are interested in the underresolved case, where $k = O(h) \gg \varepsilon$.

The random projection method is a fractional step method that solves the homogeneous equation

$$u_t + f(u)_x = 0 \tag{5}$$

by a standard shock capturing method S_c for one time step, followed by a projection step. In the projection step, we replace the unstable equilibrium, or the critical value, α, by a uniformly distributed random sequence $\theta_n \in (-1, 1)$. Let $u^* = S_c(k)u^n$. We define the **random projection** operator $S_\theta(k)$ by

$$S_\theta(k): \quad u_j^{n+1} = \begin{cases} 1, & \text{if } u_j^* > \theta_n, \\ -1, & \text{if } u_j^* \le \theta_n, \end{cases} \quad \text{for all } j. \tag{6}$$

In this method, one random value of θ_n will be selected per time step. The combination of the two steps gives the numerical scheme

$$S_{gp}(k): \qquad u^{n+1} = S_\theta(k) S_c(k) u^n. \tag{7}$$

The stability condition for this method is the usual CFL condition determined by the convection term.

Here and in our practical computations, we always use van der Corput's sampling sequence for $\{\theta_n\}$. The merit of this sequence is that it produces an equidistributed sequence on the interval $[0, 1]$, and among all known uniformly distributed sequences its deviation is minimal [12]. Let $1 \le n = \sum_{k=0}^{m} i_k 2^k$, $i_k =$

0, 1, be the binary expansion of the integer n. Then the van der Corput's sequence is defined on $[0, 1]$ as:

$$\vartheta_n = \sum_{k=0}^{m} i_k 2^{-(k+1)}, \qquad n = 1, 2, \cdots. \tag{8}$$

We rescale it in order to get a random number generator θ_n on $[-1, 1]$:

$$\theta_n = 2\vartheta_n - 1, \qquad n = 1, 2, \cdots. \tag{9}$$

The sequence $\{\theta_n,\ n = 1, 2, \cdots\}$ is equidistributed on the interval $[-1, 1]$.

We have proved [1] that the random projection method (7), when a monotonicity-preserving method S_c – which includes all TVD methods – is used for the convection term, can capture the correct location of discontinuities for the scalar model problem (1)-(2) with a first order accuracy. Let $l(n) - l(0)$ be the number of grid points that the discontinuity has traveled after n time steps.

Theorem 2.1. [1] *Given $T > 0$. Let S_c be a monotonicity preserving method. The difference between the shock location of the exact solution, $x_0 + st_n$, and the numerical one, $l(n)h$, as determined by the random projection method (7), has the following estimate:*

$$|x_0 + st_n - l(n)h| \le C(T)h|\ln h|, \tag{10}$$

for any $0 < t_n \le T$, and fixed $\lambda = sk/h$, where $C(T)$ is a positive constant depending on $|\ln(T)|$.

This shows that the shock location is captured with a first order accuracy for all time, although the reaction time ε is not resolved numerically. In [1], [2] this method was generated to two-species stiff detonation wave computation, where the key idea is to randomize the ignition temperature in a suitable domain. In [3] the method is extended for multi-species detonation computation.

3. The multi-species detonation capturing

In this section, we review the random projection method for one dimensional multi-species detonation capturing. The key idea is to **make the ignition temperature random.**

Consider the Euler equations that model the time-dependent flow of an inviscid, compressible, reactive gas with multi-species in one space dimension [17]

$$U_t + F(U)_x = \frac{1}{\varepsilon}\Psi(U), \tag{1}$$

$$U = \begin{pmatrix} \rho \\ \rho u \\ e \\ \rho z_1 \\ \rho z_2 \\ \cdots \\ \rho z_N \end{pmatrix}, \quad F(U) = \begin{pmatrix} \rho u \\ \rho u^2 + p \\ (e+p)u \\ \rho u z_1 \\ \rho u z_2 \\ \cdots \\ \rho u z_N \end{pmatrix}, \quad S(U) = \begin{pmatrix} 0 \\ 0 \\ 0 \\ w_1 \\ w_2 \\ \cdots \\ w_N \end{pmatrix}; \tag{2}$$

with

$$w_m = W_m \sum_{k=1}^{M} (\nu''_{mk} - \nu'_{mk})\, B_k T^{\alpha_k} e^{-T_k/T} \prod_{j=1}^{N} \left(\frac{\rho z_j}{W_j}\right)^{\nu'_{jk}}, \qquad 1 \le m \le N, \tag{3}$$

and

$$\sum_{m=1}^{N} z_m = 1. \tag{4}$$

The dependent variables $\rho(x,t)$, $u(x,t)$, $e(x,t)$, $z_1(x,t)$, $\cdots$, $z_N(x,t)$ are the density, velocity, total energy and the mass fraction of 1-st, $\cdots$, N-th species respectively. The pressure for ideal gas is given by

$$p = (\gamma - 1)\left(e - \frac{1}{2}\rho u^2 - q_1 \rho z_1 - q_2 \rho z_2 - \cdots - q_N \rho z_N\right), \tag{5}$$

the temperature is defined as $T = p/\rho$. The parameters B_k, M, N, q_m, T_k, w_m, W_m, α_k, γ, ν'_{mk} and ν''_{mk} correspond to frequency factor in the k-th reaction, the number of reactions, the number of species, the heat of formation for the m-th species, ignition temperature for the k-th reaction, m-th species source term, molecular weight of the m-th species, exponent of the temperature dependence of the frequency factor for the k-th reaction, c_p to c_v ratio, stoichiometric coefficient for the m-th species appearing as a reactant and product in the k-th reaction.

Equations (1)-(4) are referred to as the multi-species reactive Euler equations with Arrhenius kinetics. We will also consider (1)-(4) with

$$B_k T^{\alpha_k} e^{-T_k/T}$$

replaced by the Heaviside kinetics

$$B_k T^{\alpha_k} H(T - T_k),$$

where $H(x) = 1$ for $x > 0$ and $H(x) = 0$ for $x < 0$.

Now we will describe the random projection method to the problem (1) with the Riemann initial data

$$\begin{aligned} &(\rho(x,0), u(x,0), p(x,0), z_1(x,0), z_2(x,0), \cdots, z_N(x,0)) \\ &= \begin{cases} (\rho_l, u_l, p_l, (z_1)_l, (z_2)_l, \cdots, (z_N)_l), & \text{if } x \le x_0, \\ (\rho_r, u_r, p_r, (z_1)_r, (z_2)_r, \cdots, (z_N)_r), & \text{if } x > x_0; \end{cases} \end{aligned} \tag{6}$$

where x_0 is a given point and $\sum_{m=1}^{N}(z_m)_l = \sum_{m=1}^{N}(z_m)_r = 1$. In addition the initial mass fraction should satisfy the following condition:

$$\sum_{m=1}^{N} \frac{\mu_{mn}(z_m)_l}{W_m} = \sum_{m=1}^{N} \frac{\mu_{mn}(z_m)_r}{W_m}, \qquad 1 \le n \le L; \tag{7}$$

where L, μ_{mn} correspond to the number of different kinds of atoms in all species and the number of the n-th atom in a molecule of the m-th species respectively. The above equalities guarantee the conservation of the number of atoms in the reactions. We also assume that after the reaction, at least one species disappears (fully reacted). Thus there exists an j such that $(z_j)_l = 0$ for each reaction. This implies that

$$(w_m)_l = 0, \qquad \text{for all } m = 1, 2, \cdots, N. \tag{8}$$

This assumption clearly holds outside the reaction zone. Since an underresolved method ignores the reaction zone of width $O(\frac{1}{B_k})$, each species has only two physical states, either $(z_m)_l$ or $(z_m)_r$. Without loss of generality the data in (6) are chosen such that the discontinuity, initially at $x = x_0$, moves to the right. The case when the discontinuity moves to the left can be treated similarly.

The random projection method is a fractional step method

$$S_1(k): \qquad U^{n+1} = S_p(k)S_F(k)U^n \tag{9}$$

that consists of solving the homogeneous convection

$$U_t + F(U)_x = 0, \tag{10}$$

by a standard shock capturing method, denoted by $S_F(k)$, for one time step, followed by a random projection step for the stiff chemical reaction terms

$$\rho_t = 0, \quad (\rho u)_t = 0, \quad e_t = 0, \quad (\rho z_1)_t = w_1, \quad (\rho z_2)_t = w_2, \quad \cdots, \quad (\rho z_N)_t = w_N. \tag{11}$$

Under condition (8), the above ODE collapses to a simple projection

$$(z_m)_j^{n+1} = \begin{cases} (z_m)_l, & \text{if } T_j^* \le T_k, \\ (z_m)_r, & \text{if } T_j^* < T_k, \end{cases} \qquad \text{for all} \quad j; \tag{12}$$

with $U^* = S_F(k)U^n$, $T_j^* = p_j^*/\rho_j^*$ being the values after the convection step. However, on the discrete spatial domain, due to the grid effect, this projection (referred to as the deterministic projection) yields incorrect speed, since the smeared value of the temperature T^* in the detonation layer, once above T_k, will trigger the k-th chemical reaction too early. The key idea in the random projection method is to use a randomized ignition temperature in a suitable domain. This strategy could be successful since the speed of the front does not depend on the specific value of T_k, as far as it is in the range between the equilibrium states T_l and T_r on both sides of the detonation. We now describe the details of this random projection step.

Let

$$\theta_n = (T_l - T_r)\vartheta_n + T_r, \qquad T_l = \frac{p_l}{\rho_l}, \qquad T_r = \frac{p_r}{\rho_r}; \tag{13}$$

where ϑ_n is the van der Corput's sequence given in (8). Thus θ_n is an equidistributed sequence over the domain $[T_l, T_r]$. Assume $x_0 = l(0)h$ is a grid point. Since the projection always makes z_m either $(z_m)_l$ or $(z_m)_r$ for all $1 \leq m \leq N$, therefore, at any time step t_n, there is an $l(n) = j_0$, j_0 an integer, such that

$$(z_m)_j^n = \begin{cases} (z_m)_l, & \text{if } j \leq l(n), \\ (z_m)_r, & \text{if } j > l(n), \end{cases} \qquad 1 \leq m \leq N. \tag{14}$$

Here $l(n)$ is the location of the jump for z_m^n. The random projection will be performed around $x = l(n)h$, a procedure referred to as the **local random projection** in [1].

Let d be the estimated number of smeared points in the detonation layer. Modern shock capturing methods always introduce few smeared points across the discontinuity. The information on d can easily obtained from numerical experience (normally between 1 and 5). We now move the jump of z_m according to the following algorithm:

$$\begin{aligned} S_p(k): \qquad & \rho^{n+1} = \rho^*, \qquad m^{n+1} = m^*, \qquad e^{n+1} = e^*; \\ & \text{set } l(n+1) := l(n) - 1; \\ & \text{For } l = l(n) - 1, \cdots, l(n) + d, \ \text{ do} \\ & \quad l(n+1) := l, \quad \text{if } T_l^* > \theta_n; \\ & (z_m)_j^{n+1} = \begin{cases} (z_m)_l, & \text{if } j \leq l(n+1), \\ (z_m)_r, & \text{if } j > l(n+1), \end{cases} \qquad \text{for all } m, j. \end{aligned} \tag{15}$$

In the above algorithm, only $d + 2$ points will be scanned. The stability condition for the algorithm (9) is the usual CFL condition determined from the convection step $S_F(k)$.

4. A numerical example

In order to verify the performance of the random projection method for stiff multi-species detonation computations, we present here a numerical experiment in one dimension. In our computation, the operator $S_F(k)$ is the second order relaxed scheme [14], which is a TVD scheme without the usage of Riemann solvers or local characteristic decompositions. We choose $d = 5$ in (9) and use the Heaviside kinetics in (3) in our computations in this section.

We consider a reacting model

$$A + B \rightarrow 2\,C, \qquad 2\,C + A \rightarrow 2\,D;$$

and the species E appears as a catalyst. A prototype of this model is

$$H_2 + O_2 \rightarrow 2\,OH, \qquad 2\,OH + H_2 \rightarrow 2\,H_2O;$$

and N_2 appears as a catalyst. In this problem, there are five species and two reactions. The parameter are: $M = 2$, $N = 5$, $\gamma = 1.4$, $T_1 = 2.0$, $T_2 = 10.0$, $B_1 = 10^5$, $B_2 = 2 \times 10^4$, $\alpha_1 = \alpha_2 = 0$, $q_1 = 0.0$, $q_2 = 0.0$, $q_3 = -20.0$, $q_4 = -100.0$,

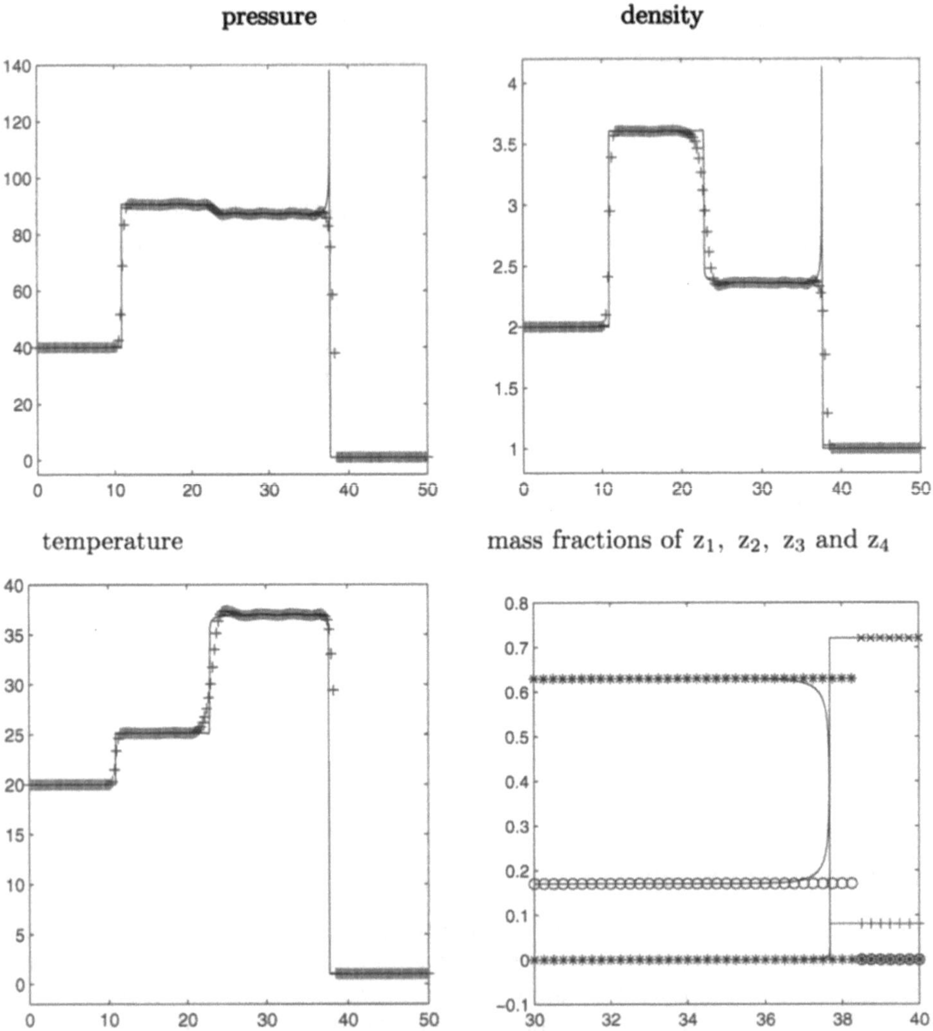

FIGURE 1. Numerical solutions at $t = 3.0$ calculated by the random projection method (9) with $h = 0.25$, $k = 0.01$. $-$: 'exact' solutions; ++: computed solutions. In the last graph, ++: z_1; xx: z_2; oo: z_3 and **: z_4.

$q_5 = 0.0$, $W_1 = 2$, $W_2 = 32$, $W_3 = 17$, $W_4 = 18$, $W_5 = 28$, $\nu'_{1,1} = 1$, $\nu'_{2,1} = 1$, $\nu'_{3,1} = 0$, $\nu'_{4,1} = 0$, $\nu'_{5,1} = 0$, $\nu'_{1,2} = 1$, $\nu'_{2,2} = 0$, $\nu'_{3,2} = 2$, $\nu'_{4,2} = 0$, $\nu'_{5,2} = 0$, $\nu''_{1,1} = 0$, $\nu''_{2,1} = 0$, $\nu''_{3,1} = 2$, $\nu''_{4,1} = 0$, $\nu''_{5,1} = 0$, $\nu''_{1,2} = 0$, $\nu''_{2,2} = 0$, $\nu''_{3,2} = 0$, $\nu''_{4,2} = 2$, $\nu''_{5,2} = 0$.

The initial data are

$$(\rho, u, p, z_1, z_2, z_3, z_4, z_5)(x, 0) = \begin{cases} (\rho_l, u_l, p_l, (z_1)_l, (z_2)_l, (z_3)_l, (z_4)_l, (z_5)_l), & \text{if } x \leq 2.5, \\ (\rho_r, u_r, p_r, (z_1)_r, (z_2)_r, (z_3)_r, (z_4)_r, (z_5)_r), & \text{if } x > 2.5; \end{cases}$$

where $p_l = 40.0$, $\rho_l = 2.0$, $u_l = 10.0$, $(z_1)_l = 0.08$, $(z_2)_l = 0.72$, $(z_3)_l = 0.0$, $(z_4)_l = 0.0$, $(z_5)_l = 0.2$; and $p_r = 1.0$, $\rho_r = 1.0$, $u_r = 0.0$, $(z_1)_r = 0.0$, $(z_2)_r = 0.0$, $(z_3)_r = 0.17$, $(z_4)_r = 0.63$, $(z_5)_r = 0.2$.

This problem is solved on the interval $[0, 50]$. The exact solution consists of a detonation wave, followed by a rarefaction wave and a shock, all moving to the right. We obtain the "exact" solution by an explicit method (the second-order relaxed scheme for the convection terms followed by an explicit Euler method for the chemical reaction terms) using a fine mesh size $h = 0.0025$ (i.e. 20001 grid points on the interval $[0, 50]$ and a small time step $k = 0.0001$). This is a resolved solution. Figure 1 shows the numerical solutions by the random projection method (9) with $h = 0.25$ (i.e. 201 grid points on the interval $[0, 50]$) and $k = 0.01$ at time $t = 3.0$. All waves are captured numerically with the correct speeds. On the other hand, spurious waves were produced when using the usual deterministic method with the same mesh size and time step.

For more numerical experiment we refer to the readers to [3].

References

[1] W. BAO AND S. JIN, *The random projection method for hyperbolic conservation laws with stiff reaction terms*, J. Comp. Phys. 163, 216-248, 2000.

[2] W. BAO AND S. JIN, *The random projection method for stiff detonation capturing*, SIAM J. Sci. Comp., to appear.

[3] W. BAO AND S. JIN, *The random projection method for stiff multi-species detonation capturing*, preprint.

[4] M. BEN-ARTZI, The generalized Riemann problem for reactive flows, J. Comput. Phys., 81 (1989), pp. 70-101.

[5] A. C. Berkenbosch, E. F. Kaasschieter and R. Klein, Detonation capturing for stiff combustion chemistry, Combust. Theory Modeling 2, 313-348 (1998).

[6] A. BOURLIOUX, A. MAJDA AND C. ROYTBURD, *Theoretical and numerical structure for unstable one-dimensional detonations*, SIAM J. Appl. Math., 51 (1991), pp. 303-343.

[7] A. J. CHORIN, *Random choice methods with applications to reacting gas flow*, J. Comput. Phys., 25 (1977), pp. 253-272.

[8] P. COLELLA, A. MAJDA AND V. ROYTBURD, *Theoretical and numerical structure for reacting shock waves*, SIAM J. Sci. Stat. Comput., 7 (1986), pp. 1059-1080.

[9] B. ENGQUIST AND B. SJOGREEN, *Numerical approximation of hyperbolic conservation laws with stiff terms*, Proc. Third Int'l Conf. on Hyperbolic Problems, 1991, pp. 848-860, Studentlitteratur, Lund.

[10] B. Engquist and B. Sjogreen, *Robust difference approximations of stiff inviscid detonation waves*, UCLA CAM Report 91-03.

[11] J. Glimm, *Solutions in the large for nonlinear hyperbolic systems of equations*, Comm. Pure Appl. Math., 18 (1965), pp. 697-715.

[12] J. M. Hammersley and D. C. Handscomb, *Monte Carlo Methods*, Methuen, London, 1965.

[13] C. Helzel, R.J. LeVeque and G. Warnecke, A modified fractional step method for the accurate approximation of detonation waves, SIAM J. Sci. Comp., 22 (2000), pp. 1489-1510.

[14] S. Jin and Z. P. Xin, *The relaxation schemes for systems of conservation laws in arbitrary space dimensions*, Comm. Pure Appl. Math., 48 (1995), pp. 235-276.

[15] R. J. LeVeque and H. C. Yee, *A study of numerical methods for hyperbolic conservation laws with stiff source terms*, J. Comput. Phys., 86 (1990), pp. 187-210.

[16] R. B. Pember, *Numerical methods for hyperbolic conservation laws with stiff relaxation. I. spurious solutions*, SIAM J. Appl. Math., 53 (1993), pp. 1293-1330.

[17] F. A. Williams, *Combustion Theory*, Addison-Wesley, Reading, MA, 1985.

W. Bao
Department of Computational Science,
National University of Singapore
bao@cz3.nus.edu.sg

S. Jin
Department of Mathematics,
University of Wisconsin,
Madison, WI 53706
jin@math.wisc.edu

International Series of Numerical Mathematics
Vol. 140, © 2001 Birkhäuser Verlag Basel/Switzerland

On the Stability of Large Amplitude Semi-discrete Shock Profiles by Means of an Evans Function in Infinite Dimensions

Sylvie Benzoni-Gavage

Abstract. The linear stability of semi-discrete (i.e. continuous in time and discrete in space) shock profiles is investigated. It is encoded in the spectrum of a retarded differential operator. In order to obtain necessary stability conditions, an Evans function based on the (infinite dimensional) eigenvalue equations and their adjoints is constructed. The method is then applied to the cases of Lax shocks and of undercompressive shocks.

1. Introduction

Sophisticated numerical schemes have been developed for hyperbolic systems of conservation laws, which often provide excellent results. However, their analysis (stability, convergence) is mostly limited to scalar conservation laws (apart from Glimm-type schemes). Regarding their qualitative properties, we often argue from analogy with viscous conservation laws. But very few results are really demonstrated (in the case of systems). In particular, the issue of discrete shocks is widely open, despite great efforts in the last decades. On one hand, discrete shocks of rational speed have been well understood since the work of Majda and Ralston [17]. On the other hand, the existence and stability of discrete shocks of irrational Diophantian speed have been shown by Liu and Yu [14, 15], for the (modified) Lax-Friedrichs scheme. But such restrictions on the discrete shock speed seem irrelevant to practical applications. We propose an alternative approach to tackle the problem. It is based on the observation that properties of a conservative scheme are mostly encoded in its flux, which corresponds to the discretization in space of the exact flux. The discretization in time is usually of Euler type, and should not lead to discriminate one scheme among others. This motivates the analysis of semi-discrete schemes, which seems more accessible than that of fully-discrete schemes. This approach is encouraged by the recent work of Chow, Mallet-Paret and Shen [7] on **Lattice Dynamical Systems**. They have shown that sufficiently stable semi-discrete traveling waves give rise to fully-discrete traveling waves, no matter the nature of the discrete speed (Theorem B in [7]). Translating to our context, the existence of spectrally stable semi-discrete shocks would show the existence of discrete shocks of arbitrary speed, for small enough Courant numbers.

Here we shall focus on the very simple "upwind" scheme, which applies to systems with characteristic speeds of one sign only. The reason for this (restrictive) choice will appear below. Semi-discrete upwind profiles can be viewed as heteroclinic orbits of a system of **Retarded Differential Equations** [11, 8]. We proved their existence in an earlier work [2], for Lax shocks of small strength. For the existence of semi-discrete profiles associated with more general schemes, which involve differential equations with both delay and advance, we refer to the recent work of Huot [5]. Assuming the existence of a profile, we are concerned here by its stability properties. Let us note that, by the translation invariance, one cannot expect asymptotic stability but only orbital stability (that is, with an asymptotic shift). This also holds true for the more classical viscous profiles, associated with a parabolic regularization of the conservation laws.

We concentrate on the **linearized stability** analysis, for shocks of arbitrary amplitude. The idea is to construct an **Evans function** and to proceed as Gardner and Zumbrun did for viscous profiles [10]. This function, introduced by Evans in the 70's [9], was developed and used by many authors in several contexts (see [1], [18], [12] and the references therein). It is an analytic function, usually denoted by D, that plays the role of the characteristic polynomial for infinite dimensional operators. It cannot be computed explicitly in general. However, in favorable cases, its low/high-frequency behaviors can be related to each other, in a way that determines the parity of the number of unstable modes. In case of translational invariance we always have $D(0) = 0$, hence the low-frequency behavior depends on $D'(0)$. We point out that the Evans function, especially its low-frequency behavior, is also a building block for the nonlinear stability analysis. We refer to the important work by Zumbrun and Howard concerning viscous profiles [19]. Furthermore, the requirement $D'(0) \neq 0$ is related to one of the basic assumptions made by Chow, Mallet-Paret and Shen [7], – the relationship between their approach and ours is explained in [4], Section 4.

Section 2 is devoted to the construction of an Evans function for semi-discrete profiles. It relies on a dynamical system formulation of the eigenvalue equations for the linearized operator about the wave. This looks like the classical approach, but there is an additional difficulty. In the work by Gardner and Zumbrun, the operator about the wave is a standard second order differential operator, and the dynamical system is just the first order version of a second order ODE. In our case, the operator about the wave is a **retarded differential operator** and yields an **infinite dimensional** dynamical system. However, we shall take advantage of the fact that the **unstable manifold** of this dynamical system is **finite dimensional**. This is where the special choice of the scheme is involved. We hope that a finite dimension reduction can also be performed for more general schemes but we do not see how at the moment. We construct a "mixed type" Evans function by using the adjoint dynamical system (see [6] for a detailed presentation of a similar Evans function in the context of viscous profiles).

In Section 3, we shall derive stability conditions from this Evans function. This will be performed by relating the adjoint dynamical system, which is rather unusual and difficult to handle, to the adjoint eigenvalue equations, which are nothing but **advanced differential equations**. For an extreme Lax shock, the stability condition that we obtain is the same as for viscous profiles associated with the "artificial" (or numerical) viscosity. This is in accordance with the common analogy. We also obtain a stability condition for undercompressive shocks. It is less explicit than for viscous profiles, due to the infinite dimensions. In this case, the analogy between semi-discrete and viscous profiles is not so clear.

This work was announced in [3]. An extended version can be found in [4].

2. The Evans function

2.1. Description of profiles

We consider a hyperbolic system of conservation laws

$$\partial_t u + \partial_x f(u) = 0, \tag{1}$$

where $u(x,t) \in \Omega \subset \mathbb{R}^N$, $f \in \mathcal{C}^2(\Omega; \mathbb{R}^N)$, and the semi-discrete system

$$\partial_t u + \frac{f(u(x,t)) - f(u(x-\Delta x, t))}{\Delta x} = 0, \tag{2}$$

where Δx is a fixed space step. A semi-discrete profile is, by definition, a traveling wave $u(x,t) = U\left(\frac{x-\sigma t}{\Delta x}\right)$, solution of Equation (2), attaining some limits $u_\pm$ at $\pm\infty$. Thus the function U must belong to $\mathcal{C}^1(\mathbb{R}; \Omega)$ and satisfy the retarded differential equation

$$\sigma U'(y) = f(U(y)) - f(U(y-1)). \tag{3}$$

We see by integrating once Equation (3), that the existence of a semi-discrete profile implies the Rankine-Hugoniot condition

$$f(u_+) - f(u_-) = \sigma(u_+ - u_-). \tag{4}$$

>From now on, we fix $(u_-, u_+, \sigma) \in \Omega \times \Omega \times \mathbb{R}$ such that (4) holds. Furthermore, we assume that the **constant states** $u_\pm$ are linearly **stable solutions** of both (1) and (2). This requires the hyperbolicity of the flux f in (1) and positive characteristic speeds. For simplicity we shall assume strict hyperbolicity.

(H): The jacobian matrices $A^\pm := df(u_\pm)$ have distinct positive eigenvalues,

$$0 < a_1^\pm < \cdots < a_N^\pm . \tag{5}$$

Equation (3) is a rather simple retarded differential equation, the occurrence of the delay only being local (or pointwise) (see [11, 8] for more general equations and related theory). As we shall see below, we may easily regard it as a dynamical system, which is nevertheless infinite dimensional. A semi-discrete profile can then be viewed as a heteroclinic connection. Because of the translational invariance, the corresponding fixed points are not hyperbolic. But, using the Rankine-Hugoniot condition (4) and the integrated form,

$$\sigma U(x) - \int_{-1}^{0} f(U(x+\theta))\, d\theta = \sigma u_\pm - f(u_\pm), \tag{6}$$

of Equation (3) we can prove the exponential decay of a profile towards the end-states. This will be important for the low frequency analysis of the Evans function, in particular to show the analogous of the gap lemma of Gardner and Zumbrun [10].

Lemma 2.1. *If U is a semi-discrete profile, that is a global solution of* (3) *such that*

$$\lim_{x\to\pm\infty} U(x) = u_\pm,$$

with $u_\pm$ satisfying (4) *and* (H), *and if there exist p, q, such that*

$$a^-_{p-1} < \sigma < a^-_p\,, \quad a^+_q < \sigma < a^+_{q+1}\,, \tag{7}$$

then there exist $C > 0$, $\alpha > 0$, such that

$$|U(x) - u_-| \leq C e^{\alpha x} \quad \forall x \leq 0\,, \qquad |U(x) - u_+| \leq C e^{-\alpha x} \quad \forall x \geq 0\,. \tag{8}$$

If $u_\pm$ are close enough, the case $p = q$ in (7) is known as the Lax shock case. A center manifold argument shows the existence of semi-discrete profiles in this case (and the exponential decays (8) are consequences of the proof), see [2]. In the following, we shall assume the existence of a profile and the inequalities (7), disregarding the amplitude and the nature of the discontinuity. Thus we shall treat undercompressive shocks (which are typically not of small strength) as well as Lax shocks.

2.2. The linearized operator about a profile

As usual in the analysis of traveling waves, we look at the stability in a moving frame. We also perform a rescaling. Through the change of variables $(x,t) \mapsto (\tilde{x} := (x - \sigma t)/\Delta x, \tilde{t} := t/\Delta x)$, the semi-discretized system (2) becomes

$$\partial_t u - \sigma\, \partial_x u + f(u(x,t)) - f(u(x-1,t)) = 0\,, \tag{9}$$

where we have dropped the tildas for convenience. The profile U is then a stationary solution of (9). Linearizing Equation (9) about U yields

$$\partial_t w = L \cdot w := \sigma\, \partial_x w - A(x) w(x,t) + A(x-1) w(x-1,t)\,, \tag{10}$$

where $A(x) := df(U(x))$. We emphasize that, provided that (5) and (7) hold true, we have $\sigma > 0$. Without loss of generality we may even assume that $\sigma = 1$. The retarded differential operator L is then a closed unbounded operator on $L^2(\mathbb{R})$, with dense domain $H^1(\mathbb{R})$. We are concerned with the spectrum of L, which should lie in the left half-plane in order to ensure stability of the profile U.

2.3. Essential spectrum of L

We recall the definition $\sigma_{\text{ess}}(L) := \mathbb{C}\backslash\{\lambda\,;\ L-\lambda \text{ is Fredholm, } \text{ind}(L-\lambda)=0\}$. Like for classical differential operators, $\sigma_{\text{ess}}(L)$ is linked to the spectra of the asymptotic operators, defined by $(L_\pm \cdot w)(x) := w'(x) - A^\pm(w(x) - w(x-1))$. We easily show by means of a Fourier transform that

$$\sigma(L_\pm) = \{\,\lambda \in \mathbb{C}\,;\ \exists \xi \in \mathbb{R}\,,\ \det(i\xi - A^\pm(1-e^{-i\xi}) - \lambda) = 0\,\}.$$

Furthermore, elementary arguments yield the following.

Lemma 2.2. *Assuming* (H), *for all* $\lambda \in \mathbb{C}$ *such that Re* $\lambda > 0$, *the "characteristic equations"*

$$\det(\mu - A^\pm(1-e^{-\mu}) - \lambda) = 0 \tag{11}$$

do not have imaginary solutions $\mu \in i\mathbb{R}$, *and have exactly* N *solutions of positive real part, counting with multiplicity.*

As a consequence of Lemma 2.2 and the results in [13, 16], we infer

Theorem 2.3. *If* (H) *holds and* $\sigma > 0$, *then for all* $\lambda \in \mathbb{C}$ *such that Re* $\lambda > 0$, $L-\lambda$ *is Fredholm of index 0, that is,* $\lambda \notin \sigma_{ess}(L)$.

2.4. Eigenvalues

We are now concerned with possible unstable eigenvalues of L. For $\lambda \in \mathbb{C}$, λ is an eigenvalue of L if and only if there exists $w \in H^1(\mathbb{R};\mathbb{C}^N)$, $w \not\equiv 0$, such that

$$w'(x) \;=\; (A(x)+\lambda)\,w(x) \;-\; A(x-1)\,w(x-1) \tag{12}$$

for all $x \in \mathbb{R}$. Equation (12) is a (simple) variational retarded differential equation, which can be formulated as a dynamical system. This is the basic idea to construct an Evans function. We consider the Hilbert space

$$H_0 := L^2((-1,0);\mathbb{C}^N) \times \mathbb{C}^N\,,$$

and the dense subspace

$$H_1 := \left\{ \begin{pmatrix} \phi \\ c \end{pmatrix} \in H_0\,;\ \phi \in H^1((-1,0);\mathbb{C}^N) \text{ and } \phi(0) = c \right\}.$$

Associating to a (smooth enough) function w and $x \in \mathbb{R}$ the element of H_1,

$$W(x) := \begin{pmatrix} w_x : \theta \mapsto w(x+\theta) \\ w(x) \end{pmatrix},$$

we obtain the following equivalence.

Proposition 2.4. *Let* $\lambda \in \mathbb{C}$ *be of positive real part. There exists a solution* $w \in H^1(\mathbb{R};\mathbb{C}^N)$, $w \not\equiv 0$, *of (12) if and only if there exists a solution* $W \in C(\mathbb{R};H_1) \cap C^1(\mathbb{R};H_0)$, $W \not\equiv 0$, *of the dynamical system*

$$\frac{dW}{dx} = \mathbb{A}(x;\lambda)\cdot W\,, \tag{13}$$

where

$$\mathbb{A}(x;\lambda): \quad H_1 \to H_0, \qquad \begin{pmatrix} \phi \\ c \end{pmatrix} \mapsto \begin{pmatrix} d\phi/d\theta \\ (A(x)+\lambda)\,c \,-\, A(x-1)\,\phi(-1) \end{pmatrix}, \tag{14}$$

with W tending to zero at $\pm\infty$.

Lemma 2.2 actually shows that both asymptotic operators

$$\mathbb{A}^{\pm}(\lambda): \quad H_1 \to H_0, \qquad \begin{pmatrix} \phi \\ c \end{pmatrix} \mapsto \begin{pmatrix} d\phi/d\theta \\ (A^{\pm}+\lambda)\,c \,-\, A^{\pm}\,\phi(-1) \end{pmatrix} \tag{15}$$

have unstable manifolds of dimension N for $\mathrm{Re}\,\lambda > 0$. This suggests that the "unstable manifold" of (13) is of dimension N and that the "stable manifold" is of co-dimension N. This is the purpose of Theorem 2.5 below, where the "stable manifold" is characterized by means of the adjoint dynamical system.

2.5. Definition of the Evans function

Our basic result concerning the variational system (13) is the following.

Theorem 2.5. *We assume that* (H) *and (7) hold true with $\sigma = 1$. If $\lambda \in \mathbb{C}$ is of positive real part, there exist solutions $W_1(\cdot;\lambda), \cdots, W_N(\cdot;\lambda)$ of (13) and solutions $Z_1(\cdot;\lambda), \cdots, Z_N(\cdot;\lambda)$ of the adjoint dynamical system*

$$\frac{dZ}{dx} = -\mathbb{A}(x;\lambda)^* \cdot Z, \tag{16}$$

such that $W \in C(\mathbb{R};H_1)\cap C^1(\mathbb{R};H_0)$ is a solution of (13) decaying to 0 at $\pm\infty$ if and only if

$$W(x) \in \ \mathit{Span}\ \{W_1(x;\lambda), \cdots, W_N(x;\lambda)\} \ \cap \tag{17}$$

$$\{\Phi \in H_0\,;\ \langle Z_i(x;\lambda), \Phi\rangle = 0 \quad \forall i = 1, \cdots, N\}$$

for some $x \in \mathbb{R}$. Equivalently, $D(\lambda) = 0$, where D is defined by

$$D(\lambda) := \det(\langle Z_i(x;\lambda), W_j(x;\lambda)\rangle)_{1\le i,j\le N}\,. \tag{18}$$

The functions W_j, Z_i, and thus also D, depend analytically on λ on any compact subset of $\{\lambda;\ \mathrm{Re}\,\lambda > 0\}$. Furthermore, they have analytic extensions to a neighborhood of $\lambda = 0$.

Note that the inner products $\langle Z_i(x;\lambda), W_j(x;\lambda)\rangle$ do not depend on x. The proof of this theorem is based on technical constructions, inspired from [10, 19], as well as on the notion of **exponential dichotomy** for solution operators (in infinite dimensions, see [13]). More specifically, the fundamental solutions W_j are constructed in such a way that, for $\mathrm{Re}\,\lambda > 0$, we have

$$\lim_{x\to-\infty} e^{-\mu_j^-(\lambda)x} W_j(x;\lambda) \;=\; \Phi_j^-(\lambda) := \begin{pmatrix} \phi_j(\lambda) : \theta \mapsto e^{\mu_j^-(\lambda)\theta} r_j^- \\ r_j^- \end{pmatrix},$$

where the vector r_j^- denotes as usual the eigenvector of A^- associated with a_j^-. The element $\Phi_j^-(\lambda)$ of H_1 is an eigenvector of $\mathbb{A}^-(\lambda)$ associated with $\mu_j^-(\lambda)$, which is defined as the unique root of $\mu - a_j^- - \lambda + a_j^- e^{-\mu}$ of positive real part. Similarly, the fundamental solutions Z_i are characterized by

$$\lim_{x \to +\infty} e^{\overline{\mu_i^+(\lambda)} x} Z_i(x;\lambda) \;=\; \Psi_i(\lambda)\,,$$

where $\Psi_i(\lambda)$ are respectively eigenvectors of $\mathbb{A}^+(\lambda)^*$ associated with $\overline{\mu_i^+(\lambda)}$, $\mu_i^+(\lambda)$ being the root of positive real part of $\mu - a_i^+ - \lambda + a_i^+ e^{-\mu}$. The important point to note for the extension to $\lambda = 0$ is that, in view of (7), we have in the limit

$$\mu_j^-(0) \;=\; 0 \quad \forall j \leq p-1\,, \quad \mu_i^+(0) \;=\; 0 \quad \forall i \leq q\,.$$

This extension is shown similarly as the gap lemma in [10] (also see [19] for a more "practical" treatment).

3. Stability conditions

Once the Evans function is well defined, the derivation of stability conditions follows the same lines as for viscous profiles [6].

We require that $\operatorname{sgn} D'(0) \operatorname{sgn} D(+\infty) \;\geq\; 0$, otherwise the (real valued) Evans function necessarily vanishes on the real axis. The aim is thus to compute $D'(0)$ and $\operatorname{sgn} D(\lambda)$ for $\lambda \gg 1$. For this we need to properly handle the adjoint dynamical system (16), which is somewhat weird. In particular, we easily check that the domain of $\mathbb{A}(x;\lambda)^*$ depends on x. However, things are restored through the following remark. There is a one-to-one relation between solutions of (16) and solutions of the **adjoint eigenvalue equation** $L^* \cdot z = \bar{\lambda} z$, which equivalently reads

$$z'(x) \;=\; z(x+1)\,A(x) \;-\; z(x)\,(A(x) + \bar{\lambda})\,. \tag{19}$$

The proof of this fact is based on the observation that, if w is a solution of (12) and z is a solution of (19) then the quantity

$$\{z, w\}(x) \;:=\; z(x)\,\overline{w(x)} \;-\; \int_{-1}^{0} z(x+1+\theta)\,A(x+\theta)\,\overline{w(x+\theta)}\,d\theta$$

actually does not depend on x.

3.1. Lax shocks

Theorem 3.1. *Assuming (H), (4), and (7) with $p = q = N$, a necessary condition for a semi-discrete profile U associated with (u_-, u_+, σ) to be stable reads*

$$\ell_N^- \cdot (u_+ \;-\; u_-) \geq 0\,, \tag{20}$$

where ℓ_N^- is a left eigenvector of A^- associated with a_N^- such that $\ell_N^- \cdot U'(x) > 0$ for $-x \gg 1$.

The condition (20) is the same as for a viscous profile associated with the numerical viscosity matrix (see [10, 6]). It is always satisfied by shocks of small strength. On the other hand, if the Majda-Liu determinant, that is $\ell_N^- \cdot (u_+ - u_-)$ vanishes at some N-shock, then generically there are neighboring shocks that are strongly unstable (see [6], Section 5).

3.2. Undercompressive shocks

Theorem 3.2. *We assume (H), (4), and (7) with $p = q + 1 = N$ and $\sigma = 1$, supplemented with the transversality condition*

$$Span\,\{r_1^-, \cdots, r_{N-1}^-, r_N^+\} = \mathbb{R}^N . \tag{21}$$

A necessary condition for a semi-discrete profile U associated with (u_-, u_+, σ) to be stable reads

$$\gamma\,\ell_N^- \cdot r_N^+ \geq 0\,, \tag{22}$$

where the orientation of ℓ_N^- is determined by the condition $\ell_N^- \cdot U'(x) > 0\,, -x \gg 1$, and γ is a "transversality coefficient". It is defined by

$$\gamma := \lim_{x \to +\infty} \{z_N, v\}(x), \tag{23}$$

where z_N is the unique solution (up to a positive constant factor) of

$$z_N'(x) = z_N(x+1)A(x) - z_N(x)A(x)$$

such that $\lim_{x\to+\infty} z_N(x) = 0$ and $z_N(x) \cdot r_N^+ > 0$ for $x \gg 1$, and v satisfies

$$\begin{aligned} v(x) - \int_{-1}^{0} A(x+\theta)v(x+\theta)\,d\theta &= U(x) - u_- - \sum_{j=1}^{N-1} \gamma_j (1 - a_j^-) r_j^- \,, \\ &= U(x) - u_+ + \gamma_N r_N^+ \end{aligned}$$

the coefficients γ_j being uniquely defined because of (21) (and (7)).

References

[1] J.C. Alexander, R. Gardner, and C.K.R.T. Jones. A topological invariant arising in the stability analysis of travelling waves. *J. Reine Angew. Math.*, 410:167–212, 1990.

[2] S. Benzoni-Gavage. Semi-discrete shock profiles for hyperbolic systems of conservation laws. *Physica D Nonlinear Analysis*, 115:109–123, 1998.

[3] S. Benzoni-Gavage. On the stability of semi-discrete shock profiles by means of an Evans function in infinite dimension. *C. R. Acad. Sci. Paris, Série I*, 329:377–382, 1999.

[4] S. Benzoni-Gavage. Stability of semi-discrete shock profiles by means of an Evans function in infinite dimensions. Postscript file available at http://www.umpa.ens-lyon.fr/ benzoni/sd.ps, 2001.

[5] S. Benzoni-Gavage and P. Huot. Existence of semi-discrete shocks. *Discrete Contin. Dynam. Systems*, To appear

[6] S. Benzoni-Gavage, D. Serre, and Zumbrun K. Alternate Evans functions and viscous shock waves. *SIAM Journal on Mathematical Analysis*, 32(5):929–962, 2001.

[7] S.-N. Chow, J. Mallet-Paret, and W. Shen. Traveling waves in lattice dynamical systems. *J. Differential Equations*, 149(2):248–291, 1998.

[8] O. Diekmann, S.A. van Gils, S.M. Verduyn Lunel, and H.-O. Walther. *Delay equations*. Springer-Verlag, New York, 1995.

[9] J.W. Evans. Nerve axon equations. I, II, III, IV. *Indiana Univ. Math. J.*, 21:877–885, 22:75–90, 22:577–593, 24(12):1169–1190,1971/72/73/74/75.

[10] R.A. Gardner and K. Zumbrun. The gap lemma and geometric criteria for instability of viscous shock profiles. *Comm. Pure Appl. Math.*, 51(7):797–855, 1998.

[11] J.K. Hale and S.M.V. Lunel. *Introduction to Functional Differential Equations*. Springer Verlag, 1993.

[12] T. Kapitula. The Evans function and generalized Melnikov integrals. *SIAM J. Math. Anal.*, 30(2):273–297 (electronic), 1999.

[13] X.-B. Lin. Exponential dichotomies and homoclinic orbits in functional-differential equations. *J. Differential Equations*, 63(2):227–254, 1986.

[14] T.-P. Liu and S.-H. Yu. Continuum shock profiles for discrete conservation laws, I. Construction. *Comm. Pure Appl. Math.*, 52(1):85–127, 1999.

[15] T.-P. Liu and S.-H. Yu. Continuum shock profiles for discrete conservation laws. II. Stability. *Comm. Pure Appl. Math.*, 52(9):1047–1073, 1999.

[16] J. Mallet-Paret. The Fredholm alternative for functional differential equations of mixed type. *J. Dynam. Differential Equations*, 11(1):1–47, 1999.

[17] A. Majda and J. Ralston. Discrete shock profiles for systems of conservation laws. *Comm. Pure Appl. Math.*, 32(4):445–482, 1979.

[18] R.L. Pego and M.I. Weinstein. Eigenvalues, and instabilities of solitary waves. *Philos. Trans. Roy. Soc. London Ser. A*, 340(1656):47–94, 1992.

[19] K. Zumbrun and P. Howard. Pointwise semigroup methods and stability of viscous shock waves. *Indiana Univ. Math. J.*, 47(3):741–871, 1998.

S. Benzoni-Gavage
UMPA,
École Normale Supérieure de Lyon,
46, allée d'Italie,
F- 69364 Lyon cedex 07, France
`benzoni@umpa.ens-lyon.fr`

International Series of Numerical Mathematics
Vol. 140, © 2001 Birkhäuser Verlag Basel/Switzerland

Viscosity Solutions for Hyperbolic Systems where Shock Curves are Straight Lines

Stefano Bianchini and Alberto Bressan

1. Introduction

Consider a strictly hyperbolic $n \times n$ system of conservation laws in one space dimension:

$$u_t + f(u)_x = 0. \tag{1}$$

Assuming that the initial data has small total variation, the global existence of weak solutions was proved by Glimm [9], while the uniqueness and stability of entropy admissible BV solutions was recently established in a series of papers [2, 4, 5, 6, 7, 13]. See also [3] for a comprehensive presentation of these results. A long standing open question is whether these discontinuous solutions can be obtained as vanishing viscosity limits. More precisely, given a smooth initial data $\bar{u} : \mathbb{R} \mapsto \mathbb{R}^n$ with small total variation, consider the parabolic Cauchy problem

$$u_t + A(u)u_x = \varepsilon\, u_{xx}, \tag{2}$$

$$u(0,x) = \bar{u}(x). \tag{3}$$

Here $A(u) \doteq Df(u)$ denotes the Jacobian matrix of f and $\varepsilon > 0$. In the light of the uniform stability result [4, 7] valid for solutions of (1), one conjectures that the solutions of the viscous system (2) also satisfy global BV bounds and are uniformly stable. Moreover, we expect that, as $\varepsilon \to 0$, the solutions u^ε of (2)-(3) converge to the unique entropy weak solution u of (1)-(3). In the case of a scalar conservation law, all these results were established in the classical paper by Kruzhkov [11]. On the other hand, for $n \times n$ systems, no general theorem in this direction is yet available.

For various 2×2 systems, relying on the method of compensated compactness [8, 14], one can show that some sequence of solutions u^ε of (2) converges weakly to a weak solution of (1). However, the uniqueness, stability and qualitative properties of these limit solutions are still not known.

For general $n \times n$ systems, the stability of some particular solutions has been analysed in detail. In particular, this refers to viscous travelling waves [13], viscous rarefactions [16] or solutions of the Riemann problem with viscosity [17]. In addition, the convergence as $\varepsilon \to 0$ has been established in the special case where the limit solution u of (1) is piecewise smooth [10].

In the past two years, we also became interested in the stability and convergence of viscous approximations. While the eventual goal is always to understand what happens for general BV solutions of $n \times n$ systems, our strategy is in a sense opposite to the one followed in previous literature. Indeed, the papers [10, 13, 16, 17, 10] deal with general hyperbolic systems, but only consider very particular solutions. On the other hand, we plan to initially study very special hyperbolic systems, and later extend the results to broader classes of systems. At every step, however, we want to establish the uniform stability and convergence for <u>all</u> solutions with small BV data.

A first result in this direction, presented in this note, is concerned with hyperbolic systems of the form (2), not necessarily in conservation form, where all characteristic curves are straight lines. More precisely, we assume that each matrix $A(u)$ has n real distinct eigenvalues, say $\lambda_1(u) < \cdots < \lambda_n(u)$. In this case, one can find left and right eigenvectors $l^1, \ldots, l^n$, $r_1, \ldots, r_n$ normalized so that

$$\big|r_i(u)\big| \equiv 1, \qquad l^i(u) \cdot r_j(u) = \begin{cases} 1 & \text{if} \quad i = j, \\ 0 & \text{if} \quad i \neq j. \end{cases} \tag{4}$$

Our key assumption is that all integral curves of the right eigenvectors are straight lines, i.e.:

$$(Dr_i)\, r_i \equiv 0 \qquad i = 1, \ldots, n. \tag{5}$$

Performing the rescalings $t \mapsto t/\varepsilon$, $x \mapsto x/\varepsilon$, the Cauchy problem (2)-(3) becomes

$$u_t + A(u)u_x = u_{xx}, \tag{6}$$

$$u(0,x) = \bar u(\varepsilon x). \tag{7}$$

Observe that, as $\varepsilon \to 0$, the initial data $u(0,\cdot)$ has constant total variation, all of its derivatives approach zero, but its $\mathbf{L}^1$-norm approaches infinity. We thus need estimates on the total variation of a solution $u(t,\cdot)$ of (6) which are independent of the $\mathbf{L}^1$-norm of the initial data. Our first result establishes uniform BV bounds and stability for solutions of (6).

Theorem 1.1. *Let the system (6) have smooth coefficients, defined for u in an open set $\Omega \subseteq I\!R^n$ containing the origin. Assume that all matrices $A(u)$ are strictly hyperbolic and that their normalized eigenvectors r_i satisfy the straight-line assumption (5). Then, for every $\delta_1 > 0$ there exist constants $L, L', \delta_0 > 0$ so that the following holds. For every initial data $\bar u$ such that*

$$\text{Tot.Var.}\{\bar u\} \le \delta_0, \qquad \|\bar u\|_{\mathbf{L}^\infty} \le \delta_0, \tag{8}$$

the Cauchy problem (6)-(3) has a unique, global solution u, which satisfies

$$\text{Tot.Var.}\big\{u(t,\cdot)\big\} \le \delta_1 \qquad \forall t \ge 0. \tag{9}$$

Moreover, for every two initial data $\bar u, \bar w$ satisfying (8), the corresponding solutions satisfy

$$\big\|u(t) - w(s)\big\|_{\mathbf{L}^1} \le L'\big(|t-s|^{1/2} + |t-s|\big) + L\,\big\|\bar u - \bar w\big\|_{\mathbf{L}^1} \qquad \forall t, s \ge 0. \tag{10}$$

We remark that the bound (9) depends only on the total variation of the initial data $\bar{u}$, not on its $\mathbf{L}^1$-norm, which may well be infinite. The above result thus yields an a priori bound on the total variation of solutions of (2)-(3), independent of the parameter ε. Our second result shows that, as $\varepsilon \to 0$, these solutions u^ε converge to a unique limit u, which provides the unique entropy weak solution to the Cauchy problem (1)-(3). We recall that the solutions $t \mapsto u(t,\cdot) = S_t\bar{u}$ of (1) determine a Lipschitz continuous semigroup, defined on a closed domain $\mathcal{D}$ of functions with small total variation [7].

Theorem 1.2. *Assume that all matrices $A(u)$ in (2) are strictly hyperbolic and that their eigenvectors satisfy (5). Then there exist constants $L, L', \delta_0 > 0$, a closed domain $\mathcal{D} \subset \mathbf{L}^1_{\mathrm{loc}}$ and a continuous semigroup $S : \mathcal{D}\times[0,\infty[\mapsto \mathcal{D}$ with the following properties.*

(i) *Every function $\bar{u}$ satisfying (8) lies in the domain $\mathcal{D}$ of the semigroup.*
(ii) *For every $\bar{u}, \bar{w} \in \mathcal{D}$ with $\bar{u} - \bar{w} \in \mathbf{L}^1$ and every $t, s \geq 0$ one has*

$$\|S_t\bar{u} - S_s\bar{w}\|_{\mathbf{L}^1} \leq L'|t-s| + L\|\bar{u} - \bar{w}\|_{\mathbf{L}^1} . \tag{11}$$

(iii) *For every $\bar{u} \in \mathcal{D}$, the trajectory $u(t,\cdot) = S_t\bar{u}$ is the unique limit in $\mathbf{L}^1_{\mathrm{loc}}$ of the corresponding solutions $u^\varepsilon(t,\cdot)$ of the viscous Cauchy problem (2)-(3), as $\varepsilon \to 0+$. In the conservative case where $A(u) = Df(u)$, this trajectory coincides with the unique entropy weak solution of (1)-(3).*

An outline of the proof of the above theorems will be given in the following sections. For all details we refer to [1].

2. The BV bounds

The main step is the proof of uniform BV bounds for a solution of (6). In the following, the directional derivative of a function $\phi = \phi(u)$ in the direction of the eigenvector r_i is written

$$r_i \bullet \phi(u) \doteq \lim_{h\to 0} \frac{\phi(u + hr_i(u)) - \phi(u)}{h}.$$

Moreover, by $u_x^i \doteq l^i(u) \cdot u_x$ we denote the i-th component of the gradient u_x w.r.t. the basis of right eigenvectors $\{r_1, \ldots, r_n\}$. Recalling (4), this implies

$$u_x = \sum_i u_x^i r_i. \tag{12}$$

With the above notations, the system (6) takes the form

$$u_t + \sum_i \lambda_i u_x^i r_i = \sum_i (u_x^i r_i)_x = \sum_i (u_x^i)_x r_i + \sum_{i,j} u_x^i u_x^j (r_j \bullet r_i). \tag{13}$$

Differentiating (12) w.r.t. t and (13) w.r.t. x and equating the results one obtains

$$\begin{aligned}\sum_i (u_x^i)_t r_i + \sum_i (\lambda_i u_x^i)_x r_i + \sum_{j\neq k} \lambda_j u_x^j u_x^k [r_k, r_j] \\ = \sum_i (u_x^i)_{xx} r_i + 2\sum_{i,j} (u_x^i)_x u_x^j (r_j \bullet r_i) \\ + \sum_{i,j,k} u_x^i u_x^j u_x^k \Big(r_k \bullet (r_j \bullet r_i) - (r_j \bullet r_k) \bullet r_i \Big).\end{aligned} \tag{14}$$

Here $[r_k, r_j] \doteq r_k \bullet r_j - r_j \bullet r_k$ is the usual Lie bracket. Taking the inner product of (14) with $l^i(u)$ one obtains

$$\begin{aligned}(u_x^i)_t + (\lambda_i u_x^i)_x - (u_x^i)_{xx} = & \sum_{j\neq k} G_{j,k}^i(u) u_x^j u_x^k + \sum_{j,k} H_{j,k}^i(u)(u_x^j)_x u_x^k \\ & + \sum_{j,k,\ell} K_{j,k,\ell}^i(u) u_x^j u_x^k u_x^\ell,\end{aligned} \tag{15}$$

where the functions G, H, K are easily recovered from (14). Since u is smooth, a BV bound on u is clearly equivalent to an estimate on the $\mathbf{L}^1$ norm of its gradient components u_x^i. Calling $v^i \doteq u_x^i$ these components, we thus seek an $\mathbf{L}^1$ estimate for solutions of

$$\begin{aligned}v_t^i + \big(\lambda_i(u) v^i\big)_x - v_{xx}^i \\ = \sum_{j\neq k} G_{j,k}^i(u) v^j v^k + \sum_{j,k} H_{j,k}^i(u) v_x^j v^k + \sum_{j,k,\ell} K_{j,k,\ell}^i(u) v^j v^k v^\ell \\ \doteq \phi^i(u, v^1, \ldots, v^n).\end{aligned} \tag{16}$$

We regard (16) as a parabolic system of n scalar equations, coupled through the terms G, H, K. These coupling terms can be split in two groups:

- *Transversal terms* involving at least two distinct components, such as $v^j v^k$, $v_x^j v^k$, $v^j v^k v^\ell$ with $j \neq k$,
- *Non-transversal terms* involving one single component, such as $v^j v_x^j$ and $v^j v^j v^j$.

Observe that, by (14), our key assumption (5) implies that all coefficients of non-transversal terms vanish identically. In this case, we can show that the coupling terms ϕ^i in (16) are integrable over the t-x plane. Since the left hand side in (16) is in conservation form, this will provide the desired $\mathbf{L}^1$ bounds on $u_x(t,\cdot)$, uniformly for all $t \geq 0$. Details of the proof are long and technical. We give here only a couple of lemmas which constitute the key building blocks. The first lemma provides an estimate on the total interaction between solutions of two independent linear diffusion equations having strictly different drifts.

Lemma 2.1. *Consider the two diffusion equations:*

$$\begin{cases} z_t + \big[\lambda(t,x) z\big]_x - z_{xx} &= 0, \\ z_t^* + \big[\lambda^*(t,x) z^*\big]_x - z_{xx}^* &= 0. \end{cases} \tag{17}$$

Assume that there exists $c > 0$ such that

$$\inf_{t,x} \lambda^*(t,x) - \sup_{t,x} \lambda(t,x) \geq c.$$

Then, for every initial data $\bar z$, $\bar z^* \in \mathbf{L}^1$, the corresponding solutions z, z^* of (17) satisfy

$$\int_0^\infty \int_{I\!R} \big|z(t,x)\big| \cdot \big|z^*(t,x)\big| \, dxdt \leq \frac{1}{c} \int_{I\!R} \big|\bar z(x)\big| \, dx \cdot \int_{I\!R} \big|\bar z^*(x)\big| \, dx. \tag{18}$$

Proof. Define the kernel

$$P(y) \doteq \begin{cases} e^{-cy/2} & \text{if} \quad y > 0, \\ 1 & \text{if} \quad y \leq 0, \end{cases} \tag{19}$$

observing that its derivatives satisfy

$$2P'' - cP' = -c\delta_0, \tag{20}$$

where δ_0 denotes the unit Dirac mass at the origin. Now consider the interaction functional

$$Q(z, z^*) \doteq \iint_{\mathbb{R}^2} P(x_2 - x_1)|z(x_1)|\,|z^*(x_2)|\,dx_1dx_2. \tag{21}$$

For any couple of solutions z, z^* of (17), set $Q(t) \doteq Q\big(z(t),\, z^*(t)\big)$. Observing that

$$\begin{cases} |z|_t + \big(\lambda(t,x)|z|\big)_x - |z|_{xx} & \leq 0, \\ |z^*|_t + \big(\lambda^*(t,x)|z^*|\big)_x - |z^*|_{xx} & \leq 0, \end{cases}$$

by direct computations we obtain

$$\begin{aligned}
&\tfrac{d}{dt}Q(t) \\
&\quad = \textstyle\iint_{\mathbb{R}^2} P(x_2 - x_1)\Big\{\big|z(t,x_1)\big|_t\,\big|z^*(t,x_2)\big| + \big|z(t,x_1)\big|\,\big|z^*(t,x_2)\big|_t\Big\}\,dx_1dx_2 \\
&\quad \leq \textstyle\iint_{\mathbb{R}^2} P(x_2 - x_1)\Big\{\Big(\big|z(t,x_1)\big|_{x_1x_1} - \big(\lambda(t,x_1)|z(t,x_1)|\big)_{x_1}\Big)|z^*(t,x_2)| \\
&\qquad\qquad + \big|z(t,x_1)\big|\,\Big(\big|z^*(t,x_2)\big|_{x_2x_2} - \big(\lambda^*(t,x_2)|z^*(t,x_2)|\big)_{x_2}\Big)\Big\}\,dx_1dx_2 \\
&\quad \leq \textstyle\iint_{\mathbb{R}^2}\Big(2P''(x_2 - x_1) + \big(\lambda(t,x_1) - \lambda^*(t,x_2)\big)P'(x_2 - x_2)\Big) \\
&\qquad\qquad \big|z(t,x_1)\big|\,\big|z^*(t,x_2)\big|\,dx_1dx_2 \\
&\quad \leq \textstyle\iint_{\mathbb{R}^2}\Big(2P''(x_2 - x_1) - cP'(x_2 - x_1)\Big)\big|z(t,x_1)\big|\,\big|z^*(t,x_2)\big|\,dx_1dx_2.
\end{aligned} \tag{22}$$

From (20) and (22) it follows

$$\frac{d}{dt}Q(t) \leq -c\int_{\mathbb{R}} \big|z(t,x)\big|\,\big|z^*(t,x)\big|\,dx. \tag{23}$$

We thus conclude

$$\int_0^\infty \int_{\mathbb{R}} \big|z(t,x)\big|\,\big|z^*(t,x)\big|\,dxdt \leq \frac{Q(0)}{c} \leq \frac{1}{c}\int_{\mathbb{R}} \big|\bar{z}(x)\big|\,dx \cdot \int_{\mathbb{R}} \big|\bar{z}^*(x)\big|\,dx, \tag{24}$$

proving the lemma. □

Remark 1. Consider the corresponding system without viscosity:

$$\begin{cases} z_t + \big[\lambda(t,x)z\big]_x & = 0, \\ z_t^* + \big[\lambda^*(t,x)z^*\big]_x & = 0. \end{cases} \tag{25}$$

In this case, the estimates (23), (24) and (18) remain valid, provided that we define

$$Q(z, z*) \doteq \iint_{x_1 > x_2} \big|z(t,x_1)\big|\,\big|z^*(t,x_2)\big|dx_1dx_2.$$

This corresponds to the kernel

$$P(y) \doteq \begin{cases} 0 & \text{if} \quad y > 0, \\ 1 & \text{if} \quad y \leq 0. \end{cases}$$

The proof of Lemma 1 thus indicates that some control of the interaction between viscous waves of different families can be achieved by means of a Glimm-type interaction functional, with the suitably modified kernel (19).

Using Lemma 1, one can show that all terms of the form $v^i v^j$ $(i \neq j)$ in (16) have a bounded integral over the t-x plane. Thanks to the $\mathbf{L}^\infty$ bound on all gradient components, the same arguments apply to the transversal terms $v^i v^j v^k$. The estimation of terms of the form $v^i v^j_x$ requires more care. Indeed, their integral cannot be controlled directly by an interaction potential as in (21). For a detailed proof of this estimate we refer to [1].

In the following, we shall prove a much simpler estimate, which however captures some essential ingredients of the actual proof. Consider the scalar, viscous conservation law

$$u_t + f(u)_x = u_{xx}, \qquad u(0,x) = \bar{u}(x), \tag{26}$$

where $\bar{u} \in \mathbf{L}^1$ is a smooth function with bounded variation. Call $\lambda(u) \doteq f'(u)$ the wave speed, and introduce the variable $v \doteq u_x$. Differentiating (26) we obtain

$$v_t + \big[\lambda(u)v\big]_x = v_{xx} \qquad v(0) = \bar{u}_x, \tag{27}$$

We seek an estimate on the total variation of $u_x(t,\cdot)$, uniform in time. We thus define

$$E(t) \doteq \int \big|u_{xx}(t,x)\big|\, dx = \big\|v_x(t)\big\|_{\mathbf{L}^1}. \tag{28}$$

By possibly performing the change of coordinates $y = x - t\cdot\lambda(0)$, it is not restrictive to assume that $\lambda(0) = 0$. Calling

$$\psi(t,x) \doteq -\lambda\big(u(t,x)\big)\cdot v(t,x),$$

the solution of (27) can be written in terms of a convolution with the standard Gauss kernel G. Namely, for every $t \geq h \geq 0$ we have

$$v(t) = G(h) * v(t-h) + \int_{t-h}^{t} G(t-s) * \psi_x(s)\, ds. \tag{29}$$

Therefore,

$$\begin{aligned} \big\|v_x(t)\big\|_{\mathbf{L}^1} &\leq \big\|G_x(h) * v(t-h)\big\|_{\mathbf{L}^1} + \textstyle\int_{t-h}^{t} \big\|G_x(t-s) * \psi_x(s)\big\|_{\mathbf{L}^1}\, ds \\ &\leq \big\|G_x(h)\big\|_{\mathbf{L}^1}\cdot\big\|v(t-h)\big\|_{\mathbf{L}^1} + \textstyle\int_{t-h}^{t} \big\|G_x(t-s)\big\|_{\mathbf{L}^1}\cdot\big\|\psi_x(s)\big\|_{\mathbf{L}^1}\, ds. \end{aligned} \tag{30}$$

Using the estimates

$$\big\|v(t)\big\|_{\mathbf{L}^1} \leq \big\|\bar{v}\big\|_{\mathbf{L}^1}, \qquad \big\|v(t,\cdot)\big\|_{\mathbf{L}^\infty} = \mathcal{O}(1)\cdot E(t),$$

$$\big\|\lambda(u(t,\cdot))\big\|_{\mathbf{L}^\infty} = \mathcal{O}(1)\cdot\big\|u(t)\big\|_{\mathbf{L}^\infty} = \mathcal{O}(1)\cdot\big\|\bar{v}\big\|_{\mathbf{L}^1},$$

$$\big\|\psi_x(t)\big\|_{\mathbf{L}^1} \leq \big\|\lambda'(u)v^2\big\|_{\mathbf{L}^1} + \big\|\lambda(u)v_x\big\|_{\mathbf{L}^1} = \mathcal{O}(1)\cdot\big\|\bar{v}\big\|_{\mathbf{L}^1}\cdot E(t),$$

from (30) we obtain

$$E(t) \le \frac{C}{\sqrt{h}} \cdot \|\bar{v}\|_{\mathbf{L}^1} + \int_{t-h}^{t} \frac{C}{2\sqrt{t-s}} \cdot \|\bar{v}\|_{\mathbf{L}^1} \cdot E(s)\,ds \tag{31}$$

for a suitable constant C. Since (31) is valid for every h, we now choose $h \doteq (2C\|\bar{v}\|)^{-2}$ and look at the equation

$$E^* = \frac{C\|\bar{v}\|}{\sqrt{h}} + C\sqrt{h}\|\bar{v}\|E^*. \tag{32}$$

With the above choice of h, an elementary computation yields $E^* = (2C\|\bar{v}\|_{\mathbf{L}^1})^2$. By a comparison argument, if the derivatives of the initial data are small enough, from (31)-(32) we deduce the desired bound

$$E(t) \le E^* \qquad \text{for all } t \ge 0. \tag{33}$$

The arguments used to obtain (33) can be applied to the system (15), yielding the estimates

$$\|v_x^j(t)\|_{\mathbf{L}^1} = \mathcal{O}(1) \qquad \text{for all } t \ge 0.$$

This bound on the total mass of $|v_x^j|$ must now be supplemented with an analysis of where this mass is located. By showing that v_x^j travels approximately with the same speed λ_j as v^j, and recalling that different characteristic speeds are strictly separated, we eventually obtain the desired bound on the integral of $v^i v_x^j$ over the t-x plane. Finally, an iteration argument based on Duhamel's principle will show the integrability of the coupling terms ϕ^i in (16).

3. Stability and convergence

To prove the Lipschitz continuous dependence w.r.t. the initial data, we first study the linearized equation describing the evolution of an infinitesimal perturbation. Replacing u by $u+\epsilon h$ in (6), letting $\epsilon \to 0$ and retaining terms of order ϵ we obtain

$$h_t + [DA(u)\cdot h]u_x + A(u)h_x = h_{xx}, \tag{34}$$

If the total variation of the reference solution u remains small, we show that every solution $h = h(t,x)$ of the linearized system (34) satisfies

$$\|h(t)\|_{\mathbf{L}^1} \le \|h(0)\|_{\mathbf{L}^1} \qquad \text{for all } t \ge 0, \tag{35}$$

for some uniform constant L. The techniques used to prove the $\mathbf{L}^1$ bounds (35) are much the same as those used to obtain the BV bounds on u. This is quite natural, since the function $h \doteq u_x$ provides one particular solution of (34).

If now two initial data $\bar{u}, \bar{w}$ are given, following [4] we construct the smooth path

$$\theta \mapsto \bar{u}^\theta \doteq \theta\bar{u} + (1-\theta)\bar{w}, \qquad \theta \in [0,1]. \tag{36}$$

Calling $t \mapsto u^\theta(t,\cdot)$ the solution of (1.4) with initial data $\bar u^\theta$, we can write

$$\|u(t)-w(t)\|_{\mathbf{L}^1} \le \int_0^1 \left\| \frac{du^\theta(t)}{d\theta}\right\|_{\mathbf{L}^1} d\theta \le L\cdot \int_0^1 \left\| \frac{du^\theta(0)}{d\theta}\right\|_{\mathbf{L}^1} d\theta = L\cdot \|\bar u-\bar w\|_{\mathbf{L}^1}. \tag{37}$$

Indeed, the tangent vector

$$h^\theta(t,x) \doteq \frac{du^\theta}{d\theta}(t,x)$$

is a solution of the linearized Cauchy problem

$$h_t^\theta + \big[DA(u^\theta)\cdot h^\theta\big]u_x^\theta + A(u^\theta)h_x^\theta = h_{xx}^\theta,$$
$$h^\theta(0,x) = \bar h^\theta(x) = \bar u(x) - \bar w(x),$$

hence it satisfies (35) for every θ. This completes the proof of Theorem 1.

Finally, we sketch the proof of Theorem 2. By the uniform estimates (37), for each $\varepsilon>0$ we have a continuous semigroup S^ε. The dependence on the initial data is Lipschitz continuous, with a constant independent of ε. Letting $\varepsilon\to 0$, taking a suitable subsequence we obtain the convergence of the S^ε to a limit semigroup S. To prove that the limit is unique and that S is indeed the semigroup of entropy weak solutions of (1), by Corollary 9.2 in [3] it suffices to check that S yields the correct solutions, locally in time, for all piecewise constant initial data where each jump is solved by waves of one single family. But in this special case we are essentially dealing with a family of one-dimensional scalar problems, and the result is well known [11].

References

[1] S. Bianchini and A. Bressan, BV solutions for a class of viscous hyperbolic systems, *Indiana Univ. Math. J.*, **49** (2000).

[2] A. Bressan, The unique limit of the Glimm scheme, *Arch. Rational Mech. Anal.* **130** (1995), 205-230.

[3] A. Bressan, *Hyperbolic Systems of Conservation Laws. The One Dimensional Cauchy Problem.* Oxford University Press, 2000.

[4] A. Bressan, G. Crasta, and B. Piccoli, Well posedness of the Cauchy problem for $n\times n$ systems of conservation laws, *Amer. Math. Soc. Memoir*, **649** (2000).

[5] A. Bressan and P. Goatin, Oleinik type estimates and uniqueness for $n\times n$ conservation laws, *J. Differential Equations* **156** (1999), 26-49.

[6] A. Bressan and M. Lewicka, A uniqueness condition for hyperbolic systems of conservation laws, *Discr. Cont. Dynam. Syst.*, **6** (2000) 673-682.

[7] A. Bressan, T. P. Liu and T. Yang, L^1 stability estimates for $n\times n$ conservation laws, *Arch. Rat. Mech. Anal.* **149** (1999), 1-22.

[8] R. J. DiPerna, Convergence of approximate solutions to conservation laws, *Arch. Rational Mech. Anal.* **82** (1983), 27-70.

[9] J. Glimm, Solutions in the large for nonlinear hyperbolic systems of equations, *Comm. Pure Appl. Math.* **18** (1965), 697-715.

[10] J. Goodman and Z. Xin, Viscous limits for piecewise smooth solutions to systems of conservation laws, *Arch. Rational Mech. Anal.* **121** (1992), 235-265.

[11] S. Kruzhkov, First-order quasilinear equations with several space variables, *Math. USSR Sbornik* **10** (1970), 217–273.

[12] T. P. Liu, Nonlinear stability of shock waves for viscous conservation laws, *Amer. Math. Soc. Memoir* **328** (1986).

[13] T. P. Liu and T. Yang, Well posedness theory for hyperbolic conservation laws, *Comm. Pure Appl. Math.* **52** (1999), 1553-1586.

[14] D. Serre, *Systèmes de Lois de Conservation II*, Diderot Editor, Paris 1996.

[15] A. Szepessy and Z. Xin, Nonlinear stability of viscous shocks, *Arch. Rational Mech. Anal.* **122** (1993), 53-103.

[16] A. Szepessy and K. Zumbrun, stability of rarefaction waves in viscous media, *Arch. Rational Mech. Anal.* **133** (1996), 249-298.

[17] S. H. Yu, Zero-dissipation limit of solutions with shocks for systems of hyperbolic conservation laws, *Arch. Rational Mech. Anal.* **146** (1999), 275-370.

S. Bianchini
S.I.S.S.A.,
Via Beirut 4,
Trieste 34014,
Italy
bianchin@sissa.it

A. Bressan
S.I.S.S.A.,
Via Beirut 4,
Trieste 34014,
Italy
bressan@sissa.it

International Series of Numerical Mathematics
Vol. 140, © 2001 Birkhäuser Verlag Basel/Switzerland

Adaptive Finite Elements for Stationary Compressible Flows at Low Mach Number

Malte Braack

Abstract. Computation of compressible flows at low Mach number is often performed by modifications of compressible flow solvers to the low Mach number limit. In particular, time stepping schemes are developed where a Poisson problem is solved for pressure updates. In this paper, we propose a different approach for fast computations of stationary solutions. The discretization is made by stabilized finite elements. It allows the analytical construction of the Jacobian for the whole coupled system, which is used in Newton iterations for solving the nonlinear problem. No time stepping is needed. The whole process is designed for adaptive mesh refinement.

1. Introduction

In several problems arising in applications a compressible flow at low Mach number is considered, for instance in free convection, laminar flames etc. In this case, the compression is due to thermodynamical changes and not caused by hydrodynamics. This leads to numerical problems when a purely 'compressible solver' is applied to this type of problems. Therefore, several papers deal with extensions of compressible flow solvers to hydrodynamically incompressible flows. The decoupling of the density from the hydrodynamical pressure is meanwhile state of the art. The equations describing stationary flow of a fluid at low Mach number read

$$\operatorname{div}(\rho v) = 0\,, \tag{1}$$

$$\operatorname{div}(\rho v \otimes v) + \operatorname{div}\tau + \nabla p - \rho g = -\rho_0 g\,, \tag{2}$$

$$\operatorname{div}(\rho v T) - \operatorname{div}(\kappa \nabla T) = 0\,, \tag{3}$$

$$\rho = \frac{P_{th}}{RT}\,. \tag{4}$$

The velocity is denoted by v, temperature by T, density by ρ. The total pressure p_{tot} is split in three parts, $p_{tot}(x) = P_{th} + p_{hs}(x) + p(x)$. The hydrodynamic pressure is simply denoted by p. The hydrostatic pressure p_{hs} is given by a reference density ρ_0 and the scalar product of the vector of gravitation g and the coordinate x: $p_{hs}(x) = -\rho_0\langle g, x\rangle$. The thermodynamic pressure P_{th} is spatially uniform. In

This work has been supported by the German Research Foundation (DFG).

open or unbounded domains P_{th} is a constant. The tensor τ is the usual stress tensor, involving the viscosity μ, which may depend on T (κ as well). The system (1)-(4) has to be designed with appropriate boundary conditions. This system is an approximation of the compressible Navier-Stokes equations, because the hydrodynamical and hydrostatic part of the pressure are neglected in the equation of state (4). This is justified, since the thermodynamic pressure is several magnitudes bigger than $p_{hs} + p$. Due to this simplification, acoustic waves are removed from the system.

The outline of this paper is as follows. We begin with the presentation of the discretization of the considered partial differential equations by stabilized finite elements of equal order for all components. Two different formulations are presented. One discretization is globally conservative, while the conservation property of the other one is violated of order $O(h^2)$, where h denotes the mesh size. The discretization is the basis of the nonlinear iteration we perform, avoiding any kind of time stepping for a wide class of stationary problems. Numerical results of a benchmark problem for low Mach number flow are given in Section 3, illustrating the performance of the solver and accuracy of the discretizations. We compare the solution with bilinear and quadratic finite elements. The aspect of adaptivity by local mesh refinement is addressed in Section 4.

2. Discretization

We begin by setting the frame for a finite element Galerkin discretization. For notational simplicity, we consider the case that the computational two dimensional domain $\Omega \subset \mathbb{R}^2$ is polygonal. Extensions for curved boundaries are possible. By $L^2(\Omega)$, we denote the Lebesgue space of square-integrable functions on Ω, equipped with the usual inner product (f, g) and norm $\|f\|$. The suggestive notation $(f, g)_Q$ and $\|f\|_Q$ is used for the L^2 inner product and norm on a subset $Q \subset \Omega$.

2.1. Variational Formulation

The variational formulation of (1)-(4) for the unknowns $u = (p, v, T)$ is obtained by multiplying the equations by appropriate test vector functions $\Phi = (\xi, \phi, \psi)$ and integrating over Ω. This leads us to introduce a semi-linear form $A(\cdot,\cdot)$ by

$$
\begin{aligned}
A(u, \Phi) \quad = \quad & (\nabla \cdot v - T^{-1} v \cdot \nabla T, \xi) \\
& + (\rho v \cdot \nabla v, \phi) - (\tau, \nabla \phi) - (p, \nabla \cdot \phi) - (\rho g, \phi) \\
& + (\rho v \cdot \nabla T, \psi) + (\kappa \nabla T, \nabla \psi) .
\end{aligned}
\tag{5}
$$

The test functions Φ are elements of the space $X = Q \times V^2 \times V$, with $Q = L^2(\Omega)$ and the Sobolev space $V = H_0^1(\Omega)$. The weak formulation reads,

$$
u \in u_0 + X : \quad A(u, \Phi) = (f, \Phi) \quad \forall \Phi \in X , \tag{6}
$$

with Dirichlet data u_0 on (parts of) the boundary $\partial\Omega$. Neumann-type boundary conditions are implicitly represented because of the partial integration of the diffusion terms. Non-homogeneous Dirichlet boundary conditions u_0 are explicitly imposed on the solution. $f = (0, -\rho_0 g, 0) \in L^2(\Omega)^{2+d}$ may be a given right-hand side. The density ρ is considered as a variable coefficient depending on temperature according to (4). The space for the hydrodynamical pressure is $L^2(\Omega)$. If Dirichlet conditions for the velocity are imposed along the entire boundary, the hydrodynamical pressure is only defined modulo a constant and the corresponding pressure space is $L^2(\Omega)/\mathbb{R}$. If an outflow boundary Γ_{out} is present, the partial integration of the pressure gradient imposes implicitly the 'natural' outflow condition $\mu\partial_n v - pn = 0$ on Γ_{out}, where n stands for the outer normal vector. Another possibility is a 'conservative formulation' where the convective fluxes are partially integrated. Instead of (5), one may consider the semi-linear form $B(\cdot,\cdot)$ given by

$$\begin{aligned} B(u,\Phi) \quad = \quad & -(\rho v, \nabla\xi) - (\rho v \otimes v + \tau, \nabla\phi) - (p, \nabla\cdot\phi) - (\rho g, \phi) \\ & -C(v, \nabla\psi) + (\kappa\nabla T, \nabla\psi) \\ & + \int_{\delta\Omega} \rho v \cdot n\,\xi\,dx + \int_{\delta\Omega} (\rho v \otimes v)\cdot n \cdot \phi\,dx + C\int_{\delta\Omega} v\cdot n\,\psi\,dx\,, \end{aligned} \tag{7}$$

with the constant $C = P_{th}/R$. The arising boundary integrals vanish if homogeneous Dirichlet conditions for the velocities on the whole boundary are imposed. In this formulation, no convective term for the temperature arises.

Our discretization uses a conforming finite element Galerkin method defined on quadrilateral (or hexahedral) meshes $\mathbf{T}_h = \{K\}$ on Ω, with cells denoted by K. The trial and test spaces $X_h \subset X$ consist of continuous, piecewise polynomial vector functions (so-called Q_r-elements) of maximal degree r for all unknowns.

The equation (6) includes as a special case the linear Stokes system. Therefore, the discretization requires a certain stability condition (so-called Babuska-Brezzi condition [8]) to be satisfied. It is well-known that the use of equal-order elements for velocity and pressure is ruled out in the normal Galerkin formulation. We introduce a stabilization term $S_h(\cdot,\cdot)$, which depends on the mesh size h. The stabilization has to serve several purposes, the treatment of the stiff pressure-velocity coupling in the low-Mach number limit, the stabilization of the advective terms, and finally the enhancement of local mass conservation.

2.2. Stabilization

For the presentation of our stabilization, we write the considered system of PDEs in operator form,

$$Lu = f, \quad u = u_0 \quad \text{on } \partial\Omega.$$

Following ideas of Hughes, et al., [10], the stabilization can be accomplished by introducing additional least-squares terms in the continuity equation and in all other equilibrium equations. The discretized systems for the use of the semi-linear form of (5) and (7) are of the form

$$u_h \in X_h : \quad A(u_h,\Phi) + S_h^A(u_h,\Phi) = (f,\Phi) \quad \forall\Phi \in X_h\,, \tag{8}$$

$$u_h \in X_h : \quad B(u_h,\Phi) + S_h^B(u_h,\Phi) = (f,\Phi) \quad \forall\Phi \in X_h\,, \tag{9}$$

where the stabilization terms $S_h^A(\cdot,\cdot)$ and $S_h^B(\cdot,\cdot)$ are based on the residuals of the strong formulation and have the form

$$S_h(u_h;\Phi) = \langle Lu_h - f, \mathbf{S}\phi\rangle_\delta\,. \tag{10}$$

For simplifying notation, we use here the weighted L^2 inner product for scalar functions

$$\langle u,v\rangle_\delta := \sum_{K\in\mathbf{T}_h} \delta_K(u,v)_K\,,$$

with cell-dependent weights δ_K depending on u_h. The modification is done in a consistent way, i.e., the solution of the partial differential equation representing (1)-(4) still satisfies these stabilized equations when u is enough regular. The definition of $\mathbf{S}$ depends on the choice of formulation with $A(\cdot,\cdot)$ or $B(\cdot,\cdot)$. While in a (pure) Galerkin-least-square technique, the stabilization is chosen as the Frechét derivative of the operator, $\mathbf{S} = L'[u]$, we choose the $\mathbf{S}$ as follows: We split the Frechét derivative of L in the differential part D and a part Z consisting of zero-order terms: $L'[u] = D + Z$. We set $\mathbf{S} = -D^*$. Now we give a justification of this choice. In order to simply the presentation, we consider the simplified stress tensor $\tau = -\mu\nabla v$. The differential part D for the considered system of PDEs, when the first formulation (8) is chosen, is given by

$$D = \begin{bmatrix} 0 & \mathrm{div} & -T^{-1}v\cdot\nabla \\ \nabla & \rho v\cdot\nabla - \nabla\cdot\mu\nabla & 0 \\ 0 & 0 & \rho v\cdot\nabla - \nabla\cdot\kappa\nabla \end{bmatrix}.$$

Using the fact that ρv is solenoidal, this choice leads to the following operator $\mathbf{S} = \mathbf{S}^A$ in (10):

$$\mathbf{S}^A = -D^* = \begin{bmatrix} 0 & \mathrm{div} & 0 \\ \nabla & \rho v\cdot\nabla + \nabla\cdot\mu\nabla & 0 \\ -T^{-1}v\cdot\nabla & 0 & \rho v\cdot\nabla + \nabla\cdot\kappa\nabla \end{bmatrix}.$$

This setting yields the following stabilization form:

$$\begin{aligned} S_h^A(u_h,\Phi) &= \langle r_p, \nabla\cdot\phi\rangle_\delta \\ &+ \langle r_v, \nabla\xi + \rho v_h\cdot\nabla\phi + \nabla\cdot(\mu\nabla\phi)\rangle_\delta \\ &+ \langle r_T, \rho v_h\cdot\nabla\psi + \nabla\cdot(\kappa\nabla\psi) - T_h^{-1}v_h\cdot\nabla\xi\rangle_\delta\,, \end{aligned} \tag{11}$$

with the component-wise residuals

$$\begin{aligned} r_p &= \nabla\cdot v_h - T_h^{-1}v_h\cdot\nabla T_h \\ r_v &= \rho v_h\cdot\nabla v_h - \nabla\cdot(\mu\nabla v_h) + \nabla p_h + (\rho_0-\rho)g \\ r_T &= \rho v_h\cdot\nabla T_h - \nabla\cdot(\kappa\nabla T_h) - f_T\,. \end{aligned}$$

We will discuss several terms arising above in order to clarify this setting.

The term $\langle r_v, \nabla\xi\rangle_\delta$ gives the pressure-velocity stabilization according to [10]. Here, the essential part for stabilization is $\langle\nabla p_h, \nabla\xi\rangle_\delta$, acting like a Laplacian on the pressure. The other terms are needed for consistency. The convective term in the momentum equation is stabilized by the streamline diffusion term $\langle r_v, \rho v_h\cdot$

$\nabla\phi\rangle_\delta$. Equivalently, in the equation for temperature, streamline diffusion is introduced by $\langle r_T, \rho v_h \cdot \nabla\psi + \nabla\cdot(\kappa\nabla\psi)\rangle_\delta$.

Hansbo and Szepessy proposed in [9] a term $\langle\nabla\cdot v_h, \nabla\cdot\phi\rangle_\delta$ for the (incompressible) Navier-Stokes equations in order to enforce the divergence free condition. Generalization to compressible flows is obtained by the term $\langle r_p, \nabla\cdot\phi\rangle_\delta$, enforcing mass conservation. The convective part in the continuity equation due to density variations is stabilized by $\langle r_T, -T_h^{-1} v_h \cdot \nabla\xi\rangle_\delta$. The cell-dependent parameters δ_K form a diagonal matrix. They are chosen cell-wise, depending on the balance between local convection $\beta = \|\rho v_h\|_{K,\infty}$ and diffusion μ and κ, $\delta_K = \delta_0 \cdot \mathrm{diag}\{\delta_p, \delta_v, \delta_v, \delta_T\}$, with

$$\delta_v = \frac{h_K^2}{6\mu + \beta h_K}, \qquad \delta_T = \frac{h_K^2}{6\kappa + \beta h_K}, \qquad \delta_p = \beta\delta_v\,.$$

The choice of δ_0 should depend on the order of the discretization. For Q_1 elements, $0.2 \leq \delta_0 \leq 0.5$ is appropriate, but should be chosen smaller for higher-order elements. The same technique is applied to the bilinear form $B(\cdot,\cdot)$ in (7).

3. A Benchmark Problem

As test case we consider the benchmark problem of Le Quere and Paillere [11], designed for measuring accuracy and performance of low-Mach-number solvers. It is an extension of earlier comparisons of incompressible flow solvers with the Boussinesq approximation by de Vahl Davis and Jones [7]. The model consists of a natural convection flow with large temperature differences in a closed cavity $\Omega = (0,1)^2$. The parameters are the Rayleigh number Ra and the non-dimensionalized temperature difference $\epsilon = 0.6$,

$$Ra = Pr\, g \left(\frac{\rho_0}{\mu_0}\right)^2 \cdot \frac{T_h - T_c}{T_0} \quad , \quad \epsilon = \frac{T_h - T_c}{T_h + T_c}\,.$$

For given ϵ and reference temperature $T_0 = 600\,K$, $\rho_0 = \rho(T_0)$ and $\mu_0 = \mu(T_0)$.

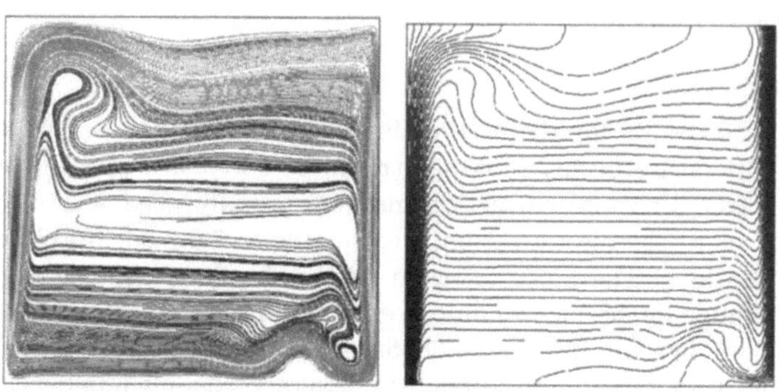

FIGURE 1. Left: Streamlines, right: isolines of T for $Ra = 10^7$

The temperature at the hot ($x = 0$) and cold walls ($x = 1$) are given by $T_h = T_0(1+\epsilon)$ and $T_c = T_0(1-\epsilon)$, respectively. At the horizontal (isolated) walls $y = 0$ and $y = 1$, Neumann type boundary condition for T are imposed. For the velocities, all boundaries are rigid walls, $v = 0$ on $\partial\Omega$. For the viscosity the Sutherland's law

$$\mu(T) = 1.68 \cdot 10^{-5} \frac{kg}{ms} \left(\frac{T}{T^*}\right)^{2/3} \frac{T^* + S}{T + S}, \quad T^* = 273\,K,\ S = 110.5\,K$$

is used. The heat conduction is given by $\kappa(T) = \mu(T)/Pr$ with constant Prandtl number $Pr = 0.71$. The reference thermodynamic pressure is $P_0 = 101325\,Pa$. In Figure 1, the solution for $Ra = 10^7$ is shown. The gas constant is $R = 287\,\frac{J}{kg\,K}$.

Furthermore, spatially uniform initial conditions $T = T_0$, $v = 0$ at initial time $t = 0$ are imposed. This leads to a further unknown in the system. The thermodynamic pressure P_{th} in a bounded vessel is not any more a constant but depends on the global temperature distribution, because the initial mass,

$$m_0 = \int_\Omega \rho_0\, dx = \frac{P_0}{R}\psi(T_0) \quad \text{with} \quad \psi(T) := \int_\Omega T^{-1} dx$$

has to be conserved. Therefore, the thermodynamic pressure for the stationary solution is defined by $P_{th} = P_0\, \psi(T_0)\, \psi(T)^{-1}$.

In this benchmark problem, the quantities of interest are the average Nusselt numbers at the hot and cold walls, Γ_h and Γ_c respectively, given by

$$\langle Nu\rangle_h = c\int_{\Gamma_h} \kappa\frac{\partial T}{\partial n}\, dx \quad \text{and} \quad \langle Nu\rangle_c = c\int_{\Gamma_c} \kappa\frac{\partial T}{\partial n}\, dx\,, \quad c = \frac{Pr}{2\mu_0 T_0 \epsilon}. \tag{12}$$

The sum of these averages is zero, because of the Neumann-type conditions at the upper and lower wall for the temperature and the non-slip conditions for the velocities,

$$\langle Nu\rangle_h + \langle Nu\rangle_c = c\int_{\partial\Omega}\left(\kappa\frac{\partial T}{\partial n} - \rho v\cdot n\right) dx = c\int_\Omega (\rho v\cdot\nabla T - \nabla\cdot(k\nabla T))\, dx = 0\,.$$

However, for the discrete solutions, one expects an error which should converge to zero under mesh refinement. We will evaluate the error for the discrete solutions by $\langle e\rangle := 1 + \langle Nu\rangle_c \langle Nu\rangle_h^{-1}$.

3.1. Numerical results

We present the results for the case of Rayleigh number $Ra = 10^6$. A boundary layer for temperature occur. It is not trivial to solve this problem with a pure stationary solver. Pressure correction methods usually need several hundreds of time steps. Due to the discretization, we are able to solve such problems in a few nonlinear iterations. The most accurate reference values fixed by the committee are for the change in thermodynamic pressure $P/P_0 = 0.9244$ and for the Nusselt number up to 5 digits is $\langle Nu\rangle_h = 8.6866$.

In Table 1 we report on the resulting average Nusselt numbers, the relative error $\langle e\rangle$ and the change of thermodynamic pressure relative to the initial pressure P_0 for the discretization (8). The thermodynamic pressure and the Nusselt numbers

	$A(\cdot,\cdot)$ with Q1				$B(\cdot,\cdot)$ with Q1		with Q2	
#dof	$\langle Nu\rangle_h$	$\langle Nu\rangle_c$	$\langle e\rangle$	P/P_0	$\langle Nu\rangle_h$	P/P_0	$\langle Nu\rangle_h$	P/P_0
1,089	9.33273	-8.21159	1.20e-1	0.900862				
4,225	9.21050	-8.33739	9.48e-2	0.931489	8.563474	0.929946	8.508630	0.938997
16,641	8.87338	-8.58009	3.31e-2	0.927150	8.615682	0.928947	8.647428	0.927672
66,049	8.74806	-8.65797	1.03e-2	0.925258	8.665559	0.925900	8.682079	0.924886
263,169	8.70491	-8.67965	2.89e-3	0.924663	8.681461	0.924843	8.686285	0.924531
1,050,625	8.69161	-8.68495	7.66e-4	0.924524	8.685401	0.924571	8.686617	0.924491
4,202,500	8.68790	-8.68619	1.96e-4	0.924495	8.686307	0.924507	8.686602	0.924487
extrapol.	8.68663	-8.68663	0	0.924485	8.686609	0.924485	8.686601	0.924488

TABLE 1. $A(\cdot,\cdot)$ with $Q1$, $B(\cdot,\cdot)$ with $Q1$ and $Q2$ elements

converge with second order respect to the mesh size. However, we observe the violation of the conservation property since $\langle e\rangle$ does not vanish (behaves also $O(h^2)$). For the 'globally conservative formulation' with $B(\cdot,\cdot)$, the accuracy in the thermodynamic pressure is comparable to the first formulation. However, the indicator of the energy lost $\langle e\rangle$ is zero up to machine accuracy and is therefore not listed. The Nusselt number converge also with second order accuracy but is more accurate than the previous 'non-conservative' discretization. A further large improved can be obtained by the use of $Q2$ elements. The order of the method with polynomials of second order gives accuracy in the Nusselt number of order $O(h^4)$, resulting in a more reliable reference solution for the same number of unknowns.

The discretized equations are solved coupled by Newton's method with an analytical Jacobian. The linear equations are solved by a multigrid V-cycle with a block ILU smoother. In each iteration, two pre- and two post-smoothing steps are performed. For details of such iterations we refer to [6, 2].

In Table 2, we show the convergence rates in the solver for computations on globally refined meshes with Q_1 elements. Here, the formulation in $A(\cdot,\cdot)$ was used, but the formulation in $B(\cdot,\cdot)$ leads to very similar results. The second column includes the convergence rate ϱ of the nonlinear iteration. For a reduction of the residual by the factor of 10^{10}, the number of needed iterations are given in the third column. Furthermore, the number of Jacobians needed are given. Instead of building the Jacobian in each nonlinear step, it is kept if the reduction of the residual of the previous step is better than a factor of 0.2. The fourth column shows the reduction rate of the linear solve in the last nonlinear step. The average number of linear iterations in each nonlinear step is given in the last column.

#dof	ϱ	Nonlinear iterations	Jacobians	rate	Linear iterations
1,089	0.37	19	17	0.256	5.1
4,225	0.30	16	10	0.182	3.4
16,641	0.23	14	9	0.178	3.2
66,049	0.13	10	3	0.115	2.8
263,169	0.10	8	1	0.073	2.0
1,050,625	0.09	8	1	0.026	2.0
4,202,500	0.07	7	1	0.012	1.8
(Q2) 4,202,500	0.11	9	4	0.139	3.0

TABLE 2. Convergence data

Under mesh refinement, the number of Newton steps decreases significantly. On the finer meshes between 1 and 4 million unknowns, only one Jacobian is assembled for reducing the residual by 10 digits. The linear convergence benefits also under mesh refinement, leading to reduction rates of better than 0.04. On the finest meshes, only two multigrid iterations are necessary for reducing the linear residual by two digits. This behavior is due to the fact that the arising matrices become more and more definite. For comparisons, we included in Table 2 the corresponding values for $Q2$ elements on a mesh with 263,169 nodes (about 4 million unknowns). In order to achieve similar robust convergence, we increased the effort in the multigrid cycle to 8 pre- and 8 post-smoothing steps.

4. Adaptive Local Refinement

Adaptive mesh refinement is well known to be an efficient method to increase the accuracy of the discrete solution respect to the number of degrees of freedom. However, the efficiency strongly depends on the refinement strategy and the information about discretization errors. The a posteriori error estimates based on weighted residuals, introduced by Becker and Rannacher [4], has already been successfully applied to various kinds of PDEs: Navier-Stokes [1], combustion problems [5, 3], elasto-plasticity. Before applying this technique on the considered benchmark problem, we shortly repeat this estimator for the error in the average Nusselt number. According to (12), the average Nusselt number can be considered as the functional

$$J(u) = c \int_{\Gamma_c} \kappa \frac{\partial T}{\partial n}\, dx \,.$$

We employ a duality argument as common in the error analysis for finite element Galerkin methods. The functional J is taken as right-hand side in a dual problem governed by the linearized adjoint bilinear form. From the solution of this dual problem, we derive the weights which describe the contribution of the local cell-residuals to the error quantity $J(u) - J(u_h)$. This information is then used for quantitive error control and as a criterion for local mesh refinement. The whole process may be viewed as a sensitivity analysis of the given continuous model with respect to the perturbations induced by the finite element discretization. The key for the efficiency of the resulting a posteriori error estimates is the Galerkin orthogonality inherent to the finite element approach, i.e., the orthogonality of the variational residual to the discrete test space.

Here, we assume J to be linear, although in the case of the Nusselt number we have a weak nonlinearity if the coefficient κ varies with temperature. However, the whole approach can be extended to nonlinear functionals (see [4]). Using the Galerkin orthogonality of the finite element scheme an error bound of the form

$$|J(u) - J(u_h)| \le \eta_\omega := \sum_{K\in\mathbf{T}_h} \Big\{ \varrho_K(\omega_K + \omega_{K,\delta}) + \varrho_{\partial K}\omega_{\partial K} \Big\}\,, \tag{13}$$

can be derived. This a posteriori error estimator consists of cell-residuals and weights

$$\varrho_K := \|f - L(u_h)\|_K, \quad \varrho_{\partial K} := h_K^{-1/2}\|[\partial_n^L u_h]\|_{\partial K},$$
$$\omega_K = \|z - z_h\|_K, \quad \omega_{K,\delta} = \delta_K\|\mathbf{S}^A(z - z_h)\|_K, \omega_{\partial K} = h_K^{-1/2}\|z - z_h\|_{\partial K}. \quad (14)$$

The 'jump terms' $\varrho_{\partial K}$ arise only in the case of Q_1 elements and vanish for higher-order elements. The residual terms ϱ_K and $\varrho_{\partial K}$ depend only on the discrete solution u_h and can be easily evaluated. The situation is different for the weighting terms ω_K, $\omega_{K,\delta}$ and $\omega_{\partial K}$. Since the system (1)–(4) is too complex for deriving analytically upper bounds for the dual solution z, we need a numerical approximation. This can be done by formulating a 'perturbed' discrete dual problem. Accordingly, we seek $z_h \in X_h$, satisfying the linear system given by the Frechét derivatives $A'[u_h]$ and $S_h^{A'}[u_h]$ of the semi-linear forms A and S_h^A previously defined:

$$A'[u_h](\phi, z_h) + S_h^{A'}[u_h](\phi, z_h) = J(\phi) \quad \forall \phi \in X_h. \quad (15)$$

Since this equation is linear, the computational cost of its solution relates to the cost of one Newton step in solving the primal problem. We note that the solution of the dual problem on the same mesh and by the same method as used for the primal problem is not mandatory. One may use a coarser mesh to save computational cost or even employ another (possibly higher-order) discretization to get the weights with enhanced accuracy. In our computations, the dual solution is always computed by the same method as used for the primal problem. Having computed an approximation to the exact dual solution z, the weights ω_K given by (14) can be determined in different ways, see [3]. We approximate the weights by second-order difference quotients of the discrete dual solution $z_h \in V_h$:

$$\omega_K \leq Ch_K^2\|\nabla^2 z\|_K \quad \approx \quad \tilde{\omega}_K := Ch_K^3|\nabla_h^2 z_h(x_K)|, \quad (16)$$

x_K is the center point of K, with an interpolation constant $C \sim 0.1$ independent of h_K, $\nabla^2 z$ is the Hessian of z and $\nabla_h^2 z_h$ a suitable difference approximation. An analogous technique is used in evaluating the two other types of weights $\omega_{\partial K}$ and $\omega_{K,\delta}$ in (13).

#nodes	$\langle Nu\rangle_c$	$J(u)-J(u_h)$
289	-7.99354	-6.93e-01
447	-7.81333	-8.73e-01
740	-8.31245	-3.74e-01
1290	-8.44896	-2.38e-01
2105	-8.54421	-1.42e-01
3650	-8.61216	-7.45e-02
5774	-8.62923	-5.74e-02
10111	-8.65286	-3.38e-02
16634	-8.66919	-1.74e-02
26947	-8.67451	-1.21e-02
45908	-8.68048	-6.15e-03

#nodes	$\langle Nu\rangle_c$	$J(u)-J(u_h)$
289	-7.99354	6.93e-01
380	-8.33295	3.54e-01
544	-8.26692	4.20e-01
814	-8.49691	1.90e-01
1397	-8.57928	1.07e-01
2500	-8.64814	3.85e-02
4000	-8.66797	1.87e-02
6940	-8.68216	4.47e-03
10831	-8.6809	5.73e-03
18155	-8.68139	5.24e-03
29567	-8.68686	-2.30e-04

TABLE 3. Left: Simple estimator η_{res}, right: weighted estimator η_ω

In Table 3, we show the obtained Nusselt numbers at the cold wall when applying the simple residual estimator $\eta_{res} = \sum_{K\in\mathbf{T}_h} h^3\rho_K$ for mesh adaptation.

Comparisons to globally refined meshes in Table 1 show that no substantial improvement can be obtained. This is due to the fact that the residual estimator η_{res} does not include any information about the quantity we look on. In contrast to this, the weighted residual estimator η_ω is especially adapted to this problem.

References

[1] R. Becker. Weighted error estimators for finite element approximations of the incompressible Navier-Stokes equations. Rapport de Recherche RR-3458, INRIA Sophia-Antipolis, 1998.

[2] R. Becker and M. Braack. Multigrid techniques for finite elements on locally refined meshes. *Numerical Linear Algebra with Applications*, 7:363–379, 2000. Special Issue.

[3] R. Becker, M. Braack, and R. Rannacher. Numerical simulation of laminar flames at low mach number with adaptive finite elements. *Combust. Theory Modelling*, 3:503–534, 1999.

[4] R. Becker and R. Rannacher. A feed-back approach to error control in finite element methods: Basic analysis and examples. *East-West J. Numer. Math.*, 4:237–264, 1996.

[5] M. Braack. An Adaptive Finite Element Method for Reactive Flow Problems. Doktorarbeit, Institut für Angewandte Mathematik, Universität Heidelberg, 1998.

[6] M. Braack and R. Rannacher. Adaptive finite element methods for low-mach-number flows with chemical reactions. In H. Deconinck, editor, *30th Computational Fluid Dynamics*, volume 1999-03 of *Lecture Series*, pages 1–93. von Karman Institute for Fluid Dynamics, 1999.

[7] G. de Vahl Davis and I.P. Jones. *Natural convection in a square cavity - A comparison exercise.* Int. J. Num. Meth. in Fluids, 3:227–248, 1983.

[8] V. Girault and P.-A. Raviart. *Finite Elements for the Navier Stokes Equations.* Springer, Berlin, 1986.

[9] P. Hansbo and A. Szepessy. A velociity-pressure streamline diffusion finite element method for the incompressible Navier-Stokes equations. *Comput. Methods Appl. Mech. Engrg.*, 84:175–192, 1990.

[10] T.J.R. Hughes, L.P. Franca, and M. Balestra. A new finite element formulation for computational fluid dynamics: V. circumvent the Babuska-Brezzi condition: A stable Petrov-Galerkin formulation for the Stokes problem accommodating equal order interpolation. *Comput. Methods Appl. Mech. Engrg.*, 59:89–99, 1986.

[11] P. Le Quere and H. Paillere. Modelling and simulation of natural convection flows with large temperature differences: a benchmark problem for low mach number solvers. *International Journal of Thermal Sciences*, 2001. to appear.

M. Braack
Institut für Angewandte Mathematik,
Universität Heidelberg, INF 294,
D-69120 Heidelberg, Germany
`malte.braack@iwr.uni-heidelberg.de`

International Series of Numerical Mathematics
Vol. 140, © 2001 Birkhäuser Verlag Basel/Switzerland

A Monge-Kantorovich Approach to the Maxwell Equations

Yann Brenier

Abstract. Our purpose is to extend to the framework of Electromagnetism the Monge-Kantorovich theory [9] which originally comes from Continuum Mechanics and has become very popular in the last ten years in the field of nonlinear PDEs, especially because of its connection with the Monge-Ampère equation [2],[3],[4], [5],[6],[7]... We construct, in the framework of Electromagnetism, an Action which is analogous to the Monge-Kantorovich (or Wasserstein) distance. From this Action, we derive a system of PDEs that look like non-linear Maxwell equations. Then, through suitable approximations, we formally recover not only the linear Maxwell equation, as expected, but, more surprisingly, an approximation to the complete (pressureless) Euler-Maxwell system.

1. A review of the Monge-Kantorovich theory

1.1. Definition of the Monge-Kantorovich distance

Given two compactly supported nonnegative functions ρ_0 and ρ_1 on $\mathbf{R}^d$, with integral equal to 1, the Monge-Kantorovich (or Wasserstein) distance (with exponent p) from ρ_0 to ρ_1 can be defined by

$$d_p(\rho_0, \rho_1)^p := \inf_M \int \frac{1}{p} |M(x) - x|^p \rho_0(x) dx, \tag{1}$$

where the infimum is performed over all (Borel) maps M from $\mathbf{R}^d$ to $\mathbf{R}^d$ such that

$$\int_{x \in A} \rho_1(x) dx = \int_{M(x) \in A} \rho_0(x) dx \tag{2}$$

for all (Borel) subset A of $\mathbf{R}^d$. From [1], we know that the MK distance can be defined in a Continuum Mechanics framework, by introducing an artificial "time" (or rather interpolation) variable $s \in [0, 1]$. More precisely, the square of the MK distance (with $p = 2$) between ρ_0 and ρ_1 can be shown to be the infimum of the following Action

$$\int_{\mathbf{R}^d} \int_0^1 K(\rho(s, x), J(s, x)) dx ds, \tag{3}$$

where

$$K(a,b) = \frac{|b|^2}{2a} \tag{4}$$

and the infimum is performed over all density and momentum fields, $\rho \geq 0$, $J \in \mathbf{R}^d$, depending on $(s,x) \in [0,1] \times \mathbf{R}^d$, subject to

$$\partial_s \rho + \nabla \cdot J = 0, \;\; \rho(0,.) = \rho_0, \;\; \rho(1,.) = \rho_1. \tag{5}$$

(Strictly speaking, (3) should be written

$$\sup_{\alpha,\beta} \int_{\mathbf{R}^d} \int_0^1 \alpha(s,x) \cdot d\rho(s,x) + \beta(s,x) \cdot dJ(s,x) dx ds, \tag{6}$$

where ρ and J are respectively real and vector-valued Radon measures subject to (5), and α, β are respectively real and vector valued continuous compactly supported functions on $[0,1] \times \mathbf{R^d}$ subject to satisfy

$$\alpha(s,x) + \frac{|\beta(s,x)|^2}{2} \leq 0 \tag{7}$$

pointwise.)

1.2. The Poisson equation in the MK framework

Before addressing the framework of Electromagnetism and the Maxwell equations, it is useful to see how the Poisson equation is related to the MK distance.

If ρ_0 and ρ_1 belong to the (homogeneous) negative Sobolev space H^{-1} the square of their distance in that space can be defined by

$$\int_{\mathbf{R^d}} |\nabla \phi(x)|^2 dx, \tag{8}$$

where ϕ solves the Poisson equation

$$-\Delta \phi = \rho_1 - \rho_0. \tag{9}$$

We can recover the H^{-1} distance, as an approximate MK distance, by minimizing (3), with constraint (5), after substituting the quadratic (and degenerate!) expression $K(a,b) = |b|^2/2$ for (4). Indeed, the Lagrangian of the corresponding minimization problem is

$$\int \{\frac{1}{2}|J|^2 - \rho \partial_s \phi - J \cdot \nabla \phi\} ds dx + \int \{\phi(1,x)\rho_1(x) - \phi(0,x)\rho_0(x)\} dx, \tag{10}$$

where $\phi(s,x) \in \mathbf{R}$ is a Lagrange multiplier for (5), and the optimality conditions are $\partial_s \phi = 0, \;\; J = \nabla \phi$. Using (5), we get

$$\rho(s,x) = (1-s)\rho_0(x) + s\rho_1(x), \;\; J = \nabla \phi, \;\; -\Delta \phi = \rho_1 - \rho_0 \tag{11}$$

and the infimum of the Action is one half of the squared L^2 norm of $\nabla \phi$, as announced.

1.3. Distance between points revisited

To extend the MK theory to the Maxwell framework, it is also worth considering formulation (3), (4), (5) in the limit case when the data are no longer functions but point masses, $\rho_0(x) = \delta(x - X_0)$, $\rho_1(x) = \delta(x - X_1)$, where X_0 and X_1 are two points. In this trivial case, the MK distance is nothing but the Euclidean distance between X_0 and X_1 and we have, of course,

$$|X_1 - X_0|^2 = \inf_X \int_0^1 |X'(s)|^2 ds, \tag{12}$$

where the infimum is taken among all curves $s \in [0,1] \to X(s) \in \mathbf{R}^d$ such that $X(1) = X_1$, $X(0) = X_0$ and achieved by the shortest path

$$X(s) = (1-s)X_0 + sX_1. \tag{13}$$

Then, it can be checked, using Jensen's inequality and the convexity of K, that the infimum of (3) is precisely achieved by the corresponding pair (ρ, J)

$$\rho(s,x) = \delta(x - X(s)), \quad J(s,x) = X'(s)\delta(x - X(s)), \tag{14}$$

which, of course, satisfies (5).

2. Extension to the Maxwell framework

2.1. Distance between curves in the Minkowski space

In the framework of Electromagnetism, instead of considering points in the Euclidean space, we consider curves in the Minkowski space, namely

$$t \in \mathbf{R} \to (t, X(t)) \in \mathbf{R}^4$$

(or more generally $\tau \in \mathbf{R} \to (T(\tau), X(\tau)) \in \mathbf{R}^4$, but this -important- generalization is not essential to our present discussion). The corresponding density and current fields are defined by

$$\rho(t,x) = \delta(x - X(t)), \quad J(t,x) = X'(t)\delta(x - X(t)) \tag{15}$$

(exactly as in (14), but now t is a genuine time variable!) and are subject to

$$\partial_t \rho + \nabla \cdot J = 0. \tag{16}$$

A natural extension of the concept of shortest path between two given points X_0, X_1 in the Euclidean space is the concept of extremal surfaces between two given curves $t \to X_0(t)$, $t \to X_1(t)$ on a given interval $[0,T]$. So, we introduce an extra "interpolation variable" $s \in [0,1]$ and look for surfaces

$$(t,s) \in [0,T] \times [0,1] \to (t, s, X(t,s)) \in \mathbf{R}^5, \tag{17}$$

such that

$$X(t, s=0) = X_0(t), \quad X(t, s=1) = X_1(t). \tag{18}$$

The area of the surface is computed in the five-dimensional space (t, s, x) (the new variable s is essential!) with the Minkowski metrics

$$(+1, -1, -1, -1, -1, -1)$$

and is given by

$$\int_0^1 \int_0^T \sqrt{1 + |\partial_s X|^2 - |\partial_t X|^2 - |\partial_s X \times \partial_t X|^2} dtds. \tag{19}$$

(Observe that with the Euclidean metrics, the expression becomes

$$\int_0^1 \int_0^T \sqrt{1 + |\partial_s X|^2 + |\partial_t X|^2 + |\partial_s X \times \partial_t X|^2} dtds \tag{20}$$

and could be used to define an elliptic version of our theory.) Notice that, by introducing (19), we have just recovered the Nambu-Goto Action used in String Theory [8].

2.2. Definition of an Action for fields

For a given surface X, it is now natural to introduce four fields

$$\rho(t, s, x) = \delta(x - X(t, s)), \quad J(t, s, x) = \partial_t X(t, s))\delta(x - X(t, s)) \tag{21}$$

$$E(t, s, x) = \partial_s X(t, s))\delta(x - X(t, s)), \tag{22}$$

$$B(t, s, x) = \partial_s X(t, s) \times \partial_t X(t, s)\delta(x - X(t, s)), \tag{23}$$

which are subject to several compatibility conditions, namely (16),

$$\partial_s \rho + \nabla \cdot E = 0, \tag{24}$$

and, less obviously,

$$\partial_t E - \partial_s J - \nabla \times B = 0. \tag{25}$$

Notice that these three compatibility conditions are not independent. Indeed, the divergence of (25) plus the s derivative of (16) is equal to the t derivative of (24). Also observe that the compatibility conditions are linear with respect to the fields. The area of X, defined by (19), can be written very simply as an Action depending on (ρ, J, E, B), namely

$$\int k(\rho(t, s, x), J(t, s, x), E(t, s, x), B(t, s, x)) dtdsdx, \tag{26}$$

where

$$k(\rho, J, E, B) = \sqrt{\rho^2 + |E|^2 - |J|^2 - |B|^2}. \tag{27}$$

Of course this definition can be extended to all fields (ρ, J, E, B) valued in $\mathbf{R} \times \mathbf{R}^3 \times \mathbf{R}^3 \times \mathbf{R}^3$, depending on $(t, s, x) \in [0, T] \times [0, 1] \times \mathbf{R}^3$, satisfying (24) and (25) as well as

$$(\rho, J)(t, s = 0, x) = (\rho_0, J_0)(t, x), \quad (\rho, J)(t, s = 1, x) = (\rho_1, J_1)(t, x), \tag{28}$$

where the density-current fields (ρ_0, J_0) and (ρ_1, J_1) are given. From now on, we forget the derivation of (ρ, J, E, B) from X through (21), (22), (23), and we only retain the compatibility conditions. In particular, the data (ρ_0, J_0) and (ρ_1, J_1)

(which must be compatible with condition (16)) are not required to be generated by curves X_0 and X_1 as in formula (21). In doing that, we are just reproducing the derivation of the MK distance betwen two given probability measures from the Euclidean distance between two given points in the Euclidean space.

2.3. Derivation of a set of nonlinear Maxwell equations

Let us now get the variational equations corresponding to (24), (25), (26), (27). The corresponding Lagrangian is given by

$$\int \{R - \rho\partial_s\phi - E\cdot\nabla\phi - E\cdot\partial_t A + J\cdot\partial_s A - B\cdot\nabla\times A\}dtdsdx, \tag{29}$$

where

$$R = \sqrt{\rho^2 + |E|^2 - |J|^2 - |B|^2} \tag{30}$$

and boundary terms have been disregarded since they do not affect the local equations we are looking for, although they play an important role to get correct boundary conditions. Varying the Lagrangian leads to

$$\rho = R\partial_s\phi, \quad J = R\partial_s A, \tag{31}$$

$$E = R(\partial_t A + \nabla\phi), \quad B = -R\nabla\times A. \tag{32}$$

Using (48) and eliminating the Lagrange multipliers in (31), (32), ϕ and A, we deduce, after elementary calculations,

$$\partial_t(JR^{-1}) - \partial_s(ER^{-1}) + \nabla(\rho R^{-1}) = 0, \tag{33}$$

$$\partial_t(BR^{-1}) + \nabla\times(ER^{-1}) = 0, \quad \nabla\cdot(BR^{-1}) = 0. \tag{34}$$

We can rewrite, in an equivalent way, this set of equations, together with (48), in terms of

$$j = JR^{-1}, \quad e = ER^{-1}, \quad b = BR^{-1}. \tag{35}$$

We get

$$\partial_t j - \partial_s e + \nabla(\rho R^{-1}) = 0, \tag{36}$$

$$\partial_t b + \nabla\times e = 0, \quad \nabla\cdot b = 0, \tag{37}$$

$$\partial_t(Re) - \partial_s(Rj) - \nabla\times(Rb) = 0, \tag{38}$$

$$\partial_s\rho + \nabla\cdot(Re) = 0, \tag{39}$$

where R is now expressed by

$$R = \sqrt{\frac{\rho^2 + |E|^2}{1 + |j|^2 + |b|^2}}. \tag{40}$$

To get a a system of time-evolution equations in $(\rho, j, E = Re, b)(t, s, x)$, we need a time evolution equation for ρ. To do that, we substitute compatibility condition (16) for (39), which leads to

$$\partial_t\rho + \nabla\cdot(Rj) = 0. \tag{41}$$

This is acceptable under compatibility conditions on the data. Notice that an additional compatibility condition can also be derived from (31), (32), namely

$$\nabla \times j + \partial_s b = 0. \tag{42}$$

Of course, all these compatibility conditions are not independent from each other and we can select those which lead to a self-consistent system of time-evolution equations.

After introducing

$$Z = \sqrt{(\rho^2 + |E|^2)(1 + |j|^2 + |b|^2)} \tag{43}$$

and noticing that

$$\frac{\partial Z}{\partial \rho} = \rho R^{-1}, \quad \frac{\partial Z}{\partial j} = Rj, \quad \frac{\partial Z}{\partial E} = ER^{-1}, \quad \frac{\partial Z}{\partial b} = Rb, \tag{44}$$

we finally get such a system, namely

$$\partial_t \rho = -\nabla \cdot (\frac{\partial Z}{\partial j}), \qquad \partial_t E = \partial_s(\frac{\partial Z}{\partial j}) + \nabla \times (\frac{\partial Z}{\partial b}), \tag{45}$$

$$\partial_t j = \partial_s(\frac{\partial Z}{\partial E}) - \nabla(\frac{\partial Z}{\partial \rho}), \qquad \partial_t b = -\nabla \times (\frac{\partial Z}{\partial E}). \tag{46}$$

From now on, we call these equations $MKMEs$ (Monge-Kantorovich Maxwell equations).

2.4. Recovery of the Maxwell equations

In the same way as the Poisson equation can be reset in the MK framework, the Maxwell equations can be recovered in the framework of (24), (25), (26) and (28), provided we substitute the degenerate and quadratic expression $k(\rho, J, E, B) = |E|^2 - |B|^2$ for (27) in the definition of the Action. Namely, the Maxwell equations can be obtained by varying

$$\int |E|^2 - |B|^2 \;\; dtdsdx, \tag{47}$$

where E, B are valued in $\mathbf{R}^3$, depend on $(t, s, x) \in [0, T] \times [0, 1] \times \mathbf{R}^3$, and are subject to

$$\partial_s \rho + \nabla \cdot E = 0, \;\; \partial_t E - \partial_s J - \nabla \times B = 0, \tag{48}$$

for some fields $\rho \in \mathbf{R}$, $J \in \mathbf{R}^3$, depending on (t, s, x) that are prescribed at $s = 0$ and $s = 1$.

2.5. An Euler-Maxwell-type system

Since the $MKMEs$ are time evolution equations, it is natural to supplement them with initial value conditions. We also need boundary conditions, at least for the interpolation variable $s \in [0, 1]$. So far, we have used condition (28). It is interesting to consider other 'natural' boundary conditions such as

$$(E, B)(t, s = 0, x) = (E, B)(t, s = 1, x) = 0. \tag{49}$$

(Notice that this boundary condition means

$$\partial_s X(t, s=0) = \partial_s X(t, s=1) = 0 \tag{50}$$

in the special case when (ρ, J, E, B) is generated from a surface X through (21), (22), (23).) Let us focus on the boundary values of the other fields (ρ, J) at $s = 0$ and $s = 1$ and denote them respectively by

$$(\rho_-, J_-)(t,x) = (\rho, J)(t, s=0, x), \quad (\rho_+, J_+)(t,x) = (\rho, J)(t, s=1, x). \tag{51}$$

Remarkable simplifications occur in the $MKMEs$ restricted at $s = 0$ and $s = 1$ for such boundary conditions. Indeed, we get, at $s = 0$ and $s = 1$, since $E = B = 0$, from (30), (40),

$$R = \sqrt{\rho^2 - |J|^2} = \frac{\rho}{\sqrt{1+|j|^2}}, \tag{52}$$

$$\frac{\rho}{R} = \sqrt{1+|j|^2}, \quad Rj = \frac{\rho j}{\sqrt{1+|j|^2}}, \quad \partial_s(Rj) = \partial_s\left(\frac{\rho j}{\sqrt{1+|j|^2}}\right). \tag{53}$$

After some calculations, we finally get at $s = 0$ and $s = 1$

$$\partial_t \rho_- + \nabla \cdot \left(\frac{\rho_- j_-}{\sqrt{1+|j_-|^2}}\right) = \partial_t \rho_+ + \nabla \cdot \left(\frac{\rho_+ j_+}{\sqrt{1+|j_+|^2}}\right) = 0, \tag{54}$$

$$\left(\partial_t + \frac{j_-}{\sqrt{1+|j_-|^2}} \cdot \nabla\right) j_- = E_- + \frac{j_-}{\sqrt{1+|j_-|^2}} \times B_-, \tag{55}$$

$$\left(\partial_t + \frac{j_+}{\sqrt{1+|j_+|^2}} \cdot \nabla\right) j_+ = E_+ + \frac{j_+}{\sqrt{1+|j_+|^2}} \times B_+, \tag{56}$$

where the 'electromagnetic' fields $(E_-, B_-)(t,x)$, $(E_+, B_+)(t,x)$ are defined by

$$E_-(t,x) = \partial_s e(t, s=0, x), \qquad E_+(t,x) = \partial_s e(t, s=1, x), \tag{57}$$

$$B_-(t,x) = \partial_s b(t, s=0, x), \qquad B_+(t,x) = \partial_s b(t, s=1, x). \tag{58}$$

These fields are linked by (37)

$$\partial_t B_- + \nabla \times E_- = \partial_t B_+ + \nabla \times E_+ = 0, \qquad \nabla \cdot B_- = \nabla \cdot B_+ = 0, \tag{59}$$

which is one half of the usual Maxwell equations. Thus, with equations (54), (55), (56) and (59), we are not far from the standard (pressureless) Euler-Maxwell system.

References

[1] J.-D. Benamou, Y. Brenier, *A Computational Fluid Mechanics solution to the Monge-Kantorovich mass transfer problem, Numerische Math. 84 (2000) 375-393.*

[2] Y.Brenier, *Polar factorization and monotone rearrangement of vector-valued functions, Comm. Pure Appl. Math. 64 (1991) 375-417.*

[3] L.A. Caffarelli, *Boundary regularity of maps with convex potentials.* Ann. of Math. *(2) 144 (1996), no. 3, 453–496.*

[4] L.C. Evans *Partial differential equations and Monge-Kantorovich mass transfer, Lecture Notes, 1998.*

[5] W. Gangbo, R.J. McCann, *The geometry of optimal transportation,* Acta Math. 177 *(1996), no. 2, 113–161.*

[6] D. Kinderlehrer, R. Jordan, F. Otto, *The variational formulation of the Fokker-Planck equation, SIAM J. Math. Anal. 29 (1998), no. 1,1–17.*

[7] F. Otto, *Evolution of microstructure in unstable porous media flow: a relaxational approach,. Comm. Pure Appl. Math. 52 (1999) 873-915.*

[8] J. Polchinski, *String theory. Vol. I.,* Cambridge University Press, *1998.*

[9] S.T. Rachev, L. Rüschendorf, *Mass transportation problems, Vol. I* and *II.* Probability *and its Applications, Springer-Verlag, New York, 1998.*

Y. Brenier
CNRS Nice Mathématiques,
Parc Valrose,
06108 Nice Cedex 02,
on leave of absence from Université Paris 6
`brenier@math.unice.fr`

International Series of Numerical Mathematics
Vol. 140, © 2001 Birkhäuser Verlag Basel/Switzerland

Convergence of the Godunov Scheme for Straight Line Systems

Alberto Bressan and Helge Kristian Jenssen

1. Introduction

Consider the Cauchy problem for an $n \times n$ system of the form

$$u_t + A(u)u_x = 0, \qquad u(x,0) = \bar{u}(x). \tag{1}$$

Here $A(u)$ is a map from a domain $\mathcal{U} \subset \mathbb{R}^n$ into $\mathbb{R}^{n\times n}$, and $(x,t) \in \mathbb{R} \times \mathbb{R}^+$. We assume strict hyperbolicity, i. e. the the matrix $A(u)$ has n real and strictly different eigenvalues for each $u \in \mathcal{U}$. In the conservative case when $A(u) = Df(u)$ for some map $f : \mathcal{U} \to \mathbb{R}^n$, Glimm [12] proved global existence of weak entropy solutions of (1) when the data has small total variation and each characteristic field is genuinely nonlinear or linearly degenerate. The recent papers [4], [6], [7], [8] have established the uniqueness and $\mathbf{L}^1$ stability of solutions obtained by the Glimm scheme or by wave-front tracking. A major open question is whether other methods of approximation, such as vanishing viscosity, relaxation and finite difference schemes, yield the same solutions. In the scalar case all these methods give the unique entropy solution. Some results have been proved for 2×2 systems [11] and $n \times n$ Temple class systems [16] using compensated compactness. For general conservative systems, convergence results for vanishing viscosity approximations are known only in the case where the limit solution is piecewise smooth [13, 17].

In general, if one can establish a priori estimates on the total variation of approximate solutions, then Helly's theorem guarantees convergence. Using the uniqueness results in [5] one can then show that this limit is the appropriate entropy weak solution. Recently such estimates have been established for systems (1) under the assumption that the integral curves of the eigenvector fields of $A(u)$ are straight lines in state space. In [3] it was proved that vanishing viscosity yield an $\mathbf{L}^1$ Lipschitz continuous semigroup which is consistent with the Riemann solver. See [2], [10] for results in the case of relaxation.

In the present note we show that, under this straight-line assumption, also the approximate solutions constructed by the Godunov scheme converge to a unique limit, the same obtained by vanishing viscosity [3]. This result has been established previously by LeVeque and Temple [15] for 2×2 straight line systems with the additional property that both fields are genuinely nonlinear. Their analysis makes essential use of Riemann invariants to bound the total variation. In contrast to this we consider $n \times n$ straight line systems with no assumptions on genuine nonlinearity or linear degeneracy and without Riemann invariants. In addition we obtain stability in L^1 by a linearization argument.

The main reason for considering these systems is that the effects of nonlinear coupling are somewhat easier to estimate. Indeed, under this straight-line assumption, new oscillations can be generated only due to the interaction of waves from *different* families.

These particular systems were observed to be very well behaved in connection with numerical computation of slowly moving shocks. In [1] Arora and Roe conjectured that if the straight line assumption is not satisfied, then any numerical scheme using a Godunov or a Roe flux will create spurious oscillations.

Our results apply to general straight-line hyperbolic systems of the form (1), not necessarily in conservation form. For simplicity we will outline the proofs only in the conservative case. For all details we refer to [9].

2. The Godunov Scheme and Main Results

Let the matrix $A(u)$ in (1) have n real and distinct eigenvalues $\lambda^k(u)$, $k = 1, \dots, n$, $u \in \mathcal{U}$. By a linear change in the t-x coordinates we can assume that

$$0 < \lambda^1(u) < \cdots < \lambda^n(u) < 1. \tag{2}$$

We also assume that these characteristic speeds are uniformly separated, i.e. there exists constants $\bar{\lambda}^0 = 0 < \bar{\lambda}^1 < \cdots < \bar{\lambda}^{n-1} < \bar{\lambda}^n = 1$ such that

$$\lambda^k(u) \in]\bar{\lambda}^{k-1}, \bar{\lambda}^k[, \quad u \in \mathcal{U}, \quad k = 1, \dots, n.$$

The corresponding right and left eigenvectors are denoted by $r^k(u)$, $l^k(u)$, respectively, and normalized such that $|r^k(u)| \equiv 1$ and $r^i(u) \cdot l^j(u) \equiv \delta_{i,j}$. The basic assumption is that the integral curves of the eigenvector fields are straight lines. This can be expressed by the relation

$$(Dr^k(u))r^k(u) = 0, \qquad u \in \mathcal{U}, \quad k = 1, \dots, n, \tag{3}$$

where Dr^k denotes the Jacobian matrix of r^k.

2.1. The Riemann Solver

Consider the Riemann problem for (1) with data

$$\bar{u}(x) = \begin{cases} u_- & \text{for } x < 0, \\ u_+ & \text{for } x > 0. \end{cases} \tag{4}$$

For $u \in \mathcal{U}$, $k = 1, \dots, n$, let $R^k(\cdot)(u)$ be the integral curve of r^k through u,

$$R^k(\sigma)(u) = u + \sigma r^k(u).$$

Given $u_-, u_+ \in \mathcal{U}$, by strict hyperbolicity there exist unique intermediate states $\omega_0 = u_-, \omega_1, \dots, \omega_n = u_+$ and wave strengths σ^k such that

$$\omega_k = R^k(\sigma^k)(\omega_{k-1}), \qquad k = 1, \dots, n.$$

For each k, define the scalar function

$$F^k(\sigma) = \int_0^\sigma \lambda^k \left(R^k(s)(\omega_{k-1})\right) ds,$$

and let $z^k(x,t)$ be the unique entropy solution to the scalar Riemann problem

$$z_t + F^k(z)_x = 0, \qquad z(x,0) = \begin{cases} 0 & \text{for } x < 0, \\ \sigma^k & \text{for } x > 0. \end{cases} \tag{5}$$

To solve the Riemann problem (u_-, u_+) we use the solution of these scalar Riemann problems to move along the integral curves. More precisely, we first define the map $\mathcal{R}(u_-, u_+) : [0,1] \to \mathcal{U}$ by

$$\mathcal{R}(u_-, u_+)(\xi) = R^k(z^k(\xi, 1))(\omega_{k-1}) \qquad \text{for } \xi \in [\bar{\lambda}^{k-1}, \bar{\lambda}^k], \quad k = 1, \ldots, n. \tag{6}$$

We then define the solution of the Riemann problem (1),(4) as

$$u(x,t) = \begin{cases} u_- & \text{for } x/t < 0, \\ \mathcal{R}(u_-, u_+)(x/t) & \text{for } x/t \in [0,1], \\ u_+ & \text{for } x/t > 1. \end{cases} \tag{7}$$

2.2. The Godunov Scheme

By the assumption (2) we can choose equal time and space step $\Delta x = \Delta t$. The constant value of the approximate solution in the i-th cell at time $j\Delta t$ is denoted by $u_{i,j}$. The Godunov scheme is now defined inductively as follows: At time $t = 0$ we let $u_{i,0}$ denote the cell average over the i-th cell of the initial data $\bar{u}$. Given the values $u_{i-1,j}$ and $u_{i,j}$ in the $(i-1)$-th and i-th cells, respectively, at time $j\Delta t$, we define the value $u_{i,j+1}$ by

$$u_{i,j+1} = \frac{1}{\Delta x} \int_0^{\Delta x} \mathcal{R}(u_{i-1,j}, u_{i,j})(x/\Delta t)\, dx = \int_0^1 \mathcal{R}(u_{i-1,j}, u_{i,j})(\xi)\, d\xi. \tag{8}$$

For each Δx the scheme gives an approximate solution $u_{\Delta x}$ of (1) defined for all $(x,t) \in \mathbb{R} \times \mathbb{R}^+$.

2.3. Main Results

We now state the main results of the paper.

Theorem 2.1. *Suppose that $A(u)$ is strictly hyperbolic for each $u \in \mathcal{U}$ and that the normalized eigenvectors satisfy (3). Then there exist constants $\delta_0, \delta_1 > 0$ such that the following holds. For each initial data $\bar{u} \in L^1$ with*

$$T.V.[\bar{u}] < \delta_0, \tag{9}$$

the corresponding solution $u_{i,j}$ of the Godunov scheme (8) is well defined for all time steps $j \in \mathbb{Z}_+$ and satisfies

$$T.V.[u(j)] < \delta_1, \qquad \text{for all } j \in \mathbb{Z}_+. \tag{10}$$

Furthermore, there exists a constant L such that for all pairs of initial data $\bar{u}, \bar{v}$ satisfying (9), the corresponding solutions satisfy

$$\|u(j) - v(j')\|_{L^1} \le L \cdot \left(|j - j'| \cdot \Delta t + \|\bar{u} - \bar{v}\|_{L^1}\right). \tag{11}$$

Theorem 2.2. *As the discretization parameter $\Delta x = \Delta t$ tends to zero, the approximate solutions given by the Godunov scheme (8) converge to the same limit as given by the method of vanishing viscosity. This limit can be characterized as the trajectory of a semigroup $S : \mathcal{D} \times \mathbb{R}^+ \to \mathcal{D}$ with the properties:*

(i) *Every initial data $\bar{u} \in L^1$ with sufficiently small total variation lies in the domain $\mathcal{D}$ of the semigroup.*

(ii) *For every $\bar{u}, \bar{v} \in \mathcal{D}$ and every $t, s \geq 0$,*

$$\|S_t\bar{u} - S_s\bar{v}\|_{L^1} \leq L \cdot \left(|t-s| + \|\bar{u} - \bar{v}\|_{L^1}\right). \tag{12}$$

(iii) *For every piecewise constant initial data $\bar{u} \in \mathcal{D}$, there exists a positive time $\tau > 0$ such that the the semigroup trajectory $S_t\bar{u}$ on $[0, \tau]$ coincides with the function obtained by patching together the solutions of the Riemann problems given by $\bar{u}$ and solved according to (7).*

In particular, for conservative systems the Godunov scheme yields the trajectories of the Standard Riemann Semigroup.

3. Wave strengths and wave speeds

We treat the conservative case with $A = Df$. The Riemann solver is then the standard one and the Godunov scheme can be written in finite difference form

$$u_{i,j+1} = u_{i,j} - [f(u_{i,j}) - f(u_{i-1,j})], \qquad i \in \mathbb{Z}, \quad j \in \mathbb{Z}_+. \tag{13}$$

Given a left state u_- and a right state u_+ connected by a single k-wave of strength σ^k, the speed of the wave is

$$\lambda^k(u_-, u_+) = \int_0^1 \lambda^k(u_- + \xi\sigma^k r^k(u_-))\, d\xi. \tag{14}$$

Given three consecutive states u_-, u_0, u_+, let $\bar{u}_-$, $\bar{u}_+$ be the resulting states,

$$\bar{u}_- = u_0 - [f(u_0) - f(u_-)], \qquad \bar{u}_+ = u_+ - [f(u_+) - f(u_0)]. \tag{15}$$

The strengths of the k-waves, $k = 1, \ldots, n$, in the three Riemann problems (u_-, u_0), (u_0, u_+), and $(\bar{u}_-, \bar{u}_+)$ are denoted by σ_-^k, σ_+^k, and $\bar{\sigma}^k$, respectively. The intermediate states in the two Riemann problems (u_-, u_0) and (u_0, u_+) are denoted by $u_-^1, \ldots, u_-^{n-1}$ and $u_+^1, \ldots, u_+^{n-1}$, respectively. Now define the map

$$\Psi^k(u_0; \sigma_-; \sigma_+) = \bar{\sigma}^k - \left(\lambda^k(u_-^{k-1}, u_-^k)\sigma_-^k + [1 - \lambda^k(u_+^{k-1}, u_+^k)]\sigma_+^k\right), \tag{16}$$

where $\sigma_\pm = (\sigma_\pm^1, \ldots, \sigma_\pm^n)$. We have the following representation.

Lemma 3.1. *Suppose that the straight-line assumption (3) holds and let Ψ^k be defined by (16). Then there exist smooth functions $A_{p,q}^k$, $B_{p,q}^k$, and $C_{p,q}^k$ depending on u_0, σ_-, σ_+ such that*

$$\Psi^k(u_0; \sigma_-; \sigma_+) = \sum_{1 \leq p \neq q \leq n} \left(A_{p,q}^k \sigma_+^p \sigma_+^q + B_{p,q}^k \sigma_+^p \sigma_-^q + C_{p,q}^k \sigma_-^p \sigma_-^q\right), \tag{17}$$

for all $u_0 \in \mathcal{U}$ and for all sufficiently weak strengths $\sigma_\pm^1, \ldots, \sigma_\pm^n$.

By a representation result for smooth maps it suffices to show that Ψ^k vanishes whenever σ_-, σ_+ contain only one wave, of the same family. That is, $\Psi^k(u_0; \sigma_-; \sigma_+) = 0$ whenever there exists an index j such that

$$\sigma_+^p = \sigma_-^p = 0 \qquad \forall\, p \neq j.$$

The above identity is a consequence of the definition of Ψ^k and the assumption (3). The lemma shows that the increase in total variation is due only to *transversal* terms. Returning to the scheme, we let $\sigma_{i,j}^k$ be the strength of the k-wave in the Riemann problem $(u_{i-1,j}, u_{i,j})$, and we let $\lambda_{i,j}^k$ be the corresponding speed given. By Lemma 3.1 the strengths satisfy the system

$$\sigma_{i,j+1}^k = \lambda_{i-1,j}^k \sigma_{i-1,j}^k + (1 - \lambda_{i,j}^k)\sigma_{i,j}^k + Q_{i,j}^k, \tag{18}$$

for $k = 1, \ldots, n$, where the quadratic coupling terms are given by

$$Q_{i,j}^k = \sum_{1 \le p \neq q \le n} \left(A_{p,q}^k \sigma_{i,j}^p \sigma_{i,j}^q + B_{p,q}^k \sigma_{i,j}^p \sigma_{i-1,j}^q + C_{p,q}^k \sigma_{i-1,j}^p \sigma_{i-1,j}^q \right). \tag{19}$$

4. A priori bounds on the total variation

We obtain the a priori bounds by bounding the total amount of waves in each family uniformly in time. That is, we establish estimates on the sums

$$\mathrm{V}^k(j) := \sum_{i=-\infty}^{\infty} |\sigma_{i,j}^k|, \qquad k = 1, \ldots, n, \tag{20}$$

that are independent of the time step j. We derive a functional relation which implies that if the initial amount of waves is sufficiently small, then it remains small for all times. The functional relation is deduced by applying Duhamel's principle, and a basic estimate on the product of two Green kernels. From (18) it follows that

$$\mathrm{V}^k(j+1) \le \mathrm{V}^k(j) + \sum_{i=-\infty}^{+\infty} |Q_{i,j}^k| \le \mathrm{V}^k(0) + \sum_{m=0}^{j} \sum_{i=-\infty}^{+\infty} |Q_{i,m}^k|. \tag{21}$$

Defining the magnitude $\mathrm{Q}^k(j)$ by

$$\mathrm{Q}^k(j) := \sum_{m=0}^{j} \sum_{i=-\infty}^{+\infty} |Q_{i,m}^k|, \tag{22}$$

and using (19), we obtain the following bound

$$\mathrm{Q}^k(j) \le C_0 \cdot \sum_{p \neq q} \sum_{m=0}^{j} \sum_{i=-\infty}^{+\infty} \left(|\sigma_{i,m}^p||\sigma_{i,m}^q| + |\sigma_{i-1,m}^p||\sigma_{i,m}^q| \right) =: E_1 + E_2, \tag{23}$$

where C_0 is a suitable constant. Let $\Gamma^k(i,j;i',j')$ denote the Green kernel corresponding to the homogeneous version of (18). By Duhamel's principle we can write the solution of the non-homogeneous system (18) as

$$\sigma^k_{i,j} = \sum_{l=-\infty}^{+\infty} \Gamma^k(i,j;l,0)\sigma^k_{l,0} + \sum_{r=0}^{j-1}\sum_{l=-\infty}^{+\infty} \Gamma^k(i,j;l,r)\mathrm{Q}^k_{l,r} =: \alpha^k_{i,j} + \beta^k_{i,j}. \tag{24}$$

We thus have,

$$\begin{aligned} E_1 &\leq C_0 \cdot \sum_{p\neq q}\sum_{m=0}^{j}\sum_{i=-\infty}^{+\infty} \Big(|\alpha^p_{i,m}||\alpha^q_{i,m}| + |\alpha^p_{i,m}||\beta^q_{i,m}| + \\ &\qquad\qquad |\beta^p_{i,m}||\alpha^q_{i,m}| + |\beta^p_{i,m}||\beta^q_{i,m}|\Big) \\ &= C_0 \cdot (S_1 + S_2 + S_3 + S_4). \end{aligned} \tag{25}$$

To proceed we must estimate the total interaction between Green kernels.

Lemma 4.1. *Assume that $0 < \lambda^p_{i,j} < \lambda < \tilde{\lambda} < \lambda^q_{i',j'} < 1$ for all $(i,j),(i',j') \in \mathbb{Z}\times\mathbb{Z}_+$. Let $\rho^p_{i,j}$ and $\rho^q_{i,j}$ denote the solutions of (18) in the homogeneous case where $Q^p_{i,j} = Q^q_{i,j} \equiv 0$ and with initial data ρ^p_0 and ρ^q_0, respectively. Then,*

$$\sum_{m=0}^{j}\sum_{i=-\infty}^{+\infty} |\rho^p_{i,m}||\rho^q_{i,m}| \leq C(p,q)\cdot\Big(\sum_{i=-\infty}^{+\infty} |\rho^p_0(i)|\Big)\cdot\Big(\sum_{i=-\infty}^{+\infty} |\rho^q_0(i)|\Big). \tag{26}$$

To prove the result one first consider the constant coefficient case and then extends by a comparison argument. Thanks to Lemma 4.1, we can now estimate each term on the right hand side of (25). By (24) and (26) we have

$$S_1 = \sum_{p\neq q}\sum_{m=0}^{j}\sum_{i=-\infty}^{+\infty} |\alpha^p_{i,m}||\alpha^q_{i,m}| \leq C_1 \cdot \sum_{p\neq q} \mathrm{V}^p(0)\cdot\mathrm{V}^q(0), \tag{27}$$

where $C_1 := \max_{p\neq q} C(p,q)$. Estimating S_2, S_3 and S_4 in the same way yields

$$E_1 \leq C_1 \cdot \sum_{p\neq q} \Big(\mathrm{V}^p(0) + \mathrm{Q}^p(j-1)\Big)\Big(\mathrm{V}^q(0) + \mathrm{Q}^q(j-1)\Big). \tag{28}$$

The quantity E_2 in (23) can be estimated similarly. From (23) we thus have

$$\mathrm{Q}^k(j) \leq 2C_1 \cdot \sum_{p\neq q} \Big(\mathrm{V}^p(0) + \mathrm{Q}^p(j-1)\Big)\Big(\mathrm{V}^q(0) + \mathrm{Q}^q(j-1)\Big). \tag{29}$$

We define,

$$\mathbf{V}(j) = \sum_{k=1}^{n} \mathrm{V}^k(j), \qquad \mathbf{Q}(j) = \sum_{k=1}^{n} \mathrm{Q}^k(j), \tag{30}$$

From (21) and (29)-(30) we conclude that

$$\mathbf{V}(j) \leq \mathbf{V}(0) + \mathbf{Q}(j-1), \quad \mathbf{Q}(j) \leq C_2 \cdot \Big(\mathbf{V}(0) + \mathbf{Q}(j-1)\Big)^2, \tag{31}$$

where $C_2 = 2n \cdot C_1$. It follows that, if the total amount of waves $\mathbf{V}(0)$ in the initial data is sufficiently small, then $\mathbf{Q}(j)$, and hence also the total amount of waves $\mathbf{V}(j)$, remains small for all time steps.

5. $\mathbf{L}^1$ Stability

Following [3],[6] we prove L^1 stability by a linearization argument. Consider two initial data $\bar{u}, \bar{v}$ which we join by the path defined by

$$\theta \mapsto \bar{u}^\theta = \theta\bar{u} + (1-\theta)\bar{v}, \qquad \theta \in [0,1]. \tag{32}$$

Let $u^\theta_{i,j}$ denote the approximate solution computed with the Godunov scheme applied to the initial data $\bar{u}^\theta$ and let $u_{i,j} = u^1_{i,j}, v_{i,j} = u^0_{i,j}$.

Suppose the left and right states of a Riemann problem depends on θ, $u_- = u^\theta_-$, $u_+ = u^\theta_+$. The resulting state $\tilde{u} = \tilde{u}^\theta$ is given by

$$\tilde{u}^\theta = u^\theta_+ - [f(u^\theta_+) - f(u^\theta_-)]. \tag{33}$$

Differentiation with respect to θ shows that the infinitesimal perturbations

$$v^\theta_- := \frac{du^\theta_-}{d\theta}, \quad v^\theta_+ := \frac{du^\theta_+}{d\theta}, \quad \tilde{v}^\theta := \frac{d\tilde{u}^\theta}{d\theta}$$

satisfy the linearized equation

$$\tilde{v}^\theta = v^\theta_+ - [A(u^\theta_+)v^\theta_+ - A(u^\theta_-)v^\theta_-]. \tag{34}$$

We decompose the perturbations along the eigenvector bases,

$$v^\theta_- = \sum_{k=1}^n \nu^{\theta,k}_- r^k(u^\theta_-), \quad v^\theta_+ = \sum_{k=1}^n \nu^{\theta,k}_+ r^k(u^\theta_+), \quad \tilde{v}^\theta = \sum_{k=1}^n \tilde{\nu}^{\theta,k} r^k(\tilde{u}^\theta),$$

Letting $\sigma^\theta = (\sigma^{\theta,1}, \dots, \sigma^{\theta,n})$ denote the strengths of the waves in the solution of the Riemann problem (u^θ_-, u^θ_+) we observe that the components $\tilde{\nu}^{\theta,k}$ are uniquely determined by u^θ_-, σ^θ and $\nu^\theta_\pm = (\nu^{\theta,1}_\pm, \dots, \nu^{\theta,n}_\pm)$. In analogy with the maps Ψ^k above we now define the maps Φ^k, $k = 1, \dots, n$, by

$$\Phi^k(u^\theta_-, \sigma^\theta, \nu^\theta_-, \nu^\theta_+) = \tilde{\nu}^{\theta,k} - \Big(\lambda(u^\theta_-)\nu^{\theta,k}_- + [1 - \lambda(u^\theta_+)]\nu^{\theta,k}_+\Big).$$

It is easily shown that

$$\Phi^k(u_-, \sigma, 0, 0) \equiv 0, \qquad \Phi^k(u_-, 0, \nu_-, \nu_+) \equiv 0.$$

It follows that Ψ^k admits a representation of the form

$$\Phi^k(u_-, \sigma, \nu_-, \nu_+) = \sum_{1 \le p,q \le n} \Big[A^k_{p,q}\sigma^p\nu^q_- + B^k_{p,q}\sigma^p\nu^q_+\Big], \tag{35}$$

where $A^k_{p,q}$ and $B^k_{p,q}$ are smooth functions of $u_-, \sigma, \nu_-, \nu_+$. Returning to the scheme we thus have the following system of equations for the components $\nu^{\theta,k}_{i,j}$ of the tangent vectors $du^\theta_{i,j}/d\theta$,

$$\nu^{\theta,k}_{i,j+1} = \lambda^k(u^\theta_{i-1,j})\nu^{\theta,k}_{i-1,j} + [1-\lambda^k(u^\theta_{i,j})]\nu^{\theta,k}_{i,j} + \mathcal{Q}^{\theta,k}_{i,j}, \tag{36}$$

where the coupling terms are of the form (35). The L^1 norm of the difference between $u_{i,j} = u^\theta_{i,j}\mid_{\theta=1}$ and $v_{i,j} = v^\theta_{i,j}\mid_{\theta=0}$ at time step $(j+1)$ is now given by

$$\begin{aligned}\|u(j+1)-v(j+1)\|_{\mathrm{L}^1} &= \sum_{i=-\infty}^{+\infty} |u_{i,j+1} - v_{i,j+1}|\Delta x \\ &\le \Delta x \cdot \left(\int_0^1 \sum_{i=-\infty}^{+\infty}\sum_{k=1}^{n} |\nu^{\theta,k}_{i,j+1}|\, d\theta\right) \\ &=: \Delta x \cdot \mathcal{N}(j+1).\end{aligned} \tag{37}$$

Using (36) we have

$$\mathcal{N}(j+1) \le \mathcal{N}(0) + \int_0^1 \sum_{i=-\infty}^{+\infty}\sum_{k=1}^{n}\sum_{s=0}^{j} \left|\mathcal{Q}^{\theta,k}_{i,s}\right| d\theta. \tag{38}$$

A simple estimate now shows that for a suitable constant C_3,

$$\int_0^1 \sum_{i=-\infty}^{+\infty}\sum_{k=1}^{n}\sum_{s=0}^{j} \left|\mathcal{Q}^{\theta,k}_{i,s}\right| d\theta \le C_3 \cdot \Big(\max_{0\le s\le j}\max_{0\le\theta\le 1} \mathbf{V}^\theta(s)\cdot\mathcal{N}(s)\Big),$$

where $\mathbf{V}^\theta(s)$ is defined by (30). From above it follows that, if the total amount of waves in the data $\bar u$, $\bar v$ are sufficiently small, then $\mathbf{V}^\theta(s)$ is uniformly bounded and small. Thus, for data with sufficiently small total variation we have

$$C_4 := C_3 \cdot \Big(\max_{0\le s\le j}\max_{0\le\theta\le 1} \mathbf{V}^\theta(s)\Big) < 1,$$

such that

$$\mathcal{N}(j+1) \le \mathcal{N}(0) + C_4 \cdot \max_{0\le s\le j} \mathcal{N}(s).$$

From (37) we conclude that there exists a Lipschitz constant L such that

$$\|u(j)-v(j)\|_{\mathrm{L}^1} \le L\cdot\|u(0)-v(0)\|_{\mathrm{L}^1}.$$

6. Convergence of the Godunov Scheme

To establish the convergence of the scheme we argue as in [3]. We define a semigroup on $\mathcal{D}$ by using the approximate solutions $u_{\Delta x}$ defined by the Godunov scheme. Given $\bar u \in \mathcal{D}$ we approximate it by functions $\bar u_\mu$ that are piecewise constant on a sequence of grids with mesh tending to zero. Let u_μ denote the corresponding Godunov approximation. By the stability result it follows that, for some subsequence $\mu \to 0$, there exists the limit

$$S_t\bar u := v(\cdot,t) = \mathrm{L}^1_{loc} - \lim_{\mu\to\infty} u_\mu(\cdot,t)$$

and this limit defines a uniformly Lipschitz semigroup. Finally, to show that the uniqueness of the limit, by Corollary 9.2 in [5], it suffices to check that the Godunov scheme provides the correct solution locally in time, in case of an initial data $\bar{u}$ which is piecewise constant with *simple jumps.* That means, we can assume that at each point x where $\bar{u}$ has a jump, the corresponding Riemann problem is solved by waves of one single family. In this case, we are thus dealing essentially with a scalar problem, and the result is well known. Indeed, the Godunov scheme converges to the unique entropy solution when applied to scalar conservation laws. This completes our sketch of the proof of convergence of Godunov approximations to a unique limit. For further details we refer to [9].

Acknowledgement. The research was supported by the European TMR Network on Hyperbolic Systems of Conservation Laws, ERBFMRXCT960033.

References

[1] M. Arora, P. L. Roe, *On postshock oscillations due to shock capturing schemes in unsteady flows,* J. Comput. Phys., **130**, (1997), 25–40.

[2] S. Bianchini, *On a Glimm type functional for a special Jin-Xin relaxation model,* Ann. Inst. Henri Poincaré - Nonlinear Analysis , to appear.

[3] S. Bianchini, A. Bressan, *BV solutions for a class of viscous hyperbolic systems,* Indiana Univ. Math. J. **49**, (2000).

[4] A. Bressan, *The unique limit of the Glimm scheme,* Arch. Rational Mech. Anal. **130**, (1995), 205–230.

[5] A. Bressan, *Hyperbolic Systems of Conservation Laws. The One Dimensional Cauchy Problem,* Oxford University Press, 2000.

[6] A. Bressan, R. M. Colombo, *The semigroup generated by* 2×2 *conservation laws,* Arch. Rational Mech. Anal. **133**, (1995), 1–75.

[7] A. Bressan, G. Crasta, B. Piccoli, *Well posedness of the Cauchy problem for* $n \times n$ *systems of conservation laws,* Amer. Math. Soc. Memoir **694**, (2000).

[8] A. Bressan, T. P. Liu, T. Yang, L^1 *stability estimates for* $n \times n$ *conservation laws,* Arch. Rational Mech. Anal. **149**, (1999), 1–22.

[9] A. Bressan, H. K. Jenssen, *On the convergence of Godunov scheme for nonlinear hyperbolic systems,* Chinese Ann. Math. , **21**, (2000), 1–16.

[10] A. Bressan, W. Shen, *BV estimates for multicomponent chromatography with relaxation,* Discrete Cont. Dyn. Syst. , **6**, (2000), 21-38.

[11] R. J. DiPerna, *Convergence of approximate solutions to conservation laws,* Arch. Rational Mech. Anal. **82**, (1983), 27–70.

[12] J. Glimm, *Solutions in the Large for Nonlinear Hyperbolic Systems of Equations,* Comm. Pure. Appl. Math. **18** (1965), 697–715.

[13] J. Goodman, Z. Xin, *Viscous limits for piecewise smooth solutions to systems of conservation laws,* Arch. Rational Mech. Anal. **121**, (1992), 235–265.

[14] D. Kröner, *Numerical Schemes for Conservation Laws*, Wiley and Teubner, 1997.

[15] R. J. LeVeque, B. Temple, *Stability of Godunov's method for a class of* 2×2 *systems of conservation laws,* SIAM J. Math. Anal. **288** (1985), 115–123.

[16] D. Serre, *Solutions with bounded variations for some hyperbolic systems of conservation laws,* J. Differential Equations **68** (1987), 137–168.

[17] S. H. Yu, *Zero dissipation limit to solutions with shocks for systems of hyperbolic conservation laws,* Arch. Rational Mech. Anal. **146**, (1999), 275–370.

A. Bressan
S.I.S.S.A.,
Via Beirut 4,
Trieste 34014, Italy
bressan@sissa.it

H.K. Jenssen
Department of Mathematics,
Indiana University,
Rawles Hall,
Bloomington, IN 47405, USA
kjenssen@indiana.edu

International Series of Numerical Mathematics
Vol. 140, © 2001 Birkhäuser Verlag Basel/Switzerland

The Convergence of Multicomponent Chromatography with Relaxation

Alberto Bressan and Wen Shen

Abstract. We consider a $2n \times 2n$ system of conservation laws with stiff relaxation terms. The system arises from the modelling of multicomponent chromatography, and the zero relaxation limit is a $n \times n$ system of conservation laws of Temple class. For initial data which are small in BV, we establish an a priori bound on the total variation of the solution, relying on a probabilistic technique. Moreover, we show that this solution depends Lipschitz continuously on the initial data in $\mathbf{L}^1$. These estimates are uniform w.r.t. the relaxation parameter. Finally, we prove the convergence towards equilibrium as the relaxation time tends to zero. This provides a first example where BV estimates and convergence are proved for general solutions to a class of $2n \times 2n$ systems with relaxation.

1. Introduction and preliminaries

In this paper we consider the system of multicomponent chromatography:

$$\begin{cases} u_t + u_x &= -\frac{1}{\varepsilon}(F(u) - v), \\ v_t &= \frac{1}{\varepsilon}(F(u) - v), \end{cases} \qquad \begin{cases} u(0,x) &= u_0(x), \\ v(0,x) &= v_0(x). \end{cases} \tag{1}$$

The system describes the following chemical process: A liquid flows with unit speed over a solid bed. Several chemical substances are partialy dissolved in the liquid which travels with unit speed, partialy deposited on the solid bed which has speed 0. The main unknowns of the system are the vectors $u = (u_1, \ldots, u_n)$ and $v = (v_1, \ldots, v_n)$, which denote the concentrations in the fluid and the solid phase respectively. A particle might pass from the fluid phase to the solid phase, and viceversa. The *relaxation time* ε measures the rate of this exchange. For the equlibrium relation $v = F(u)$ we consider the Langmuir isotherm (cf. [10])

$$F_i(u) \doteq \frac{k_i u_i}{1 + k_1 u_1 + \cdots + k_n u_n}, \qquad 0 < k_1 < \cdots < k_n. \tag{2}$$

For initial date with small total variation, as $\varepsilon \to 0$, we prove that the relaxation system (1) converges to the following system:

$$\big(u + F(u)\big)_t + u_x = 0, \tag{3}$$
$$v = F(u). \tag{4}$$

Observe that, with F defined at (2), the system (3) is of Temple class. Our results apply also to more general functions $F : \Omega \mapsto I\!R^n$ which satisfy the following assumptions:

(A1) Let $A(u) \doteq DF(u)$ be the Jacobian matrix. Then, for each $u \in \Omega$, $A(u)$ has n real, positive, distinct eigenvalues $0 < \eta_1(u) < \cdots < \eta_n(u)$.

(A2) Every characteristic field is either linearly degenerate or genuinely nonlinear.

(A3) All the integral curves of the eigenvectors of A are straight lines.

All of the above assumptions are clearly satisfied by the map F in (2).

Our first result establishes uniform BV bounds and the Lipschitz continuous dependence on the initial data.

Theorem 1.1. *For every compact set $K \subset \Omega$ and $\varepsilon > 0$, there exist constants $L, \delta_0 > 0$, a closed domain $\mathcal{D} \subset \mathbf{L}^1_{\text{loc}}(I\!R;\, I\!R^{2n})$ and a continuous semigroup S^ε : $\mathcal{D} \times [0, \infty[\, \mapsto \mathcal{D}$ with the following properties.*

(i) *The domain $\mathcal{D}$ contains every initial data (u_0, v_0) such that*

$$Tot.Var.\{u_0\} + Tot.Var.\{v_0\} \le \delta_0,$$
$$u_0(\bar{x}) \in K, \qquad \Big|F\big(u_0(\bar{x})\big) - v_0(\bar{x})\Big| \le \delta_0 \qquad \textit{for some } \bar{x} \in I\!R.$$

(ii) *Each trajectory of the semigroup: $t \mapsto \big(u(t), v(t)\big) = S^\varepsilon_t(u_0, v_0)$ is the solution of the corresponding Cauchy problem (1).*

(iii) *For every pair of initial data $(u_0, v_0), (\tilde{u}_0, \tilde{v}_0) \in \mathcal{D}$ and every $t \ge 0$ one has*

$$\big\|S^\varepsilon_t(u_0, v_0) - S^\varepsilon_t(\tilde{u}_0, \tilde{v}_0)\big\|_{\mathbf{L}^1(I\!R;I\!R^{2n})} \le L \cdot \big\|(u_0, v_0) - (\tilde{u}_0, \tilde{v}_0)\big\|_{\mathbf{L}^1(I\!R;I\!R^{2n})}.$$

The constants L, δ_0 and the domain $\mathcal{D}$ do not depend on ε.

In our second theorem we establish the limit of the trajectories of (1), as $\varepsilon \to 0$.

Theorem 1.2. *In the same setting as Theorem 1.1, for every $(u_0, v_0) \in \mathcal{D}$ and $\varepsilon > 0$, call $t \mapsto \big(u^\varepsilon(t), v^\varepsilon(t)\big)$ the corresponding solution of (1). Then, as $\varepsilon \to 0$, for each $t > 0$ one has the convergence*

$$u^\varepsilon(t) \to u(t), \qquad v^\varepsilon(t) \to v(t) \qquad \textit{in } \mathbf{L}^1_{\text{loc}}(I\!R;\, I\!R^n).$$

The limit function u can be characterized as the unique entropic solution of the system of conservation law (3) with initial data $u(0, \cdot)$ implicitly defined by

$$F\big(u(0, x)\big) = v(0, x), \qquad u(0, x) + v(0, x) = u_0(x) + v_0(x).$$

Moreover, for each $t > 0$, one has $v(t, x) = F\big(u(t, x)\big)$ for a.e. $x \in I\!R$.

The main step of the proof is the derivation of a priori estimates on the total variation of u, v, which are obtained from uniform $\mathbf{L}^1$ bounds on the derivatives u_x, v_x. The evolution equations for u_x, v_x take the form of n coupled 2×2 systems of balance laws. The key property of these systems is that the $\mathbf{L}^1$ norm of the components u_x^i, v_x^i can increase only as a result of the coupling terms, and these terms are uniformly integrable in space-time because they are only due to interactions between waves of different families. These results can be regarded as a counterpart of those proved in [1] for vanishing viscosity limits.

Hyperbolic systems of conservation laws with relaxation effects were first studied by Liu in a fundamental paper [7]. More recent results on related topic can be found in [4, 5, 8, 9] and the references therein. The limit system of chromatography was studied in detail in [10]. For the model of single component chromatography, convergence towards equilibrium and error estimates for BV solutions were studied in [12, 13] for one-dimensional cases, and in [11] for the two-dimensional cases.

The plan of the paper is as follows: In Section 2 we derive some preliminary estimates on the integral of the product of transversal terms. The main BV bounds are proved in Section 3, while the stability estimate is given in Section 4. Finally, in Section 5 we show the convergence towards the zero relaxation limit. Details of all the proofs can be found in [3].

2. A priori estimates on transversal terms

In this Section we prove the a priori estimates on the integral of the product of the transversal terms. Toward this goal, we consider the more general linear system

$$\begin{cases} \xi_t + \xi_x &= -a(t,x)\xi + b(t,x)\zeta, \\ \qquad \zeta_t &= a(t,x)\xi - b(t,x)\zeta, \end{cases} \tag{5}$$

assuming that the coefficients are uniformly bounded:

$$0 < a_- \leq a(t,x) \leq a_+, \qquad 0 < b_- \leq b(t,x) \leq b_+.$$

The system describes the following stochastic process (see [14], pp.15-17). A medium is composed by particles that can move with random speed 0 or 1. The speed of a particle can switch from 1 to 0 with rate a and from 0 to 1 with rate b. At any given time t, the functions $\zeta(t,\cdot)$ and $\xi(t,\cdot)$ yield the density of particles travelling with speed 0 and 1 respectively.

The fundamental solution $\Gamma_{ji}(t,x;t_0,x_0)$, $i,j \in \{0,1\}$ is the density of probability that a particle, which initially is at x_0 at time t_0 with speed i, reaches the point x at time $t > t_0$ with speed j. We first study the special cases where the switching rates a, b are constant

$$\begin{cases} \xi_t + \xi_x &= -\alpha\xi + \beta\zeta, \\ \qquad \zeta_t &= \alpha\xi - \beta\zeta, \end{cases} \tag{6}$$

where the fundamental solution G is invariant under time and space translation,

$$\Gamma_{ji}(t,x;t_0,x_0) = G_{ji}(t-t_0,\ x-x_0) \quad i,j \in \{0,1\}.$$

Let $P(t)$ denote the position of a particle at time t. Observe that

$$\lim_{t\to\infty} \frac{P(t)}{t} = \lambda \doteq \frac{\beta}{\alpha+\beta} \qquad \text{with probability 1.}$$

One can thus regard λ as the average speed of the random particle P. We now consider a second linear system with constant coefficients

$$\begin{cases} \xi_t + \xi_x & = -\alpha^*\xi + \beta^*\zeta, \\ \qquad \zeta_t & = \ \alpha^*\xi - \beta^*\zeta, \end{cases} \tag{7}$$

and call G^*_{ji} the corresponding kernels. We assume that the average speed λ^* of (7) is strictly smaller than that of (6), i.e.,

$$0 < \lambda^* \doteq \frac{\beta^*}{\alpha^*+\beta^*} < \lambda \doteq \frac{\beta}{\alpha+\beta} < 1. \tag{8}$$

Lemma 2.1. *If (8) holds, then the following double integral is bounded:*

$$E_0 \doteq \int_0^\infty \int_{-\infty}^\infty G_{01}(t,x)\cdot G^*_{10}(t,x)\ dxdt < \infty. \tag{9}$$

This Lemma can be intepreted as follows (see Fig. 1). Let P and P^* be the "fast" and the "slow" particles, described by the systems (6) and (7) respectively. Then, E_0 yields the expected number of crossing where P^* overtakes P, assuming P^* starts at the origin with speed 0 and P starts at the origin with speed 1. Since the average speeds are strictly different, the two particles will eventually move away from each other, and thus the number of crossing is finite. A rigorous proof of this fact can be derived from the properties of sums of independent, identically distributed random variables [6]. See [3] for details.

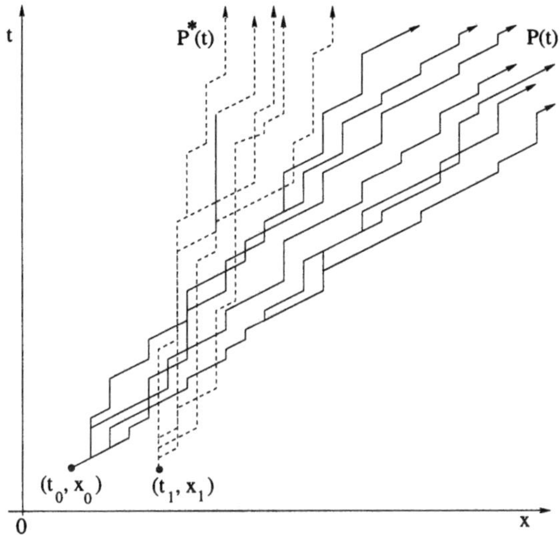

Figure 1

In the case of variable coefficients, a similar estimate can be obtained by a comparison arguement. Together with (5), we consider a second system for "slow" particles

$$\begin{cases} \xi_t + \xi_x &= -a^*(t,x)\xi + b^*(t,x)\zeta, \\ \zeta_t &= a^*(t,x)\xi - b^*(t,x)\zeta, \end{cases} \tag{10}$$

where

$$0 < a^*_- \leq a^*(t,x) \leq a^*_+, \qquad 0 < b^*_- \leq b^*(t,x) \leq b^*_+.$$

Call Γ^* the corresponding kernel. We assume that the coefficients of the systems (6), (7) and (5), (10) can be compared as follows:

$$a(t,x) \leq \alpha, \quad b(t,x) \geq \beta, \quad a^*(t,x) \geq \alpha^*, \quad b^*(t,x) \leq \beta^* \quad \text{for all } t,x. \tag{11}$$

Call Q and Q^* the particles described by (5) and (10) respectively. From the relation in (11) one sees that the particles P^* are even faster than P, since their speed is more likely to switch from 0 to 1, and less likely to switch from 1 to 0. On the other hand, the particles Q^* are even slower than the particles Q. Consequently, with the same initial states, Q and Q^* will cross each other a smaller number of times than P and P^*. This leads to the following lemma (see again [3] for details).

Lemma 2.2. *Let E_0 be the double integral defined at (9). Under the assumptions (11), for every τ_0, x_0, one has*

$$E^\sharp(\tau_0, x_0) \doteq \int_{\tau_0}^{\infty} \int_{-\infty}^{\infty} \Gamma_{01}(t,x;\tau_0,x_0) \cdot \Gamma^*_{10}(t,x;\tau_0,x_0)\, dxdt \leq E_0,$$

3. The BV bounds

In this section we obtain the bound on the total variation for the relaxation system (1). By the linear rescaling $t' = t/\varepsilon$, $x' = x/\varepsilon$, it is not restrictive to consider the case where $\varepsilon = 1$

$$\begin{cases} u_t + u_x &= -F(u) + v, \\ v_t &= F(u) - v. \end{cases} \tag{12}$$

Let $r_i(u)$, $l_i(u)$, $i = 1, \cdots, n$ be the sets of normalized right and left eigenvectors of $A(u)$

$$\big|r_i(u)\big| = 1, \qquad\qquad l_i(u) \cdot r_j(u) = \begin{cases} 1 & \text{if} \quad i = j, \\ 0 & \text{if} \quad i \neq j. \end{cases}$$

The directional derivative of a function $u \mapsto \varphi(u)$ in the direction of a vector $\mathbf{v} \in I\!R^n$ is written as

$$\mathbf{v} \bullet \varphi(u) \doteq \lim_{h \to 0} \frac{\varphi(u + h\mathbf{v}) - \varphi(u)}{h}.$$

With the above normalization, the assumption (A3) can be written in the form

$$r_i \bullet r_i \equiv 0 \qquad i = 1, \ldots, n.$$

For a given smooth solution of (12), we define the components of the gradients u_x, v_x w.r.t. the basis of eigenvectors $r_i(u)$ as

$$u_x^i \doteq l_i(u) \cdot u_x, \qquad\qquad v_x^i \doteq l_i(u) \cdot v_x, \qquad\qquad i = 1, \ldots, n.$$

Differentiating (12) we obtain

$$\begin{cases} u_{xt} + u_{xx} &= -A(u)u_x + v_x, \\ v_{xt} &= A(u)u_x - v_x. \end{cases} \tag{13}$$

After some computations, one finds that the components u_x^i, v_x^i, $i = 1, \cdots, n$ satisfy the system of $2n$ scalar equations

$$\begin{cases} (u_x^i)_t + (u_x^i)_x &= -\eta_i u_x^i + v_x^i - \sum_j G_{ij} u_x^j \\ (v_x^i)_t &= \eta_i u_x^i - v_x^i + \sum_j G_{ij} v_x^j + \sum_{j \neq k} H_{ijk} v_x^j u_x^k \end{cases} \tag{14}$$

where

$$G_{ij}(u, v) \doteq l_i \cdot \Big(\big(F(u) - v \big) \bullet r_j \Big), \qquad\qquad H_{ijk}(u, v) \doteq l_i \cdot (r_j \bullet r_k). \tag{15}$$

Note that if the terms involving G_{ij} and H_{ijk} are integrable, then by Duhamel's principle the integrals of u_x and v_x are bounded, and we achieve the BV bound of u and v. For H_{ijk}, we have the simple estimate

$$\big|H_{ijk}(u, v)\big| = \mathcal{O}(1), \tag{16}$$

where the Landau symbol $\mathcal{O}(1)$ denotes a function whose absolute value remains uniformly bounded. For G_{ij}, we get the following estimate

$$\big|G_{ij}(t,x)\big| = \mathcal{O}(1)\cdot e^{-t}\Big|\big(F(u)-v\big)(0,x)\Big| + \mathcal{O}(1)\cdot \int_0^t e^{s-t}\sum_{k\neq j}\big|u_x^k(s,x)\big|\,ds. \quad (17)$$

Motivated by (14), (15), (16) and (17), we now consider a semilinear system of $2n$ equations for the variables ξ^i, ζ^i, $i=1,\ldots,n$:

$$\begin{aligned}
\xi_t^i + \xi_x^i + \eta_i(t,x)\xi^i - \zeta^i &= C\sum_j \left(\rho_0 e^{-t} + \sum_{k\neq j}\int_0^t e^{s-t}\xi^k(s,x)\,ds\right)\xi^j,\\
\zeta_t^i - \eta_i(t,x)\xi^i + \zeta^i &= C\sum_j \left(\rho_0 e^{-t} + \sum_{k\neq j}\int_0^t e^{s-t}\xi^k(s,x)\,ds\right)\zeta^j + C\sum_{j\neq k}\xi^j\zeta^k.
\end{aligned} \quad (18)$$

If the constant C is chosen sufficiently large, and if the initial data satisfy

$$\begin{aligned}
\xi^i(0,x) = \xi_0^i(x) &\doteq \big|u_x(0,x)\big|, \qquad & \zeta^i(0,x) = \zeta_0^i(x) &\doteq \big|v_x(0,x)\big|,\\
\eta_i(t,x) &\doteq \eta_i\big(u(t,x)\big), & \rho_0 &\doteq \sup_x \Big|F\big(u(0,x)\big) - v(0,x)\Big|,
\end{aligned}$$

then a comparison argument yields

$$\big|u_x^i(t,x)\big| \leq \xi^i(t,x), \qquad \big|v_x^i(t,x)\big| \leq \zeta^i(t,x) \qquad \text{for all } t,x.$$

Therefore, the total variation of u,v is bounded by the $\mathbf{L}^1$ norm of ξ,η. Using Duhamel's principle, the $\mathbf{L}^1$ norm of ξ,η is bounded if the right hand side terms of (18) are integrable in x and t. Note that these terms involve only the product of the transversal terms where $j\neq k$, and thus they are integrable by properly applying the lemmas in Section 2. (See again [3].) This completes the proof of the uniform BV bounds of the solutions of (12).

4. Lipschitz continuous dependence

The Lipschitz continuous dependence on the initial data in $\mathbf{L}^1$ norm can be proved in a similarly way as for the BV bounds. Let u,v be a solution of (12) with small total variation. Consider a perturbation of the form $u+\varepsilon U$, $v+\varepsilon V$. Letting $\varepsilon\to 0$, the evolution of the infinitesimal perturbations U,V is described by the linearized system

$$\begin{cases} U_t + U_x &= -A(u)U + V,\\ \qquad V_t &= A(u)U - V. \end{cases} \quad (19)$$

As in Section 3, we define the components U^i, V^i as

$$U^i \doteq l_i(u)\cdot U, \qquad V^i \doteq l_i(u)\cdot V, \qquad i=1,\ldots,n.$$

After some computations we obtain a system of $2n$ scalar equations in the variables U^i, V^i:

$$\begin{cases} U^i_t + U^i_x + \eta_i U^i - V^i &= -\sum_j G_{ij} U^j, \\ V^i_t - \eta_i U^i + V^i &= -\sum_{j\neq k} H_{ijk} u^k_x V^j + \sum_j G_{ij} V^j, \end{cases} \tag{20}$$

where G_{ij} and H_{ijk} are defined in (15). By Duhamel's principle, and using the a priori estimates in Section 2, we obtain a uniform estimate on the size of the tangent vectors:

$$\|U(t)\|_{\mathbf{L}^1} + \|V(t)\|_{\mathbf{L}^1} \leq L\Big(\|U(0)\|_{\mathbf{L}^1} + \|V(0)\|_{\mathbf{L}^1}\Big). \tag{21}$$

Relying on (21), the Lipschitz continuous dependence on the initial data is obtained via a homotopy argument, as in [1]. Given any two solutions, say (u,v) and $(\tilde u, \tilde v)$, we connect them with a one-parameter family of solutions: $\theta \mapsto (u^\theta, v^\theta)$. At each time t, the $\mathbf{L}^1$ distance between the two solutions is bounded by the length of the curve $\theta \mapsto \big(u^\theta(t), v^\theta(t)\big)$. In turn, this length is estimated in terms of the size of the tangent vector

$$(U^\theta, V^\theta)(t,x) \doteq \left(\frac{du^\theta}{d\theta}, \frac{dv^\theta}{d\theta}\right)(t,x)$$

For every θ, this tangent vector is a solution of a linearized Cauchy problem of the form (19), hence it satisfies (21) for every $t \geq 0$. We thus conclude

$$\|u(t) - \tilde u(t)\|_{\mathbf{L}^1} + \|v(t) - \tilde v(t)\|_{\mathbf{L}^1} \leq L\Big(\|u(0) - \tilde u(0)\|_{\mathbf{L}^1} + \|v(0) - \tilde v(0)\|_{\mathbf{L}^1}\Big),$$

completing the proof of Theorem 1.1.

5. The convergence to the relaxation limit

We now outline the proof of Theorem 1.2. We first consider initial data (u_0, v_0) which are in equilibrium, i.e. with

$$v_0 = F(u_0) \tag{22}$$

For such initial data, for each $\varepsilon > 0$ the corresponding solution $u(t) = S^\varepsilon_t u_0$ defines a semogroup S^ε, uniformly Lipschitz continuous w.r.t. both variables t, u_0. Letting $\varepsilon \to 0$, by possibly taking a subsequence we obtain a Lipschitz semigroup $S = \lim S^\varepsilon$.

It remains to prove the uniqueness of the limit, and show that S is indeed the Standard Riemann Semigroup generated by the $n \times n$ system (3). By Corollary 9.2 in [2] it suffices to show that the correct limit is attained, locally in time, in the case where u_0 is piecewise constant and at each point of jump the corresponding Riemann problem is solved by waves of one single family. But in this case the problem is essentially scalar, and the convergence is well known [8].

Finally, the result is extended to all initial conditions (u_0, v_0), not necessarily satisfying (22). This is done observing that, at time $t = \varepsilon^{1/2}$, the the flow of (1) steers any initial data within a distance $\mathcal{O}(\varepsilon^{1/2})$ from an equilibrium configuration. For all details we again refer to [3].

References

[1] S. Bianchini and A. Bressan, BV estimates for a class of viscous hyperbolic systems, *Indiana Univ. Math. J.*, **49** (2000), 1673-1713.

[2] A. Bressan, *Hyperbolic systems of conservation laws. The one dimensional Cauchy problem*, Oxford University Press, 2000.

[3] A. Bressan and W. Shen, BV Estimates for Multicomponent Chromatography with Relaxation, *Discr. Cont. Dynam. Systems,* **6** (2000), pp. 21–38.

[4] G. Q. Chen, C. Levermore, and T. P. Liu, Hyperbolic conservation laws and stiff relaxation terms and entropy, *Comm. Pure Appl. Math.* **47** (1994), pp. 787–830.

[5] G. Q. Chen and T. P. Liu, Zero relaxation and dissipation limits for hyperbolic conservation laws, *Comm. Pure Appl. Math.* **46** (1993), pp. 755–781.

[6] M. Loève, *Probability Theory I*, fourth edition, Springer-Verlag, 1977.

[7] T. P. Liu, Hyperbolic conservation laws with relaxation, *Commun. Math. Phys.* **108** (1987), pp. 153–175.

[8] R. Natalini, Convergence to equilibrium for the relaxation approximations of conservation laws, *Comm. Pure Appl. Math.* **49** (1998), pp. 795–823.

[9] R. Natalini, Recent results on hyperbolic relaxation problems, in *Analysis of systems of conservation laws*, H. Freistühler Ed., Chapman & Hall/CRC (1998), pp. 128–198.

[10] H. K. Rhee, R. Aris, and N. R. Amundson, On the theory of multicomponent chromatography, *Philos. Trans. Royal Soc. London* **267-A** (1970), pp. 419–455.

[11] W. Shen, A. Tveito, and R. Winther, A system of conservation laws including a stiff relaxation term; the 2D case, *BIT* **36** (1996), pp. 786–813.

[12] A. Tveito and R. Winther, On the rate of convergence to equilibrium for a system of conservation laws including a relaxation term. *SIAM J. Math. Anal.*, Vol. 30, **5** (1999), pp. 1115–1135.

[13] A. Tveito and R. Winther, An error estimate for a finite difference scheme approximating a hyperbolic system of conservation laws, *SIAM J. Numer. Anal.* **30** (1993), pp. 401–424.

[14] E. Zauderer, *Partial Differential Equations of Applied Mathematics*, Wiley-Interscience, 1983.

A. Bressan
W. Shen
SISSA,
via Beirut 4,
Trieste 34014,
Italy.
bressan@sissa.it
shen@sissa.it

International Series of Numerical Mathematics
Vol. 140, © 2001 Birkhäuser Verlag Basel/Switzerland

A Strongly Degenerate Convection-diffusion Problem Modeling Centrifugation of Flocculated Suspensions

Raimund Bürger and Kenneth Hvistendahl Karlsen

Abstract. We prove existence and uniqueness of BV entropy solutions of a strongly degenerate convection-diffusion problem modeling centrifugation of flocculated suspensions. A modification of the generalized upwind method is employed to solve the initial-boundary value problem numerically, i.e., to simulate the centrifugation process.

1. Introduction

We consider the quasilinear strongly degenerate convection-diffusion equation

$$u_t + f(u,x)_x = A(u)_{xx} + g(u,x), \quad (x,t) \in Q_T := (x_1,x_2) \times (0,T), \tag{1}$$

where $A(u) := \int_0^u a(s)ds$. We assume that $f \geq 0$, $\operatorname{supp} f(\cdot,x) \subset [0,1]$, $a(u) = 0$ for $u \leq u_c$ and $u \geq 1$, and $a(u) > 0$ otherwise, i.e., equation (1) is of hyperbolic type for $u \leq u_c$ and $u \geq 1$ and of parabolic type for $u_c < u < 1$. Furthermore, we assume that $f(\cdot,x)$ is continuous and piecewise differentiable with $\|\partial_u f(\cdot,x)\| \leq M$ and that A and $g(\cdot,x)$ are Lipschitz continuous uniformly in x, and that $f(u,\cdot), g(u,\cdot) \in C^1(x_1,x_2)$ uniformly in u. In particular, the diffusion coefficient a is allowed to be discontinuous. We consider the initial and boundary conditions

$$u(x,0) = u_0(x),\ 0 \leq u_0(x) \leq 1,\ x \in [x_1,x_2], \tag{2}$$

$$f(u,x_b) - A(u(x_b,t))_x = 0,\ x_b \in \{x_1,x_2\},\ t \in (0,T]. \tag{3}$$

We define A^ε in terms of a standard C^∞ mollifier ω_ε with $\operatorname{supp}\omega_\varepsilon \subset (-\varepsilon,\varepsilon)$ via $a^\varepsilon(u) := ((a+\varepsilon) * \omega_\varepsilon)(u)$ and $A^\varepsilon(u) := \int_0^u a^\varepsilon(s)\,ds$ and assume that the initial function satisfies $u_0 \in BV(x_1,x_2)$, $u_0(x) \in [0,1]$ for all $x \in [x_1,x_2]$, and that $\mathrm{TV}_{(x_1,x_2)}(A^\varepsilon(u_0)_x)$ is uniformly bounded for all sufficiently small $\varepsilon > 0$.

Remark 1.1. *If $A \in C^1$, then it is sufficient to assume that $\mathrm{TV}_{(x_1,x_2)}(u_0') < \infty$. Moreover, the regularity assumption on $f(u,\cdot)$ can still be weakened, see [14].*

The study of strongly degenerate parabolic equations such as (1) is in part motivated by a recent theory of sedimentation-consolidation processes [8]. While the paper [4] is concerned with gravity thickening, which leads to an equation similar to (1) but without a source term, we here focus on the application of this theory to a centrifugal field in a rotating reference frame [3]. It is the purpose of this paper to briefly outline how the analysis of strongly degenerate convection-diffusion equations presented in [4, 6, 14] can be extended to the initial-boundary value problem (IBVP) (1)–(3), and to draw attention to a new application of these equations. In addition, we employ an adaptation of the generalized upwind finite difference method described in [5] to solve the IBVP numerically, that is to simulate the centrifugation process. For an overview of the analysis and numerics of (strongly) degenerate parabolic equations, we refer to [11].

In Section 2 we state the definition of entropy solutions of the IBVP. We then outline in Sections 3 and 4 the existence and uniqueness proofs, following the vanishing viscosity method and recent ideas by Carrillo [9] and Karlsen and Risebro [14], respectively. We return to the centrifugation problem in Section 5, presenting a working numerical algorithm and numerical solutions for the IBVP.

2. Entropy solutions

Due to both the degeneracy of $a(\cdot)$ and the nonlinearity of $f(\cdot,x)$, solutions of (1) are discontinuous and have to be defined as entropy solutions.

Definition 1. *A function $u \in L^\infty(Q_T)\cap BV(Q_T)$ is an* entropy solution *of the IBVP* (1)–(3) *if* (a) $A(u)_x \in L^2(Q_T)$, (b) *f.a.a.* $t\in(0,T)$, $\gamma_{x_\mathrm{b}}(f(u,\cdot)-A(u)_x)=0$, $x_\mathrm{b}\in\{x_1,x_2\}$, (c) $\lim_{t\downarrow 0} u(x,t)=u_0(x)$ *f.a.a.* $x\in(x_1,x_2)$, *and* (d)

$$\forall\varphi\in C_0^\infty(Q_T),\ \varphi\geq 0,\ \forall k\in\mathbb{R}:\iint\limits_{Q_T}\Big\{|u-k|\varphi_t+\mathrm{sgn}(u-k)\times \\ \times\Big[\big[f(u,x)-f(k,x)-A(u)_x\big]\varphi_x-\big[f_x(k,x)-g(u,x)\big]\varphi\Big]\Big\}\,dtdx\geq 0. \tag{4}$$

Here γ_{x_1} and γ_{x_2} denote the traces with respect to $x\downarrow x_1$ and $x\uparrow x_2$, respectively. Entropy inequalities like (4) go back to Kružkov [16] and Vol'pert [17] for first-order and to Vol'pert and Hudjaev [18] for second-order equations.

Remark 2.1. *The BV assumption in Definition 1 is only used to ensure the existence of the traces γ_{x_1} and γ_{x_2}. Note that we can require that the boundary conditions are satisfied in a pointwise sense a.e., whereas Dirichlet boundary conditions such as those stated in* [4] *have to be treated as entropy boundary conditions* [7]. *Finally, we point out that it is at present not known whether jump conditions for hyperbolic-parabolic equations such as those by Wu and Yin* [20] *are valid here, since these jump conditions rely on stronger regularity (for example, Lipschitz continuity) of the diffusion coefficient a than is stipulated here.*

3. Existence of entropy solutions

Existence of entropy solutions is shown by the vanishing viscosity method. To this end, consider the regularized uniformly parabolic IBVP, in which the functions f,

g and u_0 have been replaced by smooth approximations that ensure compatibility conditions and existence of smooth solutions for fixed $\varepsilon > 0$:

$$u_t^\varepsilon + f^\varepsilon(u^\varepsilon, x)_x = A^\varepsilon(u^\varepsilon)_{xx} + g^\varepsilon(u^\varepsilon, x), \quad (x,t) \in Q_T, \tag{5a}$$

$$u^\varepsilon(x,0) = u_0^\varepsilon(x), \ x \in (x_1, x_2), \tag{5b}$$

$$f^\varepsilon(u^\varepsilon, x_b) - A^\varepsilon(u^\varepsilon(x_b, t))_x = 0, \ x_b \in \{x_1, x_2\}, \ t \in (0,T]. \tag{5c}$$

Lemma 3.1. *The following uniform estimates hold for solutions u^ε of the regularized IBVP* (5), *where the constants M_1 to M_3 are independent of ε: $\|u^\varepsilon\|_{L^\infty(Q_T)} \leq M_1$; and $\|u_x^\varepsilon(\cdot,t)\|_{L^\infty(x_1,x_2)} \leq M_2$, $\|u_t^\varepsilon(\cdot,t)\|_{L^\infty(x_1,x_2)} \leq M_3$ for all $t \in (0,T]$.*

Sketch of proof. The maximum principle used to establish the first estimate is standard; we refer to the proof of Lemma 9 in [4] for details. Defining $\operatorname{sgn}_\eta(\tau) := \operatorname{sgn}(\tau)$ if $|\tau| > \eta$, $\operatorname{sgn}_\eta(\tau) := \tau/\eta$ if $|\tau| \leq \eta$ and $|x|_\eta := \int_0^x \operatorname{sgn}_\eta(\xi) d\xi$ for $\eta > 0$, we obtain by differentiating (5a) with respect to x, multiplying the result by $\operatorname{sgn}_\eta(u_x^\varepsilon)$, integrating over Q_{T_0}, $0 < T_0 \leq T$, integrating by parts and using (5b) and (5c)

$$\int_{x_1}^{x_2} \left|u_x^\varepsilon(x, T_0)\right|_\eta dx \leq \int_{x_1}^{x_2} \left|(u_0^\varepsilon)'(x)\right|_\eta dx + \int_0^{T_0} \operatorname{sgn}_\eta(u_x^\varepsilon) u_t^\varepsilon \Big|_{x_1}^{x_2} dt + \iint_{Q_{T_0}} \operatorname{sgn}_\eta'(u_x^\varepsilon) u_{xx}^\varepsilon \times$$

$$\times \left[f_u^\varepsilon(u^\varepsilon, x) - \left(a^\varepsilon(u^\varepsilon)\right)_x\right] u_x^\varepsilon \, dt dx + \iint_{Q_{T_0}} \operatorname{sgn}_\eta(u_x^\varepsilon)\left(g_u^\varepsilon(u^\varepsilon, x) u_x^\varepsilon + g_x^\varepsilon(u^\varepsilon, x)\right) dt dx.$$

To derive the second inequality of Lemma 3.1 from this inequality, we repeat the proof of part (a) of Lemma 11 in [4] to estimate the first three integrals of the right-hand side for $\eta \to 0$. The integrand of the last term can for $\eta \to 0$ be written as $|u_x^\varepsilon| g_u^\varepsilon + \operatorname{sgn}(u_x^\varepsilon) g_x^\varepsilon$. Since g_u^ε and g_x^ε are uniformly bounded, the estimate on u_x^ε can be derived from Gronwall's lemma. Analogously, the derivation of the estimate on u_t^ε in Lemma 11 of [4] also applies to the present equation with source term. ∎

From the estimates of Lemma 3.1 we may conclude that there exists a sequence $\varepsilon = \varepsilon_n \downarrow 0$ such that the sequence of solutions $\{u^{\varepsilon_n}\}$ of (5) converges in $L^1(Q_T)$ to a function $u \in L^\infty(Q_T) \cap BV(Q_T)$. We now have to show that u is indeed an entropy solution of (1)–(3). Part (a) of Definition 1 follows from the following lemma, whose short proof is a straightforward extension of that of Lemma 10 in [4]:

Lemma 3.2. *The limit function u of solutions u^ε of* (5) *satisfies $A(u)_x \in L^2(Q_T)$.*

Finally, repeating the proofs of Lemmas 5 and 12 of [4], we can show

Lemma 3.3. *The viscosity limit function u of solutions u^ε of* (5) *satisfies* (4) *and the initial and boundary conditions mentioned in Definition 1.*

Summarizing, we have:

Theorem 3.4. *The IBVP* (1)–(3) *admits an entropy solution u.*

4. Uniqueness of entropy solutions

After Carrillo's important work [9], the uniqueness proof for entropy solutions of degenerate parabolic equations has become similar to Kružkov's proof [16] for hyperbolic equations. The key proposition allowing to apply Kružkov's "doubling of the variables" to second-order equations, and to prove the main theorem of this section, is the following version (see [14] for its proof) of a lemma from [9], which identifies a certain entropy dissipation term, viz. the right-hand side of (6) below.

Lemma 4.1 (Carrillo's lemma). *Let u be an entropy solution of the IBVP* (1)–(3). *Then, for any non-negative $\varphi \in C_0^\infty(Q_T)$ and any $k \in (u_c, 1)$, we have*

$$\iint_{Q_T} \Big(|u-k|\varphi_t + \operatorname{sgn}(u-k)\Big[\big[f(u,x) - f(k,x) - A(u)_x \big]\varphi_x - \big[f_x(k,x) - g(u,x) \big]\varphi \Big] \Big)\, dtdx = \lim_{\eta \downarrow 0} \iint_{Q_T} \big(A(u)_x\big)^2 \operatorname{sgn}'_\eta\big(A(u) - A(k)\big)\varphi \, dtdx. \tag{6}$$

Theorem 4.2. *For two entropy solutions u and v of* (1)–(3), *the following inequality holds for any $\varphi \in C_0^\infty(Q_T)$, $\varphi \geq 0$:*

$$\iint_{Q_T} \Big(|v-u|\varphi_t + \operatorname{sgn}(v-u)\big[f(v,x) - f(u,x) - \big(A(v)_x - A(u)_y\big) \big]\varphi_x + \operatorname{sgn}(v-u)\big[g(v,x) - g(u,x) \big]\varphi \Big)\, dtdx \geq 0. \tag{7}$$

Sketch of proof. The argument given below relies on Lemma 4.1 and on doubling the number of dependent variables together with a penalization procedure [16]. We let $\varphi \in C^\infty(Q_T \times Q_T)$, $\varphi \geq 0$, $\varphi = \varphi(x,t,y,s)$, $v = v(x,t)$, $u = u(y,s)$, and introduce the "hyperbolic" sets $\mathcal{E}_v = \big\{(x,t) \in Q_T : v(x,t) \leq u_c \text{ or } v(x,t) \geq 1\big\}$ associated with v and $\mathcal{E}_u = \big\{(y,s) \in Q_T : u(y,s) \leq u_c \text{ or } u(y,s) \geq 1\big\}$ associated with u.

From the entropy inequalities for $v(x,t)$ (with $k = u(y,t)$) and $u(y,s)$ (with $k = v(x,t)$) and Lemma 4.1, the following inequality was derived in [14]:

$$\begin{aligned} &\iiiint_{Q_T \times Q_T} \Big(|v-u|(\varphi_t + \varphi_s) + \operatorname{sgn}(v-u)\big[f(v,x) - f(u,y) - \big(A(v)_x - A(u)_y\big) \big] \times \\ &\qquad \times (\varphi_x + \varphi_y) + \operatorname{sgn}(v-u)\big[g(v,x) - g(u,y) \big]\varphi \Big)\, dtdxdsdy + E_{\mathrm{Conv}} \\ &\geq \lim_{\eta \downarrow 0} \iiiint_{(Q_T \setminus \mathcal{E}_u) \times (Q_T \setminus \mathcal{E}_v)} \big(A(v)_x - A(u)_y\big)^2 \operatorname{sgn}'_\eta\big(A(v) - A(u)\big)\varphi \, dtdxdsdy \geq 0, \end{aligned} \tag{8}$$

where the "error term" E_{Conv} takes the form $E_{\text{Conv}} =$

$$\iiiint\limits_{Q_T \times Q_T} \operatorname{sgn}(v-u)\Big(\big[(f(u,y)-f(u,x))\varphi\big]_x - \big[(f(v,x)-f(v,y))\varphi\big]_y\Big)\, dtdxdsdy.$$

We are now on familiar ground [16] and introduce in (8) the test function

$$\varphi(x,t,y,s) = \varphi\big(\tfrac{x+y}{2}, fract + s2\big)\delta_h\big(\tfrac{x-y}{2}\big)\delta_h\big(\tfrac{t-s}{2}\big),$$

where $\{\delta_h\}_{h>0}$ is a standard regularizing sequence in $\mathbb{R}$. Observe that

$$\varphi_t + \varphi_s = \varphi_t\big(\tfrac{x+y}{2}, \tfrac{t+s}{2}\big)\delta_h\big(\tfrac{x-y}{2}\big)\delta_h\big(\tfrac{t-s}{2}\big),$$
$$\varphi_x + \varphi_y = \varphi_x\big(\tfrac{x+y}{2}, \tfrac{t+s}{2}\big)\delta_h\big(\tfrac{x-y}{2}\big)\delta_h\big(\tfrac{t-s}{2}\big).$$

Following [14] (see also [16] since $f(u,\cdot)$ is smooth), one can show that $\lim_{h\downarrow 0} E_{\text{Conv}} = 0$. Consequently, by sending $h \downarrow 0$ in (8), we get (7). ■

Define for sufficiently small $h > 0$ the functions $\rho_h(x) := \int_{-\infty}^{x} \delta_h(\xi)\, d\xi$, $\mu_h(x) := 1 - \rho_h(x - x_1 - 2h)$ and $\nu_h(x) := \rho_h(x - x_2 + 2h)$. Concerning these functions, we have the following lemma (whose proof is easy):

Lemma 4.3. *Let $u \in L^1\big(0,T; L^\infty(x_1,x_2)\big)$. If the traces $\gamma_{x_1}u := (\gamma u)(x_1,t)$ and $\gamma_{x_2}u := (\gamma u)(x_2,t)$ exist a.e. in $(0,T)$, then we have for $\varphi \in C^\infty(Q_T)$*

$$\lim_{h\downarrow 0} \iint\limits_{Q_T} \Big(\varphi\big(1-\mu_h(x)-\nu_h(x)\big)\Big)_x u\, dtdx = \int_0^T \Big(\varphi(x_1,t)\gamma_{x_1}u - \varphi(x_2,t)\gamma_{x_2}u\Big)\, dt.$$

We are now in a position to deduce from (7) the following uniqueness result:

Corollary 1. *Let v, u be entropy solutions of the IBVP (1)–(3) with initial data v_0, u_0, respectively. Then $\|v(\cdot,t)-u(\cdot,t)\|_{L^1(x_1,x_2)} \le \exp\big(t\|g\|_{\text{Lip}}\big)\|v_0-u_0\|_{L^1(x_1,x_2)}$ for all $t \in (0,T)$. In particular, the IBVP admits at most one entropy solution.*

Proof. In (7), we choose $\varphi(x,t) = (1-\mu_h(x)-\nu_h(x))\chi(t)$ with $\chi \in C_0^\infty(0,T)$ and $\chi \ge 0$. Note that φ tends to $\chi(t)$ as $h \downarrow 0$. Taking the limit $h \downarrow 0$, we obtain from Lemma 4.3 and the boundary conditions at $x = x_1, x_2$ (see Definition 1):

$$\begin{aligned}\iint\limits_{Q_T} |u-v|\chi'(t)\, dtdx &\ge -\iint\limits_{Q_T} \operatorname{sgn}(v-u)\big[g(v,x)-g(u,x)\big]\chi(t)\, dtdx \\ &\ge -\|g\|_{\text{Lip}} \iint\limits_{Q_T} |u-v|\chi(t)\, dtdx.\end{aligned} \tag{9}$$

Fixing $\tau \in (0,T)$, setting $\chi(t) := \rho_h(t) - \rho_h(t-\tau)$ in (9), then sending $h \downarrow 0$ and using Gronwall's lemma, we get the L^1 stability estimate $\|v(\cdot,\tau)-u(\cdot,\tau)\|_{L^1(x_1,x_2)} \le \exp\big(\tau\|g\|_{\text{Lip}}\big)\|v_0-u_0\|_{L^1(x_1,x_2)}$. Since τ was arbitrary, we are finished. ■

Remark 4.4. *Following [10, 13, 14], it is possible to prove that the unique entropy solution of (1)–(3) depends continuously on the nonlinearities in the problem.*

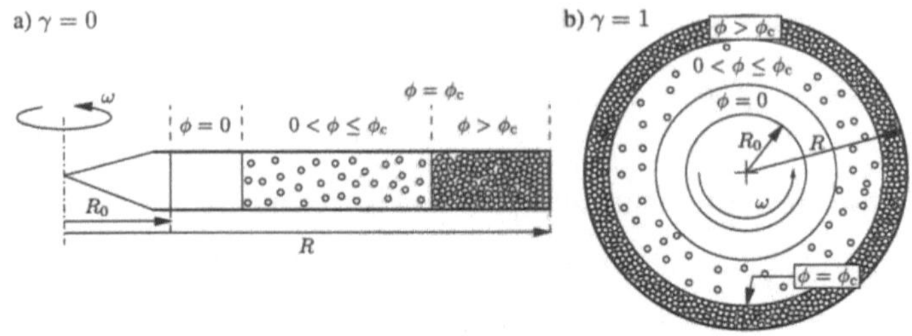

FIGURE 1. (a) Rotating tube, (b) rotating axisymmetric cylinder.

5. Batch centrifugation of flocculated suspensions

5.1. Introduction

Extending the arguments developed in [8] for a purely gravitational force to rotating systems, Bürger and Concha [3] show that the settling of a flocculated suspension in a tube with constant cross section ($\gamma = 0$, Fig. 1 (a)) or in a rotating axisymmetric cylinder ($\gamma = 1$, Fig. 1 (b)) is governed by the field equation

$$\phi_t + r^{-\gamma}\big(f_{\mathrm{ck}}(\phi)\omega^2 r^{1+\gamma}\big)_r = r^{-\gamma}\big(r^{\gamma}A(\phi)_r\big)_r, \quad 0 < R_0 < r < R,\ t > 0, \tag{10}$$

in which ϕ is the solids concentration, r the radius, f_{ck} the centrifugal Kynch flux density function, ω the angular velocity and A is defined as in (1). The function f_{ck} is a nonnegative Lipschitz continuous function with support in $[0,1]$, and the diffusion coefficient is defined by $a(\phi) := f_{\mathrm{ck}}(\phi)\sigma_{\mathrm{e}}'(\phi)/(\Delta\rho\phi)$, where $\Delta\rho > 0$ is the solid-fluid density difference and σ_{e}' is the derivative of the effective solid stress function. In Eq. (1) both the gravitational and Coriolis forces are neglected [3].

Eq. (10) inherits its degeneracy from the constitutive assumption that $\sigma_{\mathrm{e}} = 0$ while the solid flocs do not touch each other, i.e., while ϕ does not exceed a critical value ϕ_{c}, and that $\sigma_{\mathrm{e}}'(\phi) > 0$ for $\phi_{\mathrm{c}} < \phi \leq 1$. Assuming that $\operatorname{supp} f_{\mathrm{ck}} = (0,1)$, we see that $a(\phi) = 0$ and hence (10) is of hyperbolic type for $\phi \leq \phi_{\mathrm{c}}$ and $\phi \geq 1$ and that otherwise $a(\phi) > 0$, i.e., (10) is of parabolic type. Thus Eq. (10) is strongly degenerate. The included special case $a \equiv 0$ was already studied in [1].

The initial and wall boundary conditions read

$$\phi(r,0) = \phi_0(r), \quad R_0 \leq r \leq R, \tag{11}$$

$$\big(f_{\mathrm{ck}}(\phi)\omega^2 R_{\mathrm{b}} + A(\phi)_r\big)(R_{\mathrm{b}}, t) = 0, \quad t > 0,\ R_{\mathrm{b}} \in \{R, R_0\}. \tag{12}$$

Differentiating out the convection and diffusion terms in equation (10) leads to an IBVP of the type (1)–(12). The existence and uniqueness analysis therefore states that the centrifugation model admits a unique entropy solution ϕ.

5.2. Numerical algorithm

To solve the IBVP (10)–(11) numerically, we use a modification of the generalized upwind finite difference method described in [5] for gravity settling.

Let $J, N \in \mathbb{N}$, $\Delta r := (R - R_0)/J$, $\Delta t := T/N$, $r_j := R_0 + j\Delta r$, $j = 1/2, 1, \ldots, J - 1/2, J$ and $\phi_j^n \approx \phi(r_j, n\Delta t)$. The computation starts by setting $\phi_j^0 := \phi_0(r_j)$ for $j = 0, \ldots, J$. Assume then that $\phi_0^n, \ldots, \phi_J^n$ are known. To compute the values ϕ_j^{n+1}, we first compute the extrapolated values $\phi_j^{\mathrm{L}} := \phi_j^n - (\Delta r/2)s_j^n$ and $\phi_j^{\mathrm{R}} := \phi_j^n + (\Delta r/2)s_j^n$, where the slopes s_j^n can be calculated by the min-mod limiter function $M(\cdot,\cdot,\cdot)$ by $s_j^n = \mathrm{MM}(\phi_j^n - \phi_{j-1}^n, (\phi_{j+1}^n - \phi_{j-1}^n)/2, (\phi_{j+1}^n - \phi_j^n))/\Delta r$ for $j = 2, \ldots, J-2$, where $\mathrm{MM}(a,b,c) = \min\{a,b,c\}$ if $a,b,c > 0$, $\mathrm{MM}(a,b,c) = \max\{a,b,c\}$ if $a,b,c < 0$ and $\mathrm{MM}(a,b,c) = 0$ otherwise. Moreover we set $s_0^n = s_1^n = s_{J-1}^n = s_J^n = 0$. The values ϕ_j^{L} and ϕ_j^{R} are arguments of the Engquist-Osher numerical flux density function defined by $f_{\mathrm{ck}}^{\mathrm{EO}}(u,v) := f_{\mathrm{ck}}(0) + \int_0^u \max\{f_{\mathrm{ck}}'(s), 0\}\,ds + \int_0^v \min\{f_{\mathrm{ck}}'(s), 0\}\,ds$. The interior scheme for $\phi_1^n, \ldots, \phi_{J-1}^n$ then reads

$$\phi_j^{n+1} = \phi_j^n - \frac{\omega^2\Delta t}{r_j^\gamma \Delta r}\left[r_{j+1/2}^{1+\gamma} f_{\mathrm{ck}}^{\mathrm{EO}}(\phi_j^{\mathrm{R}}, \phi_{j+1}^{\mathrm{L}}) - r_{j-1/2}^{1+\gamma} f_{\mathrm{ck}}^{\mathrm{EO}}(\phi_{j-1}^{\mathrm{R}}, \phi_j^{\mathrm{L}})\right] + \frac{\Delta t}{r_j^\gamma \Delta r^2} \times$$
$$\times \left[r_{j+1/2}^\gamma\big(A(\phi_{j+1}^n) - A(\phi_j^n)\big) - r_{j-1/2}^\gamma\big(A(\phi_j^n) - A(\phi_{j-1}^n)\big)\right],\ j = 1, \ldots, J-1.$$

The boundary formulas follow by setting $j = 0$ and $j = J$ in this equation and inserting the discrete versions of the boundary conditions (12). This leads to

$$\phi_0^n = \phi_0^{n-1} - \frac{\omega^2\Delta t}{R_0^\gamma \Delta r} r_{1/2}^{1+\gamma} f_{\mathrm{ck}}^{\mathrm{EO}}(\phi_0^n, \phi_1^n) + \frac{\Delta t}{R_0^\gamma \Delta r^2} r_{1/2}^\gamma\big(A(\phi_1^n) - A(\phi_0^n)\big),$$
$$\phi_J^n = \phi_J^{n-1} + \frac{\omega^2\Delta t}{R^\gamma \Delta r} r_{J-1/2}^{1+\gamma} f_{\mathrm{ck}}^{\mathrm{EO}}(\phi_{J-1}^n, \phi_J^n) - \frac{\Delta t}{R^\gamma \Delta r^2} r_{J-1/2}^\gamma\big(A(\phi_J^n) - A(\phi_{J-1}^n)\big).$$

To ensure convergence of the scheme to the entropy solution, the CFL stability condition $R\omega^2\|f_{\mathrm{ck}}'\|_\infty(\Delta t/\Delta r) + 2\|a\|_\infty(\Delta t/\Delta r^2) \leq 1$ must be satisfied. We ensured this condition by selecting Δr freely and determining Δt appropriately. The accuracy was $J = 400$. For more details about the upwind method, see [12, 15].

5.3. Numerical example

Sambuichi et al. [19] published centrifugation experiments with three different flocculent aqueous suspensions, namely of limestone, yeast, and clay, using a cylindrical centrifuge. For each material, the measured gravitational settling rates and compression data determined the functions f_{ck} and σ_{e}, respectively. In this paper, we choose the published data referring to a limestone suspension with $\Delta\rho = 1710\,\mathrm{kg/m^3}$ (see [3] for the case of a clay suspension). Sambuichi et al. [19] approximated the measured settling rates for various concentrations by three different connecting straight segments in a logarithmic plot, which yields

$$f_{\mathrm{ck}}(\phi) = \begin{cases} (-47.923\phi^2 + 2.5474\phi) \times 10^{-5}\,\mathrm{s} & \text{for } 0 \leq \phi \leq 0.035, \\ 1.3580 \times 10^{-8}\phi^{-0.92775}\,\mathrm{s} & \text{for } 0.035 < \phi \leq 0.08, \\ 5.6319 \times 10^{-13}\phi^{-4.9228}\,\mathrm{s} & \text{for } 0.08 < \phi \leq 0.119, \\ 5.9735 \times 10^{-10}\phi^{-1.65}\,\mathrm{s} & \text{for } 0.119 < \phi \leq \phi_{\max} = 0.45, \\ 0 & \text{otherwise.} \end{cases}$$

This function does *not* satisfy all assumptions on f stated in Sect. 1 since it is cut at $\phi_{\max}$. However, due to the presence of the diffusion term, solution values are bounded away from $\phi_{\max}$, so the numerical results presented below would look

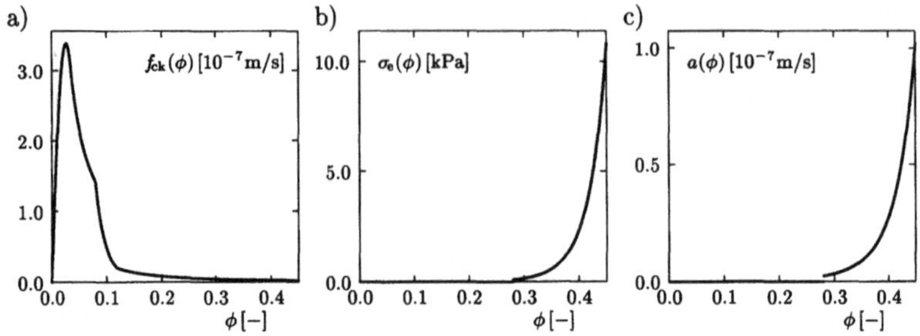

FIGURE 2. Model functions for a limestone suspension.

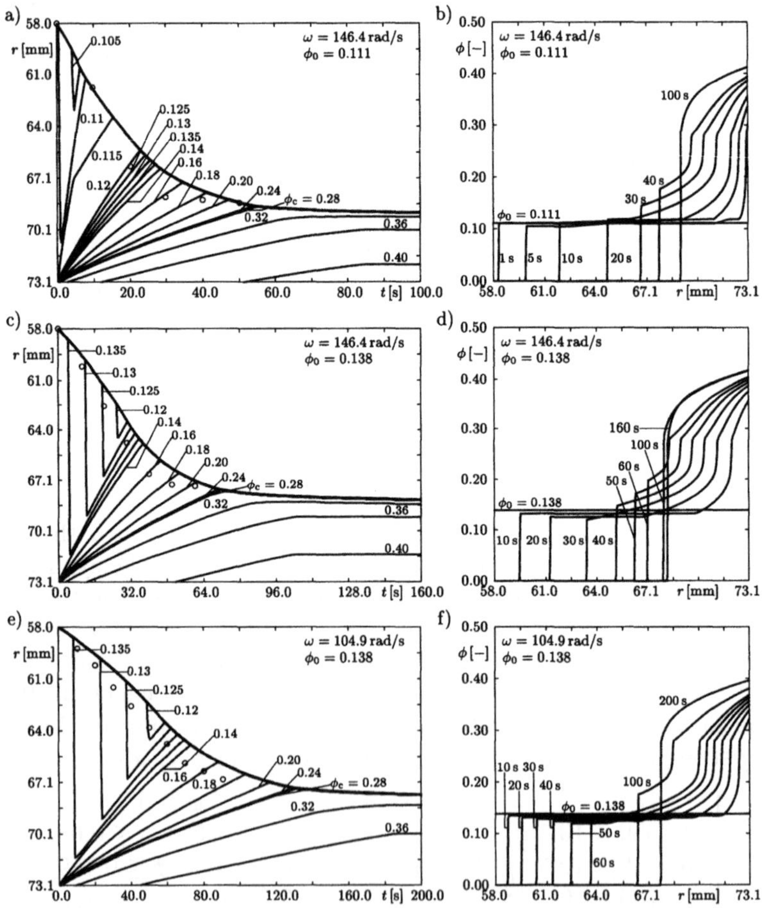

FIGURE 3. Simulation of the centrifugation of a limestone suspension.

the same if the jump of the function f_{ck} at $\phi = \phi_{\max}$ had been smoothed out in order to produce an example in which these assumptions are precisely satisfied.

The solid pressure equation suggested in [19] is equivalent to $\sigma_{\mathrm{e}}(\phi) = 0$ for $\phi \leq \phi_{\mathrm{c}} := 0.28$ and $\sigma_{\mathrm{e}}(\phi) = 0.30184(1-\phi)^{-17.544}$ Pa for $\phi > \phi_{\mathrm{c}}$. The functions f_{ck} and σ_{e} and the resulting diffusion coefficient a are plotted in Figure 2.

Figure 3 shows numerical solutions of the phenomenological model calculated with these functions in the case of a rotating cylindrical vessel ($\gamma = 1$). The left column of Figure 3 shows diagrams of numerical iso-concentration lines for selected values of ϕ, and the right column displays concentration profiles at selected times. The parameters and the data that differ in the three cases considered (indicated in Figure 3) were chosen in such a way that the simulated supernate-suspension interfaces could be compared with measurements by Sambuichi et al. [19], which are shown as open circles ($\circ$). Figure 3 thus illustrates the different effects of initial concentration and angular velocity on the dynamics of the centrifugation process.

While in the compression zone, where $\phi > \phi_{\mathrm{c}}$ and hence (10) is parabolic, the solutions are similar to those of the gravity case [2, 5], there are some distinctive features visible in the hindered settling zone ($\phi \leq \phi_{\mathrm{c}}$) where (10) is hyperbolic, due to the rotating frame of reference. Most notably, the vertical iso-concentration lines indicate that the concentration of the bulk suspension is a (decreasing) function of time, and the supernate-suspension interface has a curved trajectory. These properties have previously been found by Anestis and Schneider [1], who determined exact solutions to the centrifugation model under the assumption that $\sigma_{\mathrm{e}} \equiv 0$, i.e. $A \equiv 0$, using the method of characteristics. Of course, in the centrifugal case (in contrast to the gravitational) characteristics are *not* iso-concentration lines [1, 3].

References

[1] G. Anestis and W. Schneider, *Application of the theory of kinematic waves to the centrifugation of suspensions*, Ing.-Arch. **53** (1983), 399–407.

[2] R. Bürger and F. Concha, *Mathematical model and numerical simulation of the settling of flocculated suspensions*, Int. J. Multiphase Flow **24** (1998), 1005–1023.

[3] R. Bürger and F. Concha, *Settling velocities of particulate systems: 12. Batch centrifugation of flocculated suspensions*, Int. J. Mineral Process., to appear.

[4] R. Bürger, S. Evje and K.H. Karlsen, *On strongly degenerate convection-diffusion problems modeling sedimentation-consolidation processes*, J. Math. Anal. Appl. **247** (2000), 517–556.

[5] R. Bürger and K.H. Karlsen, *On some upwind difference schemes for the phenomenological sedimentation-consolidation model*, J. Eng. Math., to appear.

[6] R. Bürger and W.L. Wendland, *Existence, uniqueness and stability of generalized solutions of an initial-boundary value problem for a degenerating quasilinear parabolic equation*, J. Math. Anal. Appl. **218** (1998), 207–239.

[7] R. Bürger and W.L. Wendland, *Entropy boundary and jump conditions in the theory of sedimentation with compression*, Math. Meth. Appl. Sci. **21** (1998), 865–882.

[8] R. Bürger, W.L. Wendland and F. Concha, *Model equations for gravitational sedimentation-consolidation processes*, Z. Angew. Math. Mech. **80** (2000), 79–92.

[9] J. Carrillo, *Entropy solutions for nonlinear degenerate problems*, Arch. Rat. Mech. Anal. **147** (1999), 269–361.

[10] B. Cockburn and G. Gripenberg, *Continuous dependence on the nonlinearities of solutions of degenerate parabolic equations*, J. Diff. Eqns. **151** (1999), 231–251.

[11] M. S. Espedal and K.H. Karlsen, *Numerical solution of reservoir flow models based on large time step operator splitting algorithms*, In: A. Fasano and H. van Duijn, editors, *Filtration in Porous Media and Industrial Applications*, Lecture Notes in Mathematics No. 1734, Springer Verlag, Berlin 2000, 9–77.

[12] S. Evje and K.H. Karlsen. *Monotone difference approximations of BV solutions to degenerate convection-diffusion equations*, SIAM J. Num. Anal. **37** (2000), 1838–1860.

[13] S. Evje, K.H. Karlsen and N.H. Risebro, *A continuous dependence result for nonlinear degenerate parabolic equations with spatially dependent flux function*, Preprint, Department of Mathematics, University of Bergen, 2000.

[14] K.H. Karlsen and N.H. Risebro, *On the uniqueness and stability of entropy solutions of nonlinear degenerate parabolic equations with rough coefficients*, Preprint, University of Bergen, 2000.

[15] K. H. Karlsen and N. H. Risebro, *Convergence of finite difference schemes for viscous and inviscid conservation laws with rough coefficients*, Math. Modelling Numer. Anal. **35** (2001), 239–270.

[16] S.N. Kružkov, *First order quasilinear equations in several independent space variables*, Math. USSR Sb. **10** (1970), 217–243.

[17] A.I. Vol'pert, *The spaces BV and quasilinear equations*, Math. USSR Sb. **2** (1967), 225–267.

[18] A.I. Vol'pert and S.I. Hudjaev, *Cauchy's problem for degenerate second order parabolic equations*, Math. USSR Sb. **7** (1969), 365–387.

[19] M. Sambuichi, H. Nakakura and K. Osasa, *Zone settling of concentrated slurries in a centrifugal field*, J. Chem. Eng. Japan **24** (1991), 489–494.

[20] Z. Wu and J. Yin, *Some properties of functions in* BV_x *and their applications to the uniqueness of solutions for degenerate quasilinear parabolic equations*, Northeast. Math. J. **5** (1989), 395–422.

Raimund Bürger
Institute of Mathematics A,
University of Stuttgart,
Pfaffenwaldring 57,
D-70569 Stuttgart, Germany
buerger@mathematik.uni-stuttgart.de

Kenneth Hvistendahl Karlsen
Department of Mathematics,
University of Bergen,
Johs. Brunsgt. 12,
N-5008 Bergen, Norway
kennethk@mi.uib.no

International Series of Numerical Mathematics
Vol. 140, © 2001 Birkhäuser Verlag Basel/Switzerland

Weak Shock Reflection Modeled by the Unsteady Transonic Small Disturbance Equation

Sunčica Čanić, Eun Heui Kim, and Barbara Lee Keyfitz

1. Introduction

In this paper we report on our recent progress in the study of weak shock reflection modeled by the unsteady transonic small disturbance equation (UTSD)

$$\begin{aligned} u_t + uu_x + v_y &= 0 \\ -v_x + u_y &= 0\,. \end{aligned} \tag{1}$$

In contrast with our previous work [5] where we focus on shock interaction initial data, in this paper we concentrate on shock reflection initial data. Specifically, we consider here initial data that consist of an incident shock

$$S_1 : \xi = a\eta + \omega_1 \qquad \text{with} \qquad \omega_1 = \frac{1}{2} + a^2\,, \tag{2}$$

and "no flow through the wall" boundary condition

$$v(\xi, 0) = 0. \tag{3}$$

The incident shock separates the ambient (upstream) state $U_0 = (0,0)$ from the state $U_1 = (1, -a)$ behind the incident shock. The parameter a is the slope of the incident shock wave. This model was suggested in [3, 11] as an asymptotic description of the transition between Mach reflection and regular reflection for weak shocks. A numerical comparison, see [13], between the solutions of the UTSD equation and the solutions of the compressible Euler equations for weak shocks shows the same solutions structures for both equations in the region where the incident and the reflected waves interact. The motivation for the studies in [3, 11, 13] was the resolution of the so called von Neumann paradox. This refers to a weak shock reflection configuration, von Neumann reflection, that closely resembles Mach reflection, but for which shock polar analysis shows that simple Mach reflection is impossible [2, 10].

There are now numerical studies of solution structures for (1) in the parameter regimes that are believed to correspond to von Neumann reflection [3, 11, 13]. In [11] a study of the solution with $a = 0.5$ was presented. The study in [11] presented a shock polar analysis for the linearized equation and a high-resolution numerical solution showing a supersonic bubble behind the reflected wave. However, global existence of solutions to the boundary value problem for the UTSD has not yet been proved.

This paper presents an overview of the one-parameter family of initial value problems described by (1), (2) and (3) for the entire range of the parameter a. We suggest that this problem allows two qualitatively different classes of solutions. One is the well accepted *regular reflection* solution occurring for $a \geq \sqrt{2}$, and the other is a *complex reflection* occurring for $0 < a < \sqrt{2}$. We pose the corresponding free-boundary problems for the positions of the transonic shocks in each class, and use [7, 8] to suggest a strategy for proof of the global existence of a solution in the case of regular reflection. We conjecture solution structures for the complex reflection solution, basing our arguments on the well-posedness of the corresponding free-boundary problems, on shock polar analysis and on numerical simulations using the software developed in [9].

In addition, we compare our results with the solution found in [11].

2. Background

We study self-similar solutions, *i.e.*, solutions which depend on $\xi = x/t$, $\eta = y/t$. The governing equations (1) in self-similar coordinates read

$$\begin{aligned} (u-\xi)u_\xi - \eta u_\eta + v_\eta &= 0 \\ -v_\xi + u_\eta &= 0\,. \end{aligned} \tag{4}$$

The Rankine-Hugoniot equations are

$$\frac{d\xi}{d\eta} = -\frac{[v]}{[u]} = \frac{[\frac{1}{2}u^2 - \xi u]}{[v - \eta u]}\,. \tag{5}$$

Locally, at a state $U = (u, v)$, the equations are elliptic if $u > \rho = \xi + \eta^2/4$, degenerate if $u = \rho$ and hyperbolic if $u < \rho$. The characteristics in the hyperbolic region are straight lines (when the equation is linearized about a constant state) given by

$$\xi - \xi_0 + \left(\frac{\eta_0}{2} \pm \sqrt{\frac{\eta_0^2}{4} + \xi_0 - u}\right)(\eta - \eta_0) = 0. \tag{6}$$

They are tangent to the parabolas $P_u \equiv \{\xi + \eta^2/4 = u\}$. (The parabola $\rho = \xi + \eta^2/4 = u$ is the equivalent of the sonic circle in the full set of Euler equations for gas dynamics.)

It is convenient to work in the coordinate system (ρ, η); the equations become

$$\begin{aligned} (u-\rho)u_\rho - \tfrac{\eta}{2}u_\eta + v_\eta &= 0 \\ \tfrac{\eta}{2}u_\rho - v_\rho + u_\eta &= 0\,. \end{aligned} \tag{7}$$

Eliminating v yields the second-order equation

$$Q(u) \equiv \left((u-\rho)u_\rho + \frac{u}{2}\right)_\rho + u_{\eta\eta} = 0\,. \tag{8}$$

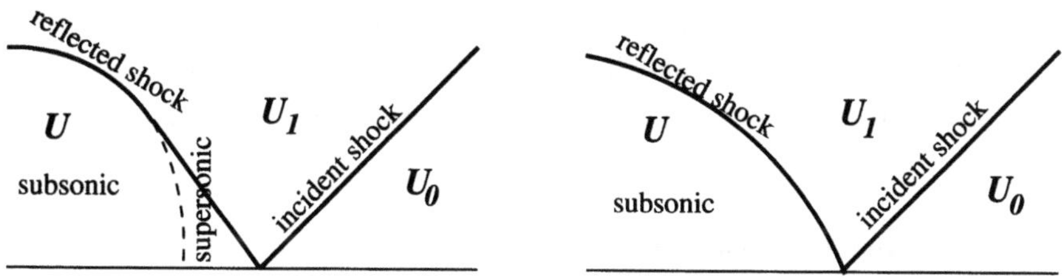

FIGURE 1. Supersonic and Transonic Regular Reflection

In the subsonic region, $u > \rho$, the equations are elliptic, with ellipticity ratio

$$\frac{\Lambda}{\lambda} = \frac{\max\{u-\rho, 1\}}{\min\{u-\rho, 1\}}.$$

Since the region where the UTSD equations are elliptic is unbounded, we introduce a cut-off boundary σ and impose far-field conditions there. To justify the far-field conditions, in [7] we develop an asymptotic expansion in which the reflected shock S_2 approaches the sonic parabola P_1. More precisely, in [7] we prove

Proposition 2.1. *The behavior $\xi(\eta) = -\frac{\eta^2}{4}+1+b(\eta)$ for the position of the reflected transonic shock S_2 as $\eta \to \infty$, with $b(\eta) \to 0$ as $\eta \to \infty$, is consistent with the Rankine-Hugoniot conditions.*

3. The Regular Reflection Solution ($a \geq \sqrt{2}$)

The regular reflection configuration describes a solution of the initial-value problem (1), (2) and (3) which consists of two shocks, the incident shock S_1 and a reflected shock S_2, meeting at a point on the wall $\Xi_a = \left(\frac{1}{2}+a^2, 0\right) = (\xi_a, 0)$. Shock polar analysis shows that regular reflection can take place only if $a \geq \sqrt{2}$ (we repeat this calculation in [5]). If $a \geq \sqrt{2}$ there are two possible reflected states at the reflection point:

$$U_R = (1 + a^2 - a\sqrt{a^2-2}, 0) \quad \text{and} \quad U_F = (1 + a^2 + a\sqrt{a^2-2}, 0). \tag{9}$$

The larger value U_F is always subsonic, and U_R is also subsonic if

$$a < a^* = \sqrt{1+\frac{\sqrt{5}}{2}} \approx 1.45.$$

The larger value U_F is known as "strong" and the other, U_R, as "weak" shock reflection. Ben-Dor, [1, page 21], calculates weak and strong regular reflections for the compressible Euler equations, and notes that the tangency at the shock polar (where the two states coalesce) is not at the sonic point, so the same phenomenon, of a small range of incident shock angles in which both weak and strong reflections are subsonic, occurs also in the full model. Neither of these solutions can be discarded on theoretical grounds [1, 12]. However, it is noted in [1] that, experimentally, only the weak shock reflection has been observed.

We pose these problems as free-boundary problems for the positions of the transonic shocks.

3.1. Transonic Regular Reflection

In transonic regular reflection the state immediately behind the reflected wave is subsonic. It was remarked in the previous subsection that this is the case for all values of $a \geq \sqrt{2}$ when strong reflection, $U = U_F$ in (9), occurs. For $\sqrt{2} \leq a \leq a^*$ both strong and weak reflection are of this type. By causality, if the state immediately behind the shock is subsonic, then it remains so, and the reflected shock remains transonic. If the horizontal flow velocity is also bounded, by $u_{\max}$ say, then the shock cannot cross the sonic parabola corresponding to $u_{\max}$, $\xi + \eta^2/4 = u_{\max}$, and hence is curved. The ellipticity ratio in this case is bounded, and the equations (7) or (8) are strictly elliptic.

We formulate the free-boundary problem in terms of the first velocity component u and recover v from the system of equations (4). To work only with u we eliminate the second velocity component v from the Rankine-Hugoniot conditions (5). This can be done in several different ways. To prove the existence of a solution by using a fixed point theorem, it is crucial to eliminate v in such a way to produce an oblique derivative boundary condition for u across the transonic shock S_2. See [7] or [8] for more details. The domain Ω in which the flow is subsonic is bounded by the reflected shock S_2, by the wall $\eta = 0$, and by the cut-off boundary σ. The free-boundary problem for the position of S_2 is as follows: Find a function u and a free boundary $\rho = \rho(\eta)$ so that

$$Q(u) \equiv \left((u-\rho)u_\rho + \tfrac{u}{2}\right)_\rho + u_{\eta\eta} = 0 \text{ in } \Omega,$$

$$\frac{d\rho}{d\eta} = -\sqrt{\rho - \frac{u+1}{2}} \quad \text{and} \quad N(u;\rho) \equiv \beta \cdot \nabla u = 0 \text{ on } S_2 \equiv \{\rho = \rho(\eta)\},$$

$$u_\eta = 0 \text{ on } \eta = 0, \quad u = f \text{ on } \sigma, \quad u(\Xi_a) = u_R, \quad \rho(0) = \xi_a.$$

Here

$$\beta = (\beta^1, \beta^2) = \left(\rho' \left[\frac{7u+1}{8} - \rho\right], \frac{5u+3}{8} - \rho\right). \tag{10}$$

This is a free-boundary problem for a quasilinear equation. In [7] we prove the existence of a *global* solution of a modified problem where we introduce a cut-off function $g(\rho - \frac{u+1}{2})$ to keep the term under the square root non-negative. We have been able to prove, so far, that the term $\rho - \frac{u+1}{2}$ is non-negative in a finite subdomain around the reflection point. Therefore, in [7] we prove

Theorem 3.1. *The elliptic free boundary problem above has a solution in a domain of finite size around the reflection point.*

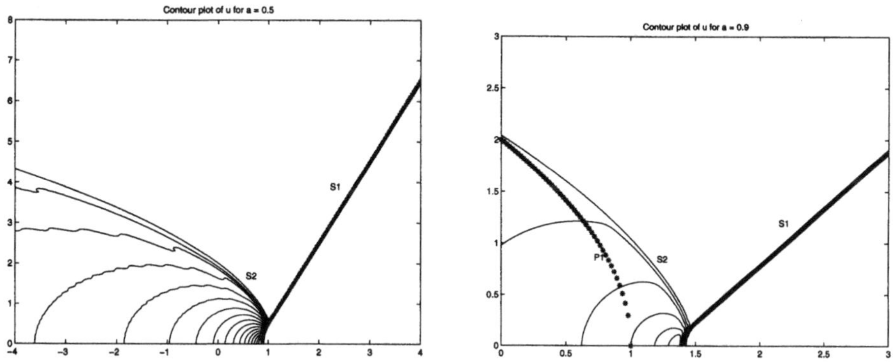

FIGURE 2. Complex reflection for $a = 0.5$ (left) and $a = 0.9$ (right)

3.2. Supersonic Regular Reflection

If $a > a^*$ the state U behind the reflected shock can be either supersonic ($U = U_R$) or subsonic ($U = U_F$). Here we pose a free boundary problem corresponding to weak reflection, in which case $U = U_R$. The solution has the constant value U_R in a region behind the reflected shock, but becomes subsonic at the parabola P_R; the solution is nonconstant and the reflected shock is transonic and curved beyond P_R; see Figure 1.

The governing equation (8) is degenerate at the sonic curve P_R. We consider a domain Ω bounded by the transonic shock S_2, by the degenerate parabola P_R, by the wall $\eta = 0$ and by the cut-off boundary σ. Equation (8) holds in the interior of the domain. The boundary conditions that hold at the free boundary S_2 are again $N(u;\rho) \equiv \beta \cdot \nabla u = 0$, where β is defined in (10), and $\frac{d\rho}{d\eta} = -\sqrt{\rho - (u+1)/2}$. The boundary condition at the sonic curve P_R is $u = \rho_R = u_R$. The sonic parabola P_R and the reflected shock S_2 meet at a fixed (known) point. The condition at the wall is $u_\eta(\xi, 0) = 0$ and the far field condition is $u = f$ on σ. This free boundary problem is different from the previous one in that the equation $Q(u) = 0$ is degenerate on P_R. Our preliminary results [6] indicate that techniques extending [7], based on fixed point theorems, elliptic regularization and upper and lower solutions will prove the existence of a solution to this problem.

4. The Complex Reflection Solution ($0 < a < \sqrt{2}$)

When $a < \sqrt{2}$ regular reflection at the wall is not possible. The structure of the global solution depends on the reflected wave, on its interaction with the incident shock, and on the subsonic flow behind the reflected wave. If the reflected wave travels faster than the incident shock, it will interact with the incident shock at a point above the wall, Ξ_T. Our numerical simulations indicate that this is the case for all $a \in (0, \sqrt{2})$. Figure 2 shows solution for the values $a = 0.5$ and $a = 0.9$.

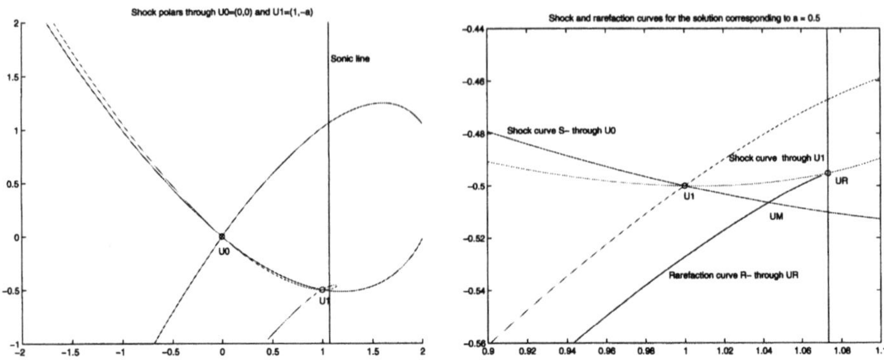

FIGURE 3. The figure on the left shows the shock loci through U_0 and U_1 for $a = 0.5$ with $\Xi_T = (1.007, 0.514)$. The figure on the right shows a magnified view of the shock loci in a neighborhood of $U_1 = (1, -a)$. The shock locus through U_0 and the rarefaction locus through U_R intersect at U_M.

To formulate the global problem we pose a quasi-one-dimensional Riemann problem at Ξ_T and solve it locally using the generalized Lax entropy condition [4, 5] to identify admissible shock waves. Then we pose a free boundary problem to complete the solution in the rest of the domain.

The local solution at the interaction point is given by the following theorem.

Theorem 4.1. *Let $0 < a < \sqrt{2}$ and let $\Xi_T = (\xi_T, \eta_T) \in S_1$ be given, with $\eta_T > 0$ and $\rho_T = \xi_T + \eta_T^2/4 > 1$. Then there exists a unique sonic state U_R that belongs to the shock locus of $U_1 = (1, -a)$ and is such that the quasi-one-dimensional Riemann problem for $U_0 = (0,0)$ and U_R has an admissible solution consisting of a (minus)-rarefaction wave from U_R to some U_M and a (plus)-shock from U_M to U_0. This solution is unique.*

The terms minus and plus refer to the sign in equation (6). A typical set of shock and rarefaction loci is shown in Figure 3. Although the interaction point Ξ_T lies outside the parabola P_1 (since U_1 must be supersonic at Ξ_T), we do not have an estimate for the position of Ξ_T.

A sketch of the solution in the physical space (ξ, η) is shown in Figure 4. In this figure we also show the sonic parabolas $P_{\rm u}$ for $u_M < u < u_R$. For a fixed sonic state U_R there is a unique space-like line L, which is tangent to the parabola P_R at Ξ_T [5]. The solution of the quasi-one-dimensional Riemann problem, with data U_0 and U_R prescribed on L at Ξ_T, consists of a rarefaction wave between U_R and U_M, and a shock between U_0 and U_M. The states in the rarefaction wave become sonic at the parabolas $P_{\rm u}$. The locus of the sonic points is given by the equation

$$P_{\rm MR} \equiv \left\{\xi - \xi_T = \frac{\eta}{2}(\eta - \eta_T)\right\}, \tag{11}$$

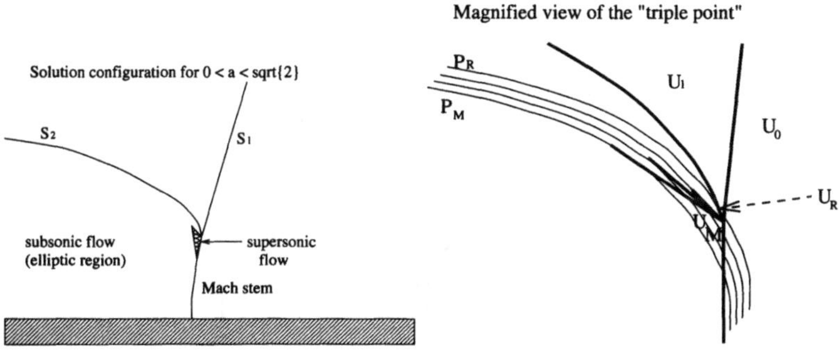

FIGURE 4. Sketch of the solution structure for $0 < a < \sqrt{2}$.

where (ξ_T, η_T) are the coordinates of the interaction point Ξ_T [5]. Near Ξ_T, the value of u behind the reflected shock S_2 increases away from the interaction point so that $u > \rho$ everywhere behind S_2 except at Ξ_T where $u = \rho$. That is, the shock S_2 is transonic everywhere except at Ξ_T. The domain in which the flow is subsonic is bounded by the Mach stem, MS, by the sonic parabola P_M defined by $\xi + \eta^2/4 = u_M$, by the sonic parabola $P_{\rm MR}$, and by the reflected shock wave, S_2. See Figure 5. Written in (ρ, η) coordinates, after eliminating the second velocity

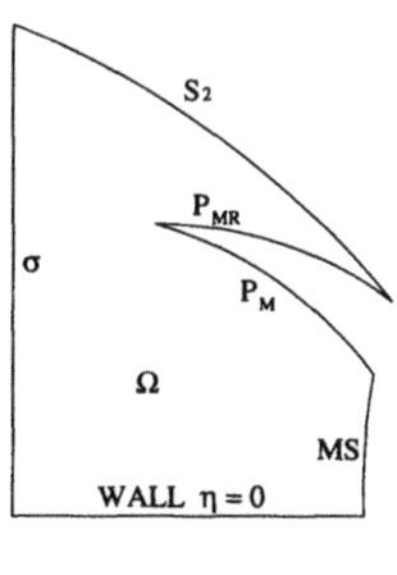

FIGURE 5. The domain of subsonic flow in complex reflection

component v, the free boundary problem for the position of the two transonic shocks S_2 and MS is the following: Find a solution u and the positions of the two free boundaries $\rho = \rho_1(\eta)$, $\rho = \rho_2(\eta)$ so that

$$Q(u) \equiv \left((u-\rho)u_\rho + \tfrac{u}{2}\right)_\rho + u_{\eta\eta} = 0 \text{ in } \Omega,$$

$$\left(\frac{d\rho}{d\eta}\right)^2 = \rho - \frac{u+1}{2} \quad \text{and} \quad N_1(u;\rho) \equiv \beta_1 \cdot \nabla u = 0 \text{ on } S_2 \equiv \{\rho = \rho_1(\eta)\},$$

$$\frac{d\rho}{d\eta} = \sqrt{\rho - \frac{u}{2}} \quad \text{and} \quad N_2(u;\rho) \equiv \beta_2 \cdot \nabla u = 0 \text{ on } MS \equiv \{\rho = \rho_2(\eta)\},$$

$$u_\eta = 0 \quad \text{on} \quad \eta = 0, \quad u = \rho \text{ on } P_{\rm MR}, \quad u = \rho_M \text{ on } P_{\rm M}, \quad \text{and} \quad u = f \text{ on } \sigma.$$

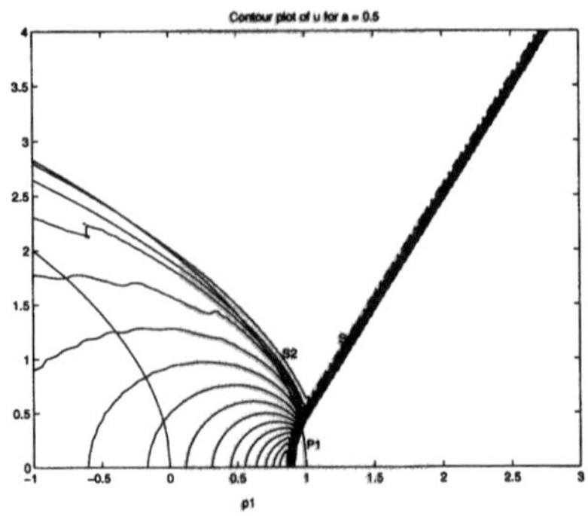

FIGURE 6. Contour plot of u for $a = 0.5$ showing that the reflected wave meets with the incident shock S_1 **outside** of parabola P_1

Here, the coefficients β_1 in the oblique derivative boundary operator $N_1(u;\rho)$ are defined by (10). The coefficients β_2 in the the oblique derivative boundary operator $N_2(u;\rho)$ are

$$\beta_2 = \left(\rho'\left[\frac{7u}{8} - \rho\right], \frac{5u}{8} - \rho\right). \tag{12}$$

We recover v from the PDE by integration. The operator Q in this case is degenerate at the portion of the boundary determined by the curves $P_{\rm MR}$ and P_M. Assuming Ξ_T is given, we foresee the following three new difficulties in this problem. First, the problem is posed on a non-convex domain. This is relevant when analyzing smoothness of the solution up to the degenerate boundary. Second, the problem consists of two free boundaries which will require proving existence of a fixed point involving two free-boundary mappings. Third, one boundary condition on the free boundary corresponding to the Mach stem loses uniform obliqueness at the point where it touches the wall; the boundary condition on S_2 also fails to be uniformly oblique, since β_2 is tangent to S_2 at the point where ρ achieves its maximum value. The first problem can be dealt with by an appropriate choice of sub- and super-solutions. The second difficulty requires synchronized estimates for the two free boundaries. The last difficulty seems to be more serious and is a subject of our current research. Finally, we must resolve the difficulty that we expect a solution to this problem to exist for only one value of Ξ_T and we must determine this value.

4.1. Conclusions

We conclude this paper by a comparison between the complex reflection solution conjectured in this paper and the corresponding solution conjectured in [11].

In [11] a solution corresponding to $a = 0.5$ was studied. It was conjectured that the solution consists of an incident shock S_1, a reflected wave S_2, a rarefaction wave and a Mach stem, all meeting at a point Ξ_T above the wall. The difference between the solution conjectured in [11] and the solution conjectured here is that

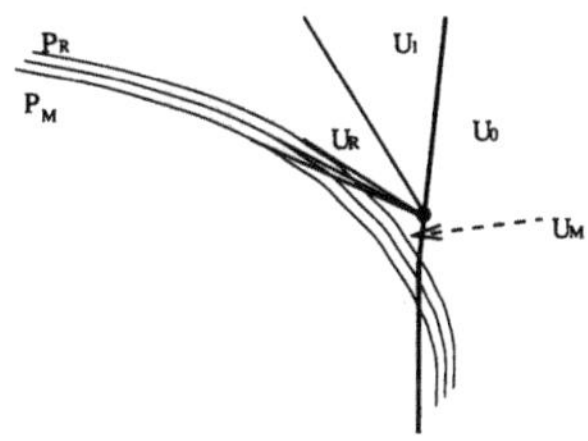

FIGURE 7. Sketch of the solution with supersonic U_R.

in this paper we argue that the state immediately behind the reflected shock wave has to be sonic, and not supersonic, as suggested in [11]. There are two reasons why we believe this to be the case. First, the solution suggested in [11] apparently violates causality, whereas the solution conjectured in this paper does not. Second, if the state immediately behind the reflected wave, call it U_R again, is supersonic, then the flow U would have to be constant ($U = U_R$) until it becomes subsonic, and the shock between U_1 and U, S_2, would have to be straight. In order for the shock S_2 to become curved it would have to become transonic. This would require the state U behind the shock to become subsonic, and this can happen only beyond the parabola P_R. However, P_R lies below the (plus) shock S_2, and so this cannot happen. See Figure 7. Therefore, S_2 would have to be a straight line all the way to infinity, and this is not what is seen in numerical simulations, see Figure 2, nor is this supported by the asymptotic analysis in [7]. Shown in Figure 7 is a sketch of the solution when the state U_R immediately behind the reflected wave is supersonic. The sketch shows the corresponding parabolas P_u, $u_M \leq u \leq u_R$ which all lie below the reflected wave S_2.

To support the conjecture presented in this paper, a global existence result for the free-boundary problem described in Section 4 is needed. This is a subject of our current and future research.

References

[1] G. Ben-Dor. *Shock Wave Reflection Phenomena.* Springer-Verlag, New York, 1992.

[2] W. Bleakney and A. H. Taub. Interaction of shock waves. *Reviews of Modern Physics*, 21:584–605, 1949.

[3] M. Brio and J. K. Hunter. Mach reflection for the two dimensional Burgers equation. *Physica D*, 60:194–207, 1992.

[4] S. Čanić and B. L. Keyfitz. Quasi-one-dimensional Riemann problems and their role in self-similar two-dimensional problems. *Archive for Rational Mechanics and Analysis*, 144:233–258, 1998.

[5] S. Čanić and B. L. Keyfitz. Riemann problems for the two-dimensional unsteady transonic small disturbance equation. *SIAM Journal on Applied Mathematics*, 58:636–665, 1998.

[6] S. Čanić, B. L. Keyfitz, and E. H. Kim. A free boundary problem for a quasilinear degenerate elliptic equation: Regular reflection of weak shock. *Communications on Pure and Applied Mathematics*, to appear, 2001.

[7] S. Čanić, B. L. Keyfitz, and E. H. Kim. Free-boundary problems for the unsteady transonic small disturbance equation: Transonic regular reflection. *Methods and Applications of Analysis*, to appear, 2000.

[8] S. Čanić, B. L. Keyfitz, and G. M. Lieberman. A proof of existence of perturbed steady transonic shocks via a free boundary problem. *Communications on Pure and Applied Mathematics*, LIII:484–511, 2000.

[9] S. Čanić and D. Mirković. A numerical study of shock reflection modeled by the unsteady transonic small disturbance equation. *SIAM Journal on Applied Mathematics*, 58:1365–1393, 1998.

[10] P. Colella and L. F. Henderson. The von Neumann paradox for the diffraction of weak shock waves. *Journal of Fluid Mechanics*, 213:71–94, 1990.

[11] J. K. Hunter and M. Brio. Weak shock reflection. *Journal of Fluid Mechanics*, 410:235–261, 2000.

[12] C. S. Morawetz. Potential theory for regular and Mach reflection of a shock at a wedge. *Communications on Pure and Applied Mathematics*, XLVII:593–624, 1994.

[13] A. R. Zakharian, M. Brio, J. K. Hunter, and G. M. Webb. The von Neumann paradox in weak shock reflection. *Journal of Fluid Mechanics*, 422:193–205, 2000.

S. Čanić
Department of Mathematics,
University of Houston,
Houston, Texas 77204-3476
canic@math.uh.edu

E.H. Kim
Department of Mathematics,
University of Houston,
Houston, Texas 77204-3476
ehkim@math.uh.edu

B. Lee Keyfitz
Department of Mathematics,
University of Houston,
Houston, Texas 77204-3476
blk@math.uh.edu

International Series of Numerical Mathematics
Vol. 140,

A Hyperbolic System of Conservation Laws in Modeling Endovascular Treatment of Abdominal Aortic Aneurysm

Sunčica Čanić and Dragan Mirković

Abstract. In this paper we present modeling of endovascular treatment of abdominal aortic aneurysm. We describe a simple, one-dimensional model, discuss the well posedness of the corresponding initial-boundary value problem, and present numerical simulations, which suggest an improvement in the design of stents used in treatment.

1. Introduction

In this paper we present a simple, one-dimensional model describing blood flow through a stented abdominal aorta exhibiting abdominal aortic aneurysm (AAA). Aneurysm is characterized by the formation of sac-like protrusions of weakened sections of blood vessels that can rupture and be fatal. Until recently, surgery has been the only treatment to prevent the aneurysm from rupturing. Since most of the patients are 55 or older and high risk, undergoing general anesthesia is not a solution.

Within the past ten years an innovative non-invasive procedure, which requires only local anesthesia, has been developed. The new procedure entails inserting a catheter (a hollow tube) into an artery and directing it to the site of the aneurysm. Placed in the catheter is a spring-like device called a stent, see Figure 1, which serves to hold open the weakened artery and to exclude the aneurysm from circulation. This lowers the probability of rupture and promotes aneurysm shrinking due to thrombosis caused by the lack of blood supply to the aneurysm tissue [2].

The procedure is still considered experimental and reports of mid- or long-term outcomes indicate structural and positional changes of the stent, including buckling, kinking and migration [21, 22]. To improve the procedure, optimal stent design depending on the patients anatomy and a study of the optimal anchoring strategies are needed. These are the objectives of our study.

The approach we take has several levels. At the first level we develop a simplified, one-dimensional model of blood flow thorough an axisymmetric, elastic tube

FIGURE 1. A sketch of a stent and of a stented aneurysm.

(blood vessel and/or stent) to detect scenarios in which complications occur. At the second level we study these scenarios in detail by considering a three-dimensional model. At the third level we plan to perform a validation of the modeling assumptions through the comparison with in vivo/in vitro measurements.

In this paper we present the one-dimensional simplified model, discuss the well posedness of the corresponding initial-boundary value problem, and present numerical simulations, which already suggest an improvement in the design of stents for endovascular AAA repair.

2. The model equations

The model equations are in the form of a hyperbolic system of conservation laws with discontinuous coefficients. The basic assumptions used to obtain the model are: the relative size of red blood cells to vessel diameter in the abdominal aorta is large so that blood can be modeled as a continuum, the **distensibility** of a vessel wall is taken into account and is assumed to dominate any effects due to the compressibility of blood; and the effects of shear thinning properties of blood in large vessels is ignored. Furthermore, the flow is assumed to be axi-symmetric.

We begin by considering the three-dimensional Navier-Stokes equations in cylindrical coordinates (x, r, Θ), with the x-axis corresponding to the vessel axial direction. We simplify the equations by assuming that angular velocity is zero, and that radial velocity is small compared to the axial velocity. The equations in terms of the *average axial velocity* $V = V(x,t)$ and *inner vessel radius* $R = R(x,t)$, obtained after integration of the simplified Navier-Stokes equations with respect to the radius r [3, 18], take the form:

$$\begin{aligned}
\frac{\partial R}{\partial t}+V\frac{\partial R}{\partial x}+\frac{R}{2}\frac{\partial V}{\partial x} &= 0\\
\frac{\partial (R^2V)}{\partial t}+\frac{\partial(\alpha R^2V^2)}{\partial x}+\frac{R^2}{\rho}\frac{\partial p}{\partial x} &= 2\nu R_0\left[\frac{\partial v_x}{\partial r}\right]_R\\
\frac{\partial p}{\partial r} &= 0.
\end{aligned}$$

Here $\alpha=\dfrac{1}{R^2V^2}\displaystyle\int_0^R 2r{v_x}^2dr$, and v_x is the axial component of the velocity. For the purpose of this paper we assume that the velocity profile v_x is independent of x and assumes the shape characteristic of oscillatory flow [18]:

$$v_x=\frac{\gamma+2}{\gamma}V\left[1-\left(\frac{r}{R_0}\right)^{\gamma}\right].$$

With this assumption α is constant, $\alpha=\dfrac{\gamma+2}{\gamma+1}$. The resulting system is similar in nature to the shallow water equations. The conservation of mass and momentum for the incompressible flow now reads

$$\frac{\partial A}{\partial t}+\frac{\partial m}{\partial x}=0, \tag{1}$$

$$\frac{\partial m}{\partial t}+\frac{\partial}{\partial x}\left(\frac{\alpha m^2}{A}\right)+\frac{1}{\rho}\frac{\partial p(A)}{\partial x}=f(A,m), \tag{2}$$

where $A(x,t)=R^2(x,t)$ is the cross-sectional area and $m(x,t)=A(x,t)V(x,t)$ is the momentum. The force term is given by $f(A,m)=f(V)=-\frac{2\nu\alpha}{\alpha-1}m/A$. To close the system we use a generalization of the pressure relation suggested in [18]

$$p(R)=G_0\left[\left(\frac{R}{R_0}\right)^{\beta}-1\right]. \tag{3}$$

Here G_0 describes tissue stiffness (pressure-strain elastic modulus), β describes linear/nonlinear behavior of the vessel or the stent (pressure-strain relationship), and R_0 is the unstressed inner vessel/stent radius. In contrast with [18], in our work we assume that G_0 and β are functions of x; more specifically, we assume that they are piecewise constant functions with the jump discontinuities at the anchoring sites of the stent. In addition, we allow the values of G_0 and β to change inside the stented region to reflect the use of multiple overlapping stents by some cardiologists [1, 10]. To account for the curved geometry, we model the effects of bending as momentum sink [9], with the sink term:

$$f_{kink}(x,m)=-mB(x)\kappa(x),$$

where $\kappa(x)$ is the curvature at x and $B(x)$ is a local empirical bending factor, $B \in [0,1]$. We solve the following initial-boundary value problem

$$\begin{aligned} \frac{\partial A}{\partial t} + \frac{\partial m}{\partial x} &= 0 \\ \frac{\partial m}{\partial t} + \frac{\partial}{\partial x}\left(\frac{\alpha m^2}{A}\right) + \frac{1}{\rho}\frac{\partial p(A)}{\partial x} &= f(A,m) + f_{kink}(x,m) \end{aligned}$$

$$p(A) = G_0(x)\left[\left(\frac{A}{A_0}\right)^{\beta(x)/2} - 1\right]$$

on the domain $Q_T = Q \times [0,T]$ where $Q = [0,\infty)$, with the initial and boundary conditions given by $V(0,t) = V_{pulsaltile}(t)$ [14], $R(x,0) = R_0(x) = 0.82$ cm [11] and $V(x,0) = 0$ m/sec. This is a hyperbolic system of conservation laws since the hyperbolicity condition

$$\alpha(\alpha-1)V^2 > -\frac{\beta G_0}{2\rho}\left(\frac{R}{R_0}\right)^{\beta}$$

is always satisfied. Because the coefficients are discontinuous and the boundary data is oscillatory, existence, uniqueness and long-time behavior of solutions is an open problem. Based on the ideas presented in [4] where oscillatory initial data was considered, and the studies in [7, 16] where scalar conservation laws with discontinuous flux functions were analyzed, our preliminary results suggest that the problem is well-posed. Analysis of the model equations is a subject of our current research.

3. The Numerical Method

We approximate system (1), (2) by Godunov's method with a second order correction implemented in a public–domain code CLAWPACK [13]. We use the *minmod* flux limiter [12] for the second order correction terms and Roe's approximate Riemann solver [17]. The source term in (2) is coupled to the homogeneous system by using "Strang splitting" [20]. The time discretization uses variable time steps controlled by the Courant number based on the maximum wave speed.

4. The Biophysical Problem and the Numerical Results

We focus our study on optimal stent design to treat abdominal aneurysm in patients with straight and curved geometries of the abdominal aorta. We consider flexible, self-expanding tubular-shaped helical vascular stents (Wallstent) shown in Figure 1. In contrast with stent grafts (graft covered stents), bare stents were shown to promote the growth of a thin layer of neointima (the innermost layer of blood vessel tissue) along the luminal surface of the stent. To further promote the growth of neointima, several cardiologist use three identical overlapping stents to

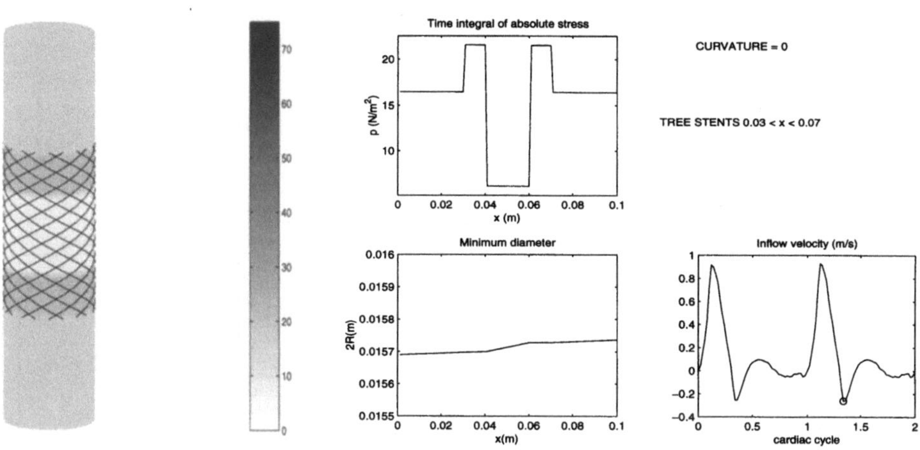

FIGURE 2. The figure on the left shows the geometry of the blood flow channel (without the aneurysmal pouch) and color-coded total pressure distribution over two cardiac cycles for the scenario **S1**. A more detailed pressure information is shown in the pressure–x diagram in the middle of the figure. Also shown is the diameter over two cardiac cycles obtained in the second diastole (shown on the right) corresponding to the minimum diameter.

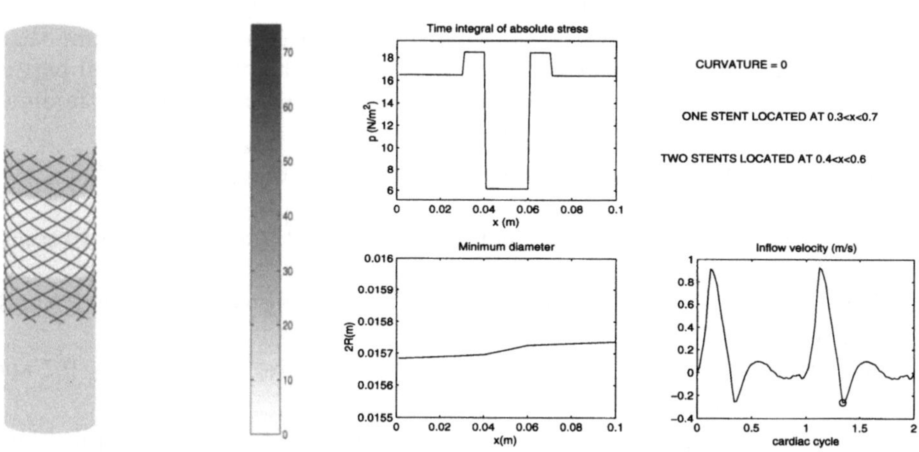

FIGURE 3. This figure shows the same flow properties as Figure 2 calculated for the scenario **S2**.

make the pores in a stent smaller. Altering the lengths of the three stents independently changes the stiffness properties of the stent. Stents with different structure cause different loads on the adjacent aorta. Since internal pressure is proportional to the circumferential tension [5] we use the pressure to measure the stress exerted

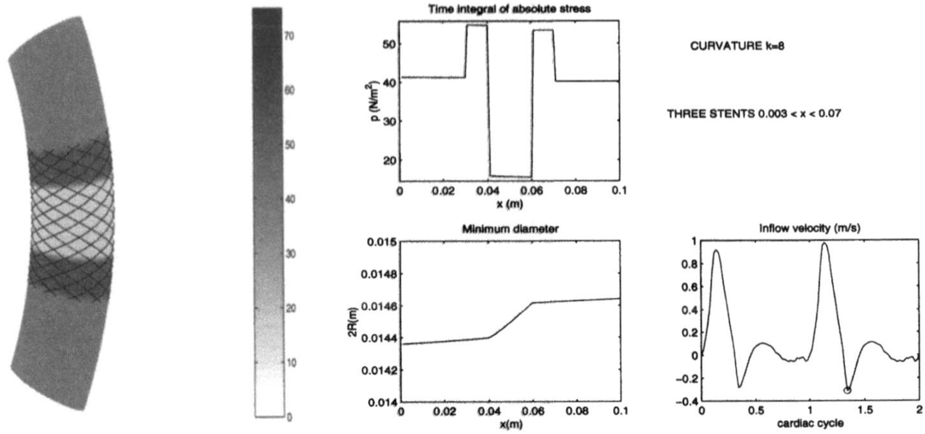

FIGURE 4. This figure shows the same flow properties as Figure 2 calculated for the scenario **S3.**

on the vessel or the stent. We integrate the absolute value of the total pressure through two cardiac cycles. The pulsaltile flow rate was obtained using trigonometric functions (FFT) to fit the physiological pulsaltile flow rate data obtained from [19]. We focus on the abdominal aorta in the region between renal and iliac arteries. We assume that this length is typically 0.1 m ($0 \leq x \leq 0.1$) and that a 4 cm stent is placed in the region $0.3 \leq x \leq 0.7$. Furthermore, we assume that, at the anchoring sites, the stent is placed 1 cm deep in the nonaneurysmal part of the aorta. See Figures 2 and 3. We calculate the total pressure and the minimal inner radius. We study the following four scenarios.

- **S 1.:** Straight vessel with three overlapping stents of equal length $0.3 \leq x \leq 0.7$.
- **S 2.:** Straight vessel with one stent in the region $0.3 \leq x \leq 0.7$ and two stents reinforcing the region $0.4 \leq x \leq 0.6$.
- **S 3.:** Curved vessel ($\kappa = 8$) with three overlapping stents of equal length $0.3 \leq x \leq 0.7$.
- **S 4.:** Curved vessel ($\kappa = 8$) with one stent in the region $0.3 \leq x \leq 0.7$ and two stents reinforcing the region $0.4 \leq x \leq 0.6$.

The results are shown in Figures 2, 3, 4 and 5. The data used to generate these pictures are: blood density $\rho = 1050kg/m^3$, viscosity $\nu = 3.2 \times 10^{-6}m^2/s$, unstressed radius of the abdominal aorta $R_0 = 0.0082m$ [11] and aortic stiffness $G_0 = 4 \times 10^4 N/m^2$ [15]. Although we have been unable to obtain the pressure-strain elastic modulus from the manufacturers of the Wallstent, our preliminary measurements indicate that $G_0 = 10^4 N/m^2$ and that the pressure-strain relationship, although nonlinear, is closer to linear (for larger strains) than that of the aorta. In our numerical simulations we used $\beta = 2$ for the aorta, $\beta = 1.2$ for the stent and $\beta = 3.2$ in the region where one stent is placed in the nonaneurismal

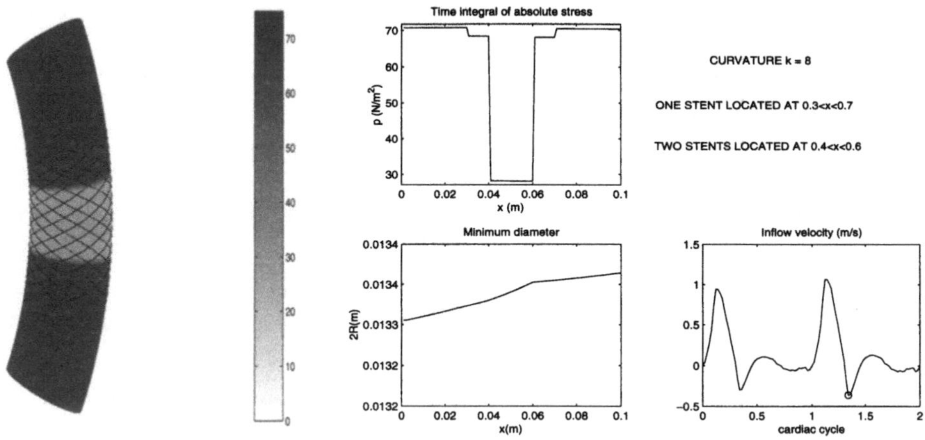

FIGURE 5. This figure shows the same flow properties as Figure 2 calculated for the scenario **S4.**

part of the aorta. Detailed measurements of the elastic properties of the Wallstent are under way by our collaborator Dr. Ravi-Chandar in the Mechanical Engineering Department at the University of Houston. Based on the results in [6], where the measured coefficient of stiffness for a double stent was twice the coefficient of stiffness for a single stent, we follow the same approach in the regions where multiple stents are placed. Small variations in the elastic properties of the stent did not produce radically different results, since the method and the problem are stable.

5. Discussion

We draw several conclusions from the results presented.

- Figures 2, 3, 4 and 5 show that the minimal diameter in the curved stented aorta is up to 26% smaller than the unstressed diameter. This confirms the observations presented in [21] which indicate that in diastole, the diameter of a stent-graft used to treat curved AAA, may decrease by as much as 30%. The observed diameter change in a nonstented vessel is typically up to 10% of the nonstressed diameter which is in good agreement with our numerical simulations which indicate 6% change for the straight, and 10% change for the curved, nonstented blood vessel.
- The total pressure on both the vessel and the stent is markedly higher in curved geometries. In addition, the pressure at the proximal end is higher than the pressure and the distal end.
- In all cases, except **S4**, the anchoring sites in the aorta are exposed to the largest inner fluid pressure. See the regions $x \in (0.03, 0.04) \cup (0.06, 0.07)$ in Figures 2, 3 and 4. This may be partly responsible for the clinically observed behavior, reported

in [23], of dislodgement of the stent attachments over time because of continued enlargement of the previously normal adjacent artery.
• We compare the performance of the two differently designed stents in two different geometries. The two stents are

STENT 1: a stent composed of three superimposed Wallstent stents of length 4 cm, and

STENT 2: a stent composed of one 4 cm-long Wallstent stent and two 2 cm-long Wallstent stents placed in the center.

The two geometries are the straight and the curved geometry with curvature $\kappa = 8$ (radius of curvature equal to 12.5 cm). The simulations presented in Figures 2, 3, 4 and 5 indicate the following.

- **STENT 2** is the preferred choice in straigh geometries because the total pressure exerted on the adjacent aorta is less than the total pressure associated with **STENT 1**.
- **STENT 1** is the preferred choice in the curved geometry with $\kappa = 8$ because the maximal total pressure exerted on the adjacent aorta is approximately 40% less than the maximal total pressure associated with **STENT 2**.
- In all of the cases, but especially in the curved geometries, the change in the total pressure inside the stent, at the points $x = 0.4$ and $x = 0.6$ (where the support from the adjacent aorta is no longer present), is drastic. The stress at those points may lead to the long-term complications reported by several cardiologists [21, 8] who detected broken axial wires, especially near the curved anchoring sites.

Movies showing the pulsation of a stented blood vessel as well as the pressure distribution over several cardiac cycles, can be viewed at:
www.math.uh.edu/˜canic/AAAstents.html.

6. Conclusions

Our results are encouraging because they seem to produce the behavior clinically observed in patients. The numerical simulations are fast and inexpensive. However, 3-dimensional simulations are needed to validate parameter regimes for which the modeling assumptions are satisfied and to provide detailed information about the solutions.

7. Acknowledgements

The authors would like to thank Dr. Zvonimir Krajcer at St. Luke's Hospital, for his support and consultations. Many thanks also go to Darryl Roy, undergraduate student at the University of Houston, for his help with pictures, for the reference search, and for many useful suggestions in writing this manuscript. Finally, the authors would like to thank Dr. Art Vailas, Provost for Research at the University of Houston, for the Undergraduate Research Support for Darryl Roy to work on this project.

References

[1] A. Achari and Z. Krajcer. A novel method for endoliminal treatment of abdominal aortic aneurysms. *Cardiovascular Interventions, Texas Heart Institute Journal*, 25:44–48, 1998.

[2] M. Aenis, A. P. Stancampiano, A. K. Wakhloo, and B. B. Lieber. Modeling of flow in a straight stented and nonstented side wall aneurysm model. *Trans. ASME*, 119:206–212, 1997.

[3] A. C. L. Barnard, W. A. Hunt, W. P. Timlake, and E. Varley. A theory of fluid flow in compliant tubes. *Biophys. J.*, 6:717–724, 1966.

[4] G-Q. Chen and H. Frid. Decay of entropy solutions of nonlinear conservation laws. *Arch. Rat. Mech. Anal*, 146:95–127, 1999.

[5] Uri Dinnar. *Cardiovascular Fluid Dynamics.* CRC Press, Boca Raton, Fla, 1981.

[6] J. F. Dyet, W. G. Watts, D. E. Ettles, and A. A. Nicholson. Mechanical properties of metallic stents: How do these properties influence the choice of stent for specific lesions? *Cardiovsc. Interv. Radiology*, 23:47–54, 2000.

[7] T. Gimse and N. H. Risebro. Solution of the Cauchy problem for a conservation law with a discontinuous flux function. *SIAM J. Math. Anal.*, 23:635–648, 1992.

[8] P. Harris, J. Brennan, J. Martin, D. Gould, A. Bakran, G. Gilling-Smith, J. Buth, E. Gevers, and D. White. Longitudinal aneurysm shrinkage following endovascular aortic aneurysm repair. *J. Endovasc. Surg.*, 6:11–16, 1999.

[9] H. Holden and N. H. Risebro. Riemann problems with a kink. *SIAM J. Math. Anal.*, 30:497–515, 1999.

[10] Z. Krajcer. Private communication.

[11] T. Länne, H. Stale, H. Bengtsson, D. Gustafsson, D. Bergqvist, B. Sonneson, H. Lecerof, and P. Dahl. Noninvasive measurement of diameter changes in the distal abdominal aorta in man. *Ultrasound in Med. & Biol.*, 18:451–457, 1992.

[12] Randall J. LeVeque. *Numerical Methods for Conservation Laws.* Birkhäuser, Boston, 1992.

[13] Randall J. LeVeque. Clawpack – a software package for solving multidimensional conservation laws. In *Proceedings of the Fifth International Conference on Hyperbolic Problems: Theory, Numerics, Applications.* World Scientific, June 1994.

[14] A. Lou and W-J Yang. A computer simulation of the blood flow at the aortic bifurcation with flexible walls. *Trans. ASME*, 115:306–315, 1993.

[15] S. T. R. MacSweeney, G. Young, R. M. Greenhalgh, and J. T. Powel. Mechanical properties of the aneurysmal aorta. *Br. J. Surg.*, 79:1281–1284, 1992.

[16] D. Ostrov. Solution of Hamilton-Jacobi equations and scalar conservation laws with discontinuous space-time dependence. Submitted.

[17] P. L. Roe. Approximate Riemann solvers, parameter vectors, and difference schemes. *J. Comp. Phys.*, 43:357–372, 1981.

[18] N. P. Smith, A. J. Pullan, and P. J. Hunter. An anatomically based model of transient coronary blood flow in the heart. Accepted in SIAM J. Appl. Math.

[19] J. C. Stettler, P. Niederer, M. Anliker, and M. Casty. Theoretical analysis of arterial hemodynamics including the influence of bifurcations, part II. *Ann. Biomed. Engineering*, 9:165–175, 1981.

[20] G. Strang. On the construction and comparison of difference schemes. *SIAM J. Num. Anal.*, 5:506–517, 1968.

[21] T. Umscheid and W. J. Stelter. Time-related alterations in shape, position, and structure of self-expanding, modular aortic stent-grafts. *J. Endovasc. Surg.*, 6:17–32, 1999.

[22] J. H. White, W. Yu, J. May, and at al. Three-year experience with the White-Yu Endovascular-GAD graft for transluminal repair of aortic and iliac aneurysm. *J. Endovasc. Surg.*, 4:124–136, 1997.

[23] J. R. Young, J. W. Olin, and J. R. Bartholomew. *Peripheral Vascular Diseases.* Mosby, Boston, 1996.

Sunčica Čanić
Department of Mathematics,
University of Houston,
Houston, Texas 77204-3476
canic@math.uh.edu

Dragan Mirković
Department of Computer Science,
University of Houston,
Houston, Texas 77204-3476
mirkovic@cs.uh.edu

International Series of Numerical Mathematics
Vol. 140, © 2001 Birkhäuser Verlag Basel/Switzerland

Study on Supersonic Flow Past a Pointed Body

Shuxing Chen

1. Main result

Study on supersonic flow around a given body is one of the main physical background of quasilinear hyperbolic system of conservation laws, as well as the main motivation of studing it. Generally speaking, when a supersonic flow attacks a given body, there will appear a shock front ahead of the body. If the body has a blunt head, the shock in front of it is detached. Otherwise, if the head of the body is sharp, then the shock can be attached on the head [3, 22, 27]. Here the main problem is to determine the location of the shock and the flow field in between the shock and the surface of the body. When the body has wing shape, the local existence of the solution and its some global property has been established in [11, 24, 14] for the wings with constant sections and in [4] for the wings with variable sections. When the body is a pointed body, a well known result is the case of a uniform supersonic flow past a circle cone with its axis parallel to the direction of the flow. However, in general case, the interaction of many factors, such as nonlinearity, multi-dimension, free boundary and the singularity caused by the tip of the pointed body let the problem become quite complicated. Therefore, the problem is open in the past 50 years, even though there has been great progress in the study of quasilinear hyperbolic system of conservation laws, particularly in one space dimensional case.

Recently, by using potential flow equation as the model to describe the motion of gas we have established the existence of the solution to such a problem under the assumption that the pointed body is a perturbation of a circle cone. Here we do not need to assume any symmetry of the body, and to assume the generators are straight neither. Our result can also be understood as the stability of the solution with a given shape of shock front in local sense.

Using the model of potential flow equation the problem of supersonic flow past a pointed body can be formulated as

$$\begin{aligned}\left(\frac{v_1^2}{a^2}-1\right)\phi_{x_1x_1}+\left(\frac{v_2^2}{a^2}-1\right)\phi_{x_2x_2}+\left(\frac{v_3^2}{a^2}-1\right)\phi_{x_3x_3}\\ +\frac{2v_1v_2}{a^2}\phi_{x_1x_2}+\frac{2v_1v_3}{a^2}\phi_{x_1x_3}+\frac{2v_2v_3}{a^2}\phi_{x_2x_3}=0,\end{aligned} \tag{1}$$

where ϕ is the potential of the flow, $v_i = \phi_{x_i}$ is the component of velocity, a is sonic speed, which is a function of $\nabla\phi$.

$$m_{x_1}\phi_{x_1} + m_{x_2}\phi_{x_2} + m_{x_3}\phi_{x_3} = 0 \tag{2}$$

on the surface of the pointed body, where $m(x_1, x_2, x_3) = 0$ is the equation of the surface.

$$\phi = \phi_-(= q_\infty x_3); \tag{3}$$

$$(\mu_{x_1}\phi_{x_1} + \mu_{x_2}\phi_{x_2} + \mu_{x_3}\phi_{x_3})H\mu_{x_3}q_\infty\rho_\infty. \tag{4}$$

the last two conditions are given on the shock with the equation $\mu(x_1, x_2, x_3) = 0$, which is unknown and will be determined with ϕ together.

The equation and the boundary conditions can also be rewitten in the cylindrical coordinates. The equation of the surface of the body can be given as $r = b(z, \theta)$. Then the differential equation and corresponding boundary conditions are

$$\begin{aligned} z^2 a_{00}\phi_{zz} + a_{11}\phi_{rr} + a_{22}\phi_{\theta\theta} + 2za_{01}\phi_{zr} + 2za_{02}\phi_{z\theta} + 2a_{12}\phi_{r\theta} \\ +a_1\phi_r + a_2\phi_\theta = 0. \end{aligned} \tag{5}$$

$$(b + zb_z)\phi_z + \frac{b_\theta}{r^2}\left(\frac{\phi_\theta}{z}\right) - (1 + b(b + zb_z))\left(\frac{\phi_r}{z}\right) = 0. \tag{6}$$

$$\begin{aligned} &\left((\frac{\phi_r}{z})^2 + \frac{1}{r^2}(\frac{\phi_\theta}{z})^2 + (\phi_z - \frac{r\phi_r}{z} - q_\infty)(\phi_z - \frac{r\phi_r}{z})\right) H \\ &-(\phi_z - \frac{r\phi_r}{z} - q_\infty)q_\infty\rho_\infty = 0. \end{aligned} \tag{7}$$

Here we omit the detailed expression of a_{ij} in (5). In the cylindrical coordinates system, our main result can be described as follows.

Theorem 1.1 *Assume that $r = b(z, \theta)$ satisfies*

$$\|b(0, \theta) - b_0\|_{C^{k_1}} \le \epsilon_0$$

$$\partial_z^k b(0, \theta) = 0 \quad \text{for } 1 \le k \le k_2$$

with k_1, k_2 being suitable integers. Assume that a supersonic flow parallel to the z-axis comes from infinity with velocity $q = q_\infty$, pressure p_∞, density ρ_∞ satisfying $q_\infty > a_\infty$ ($(\frac{\gamma p_\infty}{\rho_\infty})^{\frac{1}{2}}$). Besides, b_0 is assumed to be a constant less than the crictical value determined by $q_\infty, p_\infty, \rho_\infty$. Then the problem of supersonic flow past the given pointed body admits a weak entropy solution with a pointed shock front attached at the origin, provided ϵ_0 is small enough.

2. Approximation on level one

We seek the solution of our problem via approximation on different level. The whole process of the approximations can be briefly described as follows. The approximate solution of level zero is the background solution, which corresponds to the flow past a circle cone (the pointed body is a perturbation of this cone). On the level one, the first approximate solution corresponds to the solution of the same flow past a conical body tangential to the given pointed body. Then we use finite series expansion to modify the approximation to let it satisfy the equation and boundary conditions with error $O(z^N)$, where N is a sufficiently large integer. Based on this approximations we establish a sequence of approximate solution on level two, and finally by a further modification obtain the precise solution.

First we assume that the potential ϕ has a form of finite power expansion as

$$\phi(z,r,\theta)=\sum_{n=0}^{N} z^{n+1}\phi_n(r,\theta)+O(z^{N+2}). \tag{8}$$

By suitable choosing ϕ_n, the equation and boundary conditions can be satisfied with error $O(z^N)$. Therefore, in a neighbourhood of the origin the form of asymptotic series (8) offers a good approximation for large N. Substituting the expressions of ϕ into (5) leads to

$$a_{11}(*)\phi_{0rr}+a_{22}(*)\phi_{0\theta\theta}+2a_{12}(*)\phi_{0r\theta}+A(\phi_0,\phi_{0r},\phi_{0\theta})=0, \tag{9}$$

$$\begin{aligned}&a_{11}(*)\phi_{nrr}+a_{12}(*)\phi_{n\theta\theta}+2a_{12}(*)\phi_{nr\theta}+B_r\phi_{nr}+B_\theta\phi_{n\theta}+C\phi_n+\\ &\quad G_n(\phi_\ell,\nabla\phi_\ell,\nabla^2\phi_\ell)_{\ell<n}=0 \qquad\qquad n\geq 1\end{aligned} \tag{10}$$

where G_n are given functions depending on ϕ_ℓ and its derivatives of first and second order.

Write $b(z,\theta)$ and $s(z,\theta)$ as $\sum_{n=0}^{N} z^n b_n(\theta)$ and $\sum_{n=0}^{N} z^n s_n(\theta)$ respectively. Similar substitution gives us the boundary conditions on $r=b_0(\theta)$

$$b_0\underline{\phi}_0+\frac{1}{b_0^2}b_{0\theta}\underline{\phi}_{0\theta}-(1+b_0^2)\underline{\phi}_{0r}=0 \tag{11}$$

$$\begin{aligned}&(n+1)(\underline{\phi}_0 b_n+b_0(\underline{\phi}_n+\underline{\phi}_{0r}b_n))+\frac{1}{b_0^2}(b_{0\theta}(\underline{\phi}_{n\theta}+\underline{\phi}_{0r\theta}b_n)+\underline{\phi}_{0\theta}b_{n\theta}-2\frac{b_n}{b_0}b_{0\theta}\underline{\phi}_{0\theta})\\ &-(1+b_0^2)(\underline{\phi}_{nr}+\underline{\phi}_{0rr}b_n)-(n+2)\underline{\phi}_{0r}b_0b_n=g_n(\underline{\phi},b) \qquad (n>0),\end{aligned} \tag{12}$$

and the boundary condition on $r=s_0(\theta)$

$$(\bar{\phi}_{0r}^2+\frac{1}{r^2}\bar{\phi}_{0\theta}^2-(\bar{\phi}_0-r\bar{\phi}_{0r})r\bar{\phi}_{0r})\rho_0-r\bar{\phi}_{0r}q_\infty\rho_\infty=0, \tag{13}$$

$$\gamma_1\bar{\phi}_{nr}+\gamma_2\bar{\phi}_{n\theta}+\gamma_3\bar{\phi}_n=g_n(\bar{\phi}). \tag{14}$$

Here $\underline{\phi}_i$ means the value of ϕ_i on $r=b_0(\theta)$, $\bar{\phi}_i$ means the value of ϕ_i on $s_0(\theta)$, g_n arre given function of ϕ_ℓ or its derivatives with $\ell<n$, while many detailed expression of the coefficients are also omitted.

The solution ϕ_0 of the problem (9),(11),(13) corresponds to the solution of supersonic flow past the tangential cone with the surface $b=b_0(\theta)$. The problem can be reduced a free boundary value problem of nonlinear elliptic equation. To deal with the free boundary we perform a partial hodograph transformation T : $(r,\theta) \mapsto (p,\sigma)$ involving the unknown function ϕ_0

$$\begin{cases} \sigma = \theta \\ p = \phi_0(r,\theta). \end{cases} \tag{15}$$

Its inverse transform is T^{-1} :

$$\begin{cases} \theta = \sigma \\ r = u(p,\sigma). \end{cases} \tag{16}$$

In the new coordinates (p,σ) the function $u(p,\sigma)$ satisfies

$$\tilde{a}_{11}u_{pp} + 2\tilde{a}_{12}u_{p\sigma} + \tilde{a}_{22}u_{\sigma\sigma} + \tilde{A}(u,u_p,u_\sigma) = 0 \tag{17}$$

and the boundary conditions

$$((1+u^2) + \frac{u_\sigma^2}{u^2} - upu_p)H + upu_p\rho_\infty = 0. \tag{18}$$

on shock,

$$u = b(\sigma), \tag{19}$$

$$(1+b^2(\sigma)) + \frac{b'(\sigma)}{b^2(\sigma)}u_\sigma - b(\sigma)pu_p = 0. \tag{20}$$

on the surface of the body.

The problem (17)-(20) is still a nonlinear elliptic boundary problem with free boundary, because the potential is unknown on the surface of the body. Therefore, we introduce the domain decomposition method to consider (9) and (17) together. We decompose the domain $b_0(\theta) < r < s_0(\theta)$, where (9) is defined, to a set of overlapped rings. Correspondingly, the problem (9),(11),(13) or the problem (17)-(20) is decomposed to a set of elliptic boundary problems defined on these overlapped rings with some additional Dirichlet condition on new boundaries. Let the width of each ring be small, so that the comparison principle holds for each linearized boundary value problem on the corresponding ring. Then we can use the nonlinear Schwarz altenating iteration to establish a set of sequences $\{\phi^{(n)}\}$ (or $\{u^{(n)}\}$), which are monotonic with respect to n and uniformly bounded. Then the limit of the sequence gives the approximate solution of level one. More details can be found in [5].

The functions $\phi_n(r,\theta)$ can be determined by solving the linear boundary value problem (10),(12),(14). Here the principle part of (10) for each n is the same. Analyzing all coefficients and using the assumptions of our main theorem we can solve (10),(12),(14) for each $n \le N$ and then determine all $\phi_n(r,\theta)$, and finally obtain

$$\tilde{\phi}(z,r,\theta) = \sum_{n=0}^{N} z^{n+1}\phi_n(r,\theta),$$

which is the first approximation on level two. Here we remark that the number N, which means the degree of the approximation, is independent of the integer k_1

and k_2 in the assumption of the theorem, and only depends on the smoothness of the data.

3. Approximation on level two

Starting from the previous approximation solution we introduce another iteration scheme to modify it graduately to obtain the percise solution of the original problem. Write the unknown function ϕ as $z\chi$, and introduce a transformation

$$t = \ln z, \qquad z = e^t, \tag{21}$$

which maps the point $z = 0$ to $t = -\infty$. Then the problem (5)-(7) becomes

$$\begin{aligned} L(\chi) \equiv a_{00}(\chi_t + \chi_{tt}) + a_{11}\chi_{rr} + a_{22}\chi_{\theta\theta} + 2a_{01}\chi_{tr} + 2a_{02}\chi_{t\theta} + 2a_{12}\chi_{r\theta} \\ +(a_1 + 2a_{01})\chi_r + (a_2 + 2a_{02})\chi_\theta = 0, \end{aligned} \tag{22}$$

$$E(\chi) \equiv (b + b_t)(\chi + \chi_t) + \frac{b_\theta}{r^2}\chi_\theta - (1 + r(b + b_t))\,\chi_r = 0, \tag{23}$$

$$\begin{aligned} G(\chi) \equiv \left(\chi_r^2 + r^{-2}\chi_\theta^2 + (\chi + \chi_t - r\chi_r)(\chi + \chi_t - r\chi_r - q_\infty)\right) H \\ - (\chi + \chi_t - r\chi_r - q_\infty)\, q_\infty \rho_\infty = 0, \end{aligned} \tag{24}$$

$$\chi = q_\infty,$$

where (23) and (24) are given on the surface and the shock respectively. Since (23) is hyperbolic equation, then all method related to the ellipticity in the process of seeking the approximation of level one does not work now. Therefore, to modify the approximation once more we use Newton's interation, and to avoid the difficulty caused by the free boundary we introduce a generalized hodograph transformation

$$y_0 = t, \quad y_1 = \theta, \quad y_2 = \frac{r - b}{r - b + q_\infty - \chi}, \tag{25}$$

which transforms the boundary $r = b(z, \theta)$ and $r = s(z, \theta)$ into $y_2 = 0$ and $y_2 = 1$ respectively. In the meantime, we choose new unknown function

$$\omega(y_0, y_1, y_2) = r - b(y_0, y_1) + u_\infty - \chi(y_0, y_1, r), \tag{26}$$

where $r = r(r_0, y_1, y_2)$ is determined by (25). Under the transformation (25) involving the unknown function, the problem (22)-(24) takes the new form:

$$\mathbf{P}\omega \equiv \sum_{i,j=0,1,2} \alpha_{ij}(\omega, \nabla\omega)\omega_{y_i y_j} + R(\omega, \nabla\omega) = 0 \quad \text{in} \quad 0 < y_2 < 1 \tag{27}$$

$$\mathbf{Q_b}\omega \equiv \tilde{B}(\omega, \omega_{y_0}, \omega_{y_1}, \omega_{y_2}) \equiv \tilde{M}\nabla_y\omega + R_b(\omega) = 0 \quad \text{on} \quad y_2 = 0 \tag{28}$$

$$\mathbf{Q_s}\omega \equiv \tilde{G}(\omega, \omega_{y_0}, \omega_{y_1}, \omega_{y_2}) = 0. \quad \text{on} \quad y_2 = 1 \tag{29}$$

By direct computation we can verify that the generalized hodograph transformation (25) keeps the property of normal hyperbolicity of the equation with respect to an assigned direction. This means, the new form of the equation in (y_0, y_1, y_2) coordinates is still normal hyperbolic with respect to the image of the assigned direction under the transformation. Besides, for the nonlinear boundary conditions we can define linearized boundary operators and corresponding M-D linear stability condition for them (see [16]). The new problem under the transformation (25) still satisfies the M-D linear stability condition.

Denote P, Q_b and Q_s the linearizing operators corresponding to the nonlinear operators $\mathbf{P}, \mathbf{Q_b}$ and $\mathbf{Q_s}$ respectively, we have

$$P(\omega)\dot{\omega} \equiv \sum_{i,j=0}^{2} \alpha_{ij}(\omega)\partial_{y_i}\partial_{y_j}\dot{\omega} + \sum_{j=0}^{2} \alpha_j(\omega)\partial_{y_j}\dot{\omega} + \alpha(\omega)\dot{\omega} = f \text{ in } 0 < y_2 < 1, \quad (30)$$

$$Q_b(\omega)\dot{\omega} \equiv \sum_{j=0}^{2} \beta_{1j}\partial_{y_j}\dot{\omega} + \beta_0\dot{\omega} = g_1 \text{ on } y_2 = 0, \quad (31)$$

$$Q_s(\omega)\dot{\omega} \equiv \sum_{j=0}^{2} \gamma_{1j}\partial_{y_j}\dot{\omega} + \gamma_0\dot{\omega} = g_2 \text{ on } y_2 = 1, \quad (32)$$

For the problem (30)-(32) we can establish an energy estimate

$$|||\dot{\omega}|||^2_{s+1,\eta,T} \leq C_s \left(\frac{1}{\eta} ||f||^2_{s,\eta,T} + \langle g_1 \rangle^2_{s,\eta,T,} + \langle g_2 \rangle^2_{s,\eta,T,} \right), \quad (33)$$

where $|||\cdot|||$ and other notations in the right side of (33) are suitable weighted Sobolev norm, C_s is a constant independent of the choice of ω.

By using (31)-(33) we perform the iteration scheme as follows. The function $\tilde{\omega}(y_0, y_1, y_2)$ transformed from $\tilde{\phi}(z, r, \theta)$ approximately solves the problem (27)-(29). We take it as the first term of a sequence of approximate solutions of (27)-(29). Furthermore, we take $\dot{\omega}_0 = 0$, and define $\dot{\omega}_{k+1}$ as the solution of the linear boundary value problem:

$$P(\tilde{\omega} + \dot{\omega}_k)\dot{\omega}_{k+1} = f_k \text{ in } 0 < y_2 < 1, \quad (34)$$

$$Q_b(\tilde{\omega} + \dot{\omega}_k)\dot{\omega}_{k+1} = g_{1k} \text{ on } y_2 = 0, \quad (35)$$

$$Q_s(\tilde{\omega} + \dot{\omega}_k)\dot{\omega}_{k+1} = g_{2k} \text{ on } y_2 = 1, \quad (36)$$

where $g_{2k} = -\mathbf{Q_s}(\tilde{\omega} + \dot{\omega}_k) + Q_s(\tilde{\omega} + \dot{\omega}_k)\dot{\omega}_k, \gamma_{1j}^{(k)} = \gamma_{1j}(\tilde{\omega} + \omega_k)$ etc..

Since $\tilde{\phi}$ is the approximate solution with error $O(z^N)$, the choice of $\tilde{\omega}$ and the energy estimate (33) let the problem (34)-(36) be solvable. Namely, we can use this iteration to establish a sequence of approximate solutions on the level two. Based on the energy estimate we can further prove the boundedness in high norm and the contraction in low norm for the sequence. Obviously, by using such a fact we can confirm the convergence of the sequence of approximate solutions. Then the limit of the sequence is the precise solution as we confirmed in the main theorem.

4. Further considerations

Despite the local existence and the stability of the solution to the nonlinear problem (1)-(3) has been established. More questions remain to be answered, even restrict ourselves to the case of supersonic flow past a sharp body. Here we list some of them, and hope to attract more people's attention to this topic.

1. The global existence and asymptotic behaviour of the solution to the supersonic flow past pointed body. If the pointed body is symmetric, then the problem can be treated by using Glimm scheme (for instance, see [16]). But the results in M-D case relies on more subtle energy estimates, and much work should be done. The result on the existence of global piecewise solution in the symmetric case has been obtained in [10], and the non-symmetric case is also under consideration.

2. If the pointed body is not a small perturbation of a circle cone, then we don't have background solution, which plays important role in above discussion.. Since the linearization at background solution is the key point in many steps of the proof of Theorem 1.1, it seems to be necessary to find a new way to treat the problem.

3. If we use the model of full Euler system to describe the flow, then the problem will also be more complicated. In this case, even in the process of finding the approximation of level one we have to solve a boundary value problem of a coupled system of hyperbolic equation and elliptic equation with free boundary. Such a problem itself is also an interesting topic in the theory of partial differential equations.

4. When the coming flow is unsteady, we have to consider the time dependent problem. Could we still have existence and stability local in space and in time, or even global both in space and in time? In this case, even the body has wedge shape, the corresponding questions are also open.

Acknowledgement The work is partially supported by NNSF, Key Grant of NMST, Doctoral Programm Foundation of NEM of China and IMS of CUHK.

References

[1] S.Alinhac, *Existence d'ondes de rarefaction pour des systèmes quasi-linéaires hyperboliques multidimensionnels.* Comm. P.D.E., **14** (1989), 173-230.

[2] A. Blokhin, and A. Birkin, *Global solvability of the problem of supersonic flow around a cone.* Mat. Model, **4** (1996), 89-104.

[3] R.Courant, and K.O.Friedrichs, *Supersonic Flow and Shock Waves.* Interscience Publishers Inc., New York, 1948.

[4] S.X.Chen, *Existence of local solution to supersonic flow past a three- dimensional wing.* Advances in Appl. Math., **13** (1992), 273-304.

[5] S.X.Chen, *Non-symmetric conical supersonic flow.* International Series of Numer. Math., Birkhäuser Verlag, Basel/Switzerland, **129** (1999), 149-158.

[6] S.X.Chen, *Existence of Stationary Supersonic Flows Past a Pointed Body*, Arch. Rational Mech. Anal. **156** (2001),141-181.

[7] S.X.Chen, *Supersonic flow past a concave double wedge.* Science in China, **41** (1998), 39-47.

[8] S.X.Chen, *Asymptotic behaviour of supersonic flow past a convex combined wedge.* Chin. Ann. Math., **19B** (1998), 225-264.

[9] S.X.Chen and D.N. Li, *Supersonic flow past a symmetrically curved cone.* Indiana Univ. Math. Jour. **49** (2000).

[10] S.X.Chen, Z.P.Xin and H.C.Yin, *Global shock waves for the supersonic flow past a pointed body*, to appear.

[11] C.H.Gu *A method for solving the supersonic flow past a curved wedge* Fudan Journal (Natural Science), **7**, (1962), 11-14.

[12] O.A.Ladyzhenskaya, *The boundary value problems of mathematical physics.* Springer-Verlag, 1985, New York.

[13] D.N.Li, *Rarefaction and shock waves for multi-dimensional hyperbolic conservation laws.* Comm. in PDEs, **16** (1991), 425-450.

[14] T.T.Li, *Une remarque sur un problème à frontière libre.* C.R.Acad.Sci. Paris Sér.A **289**(1979), 99-102.

[15] T.T.Li and W.C.Yu, *Boundary value problem for quasi-linear hyperbolic systems.* Duke Univ. Math., Ser.**5** (1985).

[16] W.C.Lien and T.P.Liu, *Nonlinear stability of a self-similar 3-d gas flow* Comm. Math. Phys., **304**(1999),524-549.

[17] A.Majda, *One perspective on open problems in multi-dimensional conservation laws.* Multidimensional Hyperbolic Problems and Computation, Springer Verlag, IMA **29**(1990), 217-237.

[18] A.Majda and E.Thomann, *Multi-dimensional shock fronts for second order wave equations.* Comm. P.D.E., **12**(1987), 777-828.

[19] A.M.Meirmanov, *On a classical solution of the multidimensional Stefan problem for quasilinear parabolic equations.* Math. Sb., **112**(1980), 170-192.

[20] M.Mnif, *Probleme de Riemann pour une loi conservation scalaire hyperbolique d'order deux.* Comm. in PDEs **22** (1997), 1589-1627.

[21] C.S.Morawetz, *Potential theory for regular and Mach reflection of a shock at a wedge.* Comm. Pure Appl. Math. **47** (1994), 593-624.

[22] P.H.Osthuizen and W.E.Carscallen, *Compressible Fluid Flow.* McGraw-Hill Companies INC. (1977).

[23] A.I.Rykov, *Possible regimes of flow around pointed bodies of finite thickness for arbitrary hypersonic velocities of the incoming flow.* Journal of Appl. Math. Mech. **55** (1992), 74-78.

[24] D.G.Schaeffer, *Supersonic flow past a nearly straight wedge.* Duke Math. J., **43** (1976), 637-670.

[25] D.G.Schaeffer, *Application of the Nash-Moser theorem to a free boundary problem.* Lecture Notes in Math. **648**(1978), 129-143.

[26] W.S.Sears, *General theory of high speed aerodynamics* , Oxford University Press, London (1955).

[27] G.B.Whitham, *Linear and Nonlinear Waves* , John Wiley and Son, New York, London, Sydney, Toranto, (1974).

[28] Y.L.Zhu, *Difference methods for initial-boundary value problems and flow around bodies* Springer-Verlag, Berlin-New York; Science Press, Beijing (1988).

S. Chen
Institute of Mathematics,
Fudan University ,
Shanghai, 200433, China
sxchen@fudan.ac.cn

International Series of Numerical Mathematics
Vol. 140, © 2001 Birkhäuser Verlag Basel/Switzerland

Fine-Mesh Numerical Simulations for 2D Riemann Problems with a Multilevel Scheme

G. Chiavassa, R. Donat, and A. Marquina

1. Introduction

The numerical simulation of physical problems modeled by systems of conservation laws can be difficult due to the occurrence of discontinuities and other non-smooth features in the solution.

High-Resolution Shock Capturing (HRSC henceforth) schemes aim at obtaining numerical approximations that combine high accuracy in smooth regions with sharp and oscillation free numerical profiles at discontinuities. The power of a HRSC scheme lies usually in a 'delicate', and often very expensive, computation of the numerical flux function of the scheme, which is the main drawback of these schemes, specially in multi-dimensional computations. However, it is known that the costly numerical flux function of a HRSC scheme is strictly necessary only in a neighborhood of singularities, or in those regions where singularities will develop in the near future. In smooth regions, the highly expensive numerical flux function can, in principle, be computed with much less expensive means. This principle has been the driving force behind various types of strategies designed to reduce the computational expense of numerical schemes for conservation laws, from hybridization to Adaptive Mesh Refinement (AMR).

In [11], A. Harten proposes to reduce the cost using the smoothness information contained in a multiresolution transform of the numerical solution. In smoothness regions, the expensive numerical flux evaluation is replaced by a cheap polynomial interpolation obtained from a coarser grid. The 1D numerical experiments reported in [11] are very encouraging, and this idea has been further explored by several authors [1, 2, 4, 6, 12, 17].

In [1, 2, 4, 17] it is noted that in multidimensional computations it is the numerical divergence, instead of the numerical flux function, the quantity that should be adapted to the multilevel evaluation. These works (and also [6, 12]) use the cell-average framework for multiresolution, since in the finite-volume approach to the design of numerical schemes for conservation laws, the numerical values are considered as approximations to the cell-averages of the solution.

In [5], we developed an alternative method based on point value multiresolution transform, to be used in HRSC schemes that follow the Shu-Osher [16] formulation. The performance of this multilevel strategy has been analyzed and tested on several classical 2D configurations in gas dynamics [5]. The results in these papers allow us to conclude that the multilevel strategy can be effectively used as an acceleration tool that allows the user to perform numerical simulations on very fine meshes at an affordable cost: the cost of the basic HRSC code on a much coarser mesh.

In the present paper, the multilevel algorithm is used to obtain very high quality numerical solutions in which it is possible to observe physical effects that are undetectable when using grids with larger mesh sizes. Thanks to the large reduction in execution time procured by the method, it is feasible to run such simulations on standard personal computers.

The rest of the paper is organized as follows: In section 2 we recall the basic ingredients of the multilevel algorithm and of the underlying HRSC scheme used in our simulations. In section 3, we apply our algorithm to two particular configurations of 2D Riemann problems. We present numerical results for a (relatively) coarse grid, in order to compare our results with simulations on the same problems found in the literature. Then we present numerical results on a very fine mesh. The combination of high resolution and high order in the numerical approximation allows us to observe very local physical phenomena, like Kelvin-Helmholtz instabilities in the contact discontinuities, that are not observable at larger grid sizes.

2. The multilevel algorithm

The multilevel strategy we employ in this paper has been described an analyzed in [5] and we refer the interested reader to this paper for specific details. Here we only present the main steps and the guidelines.

2.1. General framework

We work in the finite difference context proposed by Shu and Osher [16] for the numerical solution of two dimensional systems of conservation laws:

$$\partial_t \vec{U} + \vec{f}(\vec{U})_x + \vec{h}(\vec{U})_y = \vec{0}. \tag{1}$$

The problem is discretized on a Cartesian grid G^0 using the semi-discrete formulation:

$$\frac{d\vec{U}_{ij}}{dt} + D(\vec{U})_{ij} = \vec{0}. \tag{2}$$

The numerical divergence $D(\vec{U})$ is computed as:

$$D(\vec{U})_{ij} = \frac{\vec{F}_{i+1/2,j} - \vec{F}_{i-1/2,j}}{\delta x} + \frac{\vec{H}_{i,j+1/2} - \vec{H}_{i,j-1/2}}{\delta y}, \tag{3}$$

where $\vec{F}(\vec{w}_1, .., \vec{w}_{k+m})$ and $\vec{H}(\vec{w}_1, .., \vec{w}_{k+m})$ are consistent numerical flux functions.

The numerical flux function is the trademark of the scheme. The HRSC method we shall use in our numerical simulation is Marquina's scheme [8]. The numerical flux function of this scheme involves two Jacobian evaluations at each cell boundary. High accuracy in space is achieved by using a Piecewise Hyperbolic reconstruction (PHM) in the characteristic variables at each cell boundary. High accuracy in time is achieved by using a third order TVD Runge-Kutta ODE solver as in [16] to discretize (2) in time.

The HRSC obtained in this manner has been extensively tested in [8, 9, 7], where it is shown that it leads to high quality numerical approximations, even in cases when other more classical solvers produce anomalous numerical behavior.

As explained in the introduction, the goal of the multilevel method is to *decrease the cpu time* associated to the underlying scheme by reducing the number of expensive flux evaluations. To understand the basic mechanism, let us consider, for the sake of simplicity, Euler's method applied to (2), i.e.

$$\vec{U}_{ij}^{n+1} = \vec{U}_{ij}^{n} - \delta t \; D(\vec{U})_{ij}^{n}. \tag{4}$$

If both U^n and U^{n+1} are smooth around (x_i, y_j) at time t^n, then (4) implies that the numerical divergence is also smooth and we can avoid using the numerical flux functions of the HRSC scheme in its computation. On the other hand, if a discontinuity appears during the time evolution (or when a steep gradient makes it imminent), the Riemann solver of the HRSC scheme has to be called necessarily to compute the numerical divergence if the high-resolution properties of the underlying scheme are to be maintained.

Consequently, the most important steps in the multilevel algorithm concern the smoothness analysis of U^n and U^{n+1} (which is unknown at time n) and how this information is used in the computation of $D(\vec{U})$.

2.2. Smoothness analysis

A multiresolution transform of a discrete set of data gives an equivalent discrete set composed by a coarse grid representation of the data and a sequence of *detail coefficients* (or *wavelet coefficients*) which represent the *difference in information* between two consecutive resolution levels. In this context, the different resolution levels are specified by a set of nested grids $\{G^l, \; l = 1, .., L\}$ given as follows,

$$(x_i, y_j) \; \in \; G^l \Longleftrightarrow (x_{2^l i}, y_{2^l j}) \; \in \; G^0 \tag{5}$$

where G^0 is the finest one (not necessary with fixed mesh ratio).

In Shu and Osher's framework, the numerical values can be interpreted as approximations to the point-values of the solution. Hence, in our multilevel strategy the numerical values are analyzed using the point-value framework for multiresolution, instead of the cell-average framework as in [1, 2, 4, 6, 12, 17].

In the point-value framework, the coarse grid representation of the data are simply its values at the points of the coarsest grid G^L, and the detail coefficients are interpolation errors, which are directly related to the local regularity of the data.

When applying the point-value multiresolution transform to the numerical solution, large values of the detail coefficients correspond to non-smooth zones of the solution like shocks or contact discontinuities, and it is easy to establish the relation between the local smoothness and the decay rate of the detail coefficients with respect to the scale (see [11], and also [10] or [3]).

The information about the regularity of the data contained in the multiresolution transform of the numerical solution is used to determine a *flag* vector $((b^l_{ij})_{l,i,j})$, whose value (0 or 1) will determine the choice of the procedure to evaluate D. Given a tolerance parameter ϵ, the value is obtained by applying two tests to the details coefficients $(d^l_{ij})_{l,i,j}$:

$$\begin{array}{lll} \text{if } |d^l_{ij}| & \geq & \epsilon \Longrightarrow b^l_{i-k\ j-m} = 1 \quad k,m = -2,..,2 \\ \text{if } |d^l_{ij}| & \geq & 2^r\epsilon \text{ and } l > 1 \Longrightarrow b^{l-1}_{2i-k\ 2j-m} = 1 \quad k,m = -1,0,1 \end{array} \tag{6}$$

Shock waves in the solution of a system of conservation laws form after a large gradient builds up. Once they are formed, they have a finite speed of propagation. The first test takes into account that the propagation of information in the numerical scheme is limited by the CFL number. The second test compares the decay rate of the scale coefficients to that of a smooth function and it aims at detecting the formation of a large gradient, setting the flag to 1 in regions where compression, leading to shock formation, is taking place. For specific details see [10, 4].

Remark 2.1. *In previous works* [10, 4, 5, 17], *a level-dependent tolerance was used,* $\epsilon_l = 2^{-l}\epsilon$. *This is the most adequate truncation strategy in the cell-average multiresolution context but it is unnecessarily restrictive in the point value context, where a strategy like* $\epsilon_l = \epsilon$ *is appropriate (see e.g.* [10, 3] *for details).*

2.3. Multilevel evaluation of the numerical divergence

The values stored in the flag vector are used in the computation of the numerical divergence as follows: First, $D(\vec{U})$ is computed at the points of the coarsest grid, $(x_i, y_j) \in G^L$ using the HRSC scheme. Once point-values of the divergence are known on G^l, we call them $D^l(\vec{U})$, the values on G^{l-1}, $D^{l-1}(\vec{U})$, are computed using the boolean flag:

$$\begin{array}{ll} \text{if } b^l_{ij} = 1 & \text{compute } D^{l-1}(\vec{U})_{ij} \text{ directly with Marquina's method.} \\ \text{if } b^l_{ij} = 0 & D^{l-1}(\vec{U})_{ij} = I[(x_i, y_j); D^l(\vec{U})], \end{array} \tag{7}$$

where $I[(x_i, y_j); D^l(\vec{U})]$ represents a two-dimensional polynomial interpolation of D around the point (x_i, y_j) using the values computed on G^l.
The process is repeated from $l = L$ to $l = 1$ and, once it is completed, we obtain the values of $D(\vec{U})$ on the finest grid G^0, which are needed by the ODE solver.

Remark 2.2. *In the evaluation of* $D^l(\vec{U})_{ij}$ *using the HRSC scheme, the numerical flux functions are computed using the values of* $\vec{U}$ *on* G^0, *and not on* G^l. *The finest grid values must always be available, and no memory gain can be expected from the multilevel strategy. More general strategies, which aim to obtain a fully adaptive technique are investigated in* [6, 12]. *These strategies, however, seem to suffer from* *some of the drawbacks of AMR techniques, including the need to use complicated data structures in the code.*

The 2D polynomial interpolation is the same one used for the computation of the wavelet coefficients.

This multilevel algorithm has been applied in [5] to two classical test problems in 2D, where it is shown that the quality of the numerical solution, with respect to the *reference* simulation (i.e. without multiresolution) is controlled by the tolerance parameter ϵ.

3. Numerical experiments: 2D Riemann problems

Let us consider two particular instances of the Riemann problems in two dimensions described in [14]. These test cases have been considered recently in [19, 18] from a theoretical point of view and used as test cases for numerical simulations in [15, 13].

We consider the classical compressible Euler equations for a polytropic gas (1) with $U = (\rho, \rho v_x, \rho v_y, e)$, where ρ represents the density, v_x and v_y the velocities in x and y directions and e the internal energy. The pressure P is deduced via the equation of state for an ideal gas.

Configuration A: 4-shock configuration
This configuration corresponds to N° 3 of [13] and it involves 4 constant states separated by shock waves. The computational domain is represented in Figure 1-A), and the initial values for the variables are :

$$\begin{aligned}
&\rho_1 = 1.5,\ \rho_2 = 0.5323,\ \rho_3 = 0.138,\ \rho_4 = 0.5323,\\
&v_{x_1} = 0,\ v_{x_2} = 1.206,\ v_{x_3} = 1.206,\ v_{x_4} = 0,\\
&v_{y_1} = 0,\ v_{y_2} = 0,\ v_{y_3} = 1.206,\ v_{y_4} = 1.206,\\
&p_1 = 1.5,\ p_2 = 0.3,\ p_3 = 0.029,\ p_4 = 0.3.
\end{aligned}$$

The initial geometry has been slightly changed compared to [13] to study the behavior at later times.

Configuration B: 4-contact configuration
It corresponds to the configuration N° 6 of [13] involving the interaction of 4 contact discontinuities. The domain is represented on Figure 1-B) and the values are :

$$\begin{aligned}
&\rho_1 = 1,\ \rho_2 = 2,\ \rho_3 = 1,\ \rho_4 = 3,\\
&v_{x_1} = 0.75,\ v_{x_2} = 0.75,\ v_{x_3} = -0.75,\ v_{x_4} = -0.75,\\
&v_{y_1} = -0.5,\ v_{y_2} = 0.5,\ v_{y_3} = 0.5,\ v_{y_4} = -0.5,\\
&p_1 = 1,\ p_2 = 1,\ p_3 = 1,\ p_4 = 1.
\end{aligned}$$

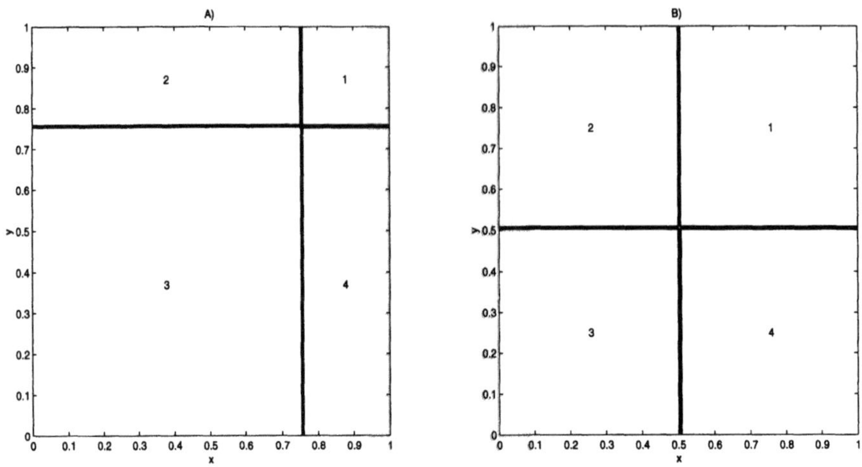

FIGURE 1. *Initial computational domains for configuration A, A), and configuration B, B).*

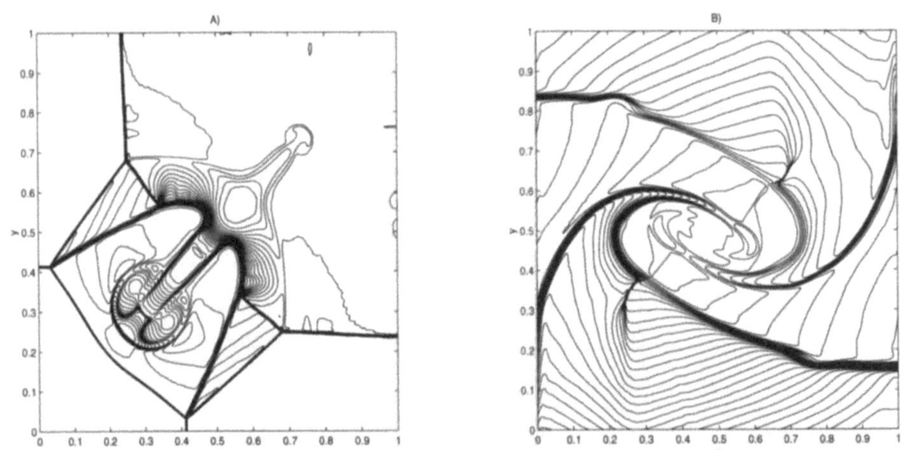

FIGURE 2. *Contour lines (30) of the density at time* $t = 0.8$ *for configuration A, A), and configuration B, B), corresponding to the grid size* 400×400.

3.1. 'Coarse' grid simulations

We first apply our multilevel algorithm to these tests using the same grid size as in [13, 15]. For both configurations, the grid size is 400×400, the number of levels in the multilevel algorithm is 4 and the tolerance parameter ϵ is set to 10^{-3}. The ratio

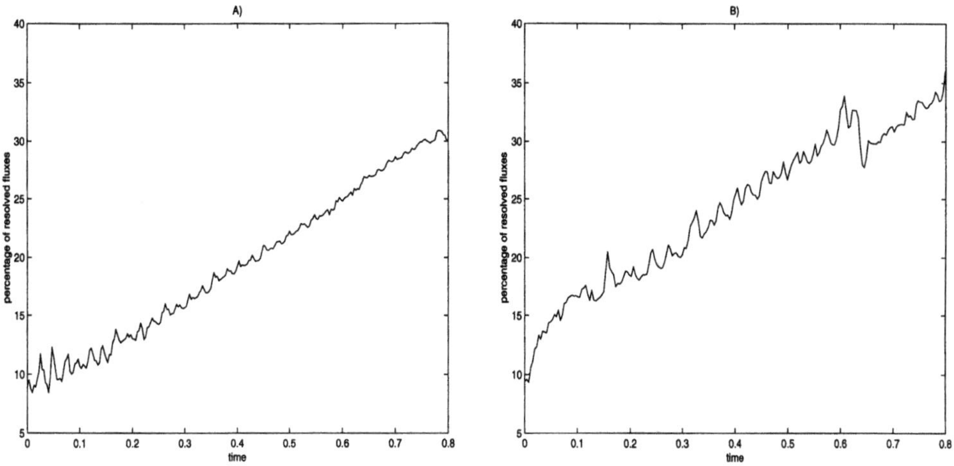

FIGURE 3. *Time evolution of the percentage of fluxes computed with the Riemann solver for both tests and for the grid size* 400 × 400. *Left configuration A, right configuration B.*

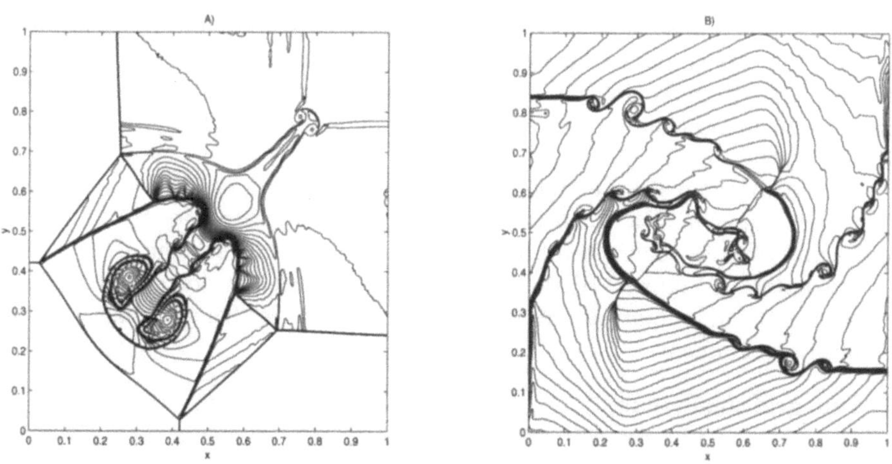

FIGURE 4. *Contour lines (30) of the density at time* $t = 0.8$ *for configuration A, A), and configuration B, B), corresponding to the grid size* 1024 × 1024.

$dt/dx = 0.25$ has been used for configuration A and $dt/dx = 0.3$ for configuration B.

The computed density is represented for both simulations in Figure 2 at time $t = 0.8$. Our numerical results can be directly compared with those of [13, 15].

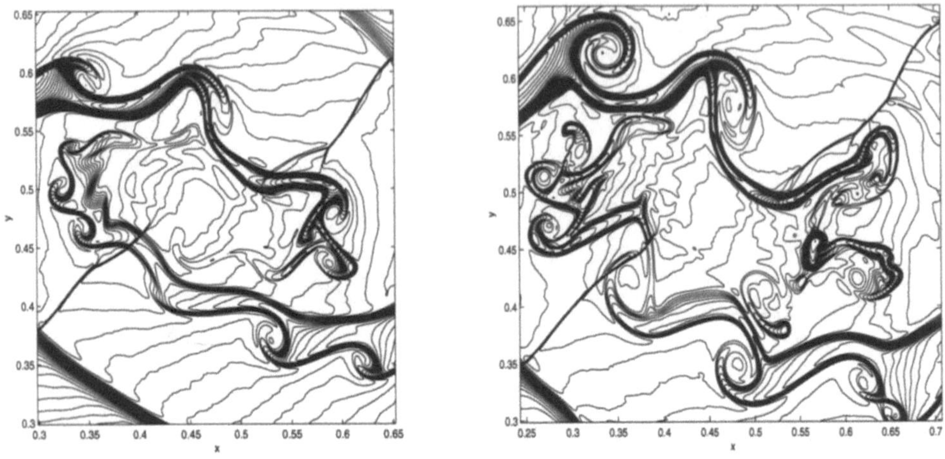

FIGURE 5. *Zoom of the central part of the density for configuration B, at time $t = 0.8$, left, and $t = 1$, right.*

In Figure 3, we represent the time evolution of the percentage of points where the HRSC scheme is used; at all other points, the numerical divergence is interpolated as described in section 2.3). Notice that the maximum percentage of numerical fluxes that are computed using the HRSC scheme is around 30%. Even though a relatively small percentage of numerical fluxes were computed with the HRSC scheme, the numerical solution obtained is what one would expect of a HRSC method of order 3.

The global cpu time of these simulations has been reduced by a factor of **2.7** for configuration A and **2.3** for configuration B, with respect to the reference simulation (without multiresolution) with the same HRSC scheme.

We notice in Figure 2-B) the two vortices trying to develop on the horizontal slip lines, but the numerical viscosity of the scheme at this grid-size prevents a complete development.

3.2. Fine grid simulations

We apply now the multilevel algorithm using a grid of 1024×1024 points. The ratios between dt and dx are unchanged, the number of levels in the multiresolution process is now 7 and the tolerance ϵ is decreased to 6.10^{-4}, to account for the increase in accuracy due to the smaller mesh size.

In Figure 4, the density for both configurations is represented at time $t = 0.8$. The increase in resolution allows us to observe several features of the flow that were undetectable in the previous simulation.

For configuration A, the main differences may be seen in the roll-up of vortices at the central jet structure due to the presence of Kelvin-Helmholtz instabilities. These instabilities develop due to a substantial difference in the size of the velocities in the internal and external parts of the jet structure. It is also important to notice the high-quality representation of the "head" of this jet with the spiraling of the slip lines.

For configuration B, many vortices can be now clearly appreciated. These are also due to the same type of instabilities. A zoom of the central area is presented on Figure 5 at time $t = 0.8$ and $t = 1$. It shows the transformation of these vortices into complicated structures after they interact with each other.

It must be noticed that the presence of all these small structures in the flow does not perturb the multilevel algorithm which remains very powerful. For configuration A, the percentage of fluxes computed with the solver grows only from 3% to 22% and for configuration B from 4% to 28%. This leads to a reduction of the global cpu time by a factor respectively of **4.2** and **3.6**.
In practice, this implies that a high resolution numerical simulation can be obtained in about three days on a Pentium-II at 350 Mhz, instead of more than ten days for the references simulations.
Cpu gain and overhead of multiresolution for other examples are carefully analyzed in [5].

4. Conclusion

In this paper we have presented numerical results concerning the application of the multilevel method proposed in [5] to 2D Riemann problems.

We combine Marquina's scheme, a powerful HRSC method, with the multilevel strategy and present numerical results on a 400×400 grid, which can be directly compared to those presented in [13, 15]. Then, thanks to the effective speed-up procured by the multilevel method, we are able to obtain high-quality, high-resolution, numerical simulations that display physical phenomena that are only observable when the viscosity of the scheme is very small.

The results reported in this paper show that coupling a HRSC scheme with a multilevel strategy leads to an efficient tool in those situations where we need high-quality and high-resolution at an affordable cost. This includes real gas simulations and three-dimensional problems. Such applications are currently investigated.

All computations have been performed on a Pentium-II at 350 Mhz.

References

[1] R. Abgrall. Multiresolution in unstructured meshes : Application to CFD. *Numerical methods for fluid dynamics, Oxford University Press*, 5, 1996.

[2] R. Abgrall, S. Lantéri, and T. Sonar. ENO schemes for compressible fluid dynamics. *ZAMM, Z. angew. Math. Mech.*, 79:3–28, 1999.

[3] F. Arandiga and R. Donat. Nonlinear multiscale decompositions: The approach of A. Harten. *Numerical Algorithms*, 23:175–216, 2000.

[4] B.L Bihari and A. Harten. Multiresolution schemes for the numerical solutions of 2D conservation laws. *SIAM J. Sci. Comp.*, 18(2):315–354, 1997.

[5] G. Chiavassa and R. Donat. Point Value Multi-scale Algorithms for 2D compressible flows. *To appear in Siam J. Sci. Comp.*, 2001.

[6] W. Dahmen, B. Gottschlich-Müller, and S. Müller. Multiresolution schemes for conservation laws. *Technical report*, Bericht Nr 159 IGMP, RWTH Aachen, 1998.

[7] R. Donat, , J.A. Font, J.M. Ibànez, and A. Marquina. A flux-split algorithm applied to relativistic flows. *J. Comput. Phys.*, 146:58–41, 1998.

[8] R. Donat and A. Marquina. Capturing shock reflections: An improved flux formula. *J. Comp. Phys.*, 125, 1996.

[9] R. Fedkiw, B. Merriman, R. Donat, and S. Osher. The penultimate scheme for systems of conservation laws: Finite difference ENO with Marquina's flux splitting. *UCLA CAM Report*, January 1997.

[10] A. Harten. Discrete multiresolution analysis and generalized wavelets. *J. Appl. Numer. Math.*, 12:153–192, 1993.

[11] A. Harten. Multiresolution algorithms for the numerical solution of hyperbolic conservation laws. *Comm. Pure Appl. Math.*, 48, No 12, 1995.

[12] S.M. Kaber and M. Postel. Finite volume schemes on triangles coupled with multiresolution analysis. *C.R. Acad. Sci., Série I, Paris*, 328:817–822, 1999.

[13] P. D. Lax and X.-D. Liu. Solution of two-dimensional Riemann problems of gas dynamics by positive schemes. *SIAM J. Sci. Comput.*, 19:319–340, 1998.

[14] C.W. Schulz-Rinne. Classification of the Riemann problem for two-dimensional gas dynamics. *SIAM J. Math Anal.*, 24:76–88, 1993.

[15] C.W. Schulz-Rinne, J. P. Collins, and H. M. Glaz. Numerical solution of the Riemann problem for two-dimensional gas dynamics. *SIAM J. Sci. Comput.*, 14:1394–1414, 1993.

[16] C. W Shu and S. J. Osher. Efficient implementation of essentially non-oscillatory shock-capturing schemes II. *J. Comp. Phys.*, 83, 1989.

[17] B. Sjögreen. Numerical experiments with the multiresolution scheme for the compressible euler equations. *J. Comp. Phys.*, 117, 1995.

[18] T. Zhang, G.-Q. Chen, and Y. Yang. On the 2D Riemann problem for the compressible Euler equations , I. interaction of shocks and rarefaction waves. *Discrete Continous Dynam. Systems*, 1:555–584, 1995.

[19] T. Zhang and Y. Zheng. Conjecture on the structure of solutions of the Riemann problem for two-dimensional gas dynamics systems. *SIAM J. Math Anal.*, 21:593–630, 1990.

G. Chiavassa
R. Donat
A. Marquina
Departamento de Matemàtica Aplicada,
Universitat de València,
Cl. Dr Moliner, 50,
València, Spain
`guillaume.chiavassa@uv.es`
`donat@uv.es`
`marquina@uv.es`

International Series of Numerical Mathematics
Vol. 140, © 2001 Birkhäuser Verlag Basel/Switzerland

Multiresolution Analysis on Triangles: Application to Gas Dynamics

Albert Cohen, Sidi Mahmoud Kaber, and Marie Postel

Abstract. Multiresolution analysis is used to improve the performances of a Finite Volumes scheme. Two schemes coupling Multiresolution and Finite Volumes are presented. One is a generalisation of Harten's original scheme for triangles. The other scheme is fully adaptive in the sense that at a given time, the solution is represented in a compressed form by a set of significant wavelets coefficients. The two schemes are applied to solve the Euler's system of gas dynamics.

1. Introduction

The main difficulties in solving nonlinear conservation laws arise from the discontinuities which the solutions of such equations may develop in finite time. The fact that the position of these discontinuities varies generally with time is of course an additional difficulty. In order to represent accurately the discontinuous solutions it is necessary to discretize the computational domain on very fine grids. Furthermore to ensure efficiency and high precision on these fine grids costly nonlinear solvers are used. In many problems the discontinuities are localized in only a small region of the total domain. The fine discretization which is necessary in the vicinity of the discontinuities is in fact a luxury everywhere else. The costly numerical fluxes could be advantageously replaced by simpler solvers wherever the solution is smooth enough. A. Harten proposed to use multiresolution analysis (MR) in order to speed up Finite Volume (FV) schemes. See [H] and the references therein, in particular [A] for applications to triangular meshes.

Let us detail the original strategy proposed in [H]. At the start one is given a FV scheme associated to a grid Ω^L which is the *finest one* in a hierarchy of nested grids Ω^ℓ for $\ell = 0, \dots, L$. At time t_n the approximate solution is represented by its averages $(\bar{u}_k^n)_k$ on the various cells of Ω^L. The values $(\bar{u}_k^{n+1})_k$ are evaluated by the FV scheme through the computations of the fluxes on Γ^L, the set of all interfaces between the cells of Ω^L. The basic idea is to use a wavelet-like multiscale decomposition of the solution at time t_n as a *smoothness indicator* in order to reduce the computations of the fluxes on Γ^L. In the regions where the details-defined below by (2)- are small (i.e. below some preassigned threshold), the flux is assumed to be smooth enough so that we can replace its exact evaluation on the finest level L by an interpolation from its values on the coarser levels (see `Algorithm 2`). This saves CPU since the precise computation of the flux is expensive, while the interpolation from coarse to fine meshes is cheap.

In section 2, we define a MR transform suited for triangulations. In section 3, we combine it with a FV scheme to obtain an extension of Harten's MR scheme to triangles. In section 4, we present an adaptive scheme that takes full advantage of the multiscale decomposition of the solution.

2. Multiresolution analysis (MR) on triangles

In [CDKP], we have defined a multiscale method adapted to cell-averages given on triangular meshes. Starting from a coarse triangulation Ω^0, we define a hierarchy of nested grids $\Omega^\ell (1 \leq \ell \leq L)$ by dividing each triangle T_k^ℓ of Ω^ℓ into 4 triangles $(T_{k,j}^{\ell+1})_{j=0}^3$ of $\Omega^{\ell+1}$. The mean value of the function u on the triangle T_k^ℓ is denoted by $\bar{u}_k^\ell = \mathfrak{A}(T_k^\ell)u = \frac{1}{|T_k^\ell|}\int_{T_k^\ell} u$ and the array of all $(\bar{u}_k^\ell)_{k=1,\ldots,N_\ell}$ is denoted $\bar{u}^\ell$.

2.1. The Multiscale decomposition

Two fundamental operators are defined.

• A *projection* operator $\mathfrak{P}_\ell^{\ell+1}$ from the resolution level $\ell+1$ to the coarser one ℓ, that maps $\bar{u}^{\ell+1}$ onto $\bar{u}^\ell$. This operator is simply defined by

$$\bar{u}_i^\ell = \frac{1}{|T_i^\ell|}\sum_{j=0}^{3} |T_{i,j}^{\ell+1}|\, \bar{u}_{i,j}^{\ell+1}.$$

• A *prediction* or *reconstruction* operator $\mathfrak{Q}_{\ell+1}^\ell$ from the resolution level ℓ to the finer level $\ell+1$. This prediction operator should satisfy the "consistency condition" $\mathfrak{P}_\ell^{\ell+1}\mathfrak{Q}_{\ell+1}^\ell = Id$. That is to say

$$\sum_{j=0}^{3} |T_{i,j}^{\ell+1}|\tilde{u}_{i,j}^{\ell+1} = |T_i^\ell|\bar{u}_i^\ell, \tag{1}$$

where $\tilde{u}_{i,j}^{\ell+1}$ are the values predicted by $\mathfrak{Q}_{\ell+1}^\ell$. We only consider here reconstruction operators for which the predicted values $\tilde{u}_{j,k}^{\ell+1}$ are given by a *local* linear combination of some immediately coarser values; $\tilde{u}_{j,k}^{\ell+1} = \sum_{m\in\mathfrak{N}(j)} \alpha_m \bar{u}_m^\ell$, where $\mathfrak{N}(j)$ corresponds to a neighborhood (in Ω^ℓ) of T_j^ℓ. Related to the prediction operator are the prediction errors or *details*

$$d_{i,j}^\ell = \bar{u}_{i,j}^{\ell+1} - \tilde{u}_{i,j}^{\ell+1} \quad \text{for } j = 0,\ldots,3. \tag{2}$$

Note that (1) implies a linear dependence for four details associated to the same triangle. This allows to compute only three details per triangle, for example $d_i^\ell := (d_{i,j}^\ell)_{j=1}^3$.

Remark. The details can be written $d^\ell_{j,k} = \langle u, \psi^\ell_k \rangle := \int_{\mathbb{R}^2} u\psi^\ell_{j,k}$, where the *wavelet* $\psi^\ell_{j,k}$ is defined by $\psi^\ell_{j,k} = \varphi^{\ell+1}_{j,k} - \sum_{m\in\mathfrak{N}(j)} \alpha_m \varphi^\ell_m$, and $\varphi^\ell_k = \frac{1}{|T^\ell_k|}\chi_{T^\ell_k}$ are the *scaling functions.* We will refer to the details as *wavelets coefficients.*
Denoting by d^ℓ the array of all the d^ℓ_i, the two representations

- $\bar{u}^L$ (all the averages on the finest level)
- $\bar{u}^0 \cup \{d^\ell, \ell = 0, \dots, L-1\}$ (averages on the coarsest level + details)

are equivalent in so far as they provide the same information and use the same memory requirement. Using the *local* structure of the projection and prediction operators, one can easily implement the multiscale transformation $\mathfrak{M} : \bar{u}^L \mapsto (\bar{u}^0, d^0, \dots, d^{L-1})$, and it's inverse $\mathfrak{M}^{-1}$ with optimal complexity.

2.2. Compression

One of the main interest of decomposing $\bar{u}^L$ into $\mathfrak{M}\bar{u}^L$ is that this new representation is more appropriate for *data compression.* Let us define

$$\Lambda := \Lambda(\varepsilon_0, \varepsilon_1, \cdots, \varepsilon_{L-1}) = \{\lambda = (\ell, j, k) \text{ s.t. } |d_\lambda := d^\ell_{j,k}| \geq \varepsilon_\ell\}$$

and the associated truncation operator $\mathfrak{T}_\Lambda$ that leaves unchanged the component d_λ if $\lambda \in \Lambda$ and replaces it by 0 otherwise. In all the computations we performed, the level depending threshold parameters are $\varepsilon_\ell := 2^{d(\ell-L)}\varepsilon$ for a fixed ε (here d is the space dimension). The parameter ε controls the truncation error resulting from the thresholding [CDKP]. Applying $\mathfrak{T}_\Lambda$ on the multiscale decomposition of $\bar{u}^L$ amounts in building a *nonlinear approximation* $\mathfrak{M}^{-1}\mathfrak{T}_\Lambda\mathfrak{M}\bar{u}^L$ in which the details in the finest scales are concentrated near the isolated singularities. Such compression properties are however related to additional assumptions on the prediction operator: *polynomial accuracy* and *multiscale stability.* Consult [CDKP, CKMP] and the references therein for a detailed presentation of this problem.

2.3. Smoothness indicators

The prediction operator $\mathfrak{Q}^\ell_{\ell+1}$ has polynomial exactness of order n if for all $u \in \mathbb{P}_n$ (polynomials of degree n), we have $\bar{u}^\ell = \tilde{u}^\ell$, i.e. $d^{\ell-1} = 0$ which expresses the orthogonality of $\psi^\ell_{j,k}$ with $\mathbb{P}_n$. Therefore, if u has local smoothness $C^s, s > 0$ (say in a neighborhood of $T^\ell_{k,j}$), then $|d^\ell_{k,j}| \leq C2^{-\ell\min(s,n+1)}$. In order to justify the use of the details as smoothness indicators, we need a converse property: small details should indicate that the encoded function is locally smooth. We also need some stability in the sense that we can control in some prescribed norm the perturbation of u resulting from thresholding the small details, see [CDKP].

3. Harten's algorithm

We now present a generalisation for triangles of the Harten's MR scheme. For cartesian grids and Finite Differences, see [CD]. The starting point is a classical FV scheme for solving the conservation law $\partial_t u + \operatorname{div} \mathbf{f}(u) = 0$ on the *finest* grid Ω^L. Here $u(x, y, t)$ is a scalar function, but generalisation to systems is straightforward.

The FV scheme computes at time t_n approximate averages $\bar{u}_k^{L,n} \simeq \mathfrak{A}(T_k^L)u(.,t_n)$ of the solution u by `Algorithm 1`.

> `Algorithm 1`: A General FV scheme.
> *Initialization:* $\bar{u}_k^{L,0} = \mathfrak{A}(T_k^L)u_0$, for $k = 1, \ldots$.
> *Iterations:* for each time step n,
> `Step 1`. Reconstruction: Use a reconstruction operator $\mathfrak{R}$ to obtain point values from cell averages.
> `Step 2`. Flux evaluation: compute $\bar{\mathfrak{D}}_k^{L,n}$, an approximation of $\mathfrak{D}_k^{L,n} = \mathfrak{A}(T_k^L)\mathrm{div}\,\mathbf{f}(\mathfrak{R})$.
> `Step 3`. Advance in time: $\bar{u}_k^{L,n+1} = \bar{u}_k^{L,n} - \Delta t \bar{\mathfrak{D}}_k^{L,n}, k = 1, \ldots, N_L$.

We only discuss here the `Step 2` of this algorithm. The flux evaluations are based on the remark that, by the divergence theorem

$$\mathfrak{D}_k^{L,n} = \frac{1}{|T_k^L|}\sum_j |\Gamma_{k,j}^L| f_{k,j}^{L,n}, \text{ with } f_{k,j}^{L,n} = \frac{1}{|\Gamma_{k,j}^L|}\int_{\Gamma_{k,j}^L} \mathbf{f}(\mathfrak{R}(\sigma)).\mathbf{n}_{k,j}\, d\sigma. \tag{3}$$

Here $\Gamma_{k,j}^L$ denotes interfaces between T_k^L and T_j^L. Therefore, $\bar{\mathfrak{D}}_k^{L,n}$ can be computed by applying (3) to some approximations $\bar{f}_{k,j}^{L,n}$ of $f_{k,j}^{L,n}$.

3.1. Multilevel computation of the fluxes

We now explain how the multiscale decomposition of the solution is used to speed up the flux evaluation, through a modification in `Step 2` of the FV algorithm. To this effect we define $\mathfrak{D}_k^{\ell,n} = \mathfrak{A}(T_k^\ell)\mathrm{div}\,\mathbf{f}(\mathfrak{R})$ for $0 \le \ell \le L$. These mean values can again be computed as in (3) and approximated by

$$\bar{\mathfrak{D}}_k^{\ell,n} = \frac{1}{|T_k^\ell|}\sum_j |\Gamma_{k,j}^\ell| \bar{f}_{k,j}^{\ell,n}, \text{ where } \bar{f}_{k,j}^{\ell,n} = \frac{1}{|\Gamma_{k,j}^\ell|}\sum_{\Gamma_m^L \subset \Gamma_{k,j}^\ell} |\Gamma_m^L| \bar{f}_m^{L,n}. \tag{4}$$

The MR representation of the solution is used to avoid the evaluation of $\bar{\mathfrak{D}}_k^{L,n}$ on the finest grid wherever it is possible. In the regions (let say around the triangle T_k^L) where the solution is smooth, $\bar{\mathfrak{D}}_k^{L,n}$ is computed by interpolation of the $\bar{\mathfrak{D}}_j^{\ell,n}$'s corresponding to some coarser levels $\ell < L$. In the region where the solution is not smooth enough, $\bar{\mathfrak{D}}_k^L$ will be computed accurately using a precise (and expensive) numerical flux. Let $\Omega^{\varepsilon,n+1}$ denote a subset of $\cup_{\ell=0}^L \Omega^\ell$ formed by all the triangles T_k^ℓ over which the numerical solution is not regular enough. This hybrid grid (builded by `Algorithm 3`) is the key ingredient of the following algorithm which is an MR version of `Step 2` of `Algorithm 1`.

`Algorithm 2`: Multilevel computation of the fluxes.
`Step 2-1.` Compute the set $\Omega^{\varepsilon,n+1}$ (see `Algorithm 5`)
`Step 2-2.` Compute the $\bar{\mathfrak{D}}^0$'s on the coarsest grid Ω^0 using (4).
`Step 2-3. For` $\ell = 1 \nearrow L$, compute the approximate $\bar{\mathfrak{D}}^\ell$'s by
`If` $d^\ell_{k,j} \in \Omega^{\varepsilon,n+1}$ `Then`
$\bar{\mathfrak{D}}^\ell_k$ is accurately computed using (4) as in `Step 2-2.`
`Else`
$\bar{\mathfrak{D}}^\ell_{k,j}$ is approximately computed by interpolation of the values $\bar{\mathfrak{D}}^{\ell-1}$, using the prediction operator $\mathfrak{Q}^{\ell-1}_\ell$.

The first test in the algorithm takes into account the propagation of information (which is limited by the CFL condition). The second test takes into account the possible appearance of discontinuities.

`Algorithm 3`: The hybrid grid.
`Step 2-1-1.` Initialize $\Omega^{\varepsilon,n+1} = \Omega^0$
`Step 2-1-2. For` $\ell = L-1 \searrow 0$
`For` $k = 1, \ldots, N_\ell$
`If` $|d^\ell_{k,j}| \geq \varepsilon_\ell$ for one j `Then`
add T^ℓ_i to $\Omega^{\varepsilon,n+1}$ if T^ℓ_i shares a vertice with T^ℓ_k.
If $|d^\ell_k| \geq 8\varepsilon_\ell$ `Then`
add $T^{\ell+1}_{k,j}$ to $\Omega^{\varepsilon,n+1}$.

3.2. Numerical tests ($2D$ Euler equations of gas dynamics)

In all the computations, we have used the following reconstruction operator (see the figure 1 for the notations) introduced in [CDKP].

$$\begin{cases} \bar{u}_{0,0} = \bar{u}_0 & \bar{u}_{0,1} = \bar{u}_0 + (\bar{u}_2 + \bar{u}_3 - 2\bar{u}_1)/6 \\ \bar{u}_{0,2} = \bar{u}_0 + (\bar{u}_1 + \bar{u}_3 - 2\bar{u}_2)/6 & \bar{u}_{0,3} = \bar{u}_0 + (\bar{u}_1 + \bar{u}_2 - 2\bar{u}_3)/6 \end{cases}$$

Reflection of a shock. We first study the reflection of an oblique shock on the lower side of a rectangular domain. The exact solution and the boundary conditions of the simulation are shown figure 2. The coarsest grid is composed of $N_0 = 200$ triangles. The MR parameters are $\varepsilon = 0.01, L = 3$. The CPU time is $T_{MR} = 0.8\,T_{FV}$. The CPU performances of Harten's scheme are disappointing on this example. To illustrate the adaptive computation of the fluxes, we plot on figure 3 what we called 'hybrid grid' consisting in all the triangles for which the flux computations are done precisely using the (supposed to be expensive) numerical solver instead of the (supposed to be cheaper) interpolation. The shape of the hybrid grid indicates that memory savings would be very interesting if the time evolution was be performed on an adaptive grid instead of the finest grid.

Mach 3 wind tunnel with a step. This model problem has been examined by several authors. The coarsest grid is composed of $N_0 = 126$ triangles. Four levels are used for the computations. The FV solution computed on the finest grid and the MR solution are displayed figure 4 (density at time $t = 4$). The CPU time for the FV run is $T_{FV} = 13944$ seconds. The MR parameters are $\varepsilon = 0.05, L = 3$. The CPU

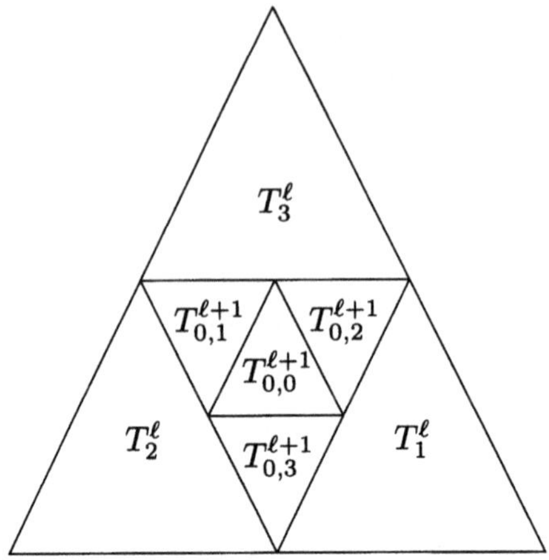

FIGURE 1. Subdivision of a triangle T_0^ℓ.

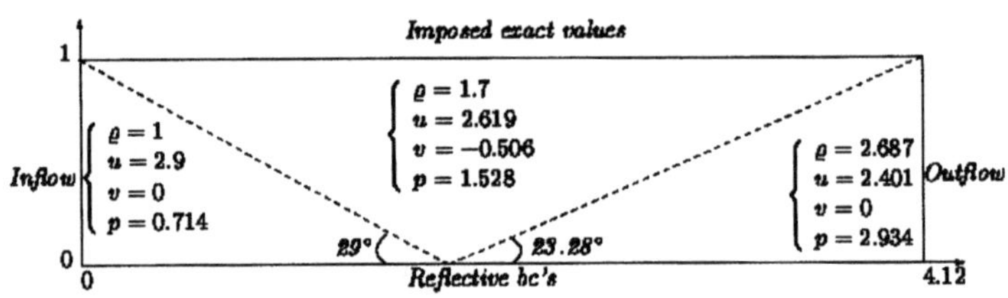

FIGURE 2. Exact solution (and boundary conditions) of the reflection problem.

time is $T_{MR} = 9931 = 0.71\, T_{FV}$. Here again the CPU gain is not tremendous, but the discontinuities are clearly localized on the hybrid grid (not displayed) which is a promising property, in view of the adaptive scheme.

3.3. Limitations of Harten's scheme

Although the flux is computed adaptively, the evolution of the solution at each time step still takes place on the finest grid Ω^L. For these reasons, the computational gain is limited. In the next section, we describe a fully adaptive algorithm for which the memory storage is proportional to the number of wavelets coefficients describing the solution. The first results using adaptive MR schemes were presented in [GMM].

4. A Fully adaptive scheme

The goal of the adaptive algorithm is to save save the maximal amount of computational time and memory space while maintaining the order of accuracy achieved by the reference FV scheme on the finest mesh.

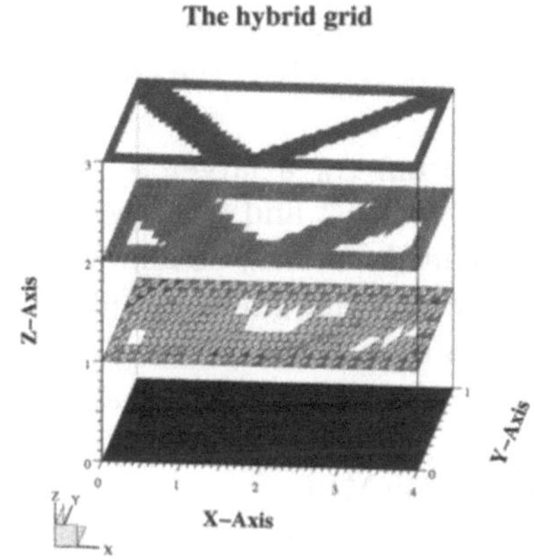

FIGURE 3. A burst view of a hybrid grid.

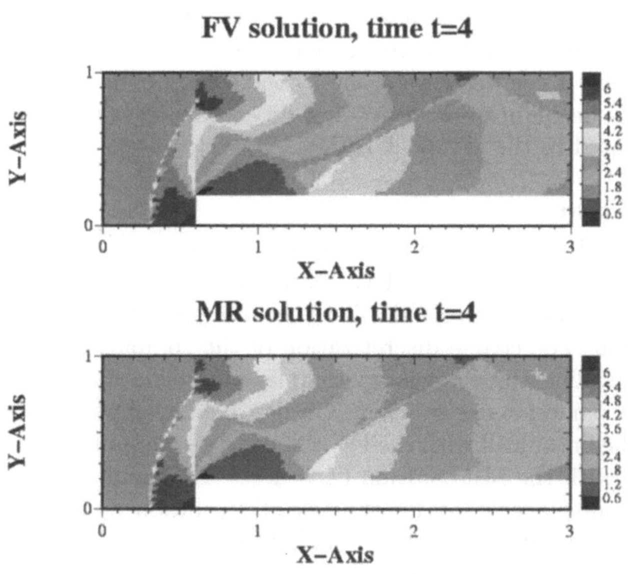

FIGURE 4. FV solution of the step problem and MR Harten's solution.

4.1. The algorithm

At a given time t_n the solution u^n is now represented on a adaptive grid Λ_n made of triangles belonging to different resolution levels. This approach gives way to new difficulties, for instance in the analysis of the stability and precision [CKMP]. Given the numerical solution represented in a compressed form by a set Λ_n of significant wavelets coefficients i.e. $u^n = \sum_{\lambda \in \Lambda_n} d_\lambda^n \psi_\lambda$, the adaptive strategy consists in three steps:

1- Refinement. Predict from Λ_n a larger set $\tilde{\Lambda}_{n+1}$ such that $\Lambda_n \subset \tilde{\Lambda}_{n+1}$, which is adapted to describe both u^n and u^{n+1} with the required accuracy. The new set of indices $\tilde{\Lambda}_{n+1}$ is defined from Λ_n as follows :

- Λ_n is padded by a "security margin" determined by the CFL condition.
- According to the magnitude of the details d_λ^n, we possibly add (at position k and scale ℓ) new wavelets coefficients located at the same space-location but with finer scales ℓ' such that $\ell < \ell' \leq \ell^* \leq L$. The precise computation of the *local refinement level* ℓ^* is given in [CKMP].

2- Evolution. Compute from u^n an intermediate solution $\tilde{u}^{n+1}$ defined from the set $\tilde{\Lambda}_{n+1}$; $\tilde{u}^{n+1} = \sum_{\lambda \in \tilde{\Lambda}_{n+1}} d_\lambda^{n+1} \psi_\lambda$. We refer to [CKP] for a complete description of this step.

3- Thresholding. Use a thresholding procedure to obtain the new set $\Lambda^{n+1} \subset \tilde{\Lambda}^{n+1}$ and the new approximation $u^{n+1} = \sum_{\lambda \in \Lambda_{n+1}} d_\lambda^{n+1} \psi_\Lambda$.

An important feature of the scheme is that the set Λ_n has a *tree* structure. This structure is crucial since it allows a one to one correspondence between the truncated multiscale decomposition of $\bar{u}^n$ and its cell averages on the adaptive triangulation associated to the set Λ_n [CKMP].

4.2. Numerical tests

In a previous work [CKP], we have shown the efficiency of the adaptive scheme for scalar $2D$ problems. The adaptive computations were (up to 10 times) faster than the pure FV ones. We consider here the $1D$ gas dynamics system.

Sod's shock tube. The coarse discretization of the computing interval consists in 200 subdivisions. The finest one is seven times finer. The standard FV solution on this fine grid is our reference solution. On figure 5, we show the density (t=0.26) computed on the finest grid with the FV scheme (this computation took $T_{FV} = 693$ seconds) along with the MR solution ($L = 4, \varepsilon = 10^{-4}$ and the adaptive grid. For $\varepsilon = 10^{-3}$ (not displayed) the finest level of resolution is used exclusively near the four singularities of the density. This induces some slight oscillations near the contact discontinuity. The L_1-error is $E_1 = .00297$ and the computing time is $T_{MR} = 86 \simeq T_{FV}/8$ seconds. For $\varepsilon = 10^{-4}$ (figure 5), the finer levels of resolution are used more often, leading to a better accuracy $E_1 = .00158$. In particular, the oscillations near the discontinuities have disappeared and the computing time remains interesting $T'_{MR} = 121 \simeq T_{FV}/5.7$.

Blast waves. We now test the code to solve the difficult problem of interacting blast waves. The results are shown on figure 6 for 5 levels of resolution. We see that the fine levels are used only where the two waves interact, even though they may have been used at intermediate times. The computing time for several runs are summarized in the following table.

	MR ($\varepsilon = 0.002$)		FV	
# levels	CPU	memory	CPU	memory
2	1329	207	1100	400
3	2966	465	3913	800
4	10677	621	18960	1600
5	38458	717	81484	3200

As far as the memory requirement is concerned, the advantages of the adaptive algorithm are indisputable. The CPU gain however is not so spectacular between the MR computation and the FV run on the corresponding finest level. This poor performance is due to the overhead costs in the MR program. We are using a *C++* prototype, which is not optimized at all and in particular uses the standard dynamic memory allocation functions. This is not a sensitive issue when the grid is created once and for all as in the standard FV scheme but becomes very costly for the time adaptive grid book-keeping.

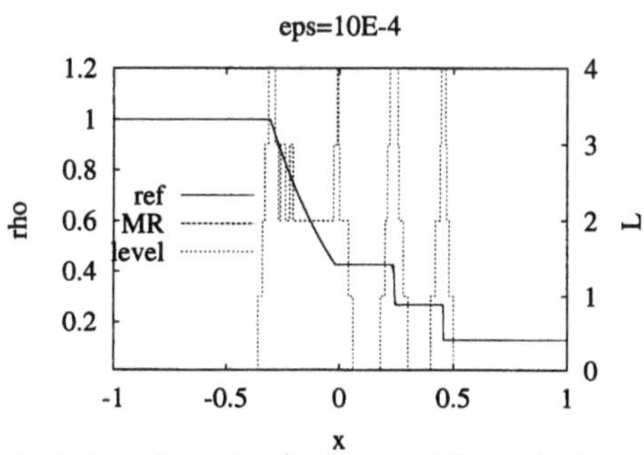

FIGURE 5. Shock tube: density (left y-axis) and adaptive grid (right).

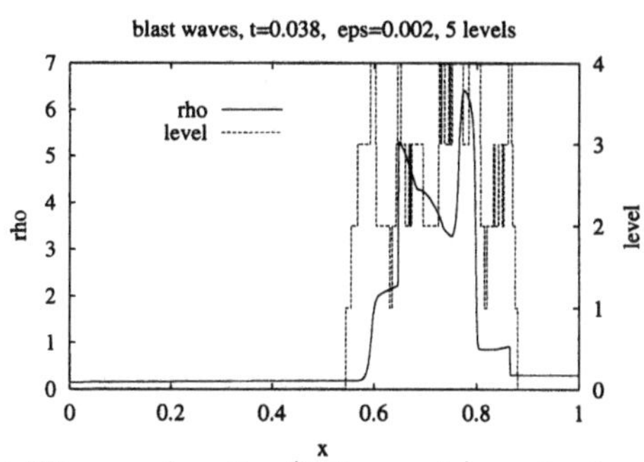

FIGURE 6. Blast Waves: density (left y-axis) and adaptive grid (right).

5. Concluding remarks

A FV Multiscale scheme has been applied to solve classical gas dynamics tests. We have shown the effectiveness of the method on triangular meshes but also its limits. Another FV Multiscale scheme is presented. This new scheme takes full advantage of the multiscale representation of the solution: at each time step, the solution is represented in a compressed form by a set of significant wavelets coefficients (or equivalently on an adaptive grid). Not only the fluxes but also the whole solution is computed adaptively. As far as the memory requirement is concerned, the advantages of the adaptive algorithm are indisputable. The improvement in terms of CPU is however not as spectacular as in the scalar case treated in [CKP] due to big overhead costs in the MR program. The big difference with the present tests, in terms of programming, is that we now handle *vectorial objects* for which the use of optimized dynamic memory allocation functions is crucial. This technical problem is currently under investigation.

References

[A] R. Abgrall. Multiresolution analysis on unstructured meshes: application to CFD, *Experimentation, modelling and combustion*, John Wiley & Sons, 1997.

[CD] G. Chiavassa, and R. Donat. Numerical Experiments with Point Value Multiresolution for 2D Compresible Flows. Submitted (1999)

[CDKP] A. Cohen, N. Dyn, S.M. Kaber, and M. Postel. Multiresolution schemes on triangles for scalar conservation laws. *J. Comp. Phys.*, 161:264–286, 2000.

[CKMP] A. Cohen, S.M. Kaber, S. Müller, and M. Postel. Fully adaptive multiresolution finite volume schemes for conservation laws.*Preprint R00009, LAN, Université Paris 6, submitted*, 2000. To appear in *Math of Comp.*

[CKP] A. Cohen, S.M. Kaber, and M. Postel. Multiresolution analysis on triangles: application to conservation laws, In *Finite volumes for complex applications. R. Vilsmeier, D. Hanel, F. Benkhaldoun Ed. Hermes Science, Paris.* 1999.

[GMM] B. Gottschlich-Müller and S. Müller, Adaptive finite volume schemes for conservation laws based on local multiresolution techniques *Proceedings of 7th International Conference on Hyperbolic Problems.* R. Jeltsch Ed. Birkhäuser-Verlag, 1998,

[H] A. Harten. Adaptive multiresolution schemes for shock computations. *J. Comp. Phys.*, 115:319–338, 1994.

A. Cohen, S.M. Kaber, M. Postel
Laboratoire d'analyse numérique,
Université Pierre & Marie Curie,
75252 Paris Cedex 05, France
cohen@ann.jussieu.fr
kaber@ann.jussieu.fr
postel@ann.jussieu.fr

International Series of Numerical Mathematics
Vol. 140, © 2001 Birkhäuser Verlag Basel/Switzerland

Propagation and Interaction of Nonlinear Waves to Quasilinear Equations

V. G. Danilov and V. M. Shelkovich

Abstract. A method for analytical description of nonlinear wave interaction is presented.

1. The weak asymptotics method

In [1] and especially in [2] a new analytical method was developed, the *weak asymptotics method*, which enables investigation of the *dynamics of propagation* of various types of singularities to quasilinear differential equations (and hyperbolic first order systems)

$$L[u] = u_t + (f(u))_x = 0, \tag{1}$$

where $f(u)$ is a smooth function, $u = u(x,t)$, $x \in R$.

The authors extended the *weak asymptotics method* [2] to the form that allows one to investigate not only the dynamics of propagation, but the *dynamics of interaction* of singularities of solutions for quasilinear equations and first order systems as well.

The Cauchy problem can also be posed for equation (1) in the case when the singularity of the initial value $u^0(x)$ or of the solution $u(x,t)$ is stronger than that of the Heaviside function, i.e., a jump. Thus, in [3], [4], [2] the problem of propagation of an infinitely narrow δ-soliton was considered, in [5], [6] problems of propagation of δ-waves and δ-shock waves are studied.

The essence of our method is as follows. If we solve the initial value problem with distribution initial data $u_\varepsilon^{*0}(x) = u_0^0(x) + \sum_{j=1}^{2} e_j^0(x) s_j(x - x_j)$, where $u_0^0(x)$, $e_j^0(x)$ are smooth functions, $s_j(\xi)$ are distributions (generalized functions) or *asymptotic distributions* (see [2] and Section 3) and x_j are constants, then the solution of this initial value problem is found in the form of the singular ansatz

The research of the first author is supported by Grant 99-01-01074 of Russian Foundation for Basic Research.
The research of the second author is supported by Grants 99-01-01074 and 99-01-00719 of Russian Foundation for Basic Research.

from algebra of asymptotic distributions:

$$u_\varepsilon^*(x,t) = u_0(x,t) + \sum_{j=1}^{2} e_j(x,t)s_j(x-\phi_j(t)) \tag{2}$$

where $u_0(x,t)$, $e_j(x,t)$, $\phi_j(t)$ are functions to be found. In the general case $e_j(x,t)$ and $s_j(\xi)$ may depend on a small parameter ε. To find the solution in the form of the singular ansatz $u_\varepsilon(x,t)$ we construct the smooth ansatz

$$u^*(x,t,\varepsilon) = u_0(x,t) + \sum_{j=1}^{2} e_j(x,t)s_j(x-\phi_j(t,\varepsilon),\varepsilon), \tag{3}$$

which is an approximation of the singular ansatz (2), $s_j(\xi,\varepsilon)$ is a smooth approximation of the distribution or asymptotic distribution $s_j(\xi)$ and $\varepsilon > 0$ is the approximation parameter.

We define a *generalized solution* (an *asymptotic solution*) of equation (1) as the weak asymptotics $u_\varepsilon^*(x,t)$ of the smooth ansatz $u(x,t,\varepsilon)$, as $\varepsilon \to +0$. Note that we consider $u_\varepsilon^*(x,t)$ as a distribution depending on the parameter t on test functions $\varphi(x) \in \mathcal{D}(R)$.

The smooth ansatz $u^*(x,t,\varepsilon)$ is substituted into equation (1) and the weak asymptotics up to $O_{\mathcal{D}'}(\varepsilon^N)$ is found of the left-hand side of this equation as $\varepsilon \to +0$. By $O_{\mathcal{D}'}(\varepsilon^\alpha)$ we denote a distribution from $\mathcal{D}'(R)$ such that $\langle O_{\mathcal{D}'}(\varepsilon^\alpha), \varphi(x)\rangle = O(\varepsilon^\alpha)$ for any test function $\varphi(x) \in \mathcal{D}(R)$ and $O(\varepsilon^\alpha)$ is understood in the ordinary sense.

Then by setting equal to zero the coefficients of the different powers of the small parameter ε and of the linearly independent distributions we obtain a system of equations (in particular, a Rankine-Hugoniot type condition) which describes the dynamics of singularities and defines the smooth functions $u_0(x,t)$, $e_j(x,t)$, $\phi_j(t)$, $j=1,2$.

We apply our method to investigate the interaction of two shock waves and of two infinitely narrow solitons. The limited space in this paper does not permit us to describe the results in detail.

2. The propagation and interaction of shock waves

1. In order to investigate the interaction of two shock waves, we have to solve the Cauchy problem with the following initial condition:

$$u^{*0}(x) = \begin{cases} u_0^0(x) + e_1^0(x) + e_2^0(x), & x < x_1^0, \\ u_0^0(x) + e_2^0(x), & x_1^0 < x < x_2^0, \\ u_0^0(x), & x > x_2^0, \end{cases} \tag{4}$$

where $u_0^0(x)$, $e_1^0(x)$, $e_2^0(x)$ are smooth functions.

In what follows we assume that $f''(u) > 0$ and that the solutions $u(x,t)$ are piecewise smooth functions satisfying the Oleinik stability conditions at every point (x,t) of a discontinuity line [7]:

$$u(x-0,t) > u(x+0,t). \tag{5}$$

Therefore we should seek a solution in the form of the singular ansatz

$$u^*(x,t) = u_0(x,t) + \sum_{k=1}^{2} \big[e_k(x,t)H(-x+\phi_k(t)\big], \tag{6}$$

and substitute into equation (1) the smooth ansatz

$$u^*(x,t,\varepsilon) = u_0(x,t) + \sum_{k=1}^{2} \big[e_k(x,t)H_k(-x+\phi_k(t,\varepsilon),\varepsilon)\big], \tag{7}$$

where $u_0(x,t)$, $e_k(x,t)$, $\phi_k(t,\varepsilon)$ are smooth functions to be found, $\lim_{\varepsilon\to+0}\phi_k(0,\varepsilon) = x_k^0$ and $H_k(\xi,\varepsilon)$ are the approximations of the Heaviside function, $k=1,2$.

Now we present the *main lemma* which gives an asymptotic expansion.

Lemma 2.1. *Let $H(-x,\varepsilon) = \omega_{01}(\frac{-x}{\varepsilon}) = \int_{-\infty}^{\frac{-x}{\varepsilon}} \omega_1(\eta)\,d\eta$, $H(-x+a,\varepsilon) = \omega_{02}(\frac{-x}{\varepsilon}) = \int_{-\infty}^{\frac{-x+a}{\varepsilon}} \omega_2(\eta)\,d\eta$ be approximations of the Heaviside functions $H(-x)$, $H(-x+a)$, respectively, the approximating functions $\omega_k(z) \in C^\infty(R)$ are nonnegative and either have compact supports or decrease sufficiently rapidly, as $|z| \to \infty$ and $\int \omega_k(z)\,dz = 1$, $k=1,2$. Let $f(u)$ be a smooth function and let $u_0(x,t)$, $e_1(x,t)$, $e_2(x,t)$ be bounded functions. Then we have the asymptotics:*

$$\begin{aligned} f\big(u_0(x,t) &+ e_1(x,t)H_1(-x,\varepsilon) + e_2(x,t)H_2(-x+a,\varepsilon)\big) = f(u_0(x,t)) \\ &+ \big[f(u_0(x,t)+e_1(x,t)) - f(u_0(x,t))\big]H(-x) \\ &+ \big[f(u_0(x,t)+e_2(x,t)) - f(u_0(x,t))\big]H(-x+a) \\ &+ B_1(x,t,\frac{a}{\varepsilon})H(-x) + B_2(x,t,-\frac{a}{\varepsilon})H(-x+a) + O_{\mathcal{D}'}(\varepsilon), \quad \varepsilon \to +0, \end{aligned} \tag{8}$$

where the estimate $O_{\mathcal{D}'}(\varepsilon)$ is uniform with respect to a. The functions

$$\begin{aligned} B_1(x,t,\rho) &= \int \Big[f'\big(u_0(x,t) + e_1(x,t)\omega_{01}(-\eta) + e_2(x,t)\omega_{02}(-\eta+\rho)\big) \\ &\qquad - f'\big(u_0(x,t) + e_1(x,t)\omega_{01}(-\eta)\big)\Big] e_1(x,t)\omega_1(-\eta)\,d\eta, \\ B_2(x,t,-\rho) &= \int \Big[f'\big(u_0(x,t) + e_1(x,t)\omega_{01}(-\eta-\rho) + e_2(x,t)\omega_{02}(-\eta)\big) \\ &\qquad - f'\big(u_0(x,t) + e_2(x,t)\omega_{02}(-\eta)\big)\Big] e_2(x,t)\omega_2(-\eta)\,d\eta \end{aligned} \tag{9}$$

are called interaction switch functions and satisfy the following relations:

$$B_1(x,t,\rho)+B_2(x,t,-\rho)=f(u_0(x,t)+e_1(x,t)+e_2(x,t))$$
$$-f(u_0(x,t)+e_1(x,t))-f(u_0(x,t)+e_2(x,t))+f(u_0(x,t)), \tag{10}$$
$$\lim_{\rho\to+\infty}B_k(x,t,\rho)=f(u_0(x,t)+e_1(x,t)+e_2(x,t))-f(u_0(x,t)+e_1(x,t))$$
$$-f(u_0(x,t)+e_2(x,t))+f(u_0(x,t)),\qquad \lim_{\rho\to-\infty}B_k(x,t,\rho)=0,\qquad k=1,2.$$

Corollary 2.2.

$$f\big(u_0(x,t)+e(x,t)H(-x,\varepsilon)\big)=f\big(u_0(x,t)\big)+\Big[f\big(u_0(x,t)+e(x,t)\big)$$
$$-f\big(u_0(x,t)\big)\Big]H(-x)+O_{\mathcal{D}'}(\varepsilon),\quad \varepsilon\to+0. \tag{11}$$

2. Generalized solutions.

Definition 2.3. We call the smooth ansatz (7) $u^*(x,t,\varepsilon)$ by a *generalized asymptotic* shock wave type solution of the equation $L[u]=0$, for $t\in[0,T]$, with initial condition $u^{0*}(x)$, if: $L[u^*(x,t,\varepsilon)]=O_{\mathcal{D}'}(\varepsilon)$, $u^*(x,0,\varepsilon)=u^{0*}(x)+O_{\mathcal{D}'}(\varepsilon)$, where the first estimate is uniform with respect to $t\in[0,T]$.

Definition 2.4. Let $u^*(x,t,\varepsilon)$ be a *generalized asymptotic* shock wave type solution of the equation $L[u]=0$ with the initial condition $u^{0*}(x)$. By a *generalized solution* of this equation we call the weak limit $u^*(x,t)=\lim_{\varepsilon\to+0}u^*(x,t,\varepsilon)$.

3. In order to investigate the propagation of the single shock wave, we have to solve the Cauchy problem in the class of piecewise smooth functions, with the initial condition (4), where $e_1^0(x)=e^0(x)$, $e_2^0(x)=0$, $x_1^0=x_2^0=x_0$. This means that we substitute the smooth ansatz (7) (approximating the singular ansatz (6)), where $e_1(x,t)=e(x,t)$, $e_2(x,t)=0$, $\phi_1(t,\varepsilon)=\phi(t)$, into the equation (1) and find the weak asymptotics up to $O_{\mathcal{D}'}(\varepsilon)$ of the left hand side of the equation, as $\varepsilon\to+0$. Then using the relation (11), we obtain:

Theorem 2.5. *Consider the Cauchy problem* (1), (4) *with* $e_1^0(x)=e^0(x)$, $e_2^0(x)=0$, $x_1^0=x_2^0=x_0$. *Consider smooth functions* $u_0^0(x)$, $e^0(x)>0$ *such that there exists* $T>0$ *such that* $-\inf_{\xi>x_0}\big[u^{0\prime}(\xi)f''(u^0(\xi))\big]<T^{-1}$, $-\inf_{\xi<x_0}\big[\big(u^0(\xi)+e^0(\xi)\big)'f''(u^0(\xi)+e^0(\xi))\big]<T^{-1}$. *Then in the sense of Definition* 2.4 *equation* (1) *for* $t\in[0,T]$ *has a discontinuous solution of the form*

$$u^*(x,t)=u_0(x,t)+e(x,t)H(-x+\phi(t)), \tag{12}$$

if and only if $u_0(x,t)$, $e(x,t)$, $\phi(t)$ *satisfy the following system of equations*

$$L[u_0(x,t)]=0,\quad x>\phi(t),\qquad L[u_0(x,t)+e(x,t)]=0,\quad x<\phi(t),$$
$$\frac{d\phi(t)}{dt}=\frac{f\big(u_0(x,t)+e(x,t)\big)-f\big(u_0(x,t)\big)}{e(x,t)}\bigg|_{x=\phi(t)}, \tag{13}$$

where $u^*(x,0)=u_0^0(x)+e^0(x)$ *for* $x<x_0$ *and* $u^*(x,0)=u_0^0(x)$ *for* $x>x_0$, $\phi(0)=x_0$.

Solution (12) *is the unique generalized solution of the same Cauchy problem in the sense of the standard definition by integral identity* [7].

To prove that the solution of the Cauchy problem in the sense of Definition 2.4 is the solution in the sense of the standard definition by the integral identity we recall that in view of Theorem 2.5, the approximation $u^*(x,t,\varepsilon)$ of solution (12) satisfies equation $L[u^*(x,t,\varepsilon)] = O_{\mathcal{D}'}(\varepsilon)$. Let us apply the left-hand and right-hand sides of this relation to an arbitrary test function $\varphi(x,t) \in \mathcal{D}(\Omega)$, $\Omega \subset R^2$. Here the discontinuity line determined by the third equation from system (13) goes through Ω. Since for $\varepsilon > 0$ the function $u^*(x,t,\varepsilon)$ is smooth, we obtain $\int_\Omega \Big[u^*(x,t,\varepsilon)\varphi_t(x,t) + f\big(u^*(x,t,\varepsilon)\big)\varphi_x(x,t)\Big]\,dxdt = \langle O_{\mathcal{D}'}(\varepsilon), \varphi(x,t)\rangle = O(\varepsilon)$. Since the limit $u^*(x,t)$ of the family $u^*(x,t,\varepsilon)$ as $\varepsilon \to +0$ in the $\mathcal{D}'$ sense is a locally integrable function, we have $\int_\Omega \Big[u^*(x,t)\varphi_t(x,t) + f\big(u^*(x,t)\big)\varphi_x(x,t)\Big]\,dxdt = 0$, as $\varepsilon \to +0$, which coincides with the usual integral identity. Due to our assumptions, the stability condition (5) is satisfied and by Theorem of O. A. Oleinik [7] this solution is unique.

4. To describe the interaction dynamics of shock waves we shall seek the phases of shock waves as functions of the fast variable $\tau = \frac{\psi_0(t)}{\varepsilon} \in R$ and the slow variable $t \geq 0$:

$$\phi_k(t,\varepsilon) \stackrel{\text{def}}{=} \widehat{\phi}_k(\tau,t) = \phi_{k0}(t) + \psi_0(t)\phi_{k1}(\tau,t). \tag{14}$$

Here $\phi_{k0}(t)$ denote the phases of non-interacting shock waves defined by the system of equations (13) from Theorem 2.5:

$$\begin{aligned} \frac{d\phi_{10}(t)}{dt} &= \frac{f(u_0(x,t)+e_1(x,t)+e_2(x,t)) - f(u_0(x,t)+e_2(x,t))}{e_1(x,t)}\bigg|_{x=\phi_{10}(t)}, \\ \frac{d\phi_{20}(t)}{dt} &= \frac{f(u_0(x,t)+e_2(x,t)) - f(u_0(x,t))}{e_2(x,t)}\bigg|_{x=\phi_{20}(t)}, \end{aligned} \tag{15}$$

for all $t \in [0,t^*)$. *The phases are extended by these equations* (15) for all $t \in [t^*, +\infty)$. The interaction time $t = t^*$ is defined as the solution of equation $\psi_0(t^*) = 0$, where $\psi_0(t) = \phi_{20}(t) - \phi_{10}(t)$ is the distance between shock wave fronts *before the interaction*. The perturbations for the phases $\phi_{k1}(\tau,t)$ are the smooth function with respect to τ. If $\tau > 0$ then $t < t^*$, i.e., the interaction has not occurred yet; if $\tau < 0$ then $t > t^*$, i.e., the interaction has occurred.

We set the following boundary conditions:

$$\phi_{k1}(\tau,t)\Big|_{\tau\to+\infty} = 0, \qquad \frac{\partial\phi_{k1}(\tau,t)}{\partial\tau}\bigg|_{\tau\to-\infty} = o(\tau^{-1}). \tag{16}$$

That is, the phases tend to zero as $\tau \to +\infty$ (i.e., before interaction). Finding the limit values of the perturbations $\phi_{k1}(\tau,t)\big|_{\tau\to-\infty} = \phi_{k1,-}(t)$, we find the phase limit values $\widehat{\phi}_k(\tau,t)\big|_{\tau\to-\infty} = \widehat{\phi}_{k,-}(t) = \phi_{k0}(t) + \psi_0(t)\phi_{k1,-}(t)$ and thus *define "the result" of the interaction of shock waves for* $t > t^*$.

Let $\psi_1(\tau,t)=\phi_{21}(\tau,t)-\phi_{11}(\tau,t)$, then the full phase difference is $\psi(t,\varepsilon)=\psi_0(t)\big(1+\psi_1(\tau,t)\big)$, the independent variable of the interaction switch function (9) has the form $\rho=\frac{\psi(t,\varepsilon)}{\varepsilon}=\tau\big(1+\psi_1(\tau,t)\big)$.

Taking into account the boundary conditions (16), we obtain the limit values of the phases and their derivatives with respect to time as $\tau\to-\infty$:

$$\widehat{\phi}_{k,-}(t)=\phi_{k0}(t)+\psi_0(t)\phi_{k1,-}(t),\qquad \Big[\frac{d}{dt}\widehat{\phi}_k(\tau,t)\Big]_-=\frac{d}{dt}\widehat{\phi}_{k,-}(t). \tag{17}$$

For the sake of simplicity we consider the case when $u_0(x,t)=u_0^0(x)=u_0$ and $e_k(x,t)=e_k^0(x)=e_k$ are *constants* in ansatz (6), where $e_k>0$ (i.e., the stability condition (5) is satisfied), though analogous results were obtained by the authors for the general case. In this case we can seek the perturbations $\phi_{k1}(\tau)$ as functions depending only on τ.

Substituting the singular ansatz (6), using the asymptotics (8), given by Lemma 2.1, and (14) we derive necessary and sufficient conditions for relation $L[u^*(x,t,\varepsilon)]=O_{\mathcal{D}'}(\varepsilon)$ to be valid:

$$\begin{aligned} e_1\Big(\frac{d\phi_{10}(t)}{dt}+\frac{d\psi_0(t)}{dt}\frac{d}{d\tau}[\tau\phi_{11}(\tau)]\Big)&=f(u_0+e_1)-f(u_0)+B_1(\rho),\\ e_2\Big(\frac{d\phi_{20}(t)}{dt}+\frac{d\psi_0(t)}{dt}\frac{d}{d\tau}[\tau\phi_{21}(\tau)]\Big)&=f(u_0+e_2)-f(u_0)+B_2(-\rho).\end{aligned} \tag{18}$$

These equations are conditions of Rankine-Hugoniot type.

Note that in the general case, when $u_0(x,t)$ and $e_k(x,t)$ are *not constants*, some new equations for these functions will be added to system (18). However, the dynamics of the interaction of shock waves is *determined by conditions of the Rankine-Hugoniot only*.

From (9), (10) if $\tau\to+\infty$ we have up to $O_{\mathcal{D}'}(\varepsilon)$ that system (18) turns into the system of equations which describes the dynamics of two noninteracting shock waves (15).

Subtracting one of the equations in system (18) from the other we obtain the following autonomous differential equation with boundary condition:

$$\frac{d\rho}{d\tau}=F(\rho),\qquad \frac{\rho}{\tau}\Big|_{\tau\to+\infty}=1, \tag{19}$$

$$F(\rho)=\frac{1}{\frac{d\psi_0(t)}{dt}}\Big[\frac{f(u_0+e_2)-f(u_0)+B_2(-\rho)}{e_2}-\frac{f(u_0+e_1)-f(u_0)+B_1(\rho)}{e_1}\Big]. \tag{20}$$

It can be shown that as $\rho\to\pm\infty$ the limit values of the right hand side $F(\rho)$ of the differential equation (19) have *opposite signs*. Then the equation $F(\rho)=0$ *has a root* ρ_0. Then the solution ρ of equation (19) tends to ρ_0, as $\tau\to-\infty$.

From here we derive a system of equations describing the evolution of shock waves after the interaction, for $t > t^*$:

$$\begin{aligned} \frac{d\phi_{10}(t)}{dt} + \phi_{11,-}\frac{d\psi_0(t)}{dt} &= \frac{f(u_0+e_1)-f(u_0)+B_1\big(\rho_0\big)}{e_1}, \\ \frac{d\phi_{20}(t)}{dt} + \phi_{21,-}\frac{d\psi_0(t)}{dt} &= \frac{f(u_0+e_2)-f(u_0)+B_2\big(-\rho_0\big)}{e_2}. \end{aligned} \tag{21}$$

Since $\rho = \tau\big(1+\psi_1(\tau)\big) \to \rho_0$ as $\tau \to -\infty$, it follows that $\psi_1(\tau,t) \to -1$, that is $\psi_1(\tau,t) = -1 + O(\tau^{-1})$. That is, up to $O(\varepsilon)$ we have

$$\widehat{\phi}_{2,-}(t) = \widehat{\phi}_{1,-}(t) \stackrel{\text{def}}{=} \widehat{\phi}_-(t). \tag{22}$$

It is clear from (21), (22) that after the interaction, i.e., for $t > t^*$, discontinuities merge constituting a new shock wave whose dynamics is determined by the equation

$$\frac{d\widehat{\phi}_-(t)}{dt} = \frac{f(u_0+e_1)-f(u_0)+B_1\big(\rho_0\big)}{e_1}. \tag{23}$$

From (21), (22) using the first relation from (10), by Lemma 2.1, we find the value $B_1(\rho_0)$ and obtain the equation which describes the propagation of the shock wave resulting from merging the initial waves:

$$\frac{d\widehat{\phi}_-(t)}{dt} = \frac{f\big(u_0+e_1+e_2\big)-f\big(u_0\big)}{e_1+e_2} \tag{24}$$

with the initial value $\widehat{\phi}_-(t^*) = \phi_{k0}(t^*) = x^*$.

5. Then, integrating systems (18), (15) and equation (19), we obtain the following Theorem 2.6

Theorem 2.6. *In the sense of Definition 2.3 for all $t \in [0,+\infty)$ there exists an asymptotic solution of equation (1): $u^*(x,t,\varepsilon) = u_0 + \sum_{k=1}^2 e_k H_k\big(-x+\phi_k(t,\varepsilon),\varepsilon\big)$ such that: (1) for all $t > 0$: $L[u^*(x,t,\varepsilon)] = O_{\mathcal{D}'}(\varepsilon)$, (2) $u^{*0}(x) - u^*(x,0,\varepsilon) = O_{\mathcal{D}'}(\varepsilon)$, where $u^{*0}(x)$ is an initial value; (3) for $t \in (0,t^*)$ $\lim_{\varepsilon\to+0} u^*(x,t,\varepsilon) = u_0 + \sum_{k=1}^2 e_k H(-x+\phi_{k0}(t))$, where $\phi_{k0}(t) = \lim_{\varepsilon\to+0}\phi_k(t,\varepsilon)$ are given by (15), $\phi_{k0}(0) = x_k^0$, $k=1,2$, $t^* = e_1e_2\frac{x_2^0-x_1^0}{e_2f(u_0+e_1+e_2)-(e_1+e_2)f(u_0+e_2)+e_1f(u_0)}$ is the shock wave merging time; (4) for $t \in (t^*,\infty)$ $\lim_{\varepsilon\to+0} u^*(x,t,\varepsilon) = u_0 + (e_1+e_2)H(-x+\widehat{\phi}_-(t))$, where $\widehat{\phi}_-(t) = \lim_{\varepsilon\to+0}\phi_k(t,\varepsilon)$ is given by formula (24) and $\widehat{\phi}_-(t^*) = \phi_{k0}(t^*) = x^*$.*

The phases $\phi_k(t,\varepsilon)$ are set in the form $\phi_k(t,\varepsilon) = \phi_{k0}(t) + \psi_0(t)\phi_{k1}(\tau)\Big|_{\tau=\frac{\psi_0(t)}{\varepsilon}}$, where $\psi_0(t) = \phi_{20}(t) - \phi_{10}(t)$ and the perturbations of phase functions are given by: $\phi_{k1}(\tau) = (-1)^{k-1}\frac{e_{3-k}}{\tau}\int_0^\tau \frac{f(u_0+e_1+e_2)-f(u_0+e_1)-f(u_0+e_2)+f(u_0)-B_1(\rho(\tau'))}{\big[f(u_0+e_1+e_2)-f(u_0+e_2)\big]e_2-\big[f(u_0+e_2)-f(u_0)\big]e_1}\,d\tau'$, $B_1(\rho)$ is given by (9), $\rho = \rho(\tau)$ is a solution of differential equation (19), (20).

This implies the following result.

Theorem 2.7. *The Cauchy problem for the equation* (1), (4) *in the sense of Definition* 2.4 *has a discontinuous solution of the form* (6)

$$u^*(x,t) = u_0 + \sum_{k=1}^{2} \Big[e_k H\big(-x+\phi_k(t)\big)\Big], \tag{25}$$

where

$$\begin{aligned}\phi_1(t) = x_1^0 + \frac{f(u_0+e_1+e_2)-f(u_0+e_2)}{e_1}t - \bigg[\frac{f(u_0+e_1+e_2)-f(u_0+e_2)}{e_1} \\ - \frac{f(u_0+e_2)-f(u_0)}{e_2}\bigg]\frac{e_2}{e_1+e_2}(t-t^*)H(t-t^*), \\ \phi_2(t) = x_2^0 + \frac{f(u_0+e_2)-f(u_0)}{e_2}t + \bigg[\frac{f(u_0+e_1+e_2)-f(u_0+e_2)}{e_1} \\ - \frac{f(u_0+e_2)-f(u_0)}{e_2}\bigg]\frac{e_1}{e_1+e_2}(t-t^*)H(t-t^*).\end{aligned} \tag{26}$$

This solution of the initial value problem (1), (4) *is the unique generalized solution of the Cauchy problem in the sense of the standard definition by the integral identity.*

Consider a domain Ω crossed by discontinuity lines. In view of Theorem 2.6, the approximation $u^*(x,t,\varepsilon)$ of solution (25) satisfies $L[u^*(x,t,\varepsilon)] = O_{\mathcal{D}'}(\varepsilon)$. By applying the left- and right-hand sides of the latter relation to an arbitrary function $\varphi(x,t) \in \mathcal{D}(\Omega)$, we obtain $\int_\Omega L[u^*(x,t,\varepsilon)]\varphi(x,t)\,dxdt = O(\varepsilon)$. By integrating by parts for $\varepsilon > 0$ and then passing to the limit as $\varepsilon \to +0$, just as in the proof of Theorem 2.5, we prove that the functions $u^*(x,t)$ determined by (25) satisfy the integral identity $\int_\Omega \Big[u^*(x,t)\varphi_t(x,t) + f(u^*(x,t))\varphi_x(x,t)\Big]\,dxdt = 0$. Since for the piecewise constant solution $u^*(x,t)$ the stability condition $e_k > 0$, $k = 1,2$, holds, this solution is unique by the Theorem of O. A. Oleinik [7].

Description of the shock wave merging process. Thus, for $t \in (0,t^*)$ there are two shock waves whose fronts move from the initial positions with the velocities (15). After interaction at the instant $t = t^*$ the shock waves merge constituting one new shock wave; for $t \in (t^*,\infty)$ this shock wave propagates from the point (x^*,t^*) with velocity (24). Up to $O_{\mathcal{D}'}(\varepsilon)$ the merging process is described by equations (18).

3. The propagation and interaction of infinitely narrow δ-solitons

It is known that the weak asymptotics of the one-soliton solution $u(x,t,\varepsilon) = \frac{3v}{2}\cosh^{-2}\big(\frac{\sqrt{v}}{2}(x-vt)/\varepsilon\big)$, $x \in R$, of the Korteweg-de Vries equation $L_{KdV}[u] = u_t + (u^2)_x + \varepsilon^2 u_{xxx} = 0$ as $\varepsilon \to +0$ up to $O_{\mathcal{D}'}(\varepsilon^2)$ represents the *infinitely narrow* δ*-soliton*

$$u_\varepsilon(x,t) = 6\sqrt{v}\varepsilon\delta(x-vt), \quad \varepsilon \to +0, \tag{27}$$

where $\delta(x)$ is the Dirac delta function.

In order to consider solutions of this type for quasilinear equations in [2] (see also [1]) we constructed an associative algebra of *asymptotic distributions* with identity and free of zero divisors generated by the linear span of (one-dimensional) associated homogeneous distributions. The elements $f_\varepsilon^*(x)$ from the algebra of asymptotic distributions will be defined as weak asymptotic expansions of elements $f^*(x,\varepsilon)$ from the linear span of approximations of one-dimensional associated homogeneous distributions as the approximation parameter ε tends to zero.

According to papers [3], [4], it is natural to consider a new type of singularity connected to infinitely narrow δ-solitons for quasilinear equations.

In order to investigate the interaction of two singularities of infinitely narrow δ-soliton type for quasilinear equations, we have to solve the Cauchy problem with the initial condition:

$$u_\varepsilon^{0*}(x) = u_0^0(x) + \sum_{k=1}^{2}\Big[g_k^0\varepsilon\delta(x-x_k) + e_k^0(x)\varepsilon H(x-x_k)\Big], \quad \varepsilon \to +0, \tag{28}$$

where $u_0^0(x)$, $e_k^0(x)$ are smooth functions and g_k^0, x_k are constants, $x_1 < x_2$.

Note that $\varepsilon\delta(x)$ and $\varepsilon H(x)$, $\varepsilon \to +0$ are *not distributions*, but *asymptotic distributions*.

In order to solve the Cauchy problem with the initial condition (28) we should seek a solution in the form of the singular ansatz (asymptotic distribution)

$$u_\varepsilon^*(x,t) = u_0(x,t) + \sum_{k=1}^{2}\Big[g_k(t)\varepsilon\delta(x-\phi_k(t)) + e_k(x,t)\varepsilon H(x-\phi_k(t))\Big] \tag{29}$$

as $\varepsilon \to +0$, where $u_\varepsilon^*(x,0) = u_\varepsilon^{0*}(x)$, and substitute into equation the smooth

$$u^*(x,t,\varepsilon) = u_0(x,t) + \sum_{k=1}^{2}\Big[g_k(t,\varepsilon)\omega_k\big(\frac{x-\phi_k(t,\varepsilon)}{\varepsilon}\big) + e_k(x,t)\varepsilon\omega_{0k}\big(\frac{x-\phi_k(t,\varepsilon)}{\varepsilon}\big), \tag{30}$$

$\varepsilon > 0$, where $u_0(x,t)$, $e_k(x,t)$, $g_k(t)$, $\phi_k(t)$ are smooth functions to be found, $\lim_{\varepsilon\to+0}\phi_k(0,\varepsilon) = x_k^0$ and $\frac{1}{\varepsilon}\omega_k(\xi/\varepsilon)$ are approximations of the delta function, $\omega_{0k}(\xi/\varepsilon)$ are approximations of the Heaviside function $H(\xi)$, $k=1,2$.

The construction of the weak asymptotics of the expression $f\big(u^*(x,t,\varepsilon)\big)$ as $\varepsilon \to +0$, where $u^*(x,t,\varepsilon)$ is the smooth ansatz of the form (30), plays the *key role* in solving the problem of interaction of infinitely narrow solitons (similar to that given in Lemma 2.1).

Lemma 3.1. *Let $f(u)$ be a smooth function and let $u_0(x,t)$, $g_k(t)$, $e_k(x,t)$ be bounded functions and $u^*(x,t,\varepsilon) = u_0 + g_1\omega_1\big(\frac{x}{\varepsilon}\big) + g_2\omega_2\big(\frac{x-a}{\varepsilon}\big) + e_1\varepsilon\omega_{01}\big(\frac{x}{\varepsilon}\big) + e_2\varepsilon\omega_{02}\big(\frac{x-a}{\varepsilon}\big)$. Then we have the asymptotics:*

$$\begin{aligned} f\big(u^*(x,t,\varepsilon)\big) &= f(u_0) + \varepsilon f'(u_0)e_1H(x) + \varepsilon f'(u_0)e_2H(x-a) + A_1(x,t)\varepsilon\delta(x) \\ &+ A_2(x,t)\varepsilon\delta(x-a) + B\big(x,t,\frac{a}{\varepsilon}\big)\big(\varepsilon\delta(x) + \varepsilon\delta(x-a)\big) + O_{\mathcal{D}'}(\varepsilon^2), \end{aligned} \tag{31}$$

as $\varepsilon \to +0$, *where the estimate* $O_{\mathcal{D}'}(\varepsilon^2)$ *is uniform with respect to* a. *Here*

$$B(x,t,\rho) = \frac{1}{2}\int \Big[f\big(u_0(x,t) + g_1(t)\omega_1(\eta) + g_2(t)\omega_2(\eta-\rho)\big) - f\big(u_0(x,t) + g_1(t)\omega_1(\eta)\big) - f\big(u_0(x,t) + g_2(t)\omega_2(\eta-\rho)\big) + f\big(u_0(x,t)\big)\Big]\,d\eta \qquad (32)$$

is the interaction switch function, where $\lim_{|\rho|\to+\infty} B(x,t,\rho) = 0$ *and* $A_k(x,t) = \int \Big[f\big(u_0(x,t) + g_k(t)\omega_k(\eta)\big) - f\big(u_0(x,t)\big)\Big]\,d\eta$, $k = 1,2$.

Note that, in contrast to the equation for shock waves, the correct limit problem describing propagation of a single δ-soliton [3], [4] arises if the additional condition $u^*(x,t,\varepsilon)L[u^*(x,t,\varepsilon)] = O_{\mathcal{D}'}(\varepsilon^2)$ is satisfied together with the condition $L[u^*(x,t,\varepsilon)] = O_{\mathcal{D}'}(\varepsilon^2)$ (for details, see [8]). The techniques of calculating weak asymptotics of left-hand sides in these relations (the same as in the preceding case) is based on Lemma 3.1.

References

[1] V. M. Shelkovich, *An associative-commutative algebra of distributions that includes multiplicators, generalized solutions of nonlinear equations*, Mathematical Notices, **57** (1995), No. 5, 765–783.

[2] V. G. Danilov, V. P. Maslov, V. M. Shelkovich, *Algebra of singularities of singular solutions to first-order quasilinear strictly hyperbolic systems*, Theor. Math. Phys., **114** (1998), No. 1, 1–42.

[3] V. P. Maslov and V. A. Tsupin, *Necessary conditions for the existence of infinitely narrow solitons in gas dynamics*, Dokl. Akad. Nauk SSSR, **246** (1979), 298–300; Engl. transl. in Soviet Phys. Dokl., **24** (1979), No. 5, 354–356.

[4] V. P. Maslov and G. A. Omel'yanov, *Asymptotic soliton-form solutions of equations with small dispersion*, Uspekhi Mat. Nauk, **36** (1981), No. 3, 63–126, Engl. transl. in Russian Math. Surveys, **36** (1981), No. 3, 73–119;

[5] F. Demengel and D. Serre, *Nonvanishing singular parts of measure valued solutions for scalar hyperbolic equations*, Commun. in Partial Diff. Eq., **16** (1991), 221–254.

[6] D. Tan, T. Zhang, Y. Zheng, *Delta-shock waves as limits of vanishing viscosity for hyperbolic systems of conservation law*, J. Diff. Eq., **112** (1994), 1–32.

[7] O. A. Oleinik, *Discontinuous solutions of nonlinear equations*, Uspekhi Mat. Nauk (N.S.), 1957, **12** (1957), No. 3, 3–73; Engl. transl. in Amer. Math. Soc. Transl. Ser., 2, **26**, 95–172.

[8] V. G. Danilov and V. M. Shelkovich, *Propagation of infinitely narrow δ-solitons*, http://arXiv.org/abs/math-ph/0012002.

V. G. Danilov
Moscow Technical University of Communication and Informatics,
Russia
danilov@amath.msk.ru

V. M. Shelkovich
St.-Petersburg State Architecture and Civil Engineering University,
Russia
shelkv@svm.abu.spb.ru

International Series of Numerical Mathematics
Vol. 140, © 2001 Birkhäuser Verlag Basel/Switzerland

MHD Instabilities Arising in Solar Physics: A Numerical Approach

Andreas Dedner, Dietmar Kröner, Christian Rohde, and Matthias Wesenberg

Abstract. Hydrodynamic instabilities are the source of many interesting physical phenomena in fluid dynamics. In this paper we consider *magnetohydrodynamic (MHD)* instabilities, in particular of Rayleigh–Taylor type. Our numerical studies are motivated by a specific application in solar physics: The development of sun spots — which can be observed from earth — is connected to magnetic field concentrations which develop in the solar convection zone. Driven by magnetic forces, these so–called flux tubes rise through the atmosphere. They are fragmented due to Rayleigh–Taylor type instabilities, and their initially simple structure is perturbed by secondary instabilities of Kelvin–Helmholtz type. An efficient numerical simulation of this complex scenario (large area with small scale structures) requires the incorporation of techniques like local adaptivity and parallelization. At the same time the code must be able to resolve the two basic instabilities in a reliable manner. We focus on these two issues and their interplay.

1. Introduction

The equations of compressible ideal magnetohydrodynamics (MHD) with gravitational source terms read

$$\partial_t \rho + \nabla \cdot (\rho \vec{u}) = 0, \tag{1a}$$
$$\partial_t (\rho \vec{u}) + \nabla \cdot (\rho \vec{u}\vec{u}^T + \mathbf{P}) = \rho \vec{g}, \tag{1b}$$
$$\partial_t \vec{B} + \nabla \times (\vec{B} \times \vec{u}) = 0, \tag{1c}$$
$$\partial_t (\rho e) + \nabla \cdot (\rho e \vec{u} + \mathbf{P}\vec{u}) = \rho \vec{u} \cdot \vec{g}, \tag{1d}$$
$$\nabla \cdot \vec{B} = 0. \tag{1e}$$

We consider a perfect gas: the gas pressure is given by $p = (\gamma - 1)\rho\varepsilon$ for a constant $\gamma > 1$; ε is the internal energy connected to the total energy by $\rho e = \rho\varepsilon + \frac{\rho}{2}|\vec{u}|^2 + \frac{1}{8\pi}|\vec{B}|^2$. The pressure tensor is defined as $P_{ij} = \left(p + \frac{1}{8\pi}|\vec{B}|^2\right)\delta_{ij} - \frac{1}{4\pi}B_i B_j \quad (i,j = 1,2,3)$, for the gravitational force vector we have $\vec{g} = (0,0,g)$. (1) can be rewritten as a system of conservation laws in the conservative variables density ρ, momentum

$\rho\vec{u} = (\rho u_x, \rho u_y, \rho u_z)$, magnetic field $\vec{B} = (B_x, B_y, B_z)$ and total energy density ρe. Suitable initial and boundary conditions have to be added.

The system (1) describes the dynamics of an ionized quasineutral plasma and serves as a mathematical model for the physics in the solar convection zone, in particular for so–called magnetic flux tubes. These are (isolated) concentrations of magnetic field which are fragmented due to magnetohydrodynamic instabilities, mainly of Rayleigh–Taylor type. Motivated by numerical studies of this phenomenon, we will at first investigate the classical Rayleigh–Taylor instability, which can be viewed as a simplified flux tube setting. Classical results of linear stability analysis [2] predict an exponential growth of small interface perturbations and a stabilizing effect of tangential magnetic fields. We will demonstrate these effects numerically and will present quantitative comparisons of two–dimensional simulations with these analytical results. Special attention will be paid to the influences of local grid adaption and higher order.

As a more realistic astrophysical test case we will also consider a Rayleigh–Taylor instability in a stratified atmosphere using the transparent boundary conditions derived in [4]. We will demonstrate the necessity of non–reflecting boundary conditions to perform accurate numerical simulations.

Finally we will return to our example from solar physics. As in the case of the investigated Rayleigh–Taylor instabilities, an additional tangential magnetic field stabilizes the magnetic flux tubes during their ascent ("twisted flux tubes"), see e.g. [11, 6].

A further aspect of the numerical discretization — the influence of numerical viscosity — is investigated in [8] for the classical hydrodynamic Rayleigh–Taylor instability: Since numerical viscosity lessens the growth of the perturbations significantly, different possibilities to minimize this effect are discussed. The best results on a fixed grid — in comparison to the unmodified finite volume scheme and an extended version with artificial compression — are obtained with the front tracking method; if the grid is refined, all three schemes lead to comparable results. Since our code is intended for simulations in different parts of the sun, covering regions with smooth solutions as well as ones evolving complicated shock interactions, we will not use the front tracking method. We will show that our higher order finite volume scheme with local grid adaption also allows efficient and accurate simulations yielding the correct growth rates without additional modifications.

2. Magnetic Flux Tubes

The simplest initial setting for a two–dimensional simulation of a magnetic flux tube in the solar convection zone consists of a static stratified background atmosphere, i.e.

$$\rho(x,z,0) = \rho_0(z), \quad \vec{u}(x,z,0) \equiv 0, \quad \vec{B}(x,z,0) \equiv 0, \tag{2a}$$

$$p(x,z,0) = p_0(z), \quad \tfrac{d}{dz} p_0(z) = \rho_0(z) g, \tag{2b}$$

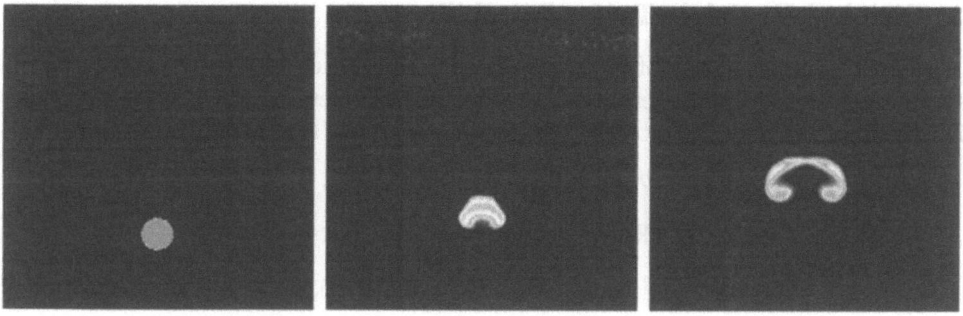

FIGURE 1. Numerical simulation of a rising magnetic flux tube: B_y at $t = 0$ (left), $t = 2$ (middle) and $t = 5$ (right).

except in a region with a strong magnetic field *normal* to the computational xz–plane (see Fig. 1 left). For the gas pressure and the density within the flux tube we use

$$p_{tube} := p_0 - \frac{1}{8\pi}|\vec{B}_{tube}|^2, \quad \rho_{tube} := p_{tube}^{\frac{1}{\gamma}}. \tag{3}$$

Since there is no exact solution and the measurement of isolated structures in the solar atmosphere is difficult, research must strongly rely on numerical simulations. Therefore the reliability of the numerical scheme is essential.

For the numerical solution of (1) we use a finite volume scheme based on the approximate Riemann solver introduced in [3]; it is (formally) extended to second order by limited linear reconstruction and Runge–Kutta time integration [5]. Since a violation of (1e) — a condition which cannot be guaranteed on the discrete level in general — may lead to severe numerical problems, see e.g. [1], we rely on the stabilization technique suggested in [9]. We use an unstructured, hierarchical grid consisting of triangles in 2D and tetrahedra in 3D, see also [5, 10]. This allows an efficient implementation of local grid refinement/coarsening and domain decomposition techniques needed for parallelization.

The numerical experiments in Fig. 1 show that the setting (2), (3) causes the flux tube to rise through the atmosphere. Due to its lowered density the magnetic flux tube is Rayleigh–Taylor unstable, and its ascent is slowed down by growing perturbations, which finally lead to a fragmentation of the tube (Fig. 1). (Note that also secondary instabilities of Kelvin–Helmholtz type contribute to the fragmentation, but we will not consider them seperately in this paper. A simulation of an isolated Kelvin–Helmholtz instability is shown in [5]). This fragmentation inhibits the flux tube from rising far enough through the atmosphere to be seen as sun spot from earth.

3. Classical Rayleigh–Taylor Instability

We start our investigation of Rayleigh–Taylor instabilities with the classical case of two superimposed fluid layers with constant positive densities in 2D; the interface between both layers is located at the height $z = 0$ and the initial conditions are

$$\rho_0(z) = \rho_t \ (z > 0), \quad \rho_0(z) = \rho_b \ (z < 0),$$
$$p_0 \in C^0(\mathbb{R}), \quad \frac{d}{dz} p_0(z) = \rho_0(z) g \ (g < 0),$$
$$\vec{u}_0(z) \equiv 0, \quad \vec{B}_0(z) \equiv (B_x, 0, 0).$$

For this equilibrium solution $\vec{U}_0$ of (1) linear stability analysis can be used to calculate the growth rate of plane wave perturbations:

Theorem (cf. [2])
Consider (1) for the incompressible case, linearized around the equilibrium solution $\vec{U}_0$. Let $\delta\vec{U}$ be a (small) perturbation of the form $\delta\vec{U}(x) = \vec{A}e^{i2\pi x}$, $\vec{A} \in \mathbb{R}^8$. Then a solution $\vec{U}$ to these equations with inital data $\vec{U}_0 + \delta\vec{U}$ exists and satisfies

$$\vec{U}(x, z, t) = \vec{U}_0 + e^{nt}\delta\vec{U}(x),$$

where — ignoring surface tension — the growth rate n fulfills

$$n^2 = 2\pi \left(|g| \frac{\rho_t - \rho_b}{\rho_t + \rho_b} - \frac{B_x^2}{\rho_t + \rho_b} \right). \tag{4}$$

From the theorem we know that a sinusoidal perturbation $\vec{A}e^{i2\pi x}$ of the initial

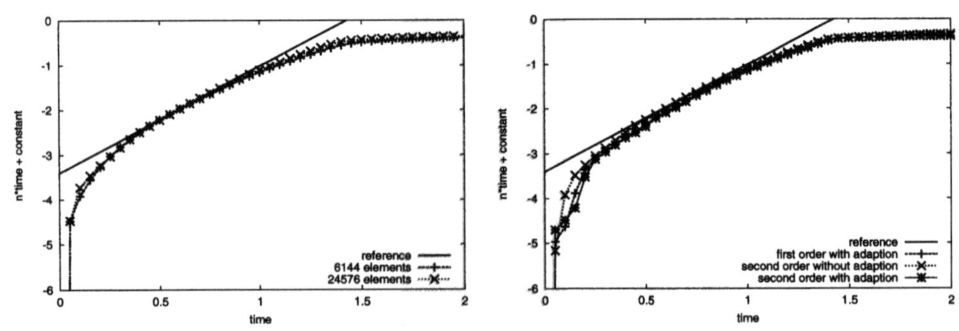

FIGURE 2. Growth rates n for the hydrodynamic setting: first order scheme without local adaption (left) and results with local adaption and second order (right).

setting leads to perturbations growing proportional to e^{nt}. In our tests we trigger the instability by introducing such a perturbation in the vertical velocity in the vicinity of the interface; numerical tests have shown that this leads to the same behaviour of the solution, cf. [7]. The growth rate n is either a non–negative real number (yielding an exponential growth) or purely imaginary (leading to a stable

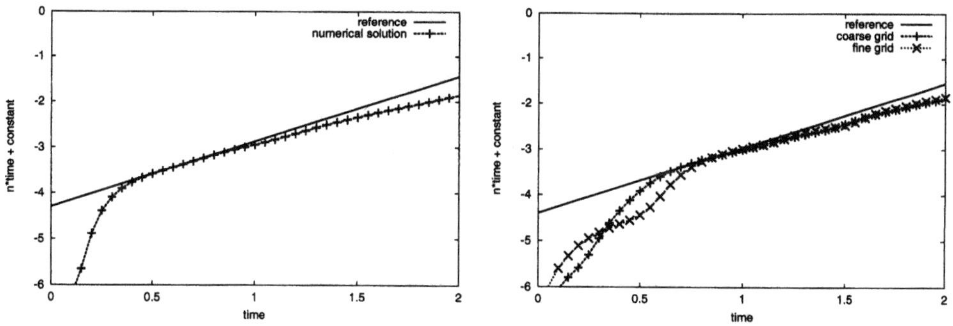

FIGURE 3. Growth rates n for the unstable magnetohydrodynamic setting: first order without local adaption (left) and second order with local adaption (right).

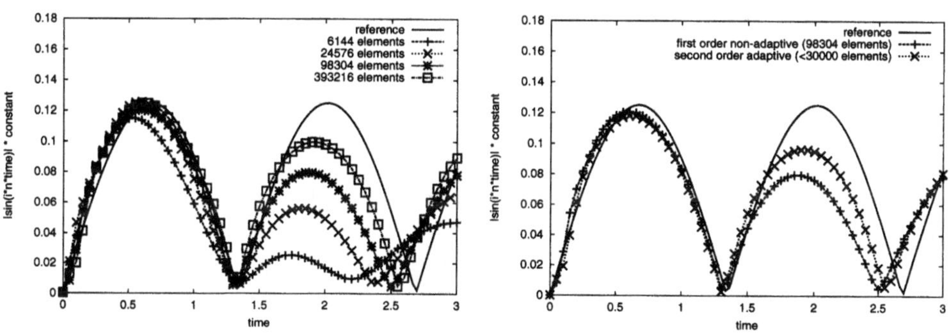

FIGURE 4. Oscillations for the stable magnetohydrodynamic setting: first order on globally refined grids (left), comparison of results with the same grid spacing on the finest level (right).

oscillation). In our simulations we have $\rho_t > \rho_b > 0$; therefore the first term in (4) is always positive and n^2 is non–negative (and n therefore real) up to the critical magnetic field strength

$$B_x^c = \sqrt{|g|(\rho_t - \rho_b)}. \tag{5}$$

This means that a tangential magnetic field stabilizes the fluid interface; for $|B_x| > B_x^c$ we have reached a stable setting. Since we consider the compressible equations (1), the theorem is not applicable directly. However, we try to minimize the effects of compressibility by choosing a large value for the pressure at the interface. The exact initial conditions for our numerical tests are taken from [7].

We have investigated how well the growth rates of our numerical solutions agree with the ones obtained from (4). Because we have found that they are quite insensitive to the method applied for calculating them, we have chosen the method suggested in [7]; our results are very similar to the ones presented there which were

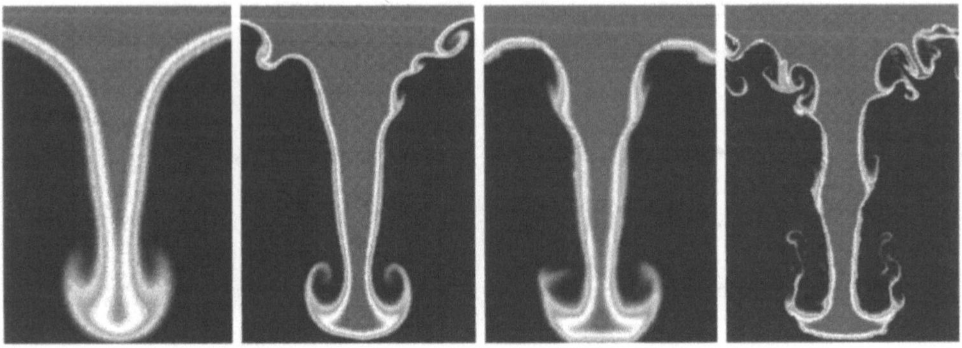

FIGURE 5. Rayleigh–Taylor instability for the hydrodynamic setting; density at time $t = 2$ (from left to right): first order non–adaptive, first order adaptive, second order non–adaptive, second order adaptive.

computed on a *cartesian grid* using *finite differences*. We have studied three different cases: a purely hydrodynamic setting ($B_x = 0$, Figs. 2 and 5), an unstable magnetohydrodynamic setting ($B_x = 0.8B_x^c$, Fig. 3) and a stable magnetohydrodynamic setting ($B_x = 1.4B_x^c$, Fig. 4). Note that only the slope (Figs. 2 and 3) or the period (Fig. 4) of the "reference solution" is given by (4), but not its absolute position (Figs. 2 and 3) nor its amplitude (Fig. 4).

Secondary Kelvin–Helmholtz instabilities can be seen in the numerical solution if the high resolution techniques are used (Fig. 5), which do not influence the corresponding growth rates (Fig. 2). Fig. 4 shows that these techniques, while using far less elements, yield results comparable with non–adaptive first order calculations. This proves that our scheme produces good results for the stable magnetohydrodynamic setting and that higher order and grid adaption allow us to capture the same behaviour of the solution with far less computational effort (the gain in computational time between the two computations compared in the r.h.s. of Fig. 4 was approximatly a factor of three). We want to stress that grid convergence as it can be seen in Fig. 4 can only be expected for the stable magnetohydrodynamic setting.

4. Atmospheric Rayleigh–Taylor Instability

In this section we want to study one key ingredient for the numerical discretization of many systems of non–linear PDEs: the boundary conditions. For the magnetic flux tube simulations we have chosen periodic boundary conditions at the vertical boundaries of the computational domain. Since all our calculations have to be restricted to a finite computational domain, we have to introduce artificial horizontal boundaries with suitable boundary conditions. Ideal conditions should be non–reflecting to avoid influences on the solution in the computational domain. In

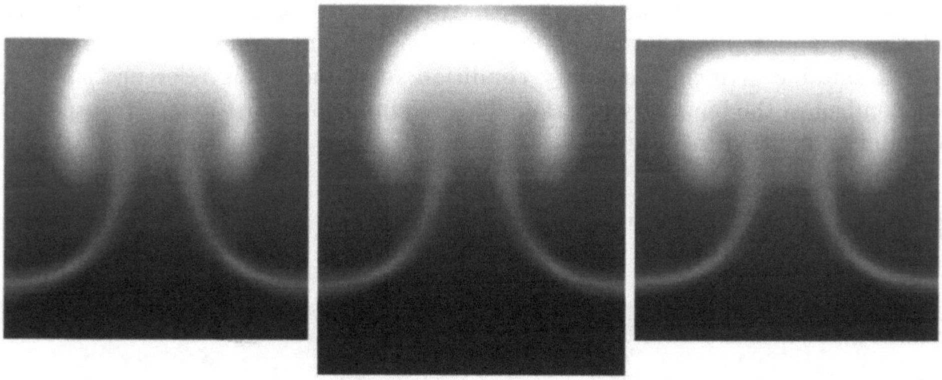

FIGURE 6. Density plot of the solution to the atmospheric Rayleigh–Taylor instablility calculated on a small domain using the transparent boundary conditions [4] (left) and Dirichlet boundary conditions (right). The middle picture shows the solution on a ten times larger domain.

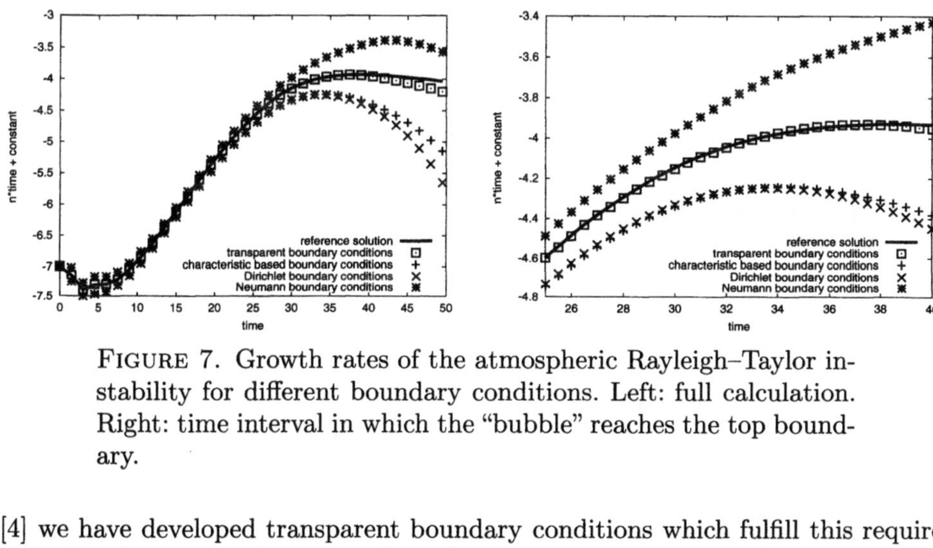

FIGURE 7. Growth rates of the atmospheric Rayleigh–Taylor instability for different boundary conditions. Left: full calculation. Right: time interval in which the “bubble” reaches the top boundary.

[4] we have developed transparent boundary conditions which fulfill this requirement for the linearized system. These boundary conditions have been derived for problems formulated in stratified, static atmospheres; the analysis was performed for atmospheres with an exponential decay of density and pressure for $z \to \infty$. The detailed analysis for these transparent boundary conditions and implementational aspects can be found in [4]. The influence of a varying tangential magnetic field strength on an atmospheric RT instability is investigated in [5].

Some results of simulations in an exponentially decaying atmosphere from [4] are shown in Fig. 6. As in the previous section, we have calculated growth rates for these numerical results. Since we are not aware of any analytical results for this setting, we compare them with growth rates obtained for a "reference solution" calculated on a far larger domain using the same grid spacing. We expect that a suitable artificial boundary condition reproduces this reference solution independent of the boundary position. This requirement is met by our transparent boundary condition, see Fig. 7.

5. Twisted Magnetic Flux Tubes

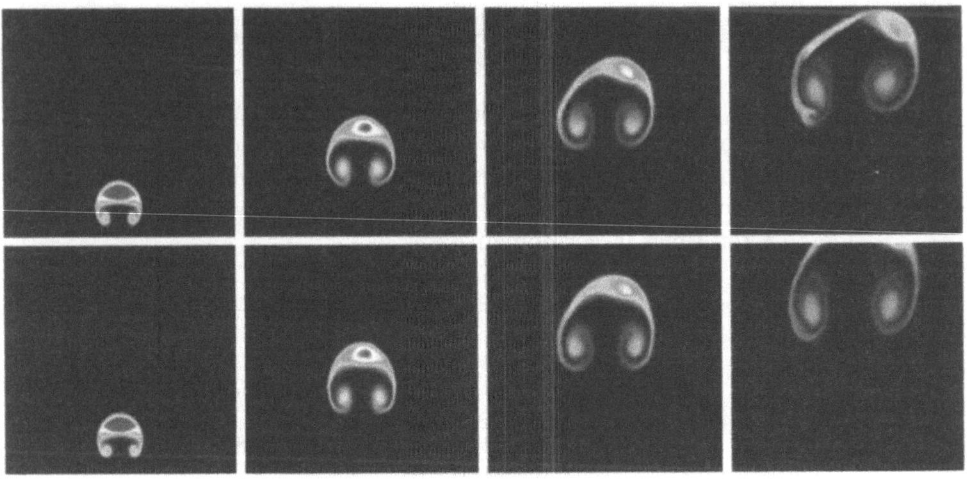

FIGURE 8. Twisted flux tubes for $t = 5$, $t = 10$, $t = 15$ and $t = 20$: results for the second order scheme on a globally refined grid (top) and on locally adapted grids (bottom) with the same minimal grid spacing.

As we have demonstrated through our numerical tests, the simplest setting for a magnetic flux tube leads to a strong fragmentation of the solution, which inhibits the tube to rise as high as expected, cf. Fig. 1. On the other hand we have seen that a tangential magnetic field stabilizes a Rayleigh–Taylor unstable interface; the same is true for Kelvin–Helmholtz instabilities [2]. This can be exploited to stabilize magnetic flux tubes by adding a tangential magnetic field, which leads to a "twist" of the magnetic field lines [6]. The stabilizing effect can be seen in Fig. 8. Again we compare simulations on a globally refined grid and on locally adapted grids. The spacing of the globally refined grid equals that of the finest adaptive grid level. At the beginning of the simulations both solutions show the same global structure of the tube (height, degree of fragmentation); at later times these global structures begin to differ. This might be due to additional perturbations introduced by the grid refinement and the reconstruction of the data, but these effects may also be caused by reflected waves coming from the boundary: in the locally adaptive calculation the damping of these waves is stronger due to the larger grid spacing far from the tube. (Note that we have not yet used non–reflecting boundary conditions in these simulations.) Since we have shown that our high resolution scheme produces good results for the stable Rayleigh–Taylor setting, we can trust our results for the twisted magnetic flux tubes — at least in the first part of the simulation.

6. Conclusions

We have illustrated how different aspects of a numerical scheme can influence the growth rate of Rayleigh–Taylor type instabilities. We have first compared results using different resolution enhancement techniques. We have shown that limited linear reconstruction and local grid adaption have only a small influence on the growth rates, while they speed up the calculations considerably. For the atmospheric Rayleigh–Taylor instability we have found a strong influence of the boundary conditions (caused by waves reflected at the boundary): the growth rate is diminished if non–transparent boundary conditions are used. Moreover it is well known that the results of *parallel* computations are dependent on the summation order (e.g. of numerical fluxes), which is usually not deterministic. While these effects are often negligible, they may become important in simulations of *unstable* problems. Especially for simulations in 3D all these problems have to be well understood in advance, since a thorough study is computationally very expensive. Therefore 2D calculations are still essential, although some properties of the instabilities are inherently three–dimensional, see [8]. A first result obtained with our 3D code is shown in Fig. 9; further work in 3D is in progress.

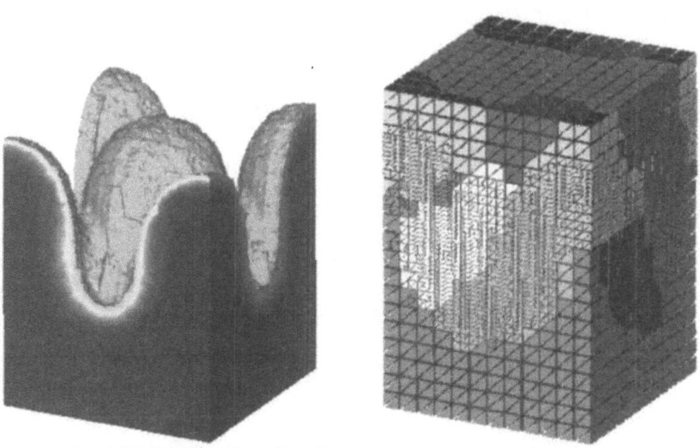

FIGURE 9. 3D MHD Rayleigh–Taylor instability ($t = 1.4$). Left: density isosurface. Right: automatically generated grid partition.

Acknowledgement

The authors were partially supported by the DFG priority research programs ANumE and DANSE.

References

[1] J. U. Brackbill and D. C. Barnes, *Note: The effect of nonzero* $\nabla \cdot \mathbf{B}$ *on the numerical solution of the magnetohydrodynamic equations*, J. Comput. Phys. **35** (1980), 426–430.

[2] S. Chandrasekhar, *Hydrodynamic and hydromagnetic stability*, Dover, New York, 1981.

[3] W. Dai and P. R. Woodward, *A simple Riemann solver and high-order Godunov schemes for hyperbolic systems of conservation laws*, J. Comput. Phys. **121** (1995), no. 1, 51–65.

[4] A. Dedner, D. Kröner, I. L. Sofronov, and M. Wesenberg, *Transparent boundary conditions for MHD simulations in stratified atmospheres*, J. Comput. Phys. **171** (2001), no. 2, 448–478.

[5] A. Dedner, C. Rohde, and M. Wesenberg, *A MHD-simulation in solar physics*, Finite Volumes for Complex Applications II (R. Vilsmeier et al., ed.), Hermès (Paris), 1999, pp. 491–498.

[6] T. Emonet and F. Moreno-Insertis, *The physics of twisted magnetic tubes rising in a stratified medium: Two-dimensional results*, Astrophys. J. **492** (1998), 804–821.

[7] B.-I. Jun, M. L. Norman, and J. M. Stone, *A numerical study of Rayleigh-Taylor instability in magnetic fluids*, Astrophys. J. **453** (1995), 332–349.

[8] X. L. Li, B. X. Jin, and J. Glimm, *Numerical study for the three-dimensional Rayleigh-Taylor instability through the TVD/AC scheme and parallel computation*, J. Comput. Phys. **126** (1996), 343–355.

[9] K. G. Powell, *An approximate Riemann solver for magnetohydrodynamics (that works in more than one dimension)*, ICASE-Report 94–24, 1994.

[10] B. Schupp, *Entwicklung eines effizienten Verfahrens zur Simulation kompressibler Strömungen in 3D auf Parallelrechnern*, PhD-thesis, Universität Freiburg, 1999; http://www.freidok.uni-freiburg.de/volltexte/68.

[11] M. Schüssler, *Magnetic buoyancy revisited — analytical and numerical results for rising flux tubes*, Astronomy and Astrophysics **71** (1979), no. 1-2, 79–91.

Institut für Angewandte Mathematik,
Albert Ludwigs Universität Freiburg,
Hermann Herder Str. 10,
D 79104 Freiburg / Breisgau, Germany
`dedner|dietmar|chris|wesenber@mathematik.uni-freiburg.de`

International Series of Numerical Mathematics
Vol. 140, © 2001 Birkhäuser Verlag Basel/Switzerland

Numerical Methods for the Real Gas MHD Equations

Andreas Dedner and Matthias Wesenberg

Abstract. In recent years many numerical schemes — mainly based on approximate Riemann solvers — for the equations of ideal magnetohydrodynamics (MHD) have been developed; their robustness and efficiency have been shown in many examples. But since the underlying equation of state (EOS) of an ideal gas is far from reality in many applications (e.g. solar physics), these numerical schemes have to be extended to cope with a more general EOS. For the Euler equations of gas dynamics two general approaches for this extension have recently been proposed. We will show that they can also be applied to MHD. Furthermore we will validate the resulting schemes and will discuss some important aspects of their behaviour.

1. Introduction

If the internal energy ε of a gas is considered to be a function of the specific volume τ and the specific entropy s, the second law of thermodynamics yields

$$p = -\partial_\tau \varepsilon(\tau, s) > 0 \quad \text{and} \quad T = \partial_s \varepsilon > 0 \tag{1}$$

for the gas pressure p and the temperature T. Therefore the function $s \mapsto \varepsilon(\tau, s)$ is invertible and we can define an *equation of state* (EOS) for p as a function of τ and ε. This function can be very complex; in some cases there is not even an analytic expression, e.g. if p is given as an implicit function or if the values of p have to be taken from tables.

The equations of magnetohydrodynamics (MHD) in the variables density ρ, velocity $\vec{u} = (u_x, u_y, u_z)$, magnetic field $\vec{B} = (B_x, B_y, B_z)$ and total energy e read:

$$\begin{aligned}
\partial_t \rho + \nabla \cdot (\rho \vec{u}) &= 0, && \text{(2a)}\\
\partial_t (\rho \vec{u}) + \nabla \cdot (\rho \vec{u}\vec{u}^T + \mathbf{P}) &= 0, && \text{(2b)}\\
\partial_t \vec{B} + \nabla \times (\vec{B} \times \vec{u}) &= 0, && \text{(2c)}\\
\partial_t (\rho e) + \nabla \cdot (\rho e \vec{u} + \mathbf{P}\vec{u}) &= 0, && \text{(2d)}\\
\nabla \cdot \vec{B} &= 0. && \text{(2e)}
\end{aligned}$$

The total energy and the pressure tensor $\mathbf{P} = (P_{ij})_{1\le i,j\le 3}$ correlate the conservative variables $\vec{U} = (\rho, \rho\vec{u}, \vec{B}, \rho e)$ to ε and p by means of

$$e = \varepsilon + \frac{1}{2}|\vec{u}|^2 + \frac{1}{8\pi\rho}|\vec{B}|^2, \quad \mathbf{P} = \left(p + \frac{1}{8\pi}|\vec{B}|^2\right)\mathbf{I} - \frac{1}{4\pi}\vec{B}\vec{B}^t. \tag{3}$$

The System (2) can be rewritten in conservation form

$$\partial_t\vec{U} + \partial_x\vec{f}_x(\vec{U}) + \partial_y\vec{f}_y(\vec{U}) + \partial_z\vec{f}_z(\vec{U}) = 0, \quad \partial_x B_x + \partial_y B_y + \partial_z B_z = 0. \tag{4}$$

Like the Euler equations of gas dynamics, the MHD equations have to be closed by a suitable EOS $p(\tau, \varepsilon)$ — for our schemes we need $\partial_\varepsilon p > 0$. In consideration of $\tau = 1/\rho$ and (3), this EOS can be written as $p(\vec{U})$.

Numerical schemes for the MHD equations for an *ideal gas*, i.e. (2) with

$$p(\vec{U}) = (\gamma - 1)\rho\varepsilon \tag{5}$$

for a constant adiabatic exponent $\gamma > 1$, have been a focus of studies in recent years, see e.g. [4, 9] and the references therein. But in many applications it is not feasible to use (5) as EOS; in this case we will talk of a "*real gas*" in the following. Real gases arise for example, if — at low temperatures and at high pressures — intermolecular forces have to be taken into account. Suitable methods for the numerical solution of the real gas MHD equations have to be derived for this class of applications. For the compressible Euler equations, i.e. (2) with $\vec{B} \equiv 0$, two methods of extending numerical fluxes to the real gas case have recently been suggested [2, 1]. We will show that both methods can be extended to MHD and give very similar results. Moreover we will discuss the restrictions on the equation of state imposed by the different solution techniques and the number of accesses to the EOS required in each time step. Additionally, we will focus on the interaction of table resolution and grid refinement in the case of a tabularized EOS. We will also include the Lax–Friedrichs scheme in our studies, since it can be used without modifications in the real gas case. For the purpose of this paper it suffices to focus on the one–dimensional equations, i.e. $\partial_y\vec{U} \equiv 0$ and $\partial_z\vec{U} \equiv 0$ in (4).

2. Direct Riemann Solver for MHD

Let $i \in \mathbb{Z}$ denote the index of the grid cells (with fixed diameter h), $n \in \mathbb{N}$ the time level and $\vec{U}_i^n$ the corresponding cell average. Then the time step $t^n \mapsto t^{n+1} := t^n + \Delta t$ of our one–dimensional finite volume scheme (cf. [4, 9]) has the form

$$\vec{U}_i^{n+1} := \vec{U}_i^n - \frac{\Delta t}{h}\left(\vec{g}(\vec{U}_i^n, \vec{U}_{i+1}^n) - \vec{g}(\vec{U}_{i-1}^n, \vec{U}_i^n)\right). \tag{6}$$

The numerical flux function $\vec{g}(\cdot,\cdot)$ is based on an approximate Riemann solver. The key ingredients for the construction of such schemes are the flux vector Jacobian and its eigensystem. Since in their derivation the EOS (5) is usually inserted directly, the first guess is that they do strongly depend on the underlying EOS. But this conjecture is not true: the approximate Riemann solver for the real gas Euler equations introduced in [1] exploits the fact that the flux Jacobian, its eigenvalues and eigenvectors are exactly the same for an *arbitrary* EOS as in the ideal gas case. The only difference is that not only p, but also γ and the sound speed c have to be interpreted as EOS–dependent *functions*:

$$c^2(\tau,\varepsilon) := -\tau^2\left(\frac{\partial p}{\partial \tau}(\tau,\varepsilon) - p(\tau,\varepsilon)\frac{\partial p}{\partial \varepsilon}(\tau,\varepsilon)\right), \quad \gamma(\tau,\varepsilon) := \frac{c^2(\tau,\varepsilon)}{\tau p(\tau,\varepsilon)}. \tag{7}$$

As the same holds for the MHD system (2), the extension of our schemes to the real gas case could be done in a straightforward manner. We follow the approach in [1] and use the primitive variables $\vec{W}(\vec{U}) := (\tau, \vec{u}, \vec{B}, p)$ in the flux implementation to minimize the number of calls to the EOS. In our MHD computations we usually use the approximate Riemann solver suggested in [3] (see e.g. [4, 9]); the main part of its real gas extension "RGDW" reads:

1. *Calculate the primitive variables* $\vec{W}_{l,r} := \vec{W}(\vec{U}_{l,r})$ *and the mean value* $\vec{W}_m := (\vec{W}_l + \vec{W}_r)/2$.
2. *Calculate for* $1 \le k \le 8$ *the eigenvalues* $\lambda_k(\vec{W}_m)$, *left and right eigenvectors* $\vec{l}_k(\vec{W}_m)$, $\vec{r}_k(\vec{W}_m)$ *and define* $\vec{W}_k^0 := \vec{W}_l$ *if* $\lambda_k(\vec{W}_m) > 0$, $\vec{W}_k^0 := \vec{W}_r$ *if* $\lambda_k(\vec{W}_m) < 0$, *respectively.*
3. *Calculate the interface state* $\vec{W}_{is} = (\vec{r}_1, \ldots, \vec{r}_8)\left(\vec{l}_1 \cdot \vec{W}_1^0, \ldots, \vec{l}_8 \cdot \vec{W}_8^0\right)^t$.
4. *Calculate the numerical flux* $\vec{g}(\vec{U}_l, \vec{U}_r) := \vec{f}_x(\vec{U}(\vec{W}_{is}))$. *Note that the transformation* $\vec{W} \mapsto \vec{U}$ *requires the function* $\varepsilon = \varepsilon(\tau, p)$.

In the full algorithm we add a small artificial viscosity for stabilization and a fix in the spirit of [5] to cope with arising low densities in special situations; both modifications do not require additional evaluations of the EOS.

3. Energy Relaxation for MHD

In each step of the time evolution (6) the energy relaxation ("ER") scheme suggested in [2] computes the fluxes by using any numerical flux function for the *ideal gas* Euler equations with an additional relaxation step for the total energy. This scheme can therefore be easily implemented as an add–on to a given finite volume scheme for an ideal gas without modification of the flux function. Analytical justification w.r.t. stability and convergence properties and the details of this scheme can be found in [2]. Due to (3) the connection between internal energy and gas pressure in the MHD system (2) is very similar to the Euler case. Therefore this method can be extended directly to MHD. If $\vec{g}^{ideal}(\cdot,\cdot;\gamma) = (g_i^{ideal}(\cdot,\cdot;\gamma))_{1\le i\le 8}$ is an arbitrary flux function for the ideal gas MHD equations, we obtain the following algorithm for two given left and right hand states $\vec{U}_{l,r} = (\rho_{l,r}, (\rho\vec{u})_{l,r}, \vec{B}_{l,r}, (\rho e)_{l,r})$:

1. *relaxation step:*
   - choose a suitable $\gamma_1 > 1$ (see below)
   - decompose $(\rho\varepsilon)_{l,r}$: $(\rho\varepsilon_1)_{l,r} := \frac{p_{l,r}}{(\gamma_1 - 1)}$, $(\rho\varepsilon_2)_{l,r} := (\rho\varepsilon)_{l,r} - (\rho\varepsilon_1)_{l,r}$

- define new left and right states $\vec{V}_{l,r} := \left(\rho_{l,r}, (\rho\vec{u})_{l,r}, \vec{B}_{l,r}, (\rho e_1)_{l,r}\right)$ with the energy density $(\rho e_1)_{l,r} := (\rho\varepsilon_1)_{l,r} + \frac{1}{2}\rho|\vec{u}_{l,r}|^2 + \frac{1}{8\pi}|\vec{B}_{l,r}|^2$

2. *flux calculation:*

$$\vec{g}(\vec{U}_l,\vec{U}_r) \;:=\; \vec{g}^{ideal}(\vec{V}_l,\vec{V}_r;\gamma_1) + \left(0,0,0,0,0,0,0,g^{corr}(\vec{U}_l,\vec{U}_r)\right).$$

Using $g^+ := \max\{g,0\}$, $g^- := \min\{g,0\}$ the flux–correction is given by

$$g^{corr}(\vec{U}_l,\vec{U}_r) := \left(g_1^{ideal}(\vec{V}_l,\vec{V}_r;\gamma_1)\right)^+ \frac{(\rho\varepsilon_2)_l}{\rho_l} + \left(g_1^{ideal}(\vec{V}_l,\vec{V}_r;\gamma_1)\right)^- \frac{(\rho\varepsilon_2)_r}{\rho_r}.$$

Here g_1^{ideal} denotes the first component of the numerical flux $\vec{g}^{ideal}$.

In [2] it has been shown that γ_1 has to fulfill the subcharacteristic condition

$$\gamma_1 > \sup_{\tau,\varepsilon}\{\gamma(\tau,\varepsilon),\Gamma(\tau,\varepsilon)\} \quad \text{with} \quad \Gamma(\tau,\varepsilon) := 1 + \tau\frac{\partial p}{\partial\varepsilon}(\tau,\varepsilon) \tag{8}$$

and $\gamma(\tau,\varepsilon)$ defined by (7). The authors also point out that it is enough to satisfy (8) *locally*, i.e. we may choose γ_1 seperately at every interface depending on $\vec{U}_l$ and $\vec{U}_r$. To our knowledge no thorough test with a locally chosen γ_1 has been published; for our implementation we use

$$\gamma_1(\vec{U}_l,\vec{U}_r) := C_1 \max\{\gamma(\tau_l,\varepsilon_l),\gamma(\tau_r,\varepsilon_r),\Gamma(\tau_l,\varepsilon_l),\Gamma(\tau_r,\varepsilon_r)\} \tag{9}$$

with $C_1 > 1$ such that (8) is still approximately fulfilled. If $\partial_\tau p$, $\partial_\varepsilon p$ are available and can be cheaply evaluated, γ_1 can be directly computed from (9). For our implementation we have decided to use central differences to approximate the derivatives of p at $\left(\frac{1}{2}(\tau_l+\tau_r),\frac{1}{2}(\varepsilon_l+\varepsilon_r)\right)$. In our ER–DW scheme we use the numerical flux from [3] for $\vec{g}^{ideal}$.

4. Demands on the Equation of State

In some applications the computation of p or its derivatives can be very time consuming; e.g. in the case of an implicitly defined EOS (like "TMV" studied in [7]) an expensive iterative procedure has to be used. It is even possible that the derivatives are not available. In this case we cannot directly use a scheme which requires the evaluation of quantities like the sound speed. Important criteria for the choice of a numerical scheme are therefore the amount of information required and the number of calls to the EOS needed per time step and computational cell. In Tab. 1 we compare the schemes w.r.t. these criteria.

5. Numerical Results

Our code is used for simulations in the solar photosphere. In this region we encounter a partially ionized plasma, which cannot be modeled by (5). In this case the relationship between pressure, internal energy and density is given by a system of nonlinear equations, which in general can only be solved numerically. To save

scheme	requirement for						no. of calls to			
	flux evaluation				time step					
	p	$\partial_\tau p$	$\partial_\varepsilon p$	ε	$\partial_\tau p$	$\partial_\varepsilon p$	p	$\partial_\tau p$	$\partial_\varepsilon p$	ε
Lax–Friedrichs	x	—	—	—	x	x	1	1	1	0
RGDW	x	x	x	x	x	x	2	1	1	1
ER, constant γ_1	x	—	—	—	—	—	1	0	0	0
ER, γ_1 from (9)	x	x	x	—	—	—	1	1	1	0
ER, FD–scheme for γ_1	x	—	—	—	—	—	2	0	0	0

TABLE 1. Comparison of the schemes w.r.t. their requirements and the number of calls to the EOS. Note that the time step in the energy relaxation ("ER") schemes is computed from the relaxation system.

computational time, we solve these equations in advance and use the resulting tables for the solution of the MHD system (2). Since exact solutions for this setting are unavailable, we have chosen to base our comparison on the far simpler EOS of van der Waals. It is an extension of the perfect gas law (5) and models also molecular forces which have to be taken into account at low temperatures and high pressures:

$$p(\tau,\varepsilon) = \frac{R\left(\varepsilon + \frac{a}{\tau}\right)}{c_v(\tau - b)} - \frac{a}{\tau^2}; \qquad \left\{ \begin{array}{lllllll} a & = & 1684.54, & b & = & 0.001692, \\ R & = & 461.5, & c_v & = & 1401.88. \end{array} \right. \tag{10}$$

The values of the constants are taken from [1]. On the one hand this EOS already admits compound waves in the solutions; on the other hand we have been able to construct an exact solution to a MHD Riemann problem. We study the behaviour of the schemes if (10) is applied directly and — to investigate the influence of tabularized pressure values — we use a table with discrete values calculated from (10).

5.1. Euler

Before we demonstrate the effectuality of the schemes' extension to MHD, we will study some aspects for two Euler Riemann problems, see Tab. 2. For a detailed study of solutions to Riemann problems for an arbitrary EOS and their properties we refer to [6].

The solution to the "Euler 1"–problem, which is a slight modification of a problem suggested in [1], consists of a rarefaction wave, a contact discontinuity and a shock (Fig. 1 left). Fig. 1 (right) shows results of the ER scheme with different choices of a global γ_1 and a local γ_1 obtained with our finite difference method: although $\gamma_1 = 4.8$ is still well below the bound given by (8), the ER scheme with the local γ_1 is already less diffusive. In Fig. 2 we compare the errors and the experimental orders of convergence (EOC) of the different schemes. It can be seen that the ER–DW scheme (now with the local γ_1) and the RGDW scheme

	Euler 1 ($t_{disp} = 0.0009$)			**Euler 2** ($t_{disp} = 0.0007$)		
	ρ	u	p	ρ	u	p
$x < 0$	333.0	0.0	37311358.0	300.0	0.0	36000000.0
$x > 0$	111.0	0.0	21770768.0	150.0	1200.0	5200000.0

TABLE 2. Initial conditions and displayed time in plots t_{disp} for Euler problems.

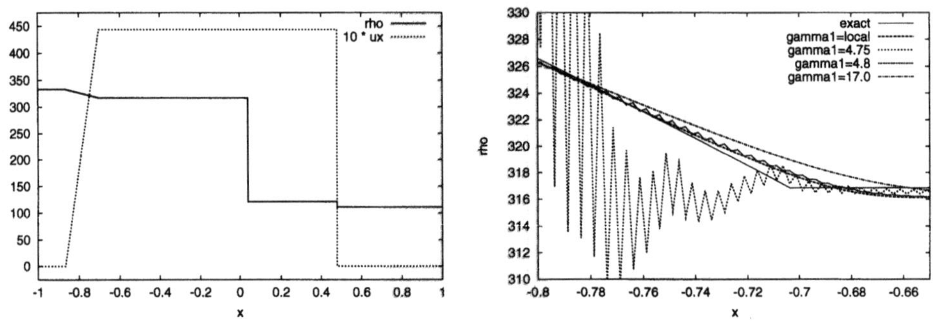

FIGURE 1. Density and velocity (left) and influence of γ_1 (right) for the Euler 1–problem.

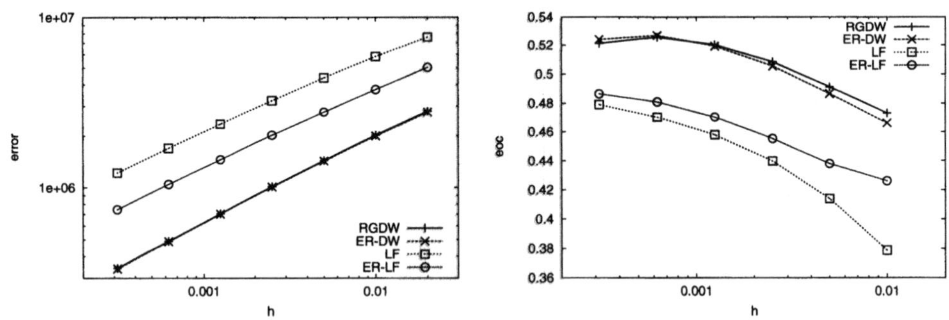

FIGURE 2. Comparison of accuracy (left) and convergence rates (right) for the Euler 1–problem. Note that the viscosity of the LF scheme can be reduced, if it is used in the ER scheme ($\hat{=}$ ER–LF).

give very similar results, while the LF (Lax–Friedrichs) scheme produces a larger error and a lower EOC. As the results of the ER–DW and the RGDW scheme have been very similar in all our tests, we have only included results for the RGDW scheme in the following. For the computations shown in Fig. 3 we have used a cartesian table with K^2 pressure values from (10). To obtain p, its derivatives and $\varepsilon(\tau, p)$ at an arbitrary location, we have introduced a bilinear interpolation of the

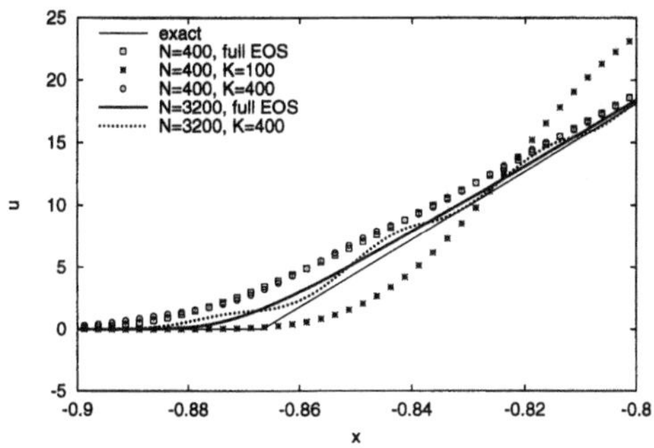

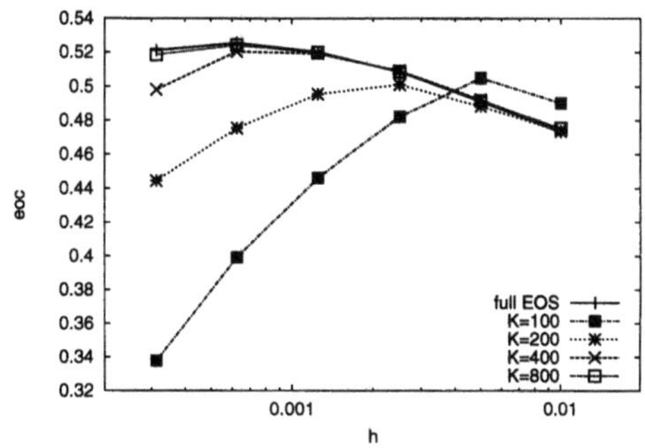

FIGURE 3. RGDW results for the Euler 1–problem with a tabularized EOS.

tabularized values. The results show that the table resolution K cannot be chosen independently of the number of grid cells $N \sim 1/h$ in the finite volume scheme: If K is fixed, grid refinement leads to oscillations in the numerical solution (Fig. 3 left) and the convergence rates decrease (Fig. 3 right).

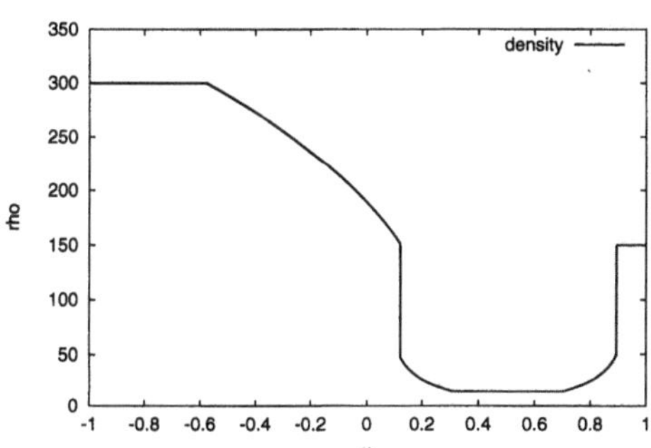

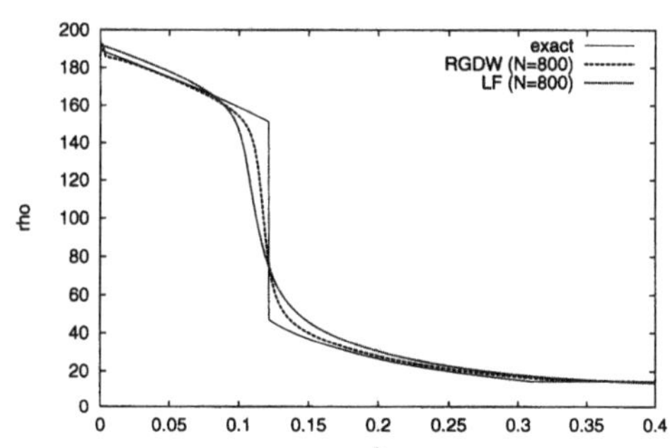

FIGURE 4. Density of the exact solution to the Euler 2–problem (left) and resolution of the left compound wave (right).

The "Euler 2"–problem has been constructed by the authors of [8], who have also provided the exact solutions for both Euler Riemann problems. In this example the maximal number of two compound waves is present, see Fig. 4 (left). Figs. 4 (right) and 5 show that the schemes are capable of resolving these compound waves. Again the LF scheme introduces much more viscosity.

5.2. MHD

We now turn our attention to the MHD Riemann problem given in Tab. 3. Since it has been constructed by attaching simple waves, we know its exact solution (Fig. 6 left). Again we have plotted numerical results (Fig. 6 right) and the L^1–error and the EOC for a number of grids (Fig. 7). The overall performance of the schemes is comparable to the Euler case.

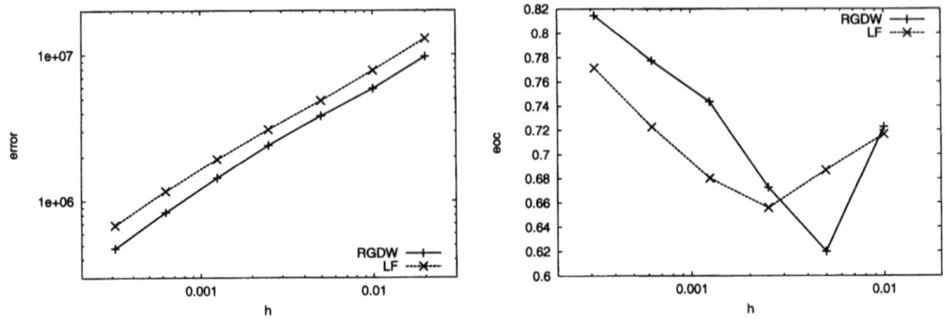

FIGURE 5. Comparison of accuracy (left) and convergence rates (right) for the Euler 2–problem.

	MHD Riemann problem ($t_{disp} = 0.001$)							
	ρ	u_x	u_y	u_z	B_x	B_y	B_z	p
$x < 0$	250.0	0.0	0.0	0.0	6175.2	5472.7	0.0	35966778.0
$x > 0$	179.0	−1.7	5.6	0.0	6175.2	4941.3	0.0	26476136.8

TABLE 3. Initial conditions and displayed time in plots t_{disp} for MHD problem.

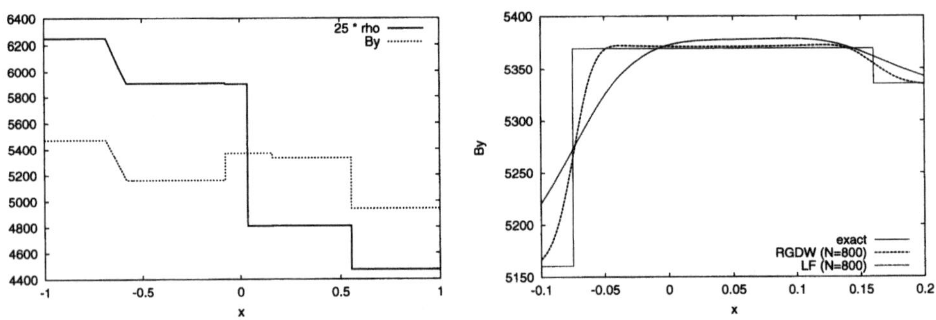

FIGURE 6. MHD Riemann problem: Density and B_y (left), section of the numerical solutions (right).

6. Conclusions

We have demonstrated that the investigated real gas MHD schemes resolve compound waves correctly and converge to the right solution. If the table and the grid of the finite volume scheme are refined simultaneously, they converge even if a tabularized EOS is used. The schemes obtained with the energy relaxation and direct Riemann solver approach yield comparable results, while they pose different constraints on the EOS. The ER scheme can be easily implemented as add–on to a given ideal gas scheme, but introduces an additional stability constraint. On

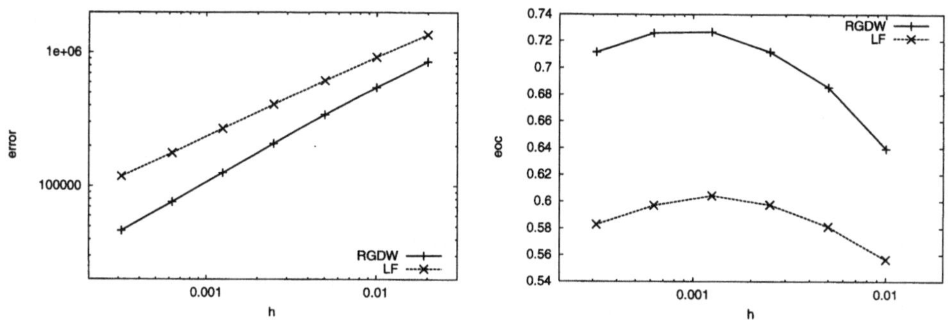

FIGURE 7. MHD Riemann problem: Comparison of accuracy (left) and convergence rates (right). Note that the results of the ER scheme are not shown, since they are very similar to those of the RGDW scheme.

the other hand the RGDW scheme is also not difficult to implement and is only subject to the usual CFL stability condition, but requires more calls to the EOS.

In a forthcoming paper, we will extend our studies of real gas MHD problems. Moreover we will focus on higher order versions of the schemes using MUSCL reconstruction. (A WENO–based extension of the ER scheme is presented in [7].) We will also study the accuracy of the schemes *relative* to their computational costs — we have refrained from including such data for the van der Waals EOS, since its evaluation is too simple to allow a *meaningful* comparison. A further issue will be the simulation of realistic astrophysical problems.

Acknowledgement

The authors were supported by the DFG priority research program ANumE.

References

[1] T. Buffard, T. Gallouët, and J. M. Herard, *A sequel to a rough Godunov scheme: Application to real gases*, Comput. Fluids (2000), no. 29, 813–847.

[2] F. Coquel and B. Perthame, *Relaxation of energy and approximate Riemann solvers for general pressure laws in fluid dynamics*, SIAM J. Numer. Anal. **35** (1998), no. 6, 2223–2249.

[3] W. Dai and P. R. Woodward, *A simple Riemann solver and high-order Godunov schemes for hyperbolic systems of conservation laws*, J. Comput. Phys. **121** (1995), no. 1, 51–65.

[4] A. Dedner, D. Kröner, C. Rohde, and M. Wesenberg, *Godunov–type schemes for the MHD equations*, to appear in Godunov Methods: Theory and Applications (edited by E. F. Toro; Kluwer), 2001.

[5] B. Einfeldt, C. D. Munz, P. L. Roe, and B. Sjögreen, *On Godunov–type methods near low densities*, J. Comput. Phys. **92** (1991), no. 2, 273–295.

[6] R. Menikoff and B. Plohr, *The Riemann problem for fluid flow of real materials*, Rev. Mod. Physics **61** (1989), no. 1, 75–120.

[7] P. Montarnal and C.-W. Shu, *Real gas computation using an energy relaxation method and high-order WENO schemes*, J. Comput. Phys. **148** (1999), no. 1, 59–80.

[8] S. Müller and A. Voß, *A Riemann solver for the Euler equations with non–convex equation of state*, Bericht 168, IGPM, RWTH Aachen, 1999.

[9] M. Wesenberg, *A note on MHD Riemann solvers*, Preprint 35, Albert–Ludwigs–Universität, Mathematische Fakultät, Freiburg, November 2000.

A. Dedner, M. Wesenberg
Institut für Angewandte Mathematik,
Albert–Ludwigs–Universität Freiburg,
Hermann–Herder–Str. 10,
D–79104 Freiburg / Breisgau, Germany
`dedner|wesenber@mathematik.uni-freiburg.de`

International Series of Numerical Mathematics
Vol. 140,

Towards a Kinetic Model of Turbulent Incompressible Fluids

P. Degond and M. Lemou

Abstract. A turbulence model for incompressible fluids is derived from kinetic theory. The kinetic model involves a relaxation time type collision operator which describes the relaxation of the probability distribution function (pdf) towards an isotropic pdf on a time scale τ. The dependence of τ upon the kinetic turbulent energy can be tuned in such a way that both the so called "viscous subrange" (dominated by molecular viscosity) and the "inertial range" (obeying the Kolmogorov law) can be described.

Key words: Boltzmann equation, Euler equation, Navier-Stokes equation, viscosity, Chapman-Enskog expansion, Turbulence.

AMS Subject classification: 41A60, 35Q20, 76P05, 35C20, 35L65, 76F99, 76N99.

Acknowledgments: This work has been supported by the TMR network No. ERB FMBX CT97 0157 on 'Asymptotic methods in kinetic theory' of the European Community.

1. Introduction: the case of inviscid fluids

The aim of this paper is to derive a turbulence model for viscous incompressible fluids from kinetic theory. In particular, we want to design a model able to take into account both the viscous subrange dominated by molecular viscosity and the inertial range in which the energy dissipation follows the Kolmogorov law. For an introduction to turbulence modeling, we refer the reader to the basic textbooks [9], [10], [8] and to the lecture notes [4]. The approach that we are going to use bears a lot of similarities with the so-called 'probability distribution function (pdf)' approach which has been widely developed in turbulence modeling. However, the usual pdf approach makes a heavy use of stochastic concepts (see the review paper [11]), which we are going to conceal in the definition of suitable collision operators. Our starting point is the following kinetic model:

$$\frac{\partial f}{\partial t} + v \cdot \nabla_x f - \nabla_x \mathcal{P} \cdot \nabla_v f = \frac{1}{\eta} Q(f, u_f) \,, \quad \int_{\Re^d} f \, dv = 1 \,, \tag{1}$$

where $f(t,x,v)$ is the probability distribution function (pdf) of fluid structures, depending on time $t \geq 0$, space $x \in \mathfrak{R}^d$, and velocity $v \in \mathfrak{R}^d$ (with $d = 1,2,3$), while $\mathcal{P}(x,t)$ is the hydrodynamic pressure of the fluid.

When the collision operator Q is set to 0, the model reduces to

$$\frac{\partial f}{\partial t} + v \cdot \nabla_x f - \nabla_x \mathcal{P} \cdot \nabla_v f = 0\,, \quad \int_{\mathfrak{R}^d} f\,dv = 1, \tag{2}$$

and has been proposed by Y. Brenier and E. Grenier [2] as a kinetic model for inviscid incompressible fluids. Indeed, from the compressible Euler equations, the fluid velocity $\bar{u}(x,t)$ and the pressure $\mathcal{P}$ satisfy:

$$\frac{\partial \bar{u}}{\partial t} + (\bar{u} \cdot \nabla_x)\bar{u} + \nabla_x \mathcal{P} = 0\,, \quad \nabla_x \cdot \bar{u} = 0\,. \tag{3}$$

Equ. (3) possesses a Lagrangean formulation: the position $x(t)$ of any given fluid element evolves according to:

$$\frac{dx}{dt} = \bar{u}(x(t),t)\,, \quad \frac{d}{dt}[\bar{u}(x(t),t)] = -\nabla_x \mathcal{P}(x(t),t)\,. \tag{4}$$

The ordinary differential system (4) appears as the system of characteristic equations of the first order PDE (2). In other words, equ. (2) is the Liouville equation of (4). One can view the connection between (2) and (3) another way, by looking at the moment system of (2). Indeed, defining the average velocity $\bar{u}$ asociated with f according to

$$\bar{u}(t,x) = \int_{\mathfrak{R}^d} v f(t,x,v)dv. \tag{5}$$

and taking the first two velocity moments of (2), we get:

$$\frac{\partial \bar{u}}{\partial t} + (\bar{u} \cdot \nabla_x)\bar{u} + \nabla_x \left(\mathcal{P} + \int_{\mathfrak{R}^d} (v-\bar{u})(v-\bar{u}) f dv \right) = 0\,, \quad \nabla_x \cdot \bar{u} = 0\,. \tag{6}$$

The model (6) can be viewed as as a turbulent Euler equation model where the tensor $R = \int (v-\bar{u})(v-\bar{u}) f dv$ is the Reynolds tensor of second order velocity correlations, expressing the influence of the probabilistic nature of the flow on the average motion of the fluid. The pressure $\mathcal{P}$ now appears as the Lagrange multiplier of the constraint $\int f\,dv = 1$.

It must be noted that model (2) is not always well-posed. There exist sufficient conditions on the initial data $f|_{t=0}$ which guarantee the well-posedness of the problem for small times but examples of initial data for which the problem is ill-posed can also be exhibitted. This question has been investigated in [3], [1]. In the present work, we shall assume that the initial data are such that the problem is well-posed.

To extend the model (2) to viscous fluids, collision operators must be added. The goal of the present paper is to propose a collision operator which to some extent describes the phenomenology of the interaction of fluid structures. It leads to the standard Navier-Stokes equation when there is no turbulence, but it incorporates appropriate modifications of the viscosity when the flow is turbulent. We present such a collision model in the next section.

2. Viscous fluids

We shall consider a collision operator which describes the following phenomenology: first, the collision operator tends to relax the distribution function towards a state of 'maximal chaos', such a state being characterized by an isotropic pdf about a mean velocity u_f, depending on f, to be specified below. Indeed, we assume that the effect of turbulence is to quickly isotropize the distribution function. Obviously, such an assumption is not valid in the case of strongly anisotropic turbulence. However, the isotropy constraint can be relaxed to some extent, as we will develop in future work. The relaxation process takes place over a certain time $\tau(\xi)$, which we shall assume dependent on the particle kinetic energy $\xi = |v - u_f|^2/2$ measured in the frame moving with velocity u_f. Finally, the collision operator is total momentum-preserving, as the interaction of fluid structures obviously obeys galilean invariance.

The simplest collision operator which satisfies the above defined constraints is of the following form:

$$Q(f, u_f) = \frac{1}{\tau} L_{u_f} f\,, \tag{7}$$

with

$$L_u f(v) = \Pi_u(f)(v) - f(v) = \frac{1}{|\mathfrak{S}^{d-1}|} \int_{\mathfrak{S}^{d-1}} f(u + |v - u|\omega')d\omega' - f(v)\,. \tag{8}$$

Π_u is the orthogonal projector (in L^2) onto the space of functions depending on $|v - u|^2/2$ only. Therefore $\Pi_u(f) = \Pi_u(f)(|v - u|^2/2)$. The collision time τ may depend functionnally on f provided that it only depends on v through $\xi = |v - u_f|^2/2$. The mean velocity u_f is defined such that the collision operator is momentum-preserving. This yields the following definition of u_f [7]:

$$\int_{\mathfrak{R}^d} (v - u_f)\, f \,\frac{dv}{\tau(|v - u_f|^2)} = 0\,, \tag{9}$$

which is actually different from the usual mean velocity $\bar{u}_f$, given by (5) or by (9) with $\tau = 1$. We also refer to [7] for the various properties of the collision operator (conservation, null-space, entropy relation, etc). In particular, it satisfies:

$$\int_{\mathfrak{R}^d} Q(f)(v) \begin{pmatrix} 1 \\ v \\ v^2 \end{pmatrix} dv = 0. \tag{10}$$

As a consequence of the first line of (10), we recover the incompressibility condition $\nabla_x \cdot \bar{u}_f = 0$ for the mean velocity $\bar{u}_f$.

The collision operator (7) is also used in the modeling of transport of cosmic rays [12], [14]. Some of its mathematical properties and associated macroscopic limits have been studied in [5], [6], [7] and [13].

In the next section, we perform the limit $\eta \to 0$ and obtain the associated macroscopic model.

3. Macroscopic limit

In the limit $\eta \to 0$, the pdf is constrained to be close to the null space of the collision operator Q and is therefore close to a function of the form $g^\eta(t,x,|v-u^\eta(x,t)|^2/2)$, where $g^\eta(t,x,\xi)$ and $u^\eta(x,t)$ satisfy a system of partial differential equations. It is easily seen that if we restrict to the leading order as $\eta \to 0$, the equation satisfied by u is nothing but the incompressible inviscid Euler equations (3). Therefore, we keep the leading and first order terms in η in the equations for g and u, using a Chapman-Enskog methodology, detailed in [7]. This gives rise to the following system, where we omit the dependences in η of the various unknowns:

$$\left\{ \begin{array}{l} \dfrac{\partial g}{\partial t} + \bar{u}\cdot\nabla_x g = \eta \Big\{ -c\cdot\nabla_x g + (\nabla_x\cdot c)\,\dfrac{2\xi}{d}\,\dfrac{\partial g}{\partial \xi} + \dfrac{2}{d}\,\xi^{1-\frac{d}{2}}\,\bar{\nabla}\left(\tau(\xi,g)\xi^{\frac{d}{2}}\bar{\nabla} g\right) \\ \qquad + \dfrac{4}{d(d+2)}\,\xi^{1-\frac{d}{2}}\,\dfrac{\partial}{\partial\xi}\left(\tau(\xi,g)\xi^{\frac{d+2}{2}}\dfrac{\partial g}{\partial \xi}\right)\,(\sigma(\bar{u}):(\nabla_x\bar{u}))\Big\}\,, \\ \dfrac{\partial \bar{u}}{\partial t} + (\nabla_x\bar{u})(\bar{u}) + \nabla_x\left(\mathcal{P} + \dfrac{2k}{d}\right) = \eta\,\nabla_x\cdot(\mu\sigma(\bar{u}))\,, \\ \nabla_x\cdot\bar{u} = 0\,, \end{array} \right. \tag{11}$$

The differential operator $\bar{\nabla}$ is a 'twisted' gradient in the extended (position-energy) space given by

$$\bar{\nabla} = \nabla_x + \nabla_x\left(\frac{2k}{d}\right)\frac{\partial}{\partial\xi}\,,$$

where k and $\sigma(u)$ respectively denote the kinetic turbulent energy and the rate-of-strain tensor:

$$k(x,t) = \int_0^\infty \xi\, g(x,\xi,t)\,dp(\xi)\,, \quad \sigma(u) = \nabla_x u + (\nabla_x u)^T, \tag{12}$$

(the exponent T meaning the transpose of a matrix). The (turbulent) viscosity μ and the vector c which measures the difference between the two mean velocities $c = (u_f - \bar{u}_f)/\eta$, are given by:

$$\mu = \mu[\tau,g] = \frac{-4}{d(d+2)}\int_0^\infty \xi^2\,\tau(\xi,g)\,\frac{\partial g}{\partial\xi}\,dp(\xi)\,, \quad c = \int_0^\infty \frac{2\xi}{d}\,\tau(\xi,g)\,\bar{\nabla} g\,dp(\xi)\,. \tag{13}$$

We recall that $dp(\xi) = |\mathfrak{S}^{d-1}||2\xi|^{(d-2)/2}d\xi$, where $|\mathfrak{S}^{d-1}|$ denotes the measure of the unit $(d-1)$-dimensional sphere and that the operator $A:B$ means the contracted product of the tensors A and B. Finally, note that the incompressibility constraint (last equation (11)) is equivalent to saying that g is a pdf:

$$\int_0^\infty g\,dp(\xi) = 1\,.$$

The structure of model (11) is best understood if one looks at the first moment equation, which is nothing but the equation for the evolution of the turbulent energy density k:

$$\frac{\partial k}{\partial t} + \bar{u}\cdot\nabla_x k = \eta\,(\,\mu\,\sigma(\bar{u}):\nabla_x\bar{u}\;-\;\nabla_x q\,)\,, \tag{14}$$

where the diffusion flux vector q is given by:

$$q = -\frac{2}{d} \int_0^\infty \xi^2 \, \tau \, \bar{\nabla} g \, dp(\xi) \, + \, \left(\frac{d+2}{d} \right) ck \, . \tag{15}$$

It is seen that the structure of the k equation is quite similar to that derived from the statistical approach of turbulence (see e.g. [10], [4]). Indeed, the turbulent energy is convected by the fluid velocity at the left-hand side of equ. (14). The first term at the right-hand side of (14) is a production term describing how turbulence is produced by the gradients in the mean-velocity of the flow. The second term is a diffusion term expressed in terms of the diffusion flux q. The major characteristics of the present model is the dependence of the viscosity μ and the diffusion flux q upon the pdf through the fairly complex expressions (13) and (15).

There is however an important omission in model so far: turbulence dissipation. Indeed, if there is no action on the system, turbulence is dissipated and the fluid is eventually brought to a laminar state. In the present model, absence of turbulence means that the turbulent energy ξ is surely 0, which in terms of the pdf is expressed by $f(v) = \delta(v - u_f)$ or $g(\xi) = \delta(\xi)$. The simplest way to introduce dissipation in the kinetic equation (1) is by adding a relaxation collision operator of the form

$$Q_{\text{diss}}(f, u_f) = -\frac{1}{\tau_D} f \, + \, \left(\int_{\Re^d} f(v) \frac{1}{\tau_D} dv \right) \delta(v - u_f) \, , \tag{16}$$

where τ_D is a relaxation time for dissipation and δ is the usual Dirac delta distribution. By construction, the operator Q_{diss} preserves mass (or equivalently the probability nature of f), i.e. $\int Q_{\text{diss}}(f, u_f) dv = 0$. In addition, it must conserve momentum, which implies a constraint of the form (9) with τ replaced by τ_D. The only way to satisfy the constraint with the same u_f for any f is to suppose that the two relaxation times are proportional. For reasons of simplicity, we therefore choose them equal: $\tau = \tau_D$.

Then, we consider the following modified kinetic equation:

$$\frac{\partial f}{\partial t} + v \cdot \nabla_x f - \nabla_x \mathcal{P} \cdot \nabla_v f = \eta \, Q_{\text{diss}}(f, u_f) + \frac{1}{\eta} Q(f, u_f). \tag{17}$$

The reason for scaling the dissipation collision operator Q_{diss} like $O(\eta)$ is that we are looking for dissipation terms in the macroscopic model of the same order as the production or diffusion terms. The g equation in model (11) is therefore modified into:

$$\frac{\partial g}{\partial t} + \bar{u} \cdot \nabla_x g = \eta \Bigg\{ -c \cdot \nabla_x g \, + \, (\nabla_x \cdot c) \, \frac{2\xi}{d} \frac{\partial g}{\partial \xi} \, + \, \frac{2}{d} \xi^{1-\frac{d}{2}} \, \bar{\nabla} \left(\tau(\xi, g) \xi^{\frac{d}{2}} \bar{\nabla} g \right)$$

$$+ \frac{4}{d(d+2)} \xi^{1-\frac{d}{2}} \frac{\partial}{\partial \xi} \left(\tau(\xi, g) \xi^{\frac{d+2}{2}} \frac{\partial g}{\partial \xi} \right) \, (\sigma(\bar{u}) : (\nabla_x \bar{u})) \, + \, \tilde{Q}_{\text{diss}}(g) \Bigg\} \, , \tag{18}$$

where $\tilde{Q}_{\text{diss}}(g)$ is given by

$$\tilde{Q}_{\text{diss}}(g) = -\frac{1}{\tau} g \, + \, \left(\int_0^\infty \frac{1}{\tau} \, g \, dp(\xi) \right) \left(\frac{dp}{d\xi} \right)^{-1} \delta(\xi). \tag{19}$$

The k equation now reads:

$$\frac{\partial k}{\partial t} + \bar{u} \cdot \nabla_x k = \eta \left(\mu \sigma(\bar{u}) : \nabla_x \bar{u} \; - \; \nabla_x q \; - \; \epsilon \right) , \tag{20}$$

where ϵ is the dissipation rate of turbulent energy, related to g by:

$$\epsilon = \int_0^\infty \frac{\xi}{\tau} \, g \, dp(\xi) \, . \tag{21}$$

The quantity ϵ plays an important role in Kolmogorov's theory (see e.g. [10]). Also, the widely used (k, ϵ) turbulence model consists in coupling the standard Navier-Stokes equation with evolution equations for k and ϵ, through a suitable expression of the turbulent viscosity. The equations for k and ϵ are derived from statistical averages of the fluctuating component of the velocity in the Navier-Stokes equation using phenomenological closure relations. The present model however provides direct expressions for k and ϵ (see (12) and (21)), provided that equ. (18) for the function g is solved.

Now, there remains to say how one can choose the collision time τ such that the model degenerates onto the standard Navier-Stokes equation with molecular viscosity ν when the fluid is laminar (i.e. not turbulent), and such that it accounts for the Kolmogorov law when the fluid is turbulent. We recall that the Navier-Stokes equation with molecular viscosity ν is written:

$$\frac{\partial \bar{u}}{\partial t} + (\bar{u} \cdot \nabla_x)\bar{u} + \nabla_x \mathcal{P} \; = \; \nu \Delta \bar{u} \, , \quad \nabla_x \cdot \bar{u} = 0 \, . \tag{22}$$

This task is achieved in the next section.

4. Model for the collision time τ

We first remark that if

$$\tau(\xi) = \frac{\alpha}{\xi}, \tag{23}$$

the viscosity μ becomes independent of g and given by $\mu = 2\alpha/(d+2)$. Therefore, if τ is given by (23) with $\alpha = (d+2)\nu/2$, ν being the molecular viscosity of the fluid, the $\bar{u}$ equation in (11) uncouples from the g equation and reduces to the standard Navier-Stokes equation (22). We denote by τ_M the function (23) with the so-determined value of α. Since the non-turbulent regime is achieved in the limit $g \to \delta(\xi)$, the classical Navier-Stokes equation will be recovered in this regime as soon as the actual collision time τ satisfies:

$$\tau(\xi) \sim \tau_M(\xi) := \frac{(d+2)\nu}{2} \, \frac{1}{\xi}, \quad \text{when} \quad \xi \to 0 \tag{24}$$

This is the first requirement we are going to impose on τ.

The second requirement on τ is that it should give rise to the Kolmogorov law in the inertial range. The statement of the Kolmogorov law, which strickly speaking is only valid in the case of homogeneous isotropic turbulence, is as follows. Let us consider the inertial range, characterized by space scales κ^{-1} (κ being the associated wave number) larger than a threshold scale κ_D^{-1}. In this range, the molecular viscosity is unable to dissipate the kinetic turbulent energy, because the space gradients are not large enough. Therefore, the spectral density of turbulent energy $E(\kappa)d\kappa$ can only depend on the wave-number κ and on the turbulent energy dissipation rate ϵ. Then by a simple dimensional analysis outcomes the Kolmogorov law:

$$E(\kappa) = C_K\, \epsilon^{2/3}\, \kappa^{-5/3}. \tag{25}$$

where C_K is a constant, experimentally close to 1.5. The Kolmogorov law has been verified in a very large number of experimental situations. For a thorough review about the Kolmogorov law, we refer to the bibliography given in section 1.

We need to connect the spectral variable κ to the energy variable ξ of our model. To that aim, we notice that the quantity $L(\xi) = \sqrt{2\xi}\tau(\xi)$ can be viewed as the largest size a fluid structure of energy ξ can achieve before being dissipated. Therefore, to this scale $L(\xi)$, we associate a wave vector $\kappa(\xi) = L^{-1}(\xi)$, i.e.

$$\kappa(\xi) = \frac{1}{\sqrt{2\xi}\,\tau(\xi)}\,. \tag{26}$$

Using this identification, we can characterize the spectral density of turbulent energy $E(\kappa)$ in terms of the pdf g according to

$$E(\kappa)\, d\kappa \;=\; \xi\, g\, dp(\xi)\,. \tag{27}$$

Inserting (26) into (27) and using the Kolmogorov law (25), we obtain:

$$C_K\, \epsilon^{2/3}\, \kappa'(\xi)\, \kappa(\xi)^{-5/3} \;=\; \xi\, g(\xi)\, \frac{dp}{d\xi}\,. \tag{28}$$

This relation, together with the assumption that in the inertial range, the size of the structure $L(\xi)$ increases with increasing energy ξ leads to the following equation:

$$L(\xi) = \sqrt{2\xi}\,\tau(\xi) = \left(\frac{2}{3C_K}\int_0^{\xi} \xi\, g(\xi)\, dp(\xi)\right)^{3/2} \frac{1}{\epsilon}\,. \tag{29}$$

Relation (29) provides a closed expression of $\tau(\xi)$ in terms of the pdf g and of the dissipation rate ϵ. We shall denote by τ_K the so-obtained value of the collision time. However, this relation is only valid in the inertial range and must be connected in some way with the value τ_M associated with molecular viscosity and prevailing in the viscous range. Note that $L_M(\xi) = \sqrt{2\xi}\tau_M(\xi) = \sqrt{2}\alpha/\sqrt{\xi}$ decreases from $+\infty$ to 0 as ξ increases from 0 to ∞ while $L_K(\xi) = \sqrt{2\xi}\tau_K(\xi)$ is an increasing function of ξ. Therefore, there is a unique point ξ_D such that $L_M(\xi_D) = L_K(\xi_D)$. For $\xi < \xi_D$, τ_M dominates over τ_K while the converse is true for $\xi > \xi_D$. Therefore ξ_D appears as a threshold energy : lower energies correspond to scales belonging to the viscous range and dominated by molecular diffusion, while larger energies are associated with scales belonging to the inertial range and well described by the Kolmogorov law. The actual collision time τ to be used must be asymptotic to τ_M for small energies and to τ_K for large energies.

There remains quite a large flexibility in the choice of τ and one can choose for instance

$$\tau_1 = \sup(\tau_M, \tau_K), \qquad \text{or} \qquad \tau_2 = \tau_M + \tau_K . \tag{30}$$

In the remainder of the paper, we shall restrict to these two choices.

It is important to notice that (30) does not give an explicit expression for τ because ϵ also depends on τ through (12). The definite value of τ is obtained through a fixed point procedure: let g and ϵ be given and define τ_1 or τ_2 according to (30). Then, compute a new value $\epsilon_1(\epsilon)$ or $\epsilon_2(\epsilon)$ by means of (12). Then, the actual value of ϵ to be used is that realizing the fixed point of the application ϵ_1 or ϵ_2 (according to which formula for τ we want to use). In the following proposition, we prove that, under some restriction on the Kolmogorov constant C_K, there exists a unique fixed point, and therefore, a unique τ_1 or τ_2 satisfying the problem.

Proposition 4.1. *If C_K satisfies the following hypothesis:*

$$C_K > \frac{2^{2/3}}{3} \tag{31}$$

Then,
(i) there exist a unique real $\xi_D > 0$ and a unique continuous function τ_1 of the form (30),
(ii) there exists a unique function τ_2 of the form (30),
such that τ_K (for $s = 1$ or 2) satisfies the following implicit equation:

$$\tau_K(\xi) = \left(\sqrt{2\xi}\int_0^\infty \frac{\xi}{\tau_s(\xi)}\, g(\xi)\, dp(\xi)\right)^{-1} \left(\frac{2}{3C_K}\int_0^\xi \xi\, g(\xi) dp(\xi)\right)^{3/2} . \tag{32}$$

We note that the experimentally determined value of the Kolmogorov constant C_K is 1.5 while the lower bound $2^{2/3}/3$ is about 0.529. This means that the present approach is consistent with the physical constraints.

Proof. We just give the proof of assertion (ii). The proof of (i) is similar but a bit longer. Let h be the following function:

$$h(\xi) = \frac{1}{\sqrt{2\xi}}\left(\frac{2}{3C_K}\int_0^\xi \xi\, g(\xi)\, dp(\xi)\right)^{3/2} . \tag{33}$$

Relation (30) for τ_2 is then equivalent to

$$\tau_2(\xi) = \frac{\alpha}{\xi} + h(\xi)\left(\int_0^{+\infty} s\, g(s)\, \tau_2^{-1}(s)\, dp(s)\right)^{-1} . \tag{34}$$

where h is defined by (33). In particular τ_2 has the form $\tau_2(\xi) = \alpha/\xi + \lambda h(\xi)$, where λ is a constant to be determined. We insert this expression into (34) and obtain that λ has to satisfy:

$$H(\lambda) := \int_0^{+\infty} \lambda\left(\lambda\, h(s) + \alpha s^{-1}\right)^{-1} s\, g(s)\, dp(s) = 1 . \tag{35}$$

Now we observe that the function H is an increasing function of $\lambda \in [0, +\infty[$, that $H(0) = 0$ and $H(+\infty) = \int s g(s) h(s)^{-1} dp(s)$. We deduce that a necessary and sufficient condition for the existence of a unique solution to (35) is $H(+\infty) > 1$. This condition is equivalent to:

$$C_K > \frac{2^{2/3}}{3} \left[\int_0^{+\infty} \xi^{3/2} g(\xi) \left(\int_0^{\xi} s\, g(s)\, dp(s) \right)^{-3/2} dp(\xi) \right]^{-2/3} . \tag{36}$$

Now we remark that the bracketed quantity in (36) is larger or equal to 1 because $g \geq 0$ and $\int g(\xi) dp(\xi) = 1$. Then, a sufficient condition for the existence and the uniqueness of a continuous function τ_1 having the form (30) is given by the lower bound (31), which concludes the proof. □

5. Conclusion

We have presented a kinetic model for a turbulent incompressible flow. The model provides a coupled system of equations for the average velocity of the fluid and for the probability distribution function of fluid structures. It relies on the determination of the typical time of interaction of fluid structures τ. This time can be calibrated in such a way that the model degenerates onto the standard Navier-Stokes equation in the case of a non turbulent fluid. At variance, when the fluid is turbulent, the model predicts values of the turbulent viscosity which are directly related to the probability distribution function. In this case, it is possible to assign values to τ that guarantee the consistency of the model with the Kolmogorov law.

References

[1] Y. Brenier, *Convergence of the Vlasov-Poisson system to the incompressible Euler equation*, to appear in Comm. PDE.

[2] Y. Brenier and E. Grenier, *Limite singulière du système de Vlasov-Poisson dans le régime de quasineutralité*, C. R. Acad. Sci. Paris, **318**, Série I, (1994), pp. 121–124.

[3] Y. Brenier and L. Corrias, *A kinetic formulation for multi-branch entropy solutions of scalar conservation laws*, Ann. Inst. H. Poincaré (Ann. Non Linéaire), **15** (1998), pp. 169–190.

[4] P. Chassaing, *Turbulence en mécanique des fluides*, lecture notes, ENSEEIHT, Toulouse, France.

[5] P. Degond, J. L. Lòpez and P. F. Peyrard, *On the macroscopic dynamics induced by a model wave-particle collision operator*. J. Cont. Mech. Therm. **10** (1998), pp. 153-178.

[6] P. Degond, J. L. Lòpez, F. Poupaud and C. Schmeiser, *Existence of solutions of a kinetic equation modeling cometary flows*, J. Stat. Phys., **96** (1999), pp. 361–376.

[7] P. Degond and M. Lemou, *On the Viscosity and thermal conduction of fluids with mutivalued internal energy.*, Eur. J. Mech. B-Fluids, **20** (2001), pp. 303-327.

[8] B. E. Launder, D. B. Spalding, *Mathematical models for turbulence*. Academic Press. 1972.

[9] M. Lesieur, *Turbulence in fluids - Stochastic and numerical modeling*, Kluwer, 1990.

[10] B. Mohammadi and O. Pironneau, *Analysis of the K-Epsilon turbulence model*, Masson and Wiley, New-York, 1993.

[11] S. B. Pope, *Turbulent Flows*, Cambridge University Press. 2000.

[12] L. L. Williams and J. R. Jokipii, *Viscosity and inertia in cosmic-ray transport: effects of an average magnetic field*, The Astrophysical Journal **371** (1991), pp. 639-647.

[13] L. Saint Raymond, *Incompressible hydrodynamic limit of a kinetic model of wave-particles interaction*, preprint LMENS-98-50, École Normale Supérieure, Paris, dec. 98.

[14] L. L. Williams, N. Schwadron, J. R. Jokipii and T. I. Gombosi, *A unified transport equation for both cosmic rays and thermal particles*, The Astrophysical Journal **405** (1993), pp. L79-L81.

P. Degond
CNRS, UMR MIP 5640, Université Paul Sabatier,
118 Route de Narbonne,
31062 Toulouse Cedex, France
`degond@mip.ups-tlse.fr`

M. Lemou
CNRS, UMR MIP 5640, Université Paul Sabatier,
118 Route de Narbonne,
31062 Toulouse Cedex, France
`lemou@mip.ups-tlse.fr`

International Series of Numerical Mathematics
Vol. 140,

Parabolic Relaxation of Semilinear Multidimensional Hyperbolic Systems

Donatella Donatelli and Pierangelo Marcati

Abstract. We study the relaxation of multi-D semilinear hyperbolic systems to parabolic systems. The singular limits are studied combining Tartar's and Gérard's generalized compensated compactness and by using the properties of the pseudodifferential symmetrizer of the system.

1. Introduction

In this note we consider the following semilinear multidimensional hyperbolic system with a small positive parameter ε

$$W_t(x,t) + \frac{1}{\varepsilon}\sum_{j=1}^{d} A_j(x)\partial_j W(x,t) = \frac{1}{\varepsilon^2}B(x,W(x,t)) + \frac{1}{\varepsilon}D(W(x,t)) \qquad (1)$$

where $W = W(x,t)$ takes values in $\mathbf{R}^N$, $x \in \mathbf{R}^d$, $t \geq 0$, $A_j(x)$, $j = 1,\ldots,d$ are $N\times N$ matrices for any $x \in \mathbf{R}^d$. Our aim is to understand the limiting behaviour of the system (1) as ε goes to zero. We will look for structure condition in order that (1) will approximate a second order parabolic system. Our interest in this problem is also motivated since it has a very strong similarity with the limiting structure appearing in the investigation of the hydrodynamic limit for the Boltzmann equation, in particular in the discrete velocity case. The Boltzmann equation describes the evolution of the density $f(x,\xi,t)$ of particles that are at time t in position x with velocity ξ and has the following form

$$\nu f_t + \xi\nabla\cdot f = \frac{1}{\varepsilon}Q(f,f)$$

where ν is the Mach number and ε the Knudsen number. By averaging $f(x,\xi,t)$ in ξ and by using higher order momenta we can define a hierarchy of macroscopic quantities. The investigation of the hydrodynamic limit regards the behaviour of those quantities (actually combined with the closure problem) as the Knudsen number goes to zero. In the case where the Mach number is of the same order of the Knudsen number our limit can be described by the Navier Stokes equation, otherwise when the Mach number is fixed and the Knudsen number tends to zero

Partially supported by the European Union TMR Project HCL #ERBFMRXCT960033 we end up with the Euler equation. The former limits can be seen as a hyperbolic to parabolic singular convergence and the latter one as an example of hyperbolic to hyperbolic limit. The relaxation problems are closely related but simpler than the hydrodynamic limits, since they do not need to overcome the difficulties of the closure problem. In particular if we deal with a discrete velocity models, the equivalent " Boltzmann equation" is a semilinear hyperbolic system. Such kind of singular limits arise also in many other situations like the porous media flow [11], the nonlinear heat conduction [12]. All of these papers, make use of the techniques of compensated compactness. Similar ideas have been applied by Marcati and Rubino [13] to show the general theory for 2×2 systems in the 1-D case and to propose a general framework that we are going to investigate here in the semilinear system case. The methods of [13] have been used in the recent paper of [8] to get the convergence of BGK type approximations to nondegenerate systems in 1-D. Preliminary results concerning semilinear systems have been obtained in [13] for systems with constant coefficients , further in [3], [4], the 1-D case with variable coefficients, has been considered . In [13], [3], the limiting process was performed using the methods of compensated compactness, while in [4] it has been necessary to use a generalization of this theory due to Tartar [17] and P.Gérard [6]. For a general theory in the pseudodifferential framework we refer to [5]

2. Multidimensional Framework

2.1. Preliminary notions

We start this section introducing the main notations and definitions used in the article. Therefore

- **(a)** $(\cdot,\cdot)$ denotes the scalar product in $\mathbf{R}^q$, $(q = 1, 2, ...)$ and $|\cdot|$ the usual norm of $\mathbf{R}^q$ $(q = 1, 2, ...)$ and $\mathcal{M}_{m\times n}$ denotes the linear space of $m \times n$ matrices,
- **(b)** $(\cdot,\cdot)_2$ denotes the scalar product in $L^2(\mathbf{R}^d)$ and $\|\cdot\|$ the norm in $L^2(\mathbf{R}^d)$,
- **(c)** $\mathcal{D}(\mathbf{R}^d \times \mathbf{R}_+)$ denotes the space of test function $C_0^\infty(\mathbf{R}^d \times \mathbf{R}_+)$, $\mathcal{D}'(\mathbf{R}^d \times \mathbf{R}_+)$ the Schwarz space of distributions and $\langle\cdot,\cdot\rangle$ the duality bracket in $\mathcal{D}'(\mathbf{R}^d \times \mathbf{R}_+)$,
- **(d)** H is a separable Hilbert space, $\mathcal{L}(H)$ the space of bounded operators, $\mathcal{K}(H)$ the space of compact operators,
- **(e)** we denote by $H^s_{loc}(\Omega, H)$ the classical local Sobolev space of order s, i.e. $u \in H^s_{loc}(\Omega, H) \Longleftrightarrow \forall\, \varphi \in C_0^\infty$, $(\widehat{\varphi u}) \in L^2(\mathbf{R}^n, (1 + |\xi|^2)^s d\xi)$.

Finally we state here our main tool to study the convergence of quadratic forms with variable coefficients: generalization of classical Compensated Compactness Theorem due to Tartar and Gérard ([17], [6])). Let H, $H^\sharp$ denote separable Hilbert spaces, $\Omega \in \mathbf{R}$, an open set. Take $m \in \mathbf{N}$, and, for every $\alpha \in \mathbf{N}^n$ such that $|\alpha| \le m$ take $a_\alpha \in C(\Omega, \mathcal{L}(H, H^\sharp))$ so that the formula

$$Pu(x) = \sum_{|\alpha|\le m} \partial^\alpha (a_\alpha(x) u(x)) \tag{2}$$

defines a differential operator $P : L^2_{loc}(\Omega) \longrightarrow H^{-m}_{loc}(\Omega, H^\sharp)$. Finally we denote by p the principal symbol of P, given by

$$p(x,\xi) = \sum_{|\alpha|=m} \xi^\alpha a_\alpha(x).$$

We have the following theorem [17], [6]

Theorem 2.1. (Compensated Compactness)
Let P defined by (2) and $\{u_k\}$ be a bounded sequence of $L^2_{loc}(\Omega, H)$, such that $u_k \rightharpoonup u$. Assume that there exists a dense subset $D \in H^\sharp$ such that, for any $h \in D$, the sequence $(\langle Pu_k, h\rangle)$ is relatively compact in $H^{-m}_{loc}(\Omega)$. Moreover, let $q \in C(\Omega, \mathcal{K}(H))$.
(i) *If $q = q^*$ and for all $(x,\xi,h) \in S^*\Omega \times H$, one has*

$$(p(x,\xi)h = 0) \Rightarrow (\langle q(x)h, h\rangle \geq 0)$$

then, for any nonnegative $\varphi \in C_0^\infty(\Omega)$

$$\liminf_{k\to\infty} \int_\Omega \varphi\langle q(x)u_k, u_k\rangle dx \geq \int_\Omega \varphi\langle q(x)u, u\rangle dx$$

(ii) *If for all $(x,\xi,h) \in S^*\Omega \times H$, one has*

$$(p(x,\xi)h = 0) \Rightarrow (\langle q(x)h, h\rangle = 0)$$

then $\langle q(x)u_k, u_k\rangle$ converges to $\langle q(x)u, u\rangle$ in $\mathcal{D}'(\Omega)$.

2.2. Formal analysis of the singular limit

We will consider the following semilinear system of equations

$$W_t(x,t) + \sum_{j=1}^d A_j(x)\partial_j W(x,t) = B(x, W(x,t)) + D(W(x,t)) \tag{3}$$

where $t \geq 0$, $x \in \mathbf{R}^d$, $W \in \mathbf{R}^N$. Moreover we assume the following hypotheses hold.

- **(A.1)** $A_j(x) \in \mathcal{M}_{N\times N}$, $j = 0, \ldots, d$,
- **(A.2)** $B(x,W), D(W) \in \mathbf{R}^N$,
- **(A.3)** the system (3) is hyperbolic, namely for all nonzero vector $\xi \in \mathbb{R}^d$, the matrix $A(x,\xi) = \sum_{j=1}^d \xi_j A_j(x)$ has real eigenvalues.

We denote by $S = span\left\{B(x,W) \mid W \in \mathbf{R}^N\right\}$ and we assume

- **(A.4)** $dim\ S = N - k$, $0 < k < N$.

As in [13] and similarly in [9] we split the vector of the conserved quantities Z^I from the remaining ones. With the previous conditions we can construct projection matrices which allow us to rewrite the system (3) in the following way

$$\begin{cases} Z_t^I + \sum_{j=1}^d M_j^{11}(x)\partial_j Z^I + \sum_{j=1}^d M_j^{12}(x)\partial_j Z^{II} = D^I(Z^I, Z^{II}) \\ Z_t^{II} + \sum_{j=1}^d M_j^{21}(x)\partial_j Z^I + \sum_{j=1}^d M_j^{22}\partial_j Z^{II} = Q(x, Z^I, Z^{II}) + D^{II}(Z^I, Z^{II}) \end{cases} \tag{4}$$

by construction $Z^I = Z^I(x,t) \in \mathbf{R}^k$, $Z^{II} = Z^{II}(x,t) \in \mathbf{R}^{N-k}$ and for any $j = 1, \dots, d$, $M_j^{11}(x) \in \mathcal{M}_{k\times k}$, $M_j^{12}(x) \in \mathcal{M}_{k\times(N-k)}$, $M_j^{21}(x) \in \mathcal{M}_{(N-k)\times k}$, $M_j^{22}(x) \in \mathcal{M}_{(N-k)\times(N-k)}$, $D^I(Z^I, Z^{II}) \in \mathbf{R}^k$, $D^{II}(Z^I, Z^{II}) \in \mathbf{R}^{N-k}$, $Q(x, Z^I, Z^{II}) \in \mathbf{R}^{N-k}$. We remark that system (4) is still hyperbolic. We also need the following relevant structural assumption

(A.5) $M_j^{11}(x) = 0$, for any $j = 1, \dots, d$,

To understand this hypothesis is necessary to make a formal power series expansions with respect to ε of the solutions. Without this condition the limit process becomes trivial and the only limit solution is zero. For more details we refer also to [9] and [13].
To analyse the relaxation process of the system (1) we rewrite it in the following way by means of the previous splitting (4)

$$\begin{cases} Z_t^I + \frac{1}{\varepsilon}\sum_{j=1}^d M_j^{12}(x)\partial_j Z^{II} = \frac{1}{\varepsilon} D^I(Z^I, Z^{II}) \\ Z_t^{II} + \frac{1}{\varepsilon}\sum_{j=1}^d M_j^{21}(x)\partial_j Z^I + \frac{1}{\varepsilon}\sum_{j=1}^d M_j^{22}(x)\partial_j Z^{II} = \frac{1}{\varepsilon^2} Q(x, Z^I, Z^{II}) + \frac{1}{\varepsilon} D^{II}(Z^I, Z^{II}) \end{cases} \tag{5}$$

By using formal asymptotics we are leaded to define

$$Z^I(x,t) = U^I(x,t), \qquad Z^{II}(x,t) = \varepsilon U^{II}(x,t). \tag{6}$$

which satisfy the following system

$$\begin{cases} U_t^I + \sum_{j=1}^d M_j^{12}(x)\partial_j U^{II} = \frac{1}{\varepsilon} D^I(U^I, \varepsilon U^{II}) \\ \varepsilon^2 U_t^{II} + \sum_{j=1}^d M_j^{21}(x)\partial_j U^I + \varepsilon\sum_{j=1}^d M_j^{22}(x)\partial_j U^{II} = \frac{1}{\varepsilon} Q(x, U^I, \varepsilon U^{II}) + D^{II}(U^I, \varepsilon U^{II}). \end{cases} \tag{7}$$

If we denote by (U^{I0}, U^{II0}) the limit profile as $\varepsilon \downarrow 0$, a formal calculation shows that the previous system relaxes to the following one

$$\begin{cases} U_t^{I0} + \sum_{j=1}^d M_j^{12}(x)\partial_j U^{II0} = D_\nu^I(U^{I0}, 0) U^{II0} \\ \sum_{j=1}^d M_j^{21}(x)\partial_j U^{I0} = Q_\nu(x, U^{I0}, 0) U^{II0} + D^{II}(U^{I0}, 0), \end{cases} \tag{8}$$

where Q_ν, D^I_ν, denote the derivative respect to the second variable respectively of Q and D^I. Under natural assumptions the system (8) is transformed into

$$\begin{cases} U^{I0}_t + \sum_{j=1}^{d} M^{12}_j(x)\partial_j U^{II0} = D^I_\nu(U^{I0},0)U^{II0} \\ U^{II0} = Q^{-1}_\nu(x,U^{I0},0)\left[\sum_{j=1}^{d} M^{21}_j(x)\partial_j U^{I0} - D^{II}(U^{I0},0)\right], \end{cases}$$

which is equivalent to (setting $U = U^{I0}$)

$$\begin{aligned} U_t + \sum_{j=1}^{d} M^{12}_j(x)\partial_j\left(Q^{-1}_\nu(x,U,0)\sum_{k=1}^{d} M^{21}_k(x)\partial_k U\right) &= \sum_{j=1}^{d} M^{12}_j(x)\partial_j(Q^{-1}_\nu(x,U,0)D^{II}(U,0)) \\ + D^I_\nu(U,0)Q^{-1}_\nu(x,U,0)\left[\sum_{k=1}^{d} M^{21}_k(x)\partial_k U - D^{II}(U,0)\right]. \end{aligned} \tag{9}$$

In the next section we will find sufficient conditions in order to justify rigorously the formal asymptotics. In particular this will be done when (9) is parabolic.

3. Estimates and convergence

In this part we consider our rescaled system (7) and we develop the rigorous theory in order to get the relaxed system (8). We want to show that, as $\varepsilon \downarrow 0$, the weak solutions of the rescaled system satisfy

$$\begin{aligned} U^I &\longrightarrow U^{I0} && \text{strongly in } L^2_{loc}(\mathbf{R}^d \times \mathbf{R}_+), \\ U^{II} &\rightharpoonup U^{II0} && \text{weakly in } L^2(\mathbf{R}^d \times [0,T]), \\ \varepsilon U^{II} &\longrightarrow 0 && \text{strongly in } L^2_{loc}(\mathbf{R}^d \times [0,T]). \end{aligned}$$

In the next subsection 3.1 we assume the existence of a symmetrizer for the system (4) whose symbol has a suitable block structure. In the definition of the symmetrizer and on its properties we follow Taylor [18], [19] but we remark that if our system is strictly hyperbolic we have automatically the existence of a symmetrizer following the construction of Kreiss [7] or of Majda and Osher in the case of boundary conditions [10]. The block structure follows also by using a deep result of Métivier [14] in the case of constant multiplicity, which includes many physical examples.

Now we are going to rewrite the system (4) into a pseudodifferential form in order to obtain "energy" type estimate by using pseudodifferential symmetrizer. Taking into account the pseudodifferential notations and setting

$$M(x,D) = OP\left\{2\pi i\sum_{j=1}^{d}\begin{pmatrix} 0 & M^{12}_j(x)\xi_j \\ M^{21}_j(x)\xi_j & M^{22}_j(x)\xi_j \end{pmatrix}\right\} \tag{10}$$

we can rewrite the system (4)in this way

$$Z_t - M(x,D)Z = Q(x,Z^I,Z^{II}) + D(Z^I,Z^{II}). \tag{11}$$

3.1. A priori estimates

The first step is to establish a priori estimates, independent of ε, for the solution of the system (7). We make on (7) the following hypotheses

(B.1) $U^I(x,0) \in \left[L^2(\mathbf{R}^d)\right]^k$, $U^{II}(x,0) \in \left[L^2(\mathbf{R}^d)\right]^{N-k}$,

(B.2) $M_j^{12}(x), M_j^{21}(x), M_j^{22}(x)$, are smooth functions for any $j = 1,\dots,d$,

(B.3) $det\left[\left(\sum_{j=1}^{d} M_j^{21}(x)\right)^T \sum_{j=1}^{d} M_j^{21}(x)\right] \neq 0$,

(B.4) $D = (D^I(Z^I,Z^{II}), D^{II}(Z^I,Z^{II})) \in C^1(\mathbf{R}^N;\mathbf{R}^k \times \mathbf{R}^{N-k})$, $D^I(Z^I,0) = 0$, $D^{II}(Z^I,0) = 0$, and D is a α-lipschitz function,

(B.5) there exists a symmetrizer $R(x,D) \in OPS^0_{1,0}$ for the operator $M(x,D)$, whose symbol has the following block form

$$R(x,\xi) = \begin{pmatrix} R_{11}(x,\xi) & 0 \\ 0 & R_{22}(x,\xi) \end{pmatrix},$$

where $R_{11}(x,\xi) \in \mathcal{M}_{k\times k}$, $R_{22}(x,\xi) \in \mathcal{M}_{(N-k)\times(N-k)}$ are symmetric positive definite matrices and $R(x,D)M(x,D)+(R(x,D)M(x,D))^* \in OPS^0_{1,0}$.

We remark that since $M_j^{21}(x) \in \mathcal{M}_{(N-k)\times k}$ $\forall j = 1,\dots,d$ from elementary linear algebra we deduce condition (B.3) is violated whenever $k > \frac{N}{2}$. Finally we state the assumptions concerning the nonhomogeneous term $Q(x,Z^I,Z^{II})$.

(D) Dissipativity condition

$Q(x,Z^I,Z^{II})$ has the following form

$$Q(x,Z^I,Z^{II}) = Q_0(x,Z^I,Z^{II}) + Q_1(x,Z^I,Z^{II})$$

and $Q(x,Z^I,0) = 0$ for any $(x,Z^I) \in \mathbf{R}^d \times \mathbf{R}^k$ moreover

(d1) $Q_0(x,Z^I,Z^{II}) \in C^1(\mathbf{R}^{N+d};\mathbf{R}^{N-k})$, $Q_{0\nu}(x,Z^I,Z^{II})$ is bounded in (x,Z^I,Z^{II}), $[Q_{0\nu}, R_{22}^{1/2}(x,D)] = 0$. There exists $\lambda_0 > 0$ such that for any $x \in \mathbf{R}^d$, $(Z^I,Z^{II}) \in \mathbf{R}^k \times \mathbf{R}^{N-k}$, $Q_{0\nu}(x,Z^I,Z^{II}) \le -\lambda_0 I$,

(d2) $Q_1(x,Z^I,Z^{II}) \in C^1(\mathbf{R}^{N+1};\mathbf{R}^{N-k})$, $Q_{1\nu}(x,Z^I,Z^{II})$ is bounded in (x,Z^I,Z^{II}) and the operator $R_{22}(x,D)Q_{1\nu}$ has norm $\|R_{22}(x,D)Q_{1\nu}\|_{\mathcal{L}(L^2)} \le \lambda_1$, $\lambda_1 > 0$, $\lambda_1 \le \lambda_0/2$.

Remark 3.1. *We have decomposed the nonhomogeneous term $Q(x,Z^I,Z^{II})$ in two parts that take into account the different dissipativity nature of Q. The first term $Q_0(x,Z^I,Z^{II})$ cares about the dissipativity of Q, and it commutes with $R_{22}(x,D)Q_{1\nu}$.*

The second one, $Q_1(x, Z^I, Z^{II})$ does not commutes but defines with $R_{22}(x, D)$ a bounded operator that doesn't affect the dissipativity of $Q_0(x, Z^I, Z^{II})$.

Remark 3.2. *The class of dissipativity terms defined in (D) is not empty. In fact it is sufficient to take Q of the following form*

$$Q(x, Z^I, Z^{II}) = C(x)Z^{II} \tag{12}$$

with $C(x) \in \mathcal{M}_{(N-k)\times(N-k)}$ for any $x \in \mathbf{R}^d$, $[R_{22}^{1/2}(x, D), C(x)] = 0$ and there exists $\gamma > 0$ such that for any $x \in \mathbf{R}^d$, $C(x) \leq -\gamma I$.

Taking advantage of the block structure of the symmetrizer and on the dissipativity condition (D) we can prove the following result

Theorem 3.3. *Let us consider the solution $\{U^I\}$, $\{U^{II}\}$ of the Cauchy problem for system (7). Assume that the hypotheses (B.1), (B.2), (B.3), (B.4), (B.5), (D) hold. Then for ε small enough, one has*

(i) *there exist $T > 0$, $M(T) > 0$, independent from ε, such that $\|U^{II}\|_{L^2(\mathbf{R}^d\times[0,T])} \leq M(T)$ and $\sup_{[0,T]} \|\varepsilon U^{II}(\cdot, t)\| \leq M(T)$,*

(ii) *$\{\varepsilon^2 U_t^{II}\}$ is relatively compact in $H_{loc}^{-1}(\mathbf{R}^d \times \mathbf{R}_+)$,*

(iii) *$\{U^I\}$ is uniformely bounded, with respect to ε, in $L^\infty(\mathbf{R}_+, L^2(\mathbf{R}^d))$, namely there exist $T > 0$, $M(T) > 0$, independent from ε, such that $\sup_{[0,T]} \|U^I(\cdot, t)\| \leq M(T)$.*

The proof is done using the previous hypotheses and the block structure of the symmetrizer in order to obtain an energy estimate which leads to (i), (ii), (iii). Details are contained in the theory developed in [5].

3.2. Strong convergence

In order to get the relaxation result we need the convergence of our sequence of solutions. We begin with a simple consequence of (i) and (ii) of Theorem (3.3).

Theorem 3.4. *Let us consider the solution $\{U^{II}\}$ of the Cauchy problem for system (7). Assume the hypotheses (B.1), (B.2), (B.3), (B.4), (B.5), (D) hold. Then there exists $U^{II0} \in [L^2(\mathbf{R}^d \times [0, T])]^{N-k}$, such that, as $\varepsilon \downarrow 0$, one has (extracting eventually subsequences)*

$$U^{II} \rightharpoonup U^{II0} \qquad \textit{weakly in } L^2(\mathbf{R}^d \times [0, T]) \tag{13}$$

$$\varepsilon U^{II} \longrightarrow 0 \qquad \textit{strongly in } L^2_{loc}(\mathbf{R}^d \times \mathbf{R}_+) \tag{14}$$

$$\{\varepsilon^2 U_t^{II}\} \longrightarrow 0 \qquad \textit{in } H^{-1}_{loc}(\mathbf{R}^d \times \mathbf{R}_+). \tag{15}$$

Our next step is to prove strong convergence for the sequence $\{U^I\}$in$L^2_{loc}(\mathbf{R}^d \times \mathbf{R}_+)$. To this end we will use only the estimate obtained in the previous paragraph and the Theorem 2.1.

Theorem 3.5. *Let us consider the solution $\{U^I\}$, of the Cauchy problem for system (7). Assume the hypotheses (B.1), (B.2), (B.3), (B.4), (B.5), (D) hold. Then there exists $U^{I0} \in [L^2(\mathbf{R}^d \times \mathbf{R}_+)]^k$, such that, as $\varepsilon \downarrow 0$, one has (extracting eventually subsequences)*

$$U^I \longrightarrow U^{I0} \qquad \textit{strongly in } L^2_{loc}(\mathbf{R}^d \times \mathbf{R}_+) \tag{16}$$

Proof. By using the hypothesis (D) and the estimate of Theorem 3.3 we conclude

$$\begin{pmatrix} U_t^I + \sum_{j=1}^d M_j^{12}(x)\partial_j U^{II} \\ \sum_{j=1}^d M_j^{21}(x)\partial_j U^I \end{pmatrix} \quad \text{is relatively compact in } (H^{-1}_{loc})^2.$$

In order to fit into the framework of the Theorem 2.1 we set

$$P\begin{bmatrix} U^I \\ U^{II} \end{bmatrix} = \begin{bmatrix} I_{k\times k} & 0 \\ 0 & 0 \end{bmatrix} \partial_t \begin{bmatrix} U^I \\ U^{II} \end{bmatrix} + \sum_{j=1}^d \begin{bmatrix} 0 & M_j^{22}(x) \\ M_j^{21}(x) & 0 \end{bmatrix} \partial_j \begin{bmatrix} U^I \\ U^{II} \end{bmatrix}$$

Let us denote by $p(x,\xi)$, for $\xi = (\xi_0, \xi') \in \mathbf{R}^{d+1}$, $|\xi| = 1$, the principal symbol of P, then we have

$$\left\{ (x,\xi,\lambda,\mu) \ \text{ such that } \ p(x,\xi)\begin{bmatrix} \lambda \\ \mu \end{bmatrix} = 0 \right\} \subset \{\lambda \mid \lambda = 0\}.$$

If we define $q(x) = \begin{bmatrix} I_{k\times k} & 0 \\ 0 & 0 \end{bmatrix}$ for all $\xi \neq 0$, $\xi = (\xi_0, \xi')$ we have that

$$p(x,\xi)\begin{bmatrix} \lambda \\ \mu \end{bmatrix} = 0 \qquad \text{implies} \qquad \langle q(x)\begin{bmatrix} \lambda \\ \mu \end{bmatrix}, \begin{bmatrix} \lambda \\ \mu \end{bmatrix} \rangle = 0$$

for all $\lambda \in \mathbf{R}^k$, $\mu \in \mathbf{R}^{N-k}$. Now we can apply the Theorem (2.1) and we conclude

$$U^I \longrightarrow U^{I0} \qquad \text{strongly in } L^2_{loc}(\mathbf{R}^d \times \mathbf{R}_+),$$

where U^{I0} denotes, in view of the Theorem (3.3) the weak limit of U^I in $L^2(\mathbf{R}^d \times \mathbf{R}_+)$. □

In this way it follows our main theorem.

Theorem 3.6. *Assume that the hypotheses of Theorems (3.4), (3.5) hold, then (U^{I0}, U^{II0}) verifies, in the sense of distributions, the following system*

$$\begin{cases} U_t^{I0} + \sum_{j=1}^d M_j^{12}(x)\partial_j U^{II0} = D_\nu^I(U^{I0},0)U^{II0} \\ \sum_{j=1}^d M_j^{21}(x)\partial_j U^{I0} = Q_\nu(x,U^{I0},0)U^{II0} + D^{II}(U^{I0},0). \end{cases} \tag{17}$$

The parabolicity of (17) follows from the relations between the blocks of the symmetrizer and the coefficents of the system (7).

References

[1] G.-Q. Chen, Levermore C.D. and T.-P. Liu, *Hyperbolic conservation laws with stiff relaxation terms and entropy*, Comm. Pure Appl. Math., **47** (1994), 787–830.

[2] C.M. Dafermos, *Hyberbolic Conservation Laws in Continuum Physics*, Grundlehren der mathematischen Wissenschaften, **325** (2000), Springer-Verlag.

[3] D. Donatelli, P. Marcati, *Relaxation of semilinear hyperbolic systems with variable coefficients*, Ricerche di Matematica, **48** (1999), suppl., 295-310.

[4] D. Donatelli, P. Marcati, $1-\mathcal{D}$ *Relaxation from hyperbolic to parabolic systems with variable coefficients*, Rendiconti dell' Istituto di Matematica dell'Università di Trieste, **31** (2000), suppl., 63–85.

[5] D. Donatelli, P. Marcati, *Convergence of singular limits for multi-D semilinear hyperbolic systems to parabolic systems*, Preprint di Matematica - n.12 della Scuola Normale Superiore di Pisa (Aprile 2000).

[6] P. Gérard, *Microlocal defect measures*, Comm. Partial Differential Equations, **16** (1991), n.11, pag. 1761-1794.

[7] H.-0. Kreiss, *Initial-boundary Value Problems for Hyperbolic systems*, Comm. on Pure and Applied Math., **23** (1970), 277-298.

[8] C. Lattanzio, R. Natalini, *Convergence of diffusive BGK approximation for parabolic systems*, Proceedings of the Royal Society of Edinburg (to appear).

[9] C. Lattanzio, W.-A. Yong, *Hyperbolic-Parabolic singular limits for first order nonlinear systems*, Comm. Partial Differential Equations (to appear)

[10] A. Majda and S. Osher, *Initial–boundary Value Problems for hyperbolic Equations with Uniformly Characteristics Boundary* , Comm. on Pure and Applied Math, **28** (1975), 607-675.

[11] P. Marcati and A. Milani, *The one-dimensional Darcy's law as the limit of a compressible Euler flow*, J. Differential Equations, **13** (1990), 129–147.

[12] P. Marcati, A. Milani, and P. Secchi, *Singular convergence of weak solutions for a quasilinear nonhomogeneous hyperbolic system*, Manuscripta Math., **60** (1988), 49–69.

[13] P. Marcati, B. Rubino, *Hyperbolic to parabolic relaxation theory for quasilinear first order systems*, J. Differential Equations, **162** (2000),No. 2, 359–399.

[14] G. Métivier, *The block structure condition for symmetric hyperbolic systems*, Bull. London Math. Soc., **32** (2000), no. 6, 689–702.

[15] L. Tartar, *Compensated compactness and applications to partial differential equations*, Research Notes in Math., **39** (1979), 136–210.

[16] L. Tartar, *The compensated compactness method applied to partial differential equations*, Systems of Nonlinear Partial Differential Equations, Reidel, Dordrecht, 1983. NATO ASI.

[17] L. Tartar, *H-measures, a new approach for studying homogenization and concentration effects in partial differential equations*, Proceedings of the Royal Society of Edinburg, **115A** (1990), 193–230.

[18] M.E. Taylor, *Pseudodifferential Operators* , Princeton mathematical series, **34** (1981), Princeton University Press, Princeton New Jersey.

[19] M.E. Taylor, *Pseudodifferential Operators and Nonlinear PDE*, Progress in Mathematics, **100** (1991), Birkäuser.

[20] W.-A. Yong, *Singular perturbations of first-order hyperbolic systems with stiff source terms*, J. Differential Equations, **155** (1999), 89-132.

Donatella Donatelli
Scuola Normale Superiore,
Piazza Cavalieri 7,
56100 Pisa, Italy
donatell@cibs.sns.it

Pierangelo Marcati
Dip. di Matematica Pura ed Applicata,
Università degli Studi dell'Aquila,
67100 L'Aquila, Italy
marcati@univaq.it

International Series of Numerical Mathematics
Vol. 140,

Large Time Asymptotics in Contaminant Transport in Porous Media with Variable Diffusion

Gema Duro and Enrique Zuazua

Abstract. This paper is devoted to the study of the large time behavior of solutions of the equations modelling contaminant transport in porous media with variable diffusion for integrable initial data.

The goal of this work is to study the large time behavior of a reactive solute which undergoes equilibrium adsorption in a porous medium with variable diffusion.

1. Description of the transport model

Let us start with a brief description of the transport model in $\mathbb{R}$. An extensive treatment of chemical reactions arising in the transport of solutes through porous media with constant diffusion is given by [3]. To be more precise we describe the absorption reactions by a Freundlich isotherm and we restrict ourselves to the case of fast reaction, whence the transport equation becomes (see for instance [2])

$$(u+u^q)_t + u_x = (au_x)_x \ \text{ in } \ (0,\infty)\times \mathbb{R} \tag{1}$$

where the a measures the hydrodynamic dispersion. We consider here the case of variable diffusion $a(x) = 1 + b(x)$.

We want to investigate the large time behaviour of solutions of the problem

$$\begin{cases} (u+u^q)_t + u_x = (a(x)u_x)_x & \text{in } (0,\infty)\times\mathbb{R} \\ u(0,x) = u_0(x). \end{cases} \tag{2}$$

On the other hand, integrating equation (2) over all $\mathbb{R}$ we deduce that the total mass of solution is conserved, that is for all $t > 0$,

$$\int_{\mathbb{R}} (u+u^q)(t,x)dx = \int_{\mathbb{R}} (u_0+u_0^q)(x)dx = M, \quad \forall t > 0.$$

To construct the asymptotic solution we make the change of variables $\xi = t - x$ which (2) the equation

$$(u+u^q)_t + u_\xi^q = (a(\xi)u_\xi)_\xi.$$

In the nondegenerate case $q > 1$, we may replace $u + u^q$ with u as t becomes large (see [3]) and obtain the nonlinear convection-diffusion equation

$$u_t + u_\xi^q = (a(\xi)u_\xi)_\xi.$$

We work on the simplified convection-diffusion problem:

$$\begin{cases} u_t - (a(x)u_x)_x + q|u|^{q-1}u_x = 0 \text{ in } (0,\infty) \times \mathbb{R} \\ u(0,x) = u_0(x) \end{cases} \tag{3}$$

where $q > 1$.

2. Main results

In this paper we study the large time behavior of solutions to convection-diffusion equations

$$\begin{cases} u_t - \operatorname{div}(a(x)\nabla u) = d \cdot \nabla(|u|^{q-1}u) \text{ in } (0,\infty) \times \mathbb{R}^N \\ u(0,x) = u_0(x) \end{cases} \tag{4}$$

with $q > 1$, $N \geq 1$ and $d \in \mathbb{R}^N$ a constant vector. By $\cdot$ we denote the scalar product in $\mathbb{R}^N$.

We are interested here in finding out how the variable diffusion and the nonlinearity affect the solution for large times.

Most of the previous work on these equations deals only with constant diffusion (see for instance [9], [8], and the references therein). However, when studying heterogeneous media we must work with variable diffusion.

For large times, a competition between the diffusive and convective effects is observed. When $a(x) = 1$, it has been established the existence of a critical value for the power q, $q_N = 1 + \frac{1}{N}$, such that:

- If $q > q_N$, the diffusion is dominant and the solution behaves like a self-similar solution with mass $M = \int u_0$ of the heat equation.
- If $q = q_N$, diffusion and convection have the same strength and the solution behaves like a selfsimilar solution with mass M of the full equation.
- If $1 < q < q_N$, the convection effects are dominant and the solution behaves like a selfsimilar solution with mass M of a reduced equation where the dissipation in some direction vanishes.

For general variable diffusion coefficients $a(x)$ it is not clear up to which point a similar description holds for the long time profiles. In this paper, we focus on two kinds of coefficients:

1. We suppose that $a(x) = 1 + b(x)$ with

$$b(x) \in L^1(\mathbb{R}^N) \cap C^{1,\alpha}(\mathbb{R}^N) \quad \text{and} \quad \|b^-(x)\|_\infty < 1, \tag{5}$$

where b^- denotes the negative part of b. We also assume that

$$|b(x)| + (1+|x|^2)^{\frac{1}{2}}|\nabla b(x)| \leq \frac{C}{(1+|x|^2)^{\frac{\delta}{2}}} \qquad \forall x \in \mathbb{R}^N \tag{6}$$

for some positive constants C and δ. This kind of coefficients will be called "*Integral coefficients*".

2. We suppose $N = 1$ and we consider that $b(x)$ is a *decreasing* function such that
$$b(x) \in C^{1,\alpha}(\mathbb{R}) \quad \text{and} \quad \|b^-(x)\|_\infty < 1, \tag{7}$$
where b^- denotes the negative part of b. Moreover, we also assume that $a(x) = 1 + b(x)$ satisfies
$$a - \alpha \in W^{1,1}(\mathbb{R}^-); \quad a - \beta \in W^{1,1}(\mathbb{R}^+), \tag{8}$$
where α and β two positive real constants. This kind of coefficients will be called "*decreasing coefficients*".

For convenience of the reader we recall some known results on the initial value problem (4) (see [6]).

For every $u_0 \in L^1(\mathbb{R}^N) \cap L^r(\mathbb{R}^N)$ with $r > \frac{Nq}{N+2}$, system (4) admits a unique solution $u \in C([0,\infty); L^1(\mathbb{R}^N))$.

This solution is smooth for $t > 0$. Integrating equation (4) over $\mathbb{R}^N$ we deduce that the total mass of solutions is conserved for all time, i. e.,
$$\int_{\mathbb{R}^N} u(t,x)dx = \int_{\mathbb{R}^N} u_0(x)dx = M, \quad \forall t > 0.$$

2.1. Integral coefficients

In order to understand the large time behavior of u solution of the nonlinear problem (4) we introduce the scaled functions $u_\lambda(t,x) = \lambda^N u(\lambda^2 t, \lambda x)$.

Let us observe that, if u solves (4), then u_λ solves
$$u_{\lambda,t} - \operatorname{div}(a(\lambda x)\nabla u_\lambda) = \lambda^{N(1-q)+1} d \cdot \nabla(|u_\lambda|^{q-1}u_\lambda).$$

On the other hand, passing to the limit on u as $t \to \infty$ can be reduced to passing to the limit on u_λ at $t = 1$ as $\lambda \to \infty$. Thus we are interested on the limit of u_λ as $\lambda \to \infty$.

Note also that $\int_{\mathbb{R}^N} u_\lambda(t,x)\,dx = \int_{\mathbb{R}^N} u_0(x)\,dx$ for all $t \geq 0$ and $\lambda > 0$.

When $q > 1 + \frac{1}{N}$, the power $N(1-q)+1$ of λ is negative. Therefore, formally (this will be made precise below), the convection term should vanish as $\lambda \to \infty$.

When $q = 1 + \frac{1}{N}$, we have that $\lambda^{N(1-q)+1} = 1$. Therefore, formally, all terms of the equation remain when passing to the limit as $\lambda \to \infty$.

On the other hand, in view of assumptions (5) we have that $a(\lambda x) - 1 \to 0$ as $\lambda \to \infty$ in $L^1(\mathbb{R}^N)$. Thus, we expect the limit diffusion to be the constant $a \equiv 1$. As a consequence of these remarks we should expect the two following results:

a) when $q = 1 + \frac{1}{N}$, the asymptotic behavior of the solutions of (4) should be given by the self-similar solutions of the full equation with $a \equiv 1$, i. e., the system should present a self-similar behavior.
b) when $q > 1 + \frac{1}{N}$, the asymptotic behavior of the solutions of (4) should be given by $MG(t)$, where M is the mass of the solution, that is, $M = \int_{\mathbb{R}^N} u_0(x)\,dx$ and G the heat kernel , i. e., the system should present a weakly nonlinear behavior.

In the study of the large time behavior with integral coefficients we have face two types of difficulties. Those coming from the nonlinearity and those coming from a linear part with a variable coefficient. As said above, we take asymptotically constant diffusion of the form $a(x) = 1 + b(x)$ with $b(x) \in L^1(\mathbb{R}^N) \cap C^{1,\alpha}(\mathbb{R}^N)$ and $\|b(x)\|_\infty < 1$. Our analysis is valid as long as the semigroup generated by $u_t - div(a(x)\nabla u) = 0$ satisfies:

$$\|\nabla T(t)(u_0)\|_p \leq C\|u_0\|_s t^{-\frac{N}{2}(\frac{1}{s}-\frac{1}{p})-\frac{1}{2}} \tag{9}$$

with $C = C(p,s)$ independent of u_0, a and t. This estimate holds when $b(x)$ (see [13]) satisfies (6) for some positive constants C and δ. Whether the same bound can be obtained under weaker decay asumptions on $b(x)$ is a deep question in theory of parabolic operators that we shall not address here. Notice that classical semigroup theory yields bounds with C dependent on a and its derivatives.

On the other hand, for this kind of coefficients we obtain the following L^p-estimates:

$$\|u(t)\|_p \leq C_p\|u_0\|_1 t^{(-N/2)(1-1/p)} \quad \forall t > 0 \tag{10}$$

$$\|\nabla u(t)\|_p \leq C_p\|u_0\|_1 t^{(-N/2)(1-1/p)-\frac{1}{2}} \quad \forall t > 1 \tag{11}$$

for every $p \in [1,\infty]$. Thanks to (9), the constant C_p does not depend on b or its derivatives.

Concerning the self-similar large time behavior of the solutions for $q = 1 + \frac{1}{N}$ we have the following result.

Theorem 1. *Suppose that a is an integral coefficient, i.e. $a = 1+b$ with b satisfying (5) and (6). Assume that $q = 1 + \frac{1}{N}$.*

Let $u_0 \in L^1(\mathbb{R}^N)$ be such that $M = \int_{\mathbb{R}^N} u_0(x)\,dx$. Then the solution $u = u(t,x)$ of (4) satisfies

$$t^{\frac{N}{2}(1-\frac{1}{p})}\|u(t) - v(t)\|_p \to 0 \quad as \quad t \to \infty \tag{12}$$

for every $p \in [1,\infty]$, where $v(t)$ is the unique solution of

$$\begin{cases} v_t - \Delta v = d \cdot \nabla(|v|^{\frac{1}{N}} v) & in \ (0,\infty) \times \mathbb{R}^N \\ v(0,x) = M\delta \end{cases} \tag{13}$$

which is self-similar: $v(t,x) = t^{\frac{-N}{2}} f_M(\frac{x}{\sqrt{t}})$ with f_M such that $\int_{\mathbb{R}^N} f_M(x)\,dx = M$.

Remark 1. *When $d = 0$ Theorem 1 guarantees that if $u_0 \in L^1(\mathbb{R}^N)$ the solutions of*

$$u_t - \operatorname{div}(a(x)\nabla u) = 0 \tag{14}$$

behave like the heat kernel as $t \to \infty$ in a first approximation.

Idea of proof of Theorem 1.

The proof uses a scaling technique. We proceed in several steps. In the first one we introduce the scaled functions. In step 2 we obtain estimates which allow us to pass to the limit in the scaled equation. In steps 3 and 4 we pass to the limit in the variational formulation. In the last step, using the uniqueness Theorem proved in [1], we complete the proof.

Step 1. Scaling.

For $\lambda > 0$, we introduce the scaled functions $u_\lambda(t,x) = \lambda^N u(\lambda^2 t, \lambda x)$ that satisfy

$$\begin{cases} u_{\lambda,t} - \operatorname{div}(a(\lambda x)\nabla u_\lambda) = d \cdot \nabla(|u_\lambda|^{\frac{1}{N}} u_\lambda) \text{ in } (0,\infty) \times \mathbb{R}^N \\ u_\lambda(0,x) = u_{\lambda,0}(x) = \lambda^N u_0(\lambda x). \end{cases} \tag{15}$$

On the other hand, as the solution v of (13) is self-similar, we observe that proving (12) is equivalent to prove that

$$u_\lambda(1) \to v(1) \quad \text{in} \quad L^p(\mathbb{R}^N) \quad \text{when} \quad \lambda \to \infty$$

for every $p \in [1,\infty]$. Thus, the convergence result in Theorem 1 consists precisely in showing that $u_\lambda(1,.)$ converges to the solution v of equation (13) at time $t = 1$ when $\lambda \to \infty$.

Step 2. Uniform estimates for the scaled solutions.

As a consequence of (10) and (11) we get:

$$\|u_\lambda(t)\|_p \leq C_p \|u_0\|_1 t^{(-N/2)(1-1/p)}, \qquad \forall t > 0 \tag{16}$$

$$\|\nabla u_\lambda(t)\|_p \leq C_p \|u_0\|_1 t^{(-N/2)(1-1/p)-\frac{1}{2}}, \qquad \forall t > 0 \tag{17}$$

for every $p \in [1,\infty]$ with $C_p > 0$ independent of $\lambda > 0$, also obtain that

$$\|u_\lambda(t,\cdot)\|_{L^1(|x|>R)} \to 0 \quad \text{as} \quad R \to \infty \tag{18}$$

uniformly in $\lambda \geq 1$ and $t \in [0,t_0]$, for any $t_0 > 0$ fixed.

By (16) and (17) we deduce that

$$\|\nabla |u_\lambda|^{\frac{1}{N}+1+r}(t)\|_1 \leq C_1 t^{\frac{-Nr}{2}-1}, \qquad \forall t > 0 \tag{19}$$

with $C_1 > 0$ independent of λ and for every $r \geq 0$.

Using (17), (19), the fact that $a(\lambda x)$ is uniformly bounded in $L^\infty(\mathbb{R}^N)$ and equation (15) we deduce that

a) $\{\partial_t u_\lambda\}$ is uniformly bounded in $L^2_{loc}((0,\infty); H^{-s}(\Omega))$ for some $s > 0$ and every bounded domain Ω of $\mathbb{R}^N$.

On the other hand, as a consequence of the estimates (16) above we have

b) $\{u_\lambda\}$ is uniformly bounded in $L^\infty_{loc}((0,\infty); L^2(\mathbb{R}^N))$.

Taking into account that $L^2(\Omega)$ is compactly embedded in $H^{-\varepsilon}(\Omega)$ for every $\varepsilon > 0$, and that $H^{-\varepsilon}(\Omega) \hookrightarrow H^{-s}(\Omega)$ with continuous embedding for every $s > \varepsilon$, combining a) and b) and applying classical compactness results (cf. [14], Corollary 4, page 85) we deduce that $\{u_\lambda\}$ is relatively compact in $C([t_1,t_2]; H^{-\varepsilon}(\Omega))$ for every $0 < t_1 < t_2 < \infty$ and $\varepsilon > 0$.

Extracting a subsequence $\lambda_n \to \infty$ (that we still denote by λ), we may assert that

$$u_\lambda \to u \quad \text{in} \quad C([t_1,t_2]; H^{-\varepsilon}(\Omega)) \tag{20}$$

After we pass to the limit in the variational formulation. Finally, taking into account, the uniqueness of the limit (see [1]) and (20), we have that the whole family $\{u_\lambda\}$ to strongly converge to u in $C([t_1,t_2];H^{-\varepsilon}(\Omega))$ as $\lambda\to\infty$. This concludes the proof of Theorem 1.

Concerning the weakly non-linear large time behavior of solutions for $q > 1+\frac{1}{N}$ we have the following result.

Theorem 2. *Suppose that a is an integral coefficient, i.e. $a = 1+b$ with b satisfying (5) and (6). Assume that $q > 1+\frac{1}{N}$.*

Let $u_0 \in L^1(\mathbb{R}^N)\cap L^r(\mathbb{R}^N)$ with $r > Nq/(N+2)$ be such that $M = \int_{\mathbb{R}^N} u_0(x)\,dx$. Then the solution $u = u(t,x)$ of (4) satisfies

$$t^{\frac{N}{2}(1-\frac{1}{p})}\|u(t)-MG(t)\|_p \to 0 \quad as \quad t\to\infty \tag{21}$$

for every $p\in[1,\infty]$, where G is the heat kernel.

Moreover, if $u_0\in L^1(\mathbb{R}^N;1+|x|)\cap L^\infty(\mathbb{R}^N)$, for every $p\in[1,\infty]$, there exists $C_p>0$ such that

$$\|u(t)-MG(t)\|_p \le C_p t^{\frac{-N}{2}(1-\frac{1}{p})}\varepsilon(t) \quad \forall\, t\ge 1 \tag{22}$$

with

$$\varepsilon(t) = \begin{cases} t^{\frac{-1}{2}} & when\ q > 1+\frac{2}{N} \\ t^{\frac{-1}{2}}\log(t+2) & when\ q = 1+\frac{2}{N} \\ t^{-(N(q-1)-1)/2} & when\ 1+\frac{1}{N} < q < 1+\frac{2}{N}. \end{cases} \tag{23}$$

Idea of proof of Theorem 2.

We prove Theorem 2 working directly on the integral equation associated with (4):

$$u(t) = T(t)[u_0] + \int_0^t d\cdot T(t-\tau)[\nabla F(u(\tau))]d\tau$$

where $F(u) = |u|^{q-1}u$ and $T(t)$ is the contraction semigroup in $L^1(\mathbb{R}^N)$ generated by the operator $Au = -\mathrm{div}(a(x)\nabla u)$.

On the other hand, by Remark 1, we have

$$t^{\frac{N}{2}(1-\frac{1}{p})}\|v(t)-MG(t)\|_p \to 0 \quad \text{as} \quad t\to\infty \tag{24}$$

for any $u_0\in L^1(\mathbb{R}^N)$ without assuming that $|x|u_0\in L^1(\mathbb{R}^N)$.

Finally, using the above estimates on the solution we prove that

$$t^{\frac{N}{2}(1-\frac{1}{p})}\left\|\int_0^t d\cdot T(t-\tau)[\nabla F(u(\tau))]\,d\tau\right\|_p \to 0 \quad \text{as} \quad t\to\infty \tag{25}$$

for every $p\in[1,\infty]$.

2.2. Decreasing coefficients

In this section, we continue the study in the most interesting case from the point of view of applications, that is $1 < q < 2$. We also study the cases $q \geq 2$ for more general coefficients that in [6].

Concerning the large time behavior of the solution for $1 < q < 2$, we have the following result:

Theorem 3. *Suppose that a is a decreasing coefficients, i. e. $a = 1 + b$ where b is a decreasing function satisfying (7) and (8). Assume that $1 < q < 2$.*

Let $u_0 \in L^1(\mathbb{R})$ be such that $M = \int_{\mathbb{R}} u_0(x)dx$. Then the solution $u = u(t,x)$ of (3) satisfies

$$t^{\frac{1}{q}(1-\frac{1}{p})}\|u(t,\cdot) - v(t,\cdot)\|_p \to 0 \quad \text{as } t \to \infty \tag{26}$$

for every $p \in [1, \infty)$, where v is the solution of problem

$$\begin{cases} v_t + (|v|^{q-1}v)_x = 0 & \text{in } (0,\infty) \times \mathbb{R} \\ v(0,x) = M\delta. \end{cases} \tag{27}$$

Remark 2. *We recall that solutions of such conservations laws may be discontinuous and that uniqueness holds only for a special class of solutions called entropy solutions, see* [10].

Remark 3. *According to* [11] *there exists only one entropy solution of (27) with initial data $M\delta(x)$ and is given by the formula*

$$v(t,x) = \begin{cases} (\frac{x}{t})^{\frac{1}{q-1}} & \text{if } 0 < x < r(t) \\ 0 & \text{otherwise} \end{cases} \tag{28}$$

with $r(t) = CM^{\frac{q-1}{q}} t^{\frac{1}{q}}$ and $C = (\frac{q}{q-1})^{\frac{q-1}{q}}$.

Notice that an immediate Corollary of Theorem 3 is that solutions of (3) decay like $t^{-1/q}$ in L^∞. Obtaining this decay rate is actually the most delicate of the proof of Theorem 3.

The main difference between the cases $1 < q < 2$ and $q \geq 2$ is due to the observed decay rates of the solutions.

In particular, for $1 < q < 2$ all nonnegative solutions of (3) satisfy the decay estimate

$$\|u(t)\|_\infty \leq Ct^{\frac{-1}{q}}. \tag{29}$$

In the range $q < 2$, $\frac{1}{q} > \frac{1}{2}$. Therefore, this decay rate is faster than the optimal decay rate for $q \geq 2$, $\|u(t)\|_\infty \leq C\, t^{\frac{-1}{2}}$.

A previous step in order to prove (29) is to obtain a bound for the derivative $(u^{q-1})_x$, (this estimate is called *entropy estimates):*

$$(u^{q-1})_x \leq \frac{1}{qt}$$

This is the main result of this section. Proving this entropy estimate for bounded variable coefficients b is an open problem even if we assume integrability hypothesis on b. To prove this estimate we need to suppose that b is a decreasing function.

Theorem 3 follows from classical scaling techniques thanks to sharp estimates on the decay of the solutions (see [4]).

Concerning the large time behavior of the solution for $q = 2$ and $q > 2$ we have the following result:

Theorem 4. *Suppose that the diffusion coefficient a satisfies (8) and b verifies (7). Assume that* $q > 2$ *and* $u_0 \in L^1(I\!R)$ *with* $M = \int_{I\!R} u_0(x)\,dx$. *Then, the solution of (3) satisfies*

$$t^{\frac{1}{2}(1-\frac{1}{p})}\|u(t) - v(t)\|_p \to 0 \quad as \quad t \to \infty \tag{30}$$

for every $p \in [1, \infty]$, *where* $v(t,x)$ *is the self-similar fundamental solution of problem*

$$\begin{cases} v_t - (\hat{a}(x)v_x)_x = 0 & in\ (0,\infty) \times I\!R \\ v(0,x) = M\delta \end{cases} \tag{31}$$

where $M = \int_{I\!R} u_0(x)dx$ *and*

$$\hat{a}(x) = \begin{cases} \alpha & if\ x \leq 0 \\ \beta & if\ x > 0. \end{cases} \tag{32}$$

Theorem 5. *Suppose that the diffusion coefficient a satisfies (8) and b verifies (7). Assume that* $q = 2$ *and* $u_0 \in L^1(I\!R)$ *with* $M = \int_{I\!R} u_0(x)\,dx$. *Then, the solution of (3) satisfies that*

$$t^{\frac{1}{2}(1-\frac{1}{p})}\|u(t) - v(t)\|_p \to 0 \quad as \quad t \to \infty \tag{33}$$

for every $p \in [1, \infty]$, *with* $v(t)$ *the solution of*

$$\begin{cases} v_t - (\hat{a}v_x)_x + (|v|v)_x = 0 & in\ (0,\infty) \times I\!R \\ v(0,x) = M\delta \end{cases} \tag{34}$$

where $\hat{a}$ *satisfies (32).*

Theorems 4 and 5 are also satisfied if we assume $a - 1 \in W^{1,1}(I\!R)$ instead of $a - \alpha \in W^{1,1}(I\!R^-)$; $a - \beta \in W^{1,1}(I\!R^+)$ with α and β two positive real constants. Therefore are more restrictive the hypothesis assumed for integral coefficients. For decreasing coefficients we do not assume that b satisfies (6).

Theorems 4 and 5 also follow from classical scaling techniques (see [4]).

References

[1] A. Carpio. *Large time behaviour in some convection-diffusion equations.* Ann. Sc. Norm. Sup. Pisa, serie IV, **XXIII**, Fasc. 3, (1996), 551-574.

[2] C. N. Dawson, C. J. Van Duijn and R. E. Grundy. *Asymptotic profiles with finite mass in one-dimensional contaminant transport through porous media: The fast reaction case.* Quart. J. Mech. Appl. Math., **47**, (1994), 69-106.

[3] C. N. Dawson, C. J. Van Duijn and R. E. Grundy. *Large time asymptotics in contaminant transport in porous media.* SIAM J. Appl. Math. **56**, 4, (1996), 965-993.

[4] G. Duro. *Asymptotic profiles in contaminant transport in porous media with variable diffusion.* To appear in Adv Math, Sci. and Appl., **11,** 2 (2001).

[5] G. Duro and A. Carpio. *Asymptotic profiles in convection-diffusion with variable diffusion.* To appear in Nonlinear Anal., (2001).

[6] G. Duro and E. Zuazua. *Large time behavior for convection-diffusion equations in $\mathbb{R}^N$ with asymptotically constant diffusion*, Comm. P.D.E, **24** (7&8), (1999), 1283-1340.

[7] G. Duro and E. Zuazua. *Large time behavior for convection-diffusion equations with periodic coefficients.* Jour. Diff. Equations, **167**, 275-315 (2000).

[8] M. Escobedo, J.L. Vázquez and E. Zuazua. *Asymptotic behavior and source-type solutions for a diffusion-convection equation.* Archives Rat. Mech. Anal., **124**, (1993), 43-65.

[9] M. Escobedo and E. Zuazua. *Large time behavior for solutions of a convection diffusion equation in $\mathbb{R}^N$.* J. Funct. Anal., **100** (1), (1991), 119-162.

[10] P. Lax. *Hyperbolic systems of conservation laws.* II, Comm. Pure Appl. Maths., **10**, (1957), 537-566.

[11] T. P. Liu and M. Pierre. *Source solutions and asymptotic behaviour in conservation laws.* Jour. Diff. Equations, **51** (1984), 419-441.

[12] P. Li and S. T. Yau. *On the parabolic Kernel of the Schrödinger operator.* Acta Mathematica, **156**, (1986), 153-201.

[13] M. Murata. *Large time asymptotics for fundamental solutions of diffusion equations.* Tohoku Mathematical Journal, **37**, (1985), 151-195.

[14] J. Simon. *Compact sets in the space $L^p(0,T;B)$.* Annali Mat. Pura Appl., CXLVI, (1987), 65-96.

G. Duro
Departamento de Análisis Económico,
Economía Cuantitativa,
Universidad Autónoma de Madrid,
28049 Madrid, Spain
gema.duro@uam.es

E. Zuazua
Departamento de Matemática Aplicada,
Universidad Complutense de Madrid,
28040 Madrid, Spain
zuazua@sunma4.mat.ucm.es

International Series of Numerical Mathematics
Vol. 140,

A Nonlinear Flux Vector Split Defect Correction Scheme for Fast Solutions of the Euler and Navier-Stokes Equations

Albrecht Eberle

Abstract. A defect correction scheme is suggested which makes use of embedding the high order accurate total numerical residuals of the dicretized EULER or NAVIER-STOKES (NS) equations into a simple Flux Vector Splitting (FVS) scheme formulated in the classic first order mode. The update of the flow equations (EULER or NS) is performed using a matrix free approach which even does not require the calculation of search directions as in Krylov subspace methods. It rather works like a point Gauss-Seidel method with immediate first order flux updates performed with the newest available flow quantities. By this concept the expensive calculation of the matrix for Newton type classic implicit solvers is avoided, and the storage requirements are those of an explicit scheme. The investigations undertaken with the present kind of implicit solution strategy revealed that a defect correction scheme should use diffusive first order accurate numerical fluxes for providing sufficient smoothing of the increments of the flow variables. For this purpose a novel non-polynomial flux vector splitting is used which exhibits extraordinary simplicity and still retains acceptable accuracy if used in the context of higher order reconstructions of the flow variables.

1. A Defect Correction Scheme

Assuming that R^μ is the total (eventually unsteady) local numerical residual of an upwinded EULER or NS solver at iteration level μ with initial values given at time level n

$$R^\mu = \frac{(VU)^\mu - (VU)^n}{\Delta t} + \oint F^\mu_{normal} dS + \Phi$$

F means the high (usually second) order accurate flux vector of an upwind scheme which may include sophisticated limiters, flux blendig, viscous terms etc. S is the surface area of the control volume under consideration. V is the volume of the control cell, U is the vector of the flow variables and t is the time. Φ is a source term which contains known informations from previous time levels necessary for a high order accurate time integration. Now, a similar residual at an outer iteration level ν is defined by

$$r^\nu = \frac{(VU)^\nu - (VU)^\mu}{\Delta t} + \oint f^\nu_{normal} dS$$

The flux vector f is a first order accurate mapping of the high order accurate flux F. It may consist of simple FVS convective fluxes and the viscous fluxes computed for the high order accurate residual R which have to be stored for use in the residual r. With these definitions a simple defect correction scheme may be constructed the following way

$$D\Delta U + r^{\nu+1,\nu} + \omega R^\mu - r^\mu = 0$$

D is the local 5x5 diagonal matrix in 3D or the 4x4 diagonal Jacobian matrix in 2D of the first order accurate FVS residual r. For simplicity the diagonal matrix may be replaced by the identity matrix multiplied by the inverse local time step

$$\frac{\sum^{nface}(|\lambda_0| + cA)}{CFL}$$

CFL is a user specified dimensionless number. λ_0 is the relative velocity normal to the cell face multiplied with its absolute area A:

$$\lambda_0 = uA_x + vA_y + wA_z - gA$$

g is the normal grid velocity being different from zero in case of a moving mesh. c is the speed of sound. ω is an under relaxation coefficient.
Key point of the method is the immediate local update of the newest available flow variables after having solved locally the system of 5 (4) equations for the update cell:

$$U^{\nu+1} = U^\nu + \Delta U$$

The first order accurate residual $r^{\nu+1,\nu}$ is computed from flow variables at the newest iteration level and the previous iteration level. The scheme works like the solution of the first order method augmented by a constant source, the latter being the difference of the high order and low order solution calculated at the iteration (time) level n:

$$D\Delta U + r^{\nu+1,\nu} = S^\mu$$

$$S^\mu = r^\mu - \omega R^\mu$$

Clearly a steady state solution can be calculated by assigning the time step Δt a very large value. Convergence for an iteration (time) step is achieved when $U^\nu = U^\mu = U^{n+1}$. The method grants faster overall convergence than that obtainable with classic matrix linearized implicit methods. It may be called a matrix free nonlinear Gauss-Seidel method. It pays off to stop the ν iterations when the dimensionless ratio $\frac{|r^\nu - S^\mu|}{|r^1 - S^\mu|}$ is less than a given constant threshold ϵ. The iteration procedure may be augmented by restarting the iterations after the implicit smoothing of the total residual

$$R^t = r^{\nu+1,\nu} + \omega R^\mu - r^\mu$$

This is achieved by a symmetric Gauss-Seidel procedure applied to

$$R_0^{tnew} = \frac{R_0^t + \beta \sum R^t_{neighbours}}{1 + neighbours * \beta}$$

with a user specified parameter β controlling the implicitness of the residual smoothing. The iterative procedure is restartet by simply putting $R = R^{tnew}$.

2. A Robust Flux Vector Splitting

The first order residual r of the defect correction scheme needs not be necessarily the first order accurate counterpart of the high order accurate residual R. It only should feature characteristica similar to the high order accurate residual. Thus, if the convective fluxes from which R is calculated are taken from an eigenvalue based upwind scheme, then r also should be based on upwind convective fluxes. If R is found from central differenced fluxes, then r also should be calculated from central differences. By try and error it was found out that the first order accurate residual should contain a considerable amount of numerical diffusion. A FVS scheme with controllable diffusion is particularly well suited for this purpose. It may be constructed from a flux vector splitting controlled by one parameter for enhancing robustness while still retaining acceptable accuracy. In contrast to conventional polynomial flux vector splitting the present approach introduces a pseudo speed of sound

$$s^2 = \alpha c^2 + q^2(1 - 2\alpha + \alpha \frac{q^2}{c^2})$$

with $q^2 = min[c^2, (\frac{\lambda_0}{A})^2]$.

To use the normal velocity $\frac{\lambda_0}{A}$ in spite of the square of the total speed $q^2 = u^2 + v^2 + w^2$ makes sense: It grants the scheme to be perfectly upwinding.
c is the speed of sound

$$c^2 = \gamma \frac{p}{\rho}.$$

The mass flux is defined as

$$f^{\pm} = \frac{\rho}{2}(\lambda_0 \pm sA)$$

The parameter α should be large enough to provide sufficient diffusion of the numerical fluxes. γ is the ratio of specific heats. $A = \sqrt{A_x^2 + A_y^2 + A_z^2}$ is the cell face area, ρ is the air density, λ_0 is the total normal velocity multiplied with A including the grid speed g normal to the cell face A:

$$\lambda_0 = uA_x + vA_y + wA_z - gA$$

The Cartesian momentum fluxes are

$$
\begin{aligned}
f_x^{\pm} &= (u \pm \frac{pA_x}{A\rho s}) f^{\pm} \\
f_y^{\pm} &= (v \pm \frac{pA_y}{A\rho s}) f^{\pm} \\
f_z^{\pm} &= (w \pm \frac{pA_z}{A\rho s}) f^{\pm}
\end{aligned}
$$

p is the gasdynamic pressure. The energy flux is defined as

$$
F^{\pm} = [\frac{c^2}{\gamma(\gamma-1)} + \frac{u^2+v^2+w^2}{2} + \frac{p}{\rho}(1 \pm \frac{g}{s})] f^{\pm}
$$

Numerical upwinding is performed by calculating the positive fluxes on the left side of the cell face under consideration. The negative fluxes are evaluated with the flow variables on the right side of the cell face. The sum of both is the total numerical flux through the cell face.

3. Extensions of the Flux Vector Splitting

The free parameter α was basically considered as a tuning coefficient for the artificial diffusion of the FVS scheme. Heiermann et al. [1, 2] found out that the parameter α can be assigned a value taken from molecular gasdynamics: it may be computed from the distribution of molecules passing through the cell face in each of the two normal directions. A short description of their concept may be found in their referenced papers.

4. Results

As a first example an inviscid steady transonic flow past the 2D-airfoil of the ONERA M6 wing is considered. The underlying high order accurate flux calculation is based on a noniterative Riemann solver the coefficients of which are taken from the exact solution of the accoustic one dimensional wave pattern normal to the cell face under consideration. A detailed description of this Riemann solver is given in [3]. The reconstruction is the classic second order MUSCL-κ-formula augmented by a very sophisticated limiter for the flow variables developed by the author. The entries of the reconstruction are the conservative flow variables thus grantig possibly lowest numerical diffusion. An overview of the airfoil flow is given in the following figure.

Next the convergence behaviour to steady state by an inexact Newton procedure is depicted. The Jacobian matrix is calculated from the FVS described herein with the simple definition of the reference speed $s = c$ incorporated, in order to keep the matrix calculation as simple as possible. The discretisation in space is first order accurate using the nearest neighbour formulation. In the following 'step' means one call of the high order accurate residual.

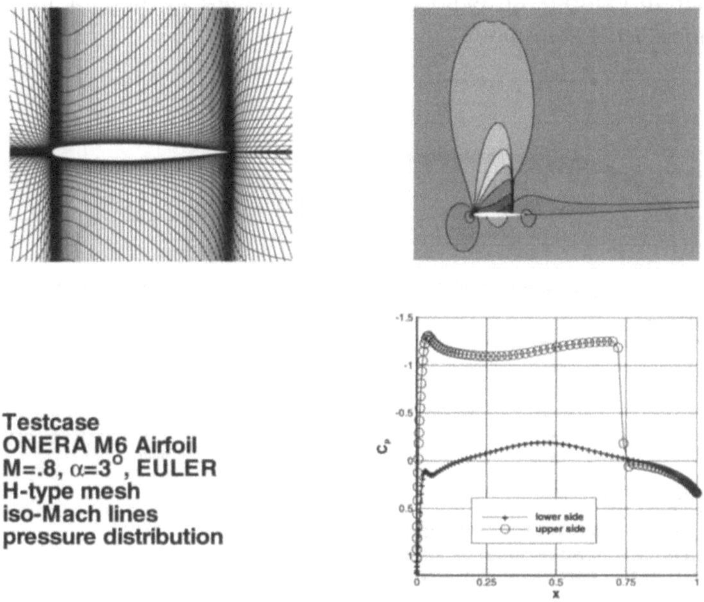

FIGURE 1

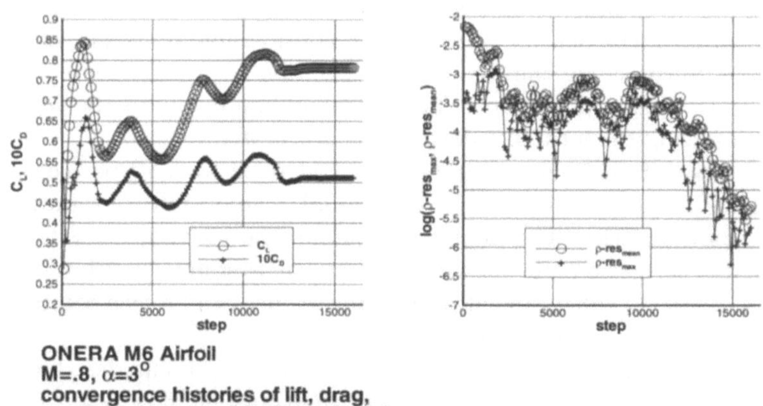

FIGURE 2

The convergence history exhibits wild variations of the lift and drag amplitudes. The number of iterations needed for convergence to the steady state is well over 10k. In contrast to the inexact Newton procedure the defect correction scheme presented herein exhibits an exponential convergence history for lift and drag if run in the local time step mode.

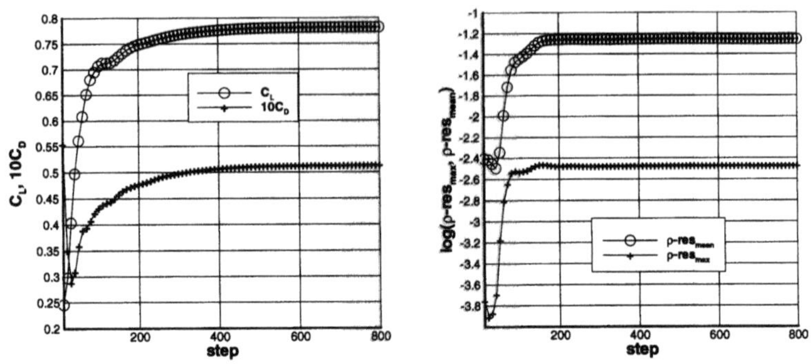

ONERA M6 Airfoil
M=.8, α=3°
convergence histories of lift, drag,
maximum/mean residual of continuity,
SGS defect correction with local time stepping

FIGURE 3

Unfortunately a large number of subiterates is needed in this case. The missing residual drop of the density equation is a considerable drawback of the local time stepping. Surprisingly the steady state solution is just the same as one obtains with other update procedures. The following figure depicts the convergence characteristics using the local 4x4 diagonal as a preconditioner.

The number of nonlinear subiterates is controlled by a 10 per cent drop of the absolute residual relative to the first residual. The overall convergence shows favourable characteristics. The next figure shows the convergence story for the restarted defect correction scheme with the implicit residual smoothing option.

Only a moderate enhancement of the convergence can be seen compared with the plain defect correction scheme.

Finally let us have a look on the performance of the present defect correction scheme applied to a 3D case. An overview of the transonic flow result is given in the next figure.

The defect correction scheme with local time stepping exhibits the same behaviour as in the 2D case. Although the integral values such as lift and drag do converge to their expected values, the residuals do not drop. The pressure distribution, however, is indistinguishible compared with the results obtained from the other update procedures. What cannot be seen in the figure: the number of explicit subiterates is intolerable large.

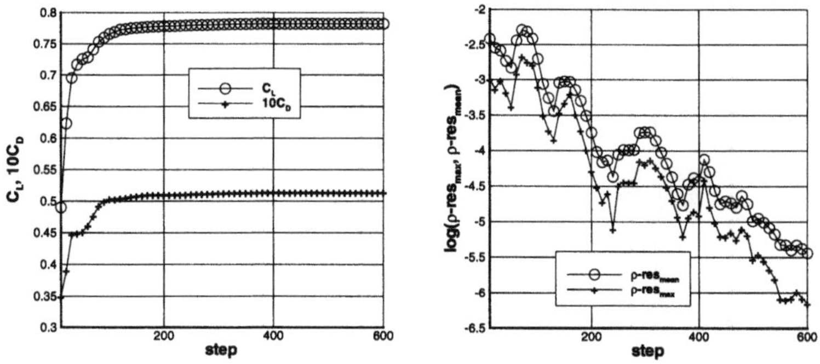

ONERA M6 Airfoil
M=.8, α=3°
convergence histories of lift, drag,
maximum/mean residual of continuity,
SGS defect correction with 4x4 diagonal matrix

FIGURE 4

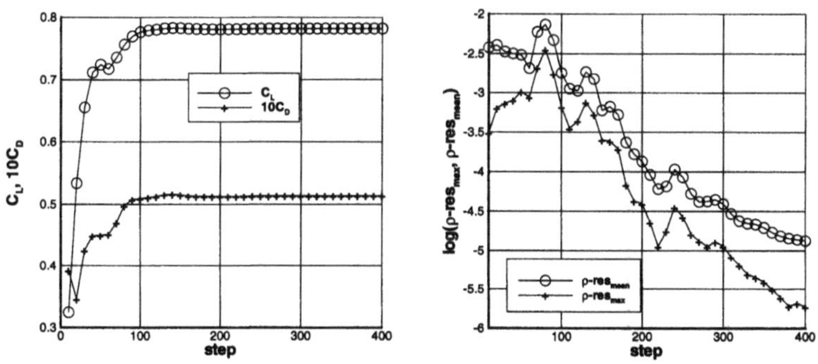

ONERA M6 Airfoil
M=.8, α=3°
convergence histories of lift, drag,
maximum/mean residual of continuity,
SGS defect correction with 4x4 diagonal matrix,
4 restarts after smoothing each 4th subiterate

FIGURE 5

Again, the preconditioning with the local 5x5 diagonal matrix leads to the expected success. In the next figure the result with full defect correction is depicted. Full defect correction means that the code is run for some iterations in the first order FVS mode which is simply achieved by putting the forcing source term on the right hand side zero.

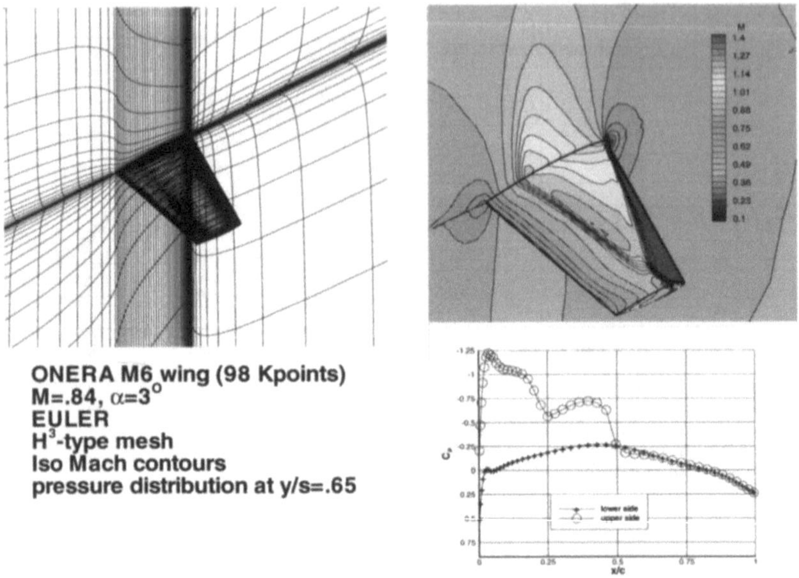

FIGURE 6

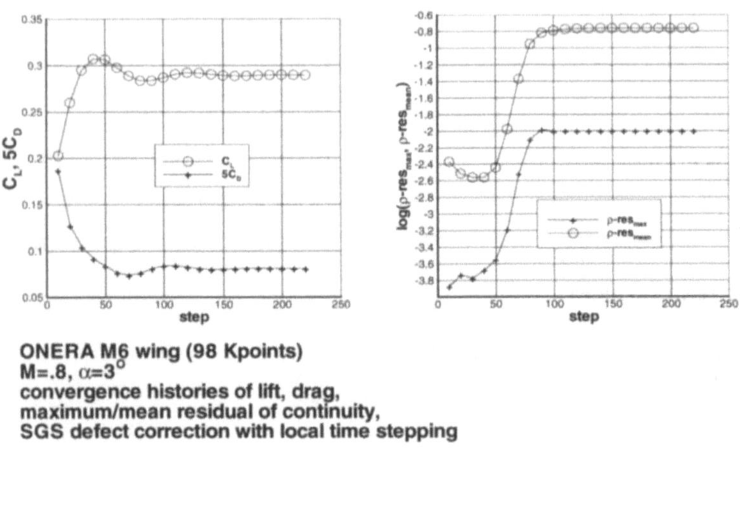

FIGURE 7

5. Conclusions

The defect correction scheme presented herein is a non linear Gauss-Seidel scheme. CPU time and number of iteration measures do reveal that this kind of algebraic acceleration beats classic incomplete Newton procedures. Local time stepping turned

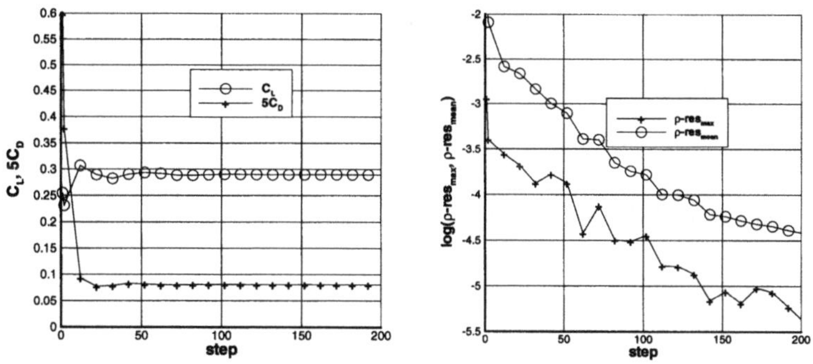

ONERA M6 wing (98 Kpoints)
M=.84, α=3°
convergence histories of lift, drag,
maximum/mean residual of continuity,
SGS full defect correction (80 first order steps)
with 5x5 diagonal matrix

FIGURE 8

out to be not a satisfying preconditioning since it always ends up in limit cycling and needs too many subiterates. Solving the diagonal matrix locally for the flow variables, however, turned out to be a powerful so called right preconditioner. Left preconditioning by implicit residual smoothing results only in a moderate enhancement of the convergence speed. Residual smoothing may be included since the computational cost is very low. The new flux vector splitting presented herein is a contribution to a robust and still very low diffusion numerical flux evaluation if used in the context of high order accurate reconstruction. It exhibits a surprising performance concerning low numerical entropy errors which hardly is achieved with other FVS formulations.

References

[1] J. Heiermann, M. Auweter-Kurtz, H.J. Kaeppeler, P.C. Sleziona, A. Eberle, U. Iben
Recent Improvements of Numerical Methods for the Simulation of MPD Thruster Flow on Adaptive Meshes
26th International Electric Propulsion Conference
Kitakyushu, Japan, Oct. 17-21, 1999

[2] J. Heiermann, M. Auweter-Kurtz, P.C. Sleziona, A. Eberle, U. Iben
Robuste hochaufloesende Methoden zur Simulation magnetoplasmadynamischer Raketentriebwerke
DGLR Tagung, Berlin, September 1999

[3] A. Eberle, A. Rizzi, E.H. Hirschel
Numerical Solutions of the Euler Equations for Steady Flow Problems
Vieweg, 1992
Notes on numerical fluid mechanics: vol. 34
ISBN 3-528-07634-8

A. Eberle
European Aeronautic Defence and Space Company,
MT63,
81663 Muenchen,
Germany
albrecht.eberle@m.eads.net

International Series of Numerical Mathematics
Vol. 140, © 2001 Birkhäuser Verlag Basel/Switzerland

A Continuous Dependence Result for Nonlinear Degenerate Parabolic Equations with Spatially Dependent Flux Function

S. Evje, K. H. Karlsen, and N. H. Risebro

Abstract. We study entropy solutions of nonlinear degenerate parabolic equations of form $u_t + \operatorname{div}(k(x)f(u)) = \Delta A(u)$, where $k(x)$ is a vector-valued function and $f(u), A(u)$ are scalar functions. We prove a result concerning the continuous dependence on the initial data, the flux function $k(x)f(u)$, and the diffusion function $A(u)$. This paper complements previous work [7] by two of the authors, which contained a continuous dependence result concerning the initial data and the flux function $k(x)f(u)$.

1. Introduction

In this paper we are concerned with entropy solutions of the initial value problem

$$u_t + \operatorname{div}(k(x)f(u)) = \Delta A(u), \quad u(x,0) = u_0(x), \tag{1}$$

for $(x,t) \in \Pi_T = \mathbb{R}^d \times (0,T)$ with $T > 0$ fixed. In (1), $u(x,t)$ is the scalar unknown function that is sought, $k(x)f(u)$ is the flux function, and $A = A(u)$ is the diffusion function. We *always* assume that $k : \mathbb{R}^d \to \mathbb{R}^d$, $f : \mathbb{R} \to \mathbb{R}$, and $A : \mathbb{R} \to \mathbb{R}$ satisfy

$$\begin{cases} k \in W^{1,1}_{\text{loc}}(\mathbb{R}^d);\ k, \operatorname{div} k \in L^\infty(\mathbb{R}^d);\ f \in \operatorname{Lip}_{\text{loc}}(\mathbb{R});\ f(0) = 0; \\ A \in \operatorname{Lip}_{\text{loc}}(\mathbb{R}) \text{ and } A(\cdot) \text{ is nondecreasing with } A(0) = 0. \end{cases} \tag{2}$$

Since $A'(\cdot)$ is allowed to be zero on an interval $[\alpha, \beta]$ (the scalar conservation law is a special case of (1)), solutions may become discontinuous in finite time even with a smooth initial function. Consequently, one needs to interpret (1) in the weak sense. However, weak solutions are in general not uniquely determined by their initial data and an entropy condition must be imposed to single out the physically correct solution.

Definition 1.1. *A measurable function $u(x,t)$ is an entropy solution of (1) if*

D.1 $u \in L^1(\Pi_T) \cap L^\infty(\Pi_T) \cap C(0,T; L^1(\mathbb{R}^d))$ *and* $A(u) \in L^2(0,T; H^1(\mathbb{R}^d))$.

D.2 *For all $c \in \mathbb{R}$ and all non-negative test functions in $C_0^\infty(\Pi_T)$,*

$$\iint_{\Pi_T} \Big(|u-c|\phi_t + \operatorname{sign}(u-c)\, k(x)\big(f(u)-f(c)\big)\cdot\nabla\phi + |A(u)-A(c)|\Delta\phi \\ - \operatorname{sign}(u-c)\operatorname{div}k(x) f(c)\phi \Big)\, dt\, dx \geq 0. \tag{3}$$

D.3 *Essentially as $t \downarrow 0$, $\|u(\cdot,t)-u_0(x)\|_{L^1(\mathbb{R}^d)} \to 0$.*

Following Kružkov [9] and the recent work of Carrillo [3], two of the authors proved in [7] a uniqueness result for entropy solutions of the more general equation

$$u_t + \operatorname{div} f(x,t,u) = \Delta A(u) + q(x,t,u), \tag{4}$$

where the flux function $f = f(x,t,u)$ may have a non-smooth spatial dependence, see [7] for the precise assumptions on f and q in (4). Moreover, in the $L^\infty(0,T;BV(\mathbb{R}^d))$ class of entropy solutions, the authors of [7] proved continuous dependence on the initial function u_0 and flux function in the case $f(x,t,u) = k(x)f(u)$. However, in [7] the question of continuous dependence with respect to the diffusion function A was left open. Recently, Cockburn and Gripenberg [4] have obtained such a result when $k(x) = 1$. Their result does *not*, however, imply uniqueness of the entropy solution (from reasons that will become apparent later). Let us also mention that results regarding continuous dependence on the flux function in scalar conservation laws ($A' \equiv 0$) have been obtained in [11, 1, 8].

The purpose of the present paper is to combine the ideas in [7] with those in [4] and prove a version of Theorem 1.3 in [7] which also includes continuous dependence on the diffusion function A. To state our continuous dependence result, let us introduce

$$v_t + \operatorname{div}\big(l(x)g(v)\big) = \Delta B(v), \qquad v(x,0) = v_0(x). \tag{5}$$

We assume that $l : \mathbb{R}^d \to \mathbb{R}$, $g : \mathbb{R} \to \mathbb{R}$, and $B : \mathbb{R} \to \mathbb{R}$ satisfy the same conditions as k, f, A, see (2). We now state our main result:

Theorem 1.1. *Let $v, u \in L^\infty(0,T;BV(\mathbb{R}^d))$ be the unique entropy solutions of* (5), (1) *with initial data $v_0, u_0 \in L^1(\mathbb{R}^d)\cap L^\infty(\mathbb{R}^d)\cap BV(\mathbb{R}^d)$, respectively. Suppose that v,u take values in in the closed interval $I \subset \mathbb{R}$ and define $V_v = \sup_{t\in(0,T)} |v(\cdot,t)|_{BV(\mathbb{R}^d)}$. Suppose $k \in \operatorname{Lip}(\mathbb{R}^d)$ and $\operatorname{div}k \in BV(\mathbb{R}^d)$. Then for almost all $t \in (0,T)$,*

$$\begin{aligned} \|v(\cdot,t)-u(\cdot,t)\|_{L^1(\mathbb{R}^d)} &\leq \|v_0-u_0\|_{L^1(\mathbb{R}^d)} \\ &+ C_{\mathrm{Conv}}t\Big(\|l-k\|_{L^\infty(\mathbb{R}^d)} + |l-k|_{BV(\mathbb{R}^d)} + \|g-f\|_{L^\infty(I)} + \|g-f\|_{\mathrm{Lip}(I)}\Big) \\ &+ C_{\mathrm{Diff}}\sqrt{t}\,\Big\|\sqrt{B'}-\sqrt{A'}\Big\|_{L^\infty(I)}, \end{aligned} \tag{6}$$

for some constants $C_{\mathrm{Conv}}, C_{\mathrm{Diff}}$. Here C_{Conv} depends on V_v, $\|k\|_{L^\infty(\mathbb{R}^d)}$, $|k|_{BV(\mathbb{R}^d)}$, $\|g\|_{L^\infty(I)}$, $\|g\|_{\mathrm{Lip}(I)}$ and C_{Diff} depends on V_v, $\|k\|_{\mathrm{Lip}(\mathbb{R}^d)}$, $|\operatorname{div}k|_{BV(\mathbb{R}^d)}$, $\|g\|_{L^\infty(I)}$, $\|g\|_{\mathrm{Lip}(I)}$. The explicit form of the constants $C_{\mathrm{Conv}}, C_{\mathrm{Diff}}$ can be traced from the proof of Theorem 1.1.

We remark that existence of $BV(\Pi_T)$ entropy solutions of (1) (or (4)) can be proved by the vanishing viscosity method provided f, A, q, u_0 are sufficiently smooth, see Vo'lpert and Hudjaev [13]. Existence of $L^\infty(0,T;BV(\mathbb{R}^d))$ entropy solutions of (1) is guaranteed if $\mathrm{div} k \in BV(\mathbb{R}^d)$. This follows from the results obtained by Karlsen and Risebro [6], who proved convergence of finite difference schemes for degenerate parabolic equations with rough coefficients. For an overview of the literature dealing with numerical methods for approximating entropy solutions of degenerate parabolic equations, we refer to [5]. In this connection, we should mention that the arguments used to prove Theorem 1.1 can be used to prove error estimates for numerical methods. This will be discussed elsewhere.

For later use, we mention that the results in [6] can be used to prove the existence of $L^\infty(0,T;BV(\mathbb{R}^d))$ entropy solutions of (1) by the vanishing viscosity method. To this end, consider the uniformly parabolic problem

$$u_t^\mu + \mathrm{div}\big(k(x)f(u^\mu)\big) = \Delta A(u^\mu) + \mu\Delta u^\mu, \qquad u^\mu(x,0) = u_0(x), \tag{7}$$

for $\mu > 0$. Provided k, f, A, u_0 are sufficiently smooth, it is well known that there exists a unique classical (and hence entropy) solution of (7) which possesses all the continuous derivatives occurring in the partial differential equation in (7). Using the space and time translation estimates derived in [6], it is not difficult to show that u^μ converges in $L^1_{\mathrm{loc}}(\Pi_T)$ as $\mu \downarrow 0$ to an entropy solution u of (1) (see also Vo'lpert and Hudjaev [13]). Convergence of the viscosity method and smoothness of the solution u^μ of (7) will be used in the proof of Theorem 1.1. Finally, to relax the smoothness assumptions on k, f, A, u_0 needed by the vanishing viscosity method to those actually required by Theorem 1.1, one can approximate k, f, A, u_0 by smoother functions and then use Theorem 1.1 to pass to the limit as the smoothing parameter tends to zero. We will not go into further details about this limiting operation but instead leave this as an exercise for the interested reader. Also, in this paper we have exclusively treated the initial value problem but it is possible to treat various initial-boundary value problems. For work in this direction, we refer to Bürger, Evje, and Karlsen [2] and Rouvre and Gagneux [12].

Before ending this section, we present an immediate corollary of Theorem 1.1 concerning the convergence rate of the viscosity method.

Corollary 1.2. *Let $u \in L^\infty(0,T;BV(\mathbb{R}^d))$ be the unique entropy solution of* (1) *with initial data $u_0 \in L^1(\mathbb{R}^d) \cap L^\infty(\mathbb{R}^d) \cap BV(\mathbb{R}^d)$ and let u^μ be the corresponding viscous approximation of u, i.e., u^μ is the unique classical solution of* (7). *Suppose $k \in \mathrm{Lip}(\mathbb{R}^d)$ and $\mathrm{div} k \in BV(\mathbb{R}^d)$. Then for almost all $t \in (0,T)$,*

$$\|u(\cdot,t) - u^\mu(\cdot,t)\|_{L^1(\mathbb{R}^d)} \le C\sqrt{\mu}, \tag{8}$$

for some non-negative constant C.

The remaining part of this paper is devoted to proving Theorem 1.1.

2. Some preliminaries

Let u be an entropy solution of (1). It easy to see from Definition 1.1 that the equality

$$\iint_{\Pi_T} \Big(u\phi_t + \big[k(x)f(u) - \nabla A(u)\big] \cdot \nabla\phi\Big)\, dt\, dx = 0 \tag{9}$$

holds for all $\phi \in L^2(0,T;H_0^1(\mathbb{R}^d)) \cap W^{1,1}(0,T;L^\infty(\mathbb{R}^d))$. Let $\langle\cdot,\cdot\rangle$ denote the usual pairing between $H^{-1}(\mathbb{R}^d)$ and $H_0^1(\mathbb{R}^d)$. From (9), we conclude that $\partial_t u \in L^2(0,T;H^{-1}(\mathbb{R}^d))$, so that the equality (9) can restated as

$$-\int_0^T \langle \partial_t u, \phi\rangle\, dt + \iint_{\Pi_T} \Big(\big[k(x)f(u) - \nabla A(u)\big] \cdot \nabla\phi\Big)\, dt\, dx = 0 \tag{10}$$

for all $\phi \in L^2(0,T;H_0^1(\mathbb{R}^d)) \cap W^{1,1}(0,T;L^\infty(\mathbb{R}^d))$. The fact that an entropy solution u satisfies (10) is important for the uniqueness proof [3, 7]. We recall that after the work of Carrillo [3], the uniquness proof for entropy solutions of degenerate parabolic equations has become very similar to the "doubling of variables" proof introduced by Kružkov [9] many years ago for first order hyperbolic equations. However, to apply the "doubling of variables" device to second order equations, one needs a version of an important lemma stated and proved first in [3]. This lemma indentifies a certain entropy dissipation term that must be taken into account if the "doubling device" is going to work.

Before stating this lemma, we introduce some notation. For $\varepsilon > 0$, set

$$\operatorname{sign}_\varepsilon(\tau) = \begin{cases} -1, & \tau < \varepsilon, \\ \tau/\varepsilon, & \varepsilon \le \tau \le \varepsilon, \\ 1 & \tau > \varepsilon. \end{cases}$$

Moreover, we let $A^{-1} : \mathbb{R} \to \mathbb{R}$ denote the unique left-continuous function satisfying $A^{-1}(A(u)) = u$ for all $u \in \mathbb{R}$. Define $E = \{r : A^{-1}(\cdot) \text{ discontinuous at } r\}$. Note that E is associated with the set of points $\{u : A'(u) = 0\}$ at which the operator $u \mapsto \Delta A(u)$ is degenerate elliptic. We can now state following lemma:

Lemma 2.1 ([7]). *Let u be an entropy solution of* (1). *Then, for any non-negative $\phi \in C_0^\infty(\Pi_T)$ and any $c \in \mathbb{R}$ such that $A(c) \notin E$, we have*

$$\iint_{\Pi_T} \Big(|u-c|\phi_t + \operatorname{sign}(u-c)\big[k(x)\big(f(u)-f(c)\big) - \nabla A(u)\big]\cdot\nabla\phi$$
$$- \operatorname{sign}(u-c)\operatorname{div}k(x)f(c)\phi\Big)\, dt\, dx = \lim_{\varepsilon\downarrow 0} \iint_{\Pi_T} \big|\nabla A(u)\big|^2 \operatorname{sign}_\varepsilon'(A(u)-A(c))\,\phi\, dt\, dx. \tag{11}$$

Note that if (1) is uniformly parabolic ($A' > 0$), then the set E is empty and a weak solution is automatically an entropy solution. The idea of the proof of Lemma 2.1 is to use $[\operatorname{sign}_\varepsilon(A(u)-A(c))\,\phi] \in L^2(0,T;H_0^1(\mathbb{R}))$ as a test function in (10) together with a "weak chain rule" to deal with the time derivative and subsequently sending $\varepsilon \downarrow 0$. We refer to [7] for details on the proof of Lemma 2.1 (see Carrillo [3] when $k(x) \equiv 1$).

Although the identification of the entropy dissipation term (i.e., the right-hand side of (11)) is the cornerstone of the uniqueness proof as well as the proof of continuous dependence on the flux function $k(x)f(u)$, it seems difficult to obtain continuous dependence on the diffusion function $A(u)$ with this form of the dissipation term. However, it is possible to derive a version of (11) in which a different (form of the) entropy dissipation term appears. But this seems possible only if u is smooth or at least belongs to $L^2(0,T;H^1(\mathbb{R}^d))$. Consequently, Theorem 1.1 does *not* yield uniqueness of the entropy solution.

Provided (1) is uniformly parabolic and hence admits a unique classical solution, the following version of Lemma 2.1 holds:

Lemma 2.2. *Suppose* (1) *is uniformly parabolic (i.e., $A' > 0$). Let u be a classical solution of* (1). *Then, for any non-negative $\phi \in C_0^\infty(\Pi_T)$ and any $c \in \mathbb{R}$, we have*

$$\iint_{\Pi_T} \Big(|u-c|\phi_t + \operatorname{sign}(u-c)\left[k(x)\big(f(u)-f(c)\big) - \nabla A(u)\right]\cdot\nabla\phi$$
$$- \operatorname{sign}(u-c)\operatorname{div} k(x) f(c)\phi \Big)\, dt\, dx = \lim_{\varepsilon\downarrow 0}\iint_{\Pi_T} A'(u)\left|\nabla u\right|^2 \operatorname{sign}_\varepsilon'(u-c)\,\phi\, dt\, dx. \tag{12}$$

This lemma can proved by using $[\operatorname{sign}_\varepsilon(u-c)\,\phi]$ as a test function (this is indeed a test function since u is smooth) in (10) and then sending $\varepsilon \downarrow 0$. The proof of Lemma 2.2 is similar to the proof of Lemma 2.1 and it is therefore omitted. Notice the difference between the entropy dissipation terms in (11) and (12).

3. Proof of Theorem 1.1

We are now interested in estimating the L^1 difference between the entropy solution v of (5) and the entropy solution u of (1). In view of the discussion in Section 1, we will prove Theorem 1.1 under the assumption that $B', A' > 0$ so that (5) and (1) become uniformly parabolic problems and therefore admit unique classical solutions. To treat the degenerate parabolic case ($B', A' \geq 0$), we proceed via the vanishing viscosity method, i.e., we replace $\Delta A(u)$ and $\Delta B(v)$ in (1) and (5) by $\Delta A(u) + \eta\Delta u$ and $\Delta B(v) + \eta\Delta v$, respectively, and then send $\eta \downarrow 0$.

The argument given below uses Lemma 2.2 and Kružkov's idea of doubling the number of dependent variables together with a penalization procedure. It is inspired by Carrillo [3] and Cockburn and Gripenberg [4]. Strictly speaking, we could have carried out the argument below under the assumptions that the (entropy) solutions v, u belong to $L^2(0,T;H^1(\mathbb{R}^d))$, i.e., v, u need not be (entirely!) classical solutions. Although we do not prove this here, all functions below that are integrated can be proven to be integrable!

Following [9, 10], we now specify a non-negative test function $\phi \in C_0^\infty(\Pi_T \times \Pi_T)$. To this end, introduce a nonnegative function $\delta \in C_0^\infty(\mathbb{R})$ which satisfies $\delta(\sigma) = \delta(-\sigma)$, $\delta(\sigma) \equiv 0$ for $|\sigma| \geq 1$, and $\int_{\mathbb{R}} \delta(\sigma)\, d\sigma = 1$. For $\rho_0 > 0$, let $\delta_{\rho_0}(\sigma) = \frac{1}{\rho_0}\delta(\frac{\sigma}{\rho_0})$. Pick two (arbitrary but fixed) points $\nu, \tau \in (0, T)$. For any $\alpha_0 > 0$, we define $W_{\alpha_0}(t) = H_{\alpha_0}(t-\nu) - H_{\alpha_0}(t-\tau)$, where $H_{\alpha_0}(t) = \int_{-\infty}^t \delta_{\alpha_0}(s)\, ds$. With $0 < \rho_0 < \min(\nu, T-\tau)$ and $\alpha_0 \in (0, \min(\nu - \rho_0, T - \tau - \rho_0))$, we define the test function $\phi(x,t,y,s) = W_{\alpha_0}(t)\delta_\rho(x-y)\delta_{\rho_0}(t-s)$, $\rho > 0$. One should observe that $\phi_t + \phi_s = \big[\delta_{\alpha_0}(t-\nu) - \delta_{\alpha_0}(t-\tau)\big]\delta_\rho(x-y)\delta_{\rho_0}(t-s)$ and $\nabla_x\phi + \nabla_y\phi \equiv 0$.

Applying Lemma 2.2 with $v = v(x,t)$ and $c = u(y,s)$ and then integrating the resulting equation with respect to $(y,s) \in \Pi_T$, we get

$$
\begin{aligned}
-\iiiint\limits_{\Pi_T\times\Pi_T} \Big(& |v-u|\phi_t + \operatorname{sign}(v-u)\,\big[l(x)\big(g(v)-g(u)\big) - \nabla_x B(v)\big]\cdot\nabla_x\phi \\
& - \operatorname{sign}(v-u)\operatorname{div}_x l(x) g(u)\phi\Big)\, d\xi \qquad (13)\\
= -\lim_{\varepsilon\downarrow 0} & \iiiint\limits_{\Pi_T\times\Pi_T} B'(v)\big|\nabla_x v\big|^2 \operatorname{sign}_\varepsilon'(v-u)\,\phi\, d\xi,
\end{aligned}
$$

where $d\xi = dt\,dx\,ds\,dy$. Similarly, applying Lemma 2.2 with $u = u(y,s)$ and $c = v(x,t)$ and then integrating the resulting equation with respect to $(x,t) \in \Pi_T$, we get

$$
\begin{aligned}
-\iiiint\limits_{\Pi_T\times\Pi_T} \Big(& |u-v|\phi_t + \operatorname{sign}(u-v)\,\big[k(y)\big(f(u)-f(v)\big) - \nabla_y A(u)\big]\cdot\nabla_y\phi \\
& - \operatorname{sign}(u-v)\operatorname{div}_y k(y) f(v)\phi\Big)\, d\xi \qquad (14)\\
= -\lim_{\varepsilon\downarrow 0} & \iiiint\limits_{\Pi_T\times\Pi_T} A'(u)\big|\nabla_y u\big|^2 \operatorname{sign}_\varepsilon'(u-v)\,\phi\, d\xi.
\end{aligned}
$$

Following [7] when adding (13) and (14), we get

$$
\begin{aligned}
-\iiiint\limits_{\Pi_T\times\Pi_T} & \Big(|v-u|(\phi_t+\phi_s) + I_{\text{Conv}} - I_{\text{Diff}}^1\Big)\, d\xi \\
&= -\lim_{\varepsilon\downarrow 0} \iiiint\limits_{\Pi_T\times\Pi_T} \Big(B'(v)\big|\nabla_x v\big|^2 + A'(u)\big|\nabla_y u\big|^2\Big)\operatorname{sign}_\varepsilon'(v-u)\,\phi\, d\xi, \\
&= -\lim_{\varepsilon\downarrow 0} \iiiint\limits_{\Pi_T\times\Pi_T} \bigg(\Big(\sqrt{B'(v)}\nabla_x v - \sqrt{A'(u)}\nabla_y u\Big)^2 \qquad (15)\\
&\qquad + 2\sqrt{B'(v)}\sqrt{A'(u)}\nabla_x v\cdot\nabla_y u\bigg)\operatorname{sign}_\varepsilon'(v-u)\,\phi\, d\xi, \\
&\leq -\lim_{\varepsilon\downarrow 0} \iiiint\limits_{\Pi_T\times\Pi_T} I_{\text{Diff}}^2\, d\xi,
\end{aligned}
$$

where

$$I_{\text{Conv}} = \operatorname{sign}(v-u)\Big(\operatorname{div}_x\big[\big(k(y)f(u)-l(x)g(u)\big)\phi\big] \\ - \operatorname{div}_y\big[\big(l(x)g(v)-k(y)f(v)\big)\phi\big]\Big),$$

$$I^1_{\text{Diff}} = \operatorname{sign}(v-u)\,\nabla_x B(v)\cdot\nabla_x\phi + \operatorname{sign}(u-v)\,\nabla_y A(u)\cdot\nabla_y\phi,$$

$$I^2_{\text{Diff}} = \Big(2\sqrt{B'(v)}\sqrt{A'(u)}\nabla_x v\cdot\nabla_y u\Big)\operatorname{sign}'_\varepsilon(v-u)\,\phi.$$

By the triangle inequality, we get

$$-\iiiint_{\Pi_T\times\Pi_T} |v(x,t)-u(y,s)|\big(\phi_t+\phi_s\big)\,d\xi \le I + R^t + R^x,$$

where

$$I = -\iiiint_{\Pi_T\times\Pi_T} |v(y,t)-u(y,t)|\big[\delta_{\alpha_0}(t-\nu)-\delta_{\alpha_0}(t-\tau)\big]\delta_\rho(x-y)\delta_{\rho_0}(t-s)\,d\xi,$$

$$R^t = -\iiiint_{\Pi_T\times\Pi_T} |u(y,t)-u(y,s)|\big[\delta_{\alpha_0}(t-\nu)-\delta_{\alpha_0}(t-\tau)\big]\delta_\rho(x-y)\delta_{\rho_0}(t-s)\,d\xi,$$

$$R^x = -\iiiint_{\Pi_T\times\Pi_T} |v(x,t)-v(y,t)|\big[\delta_{\alpha_0}(t-\nu)-\delta_{\alpha_0}(t-\tau)\big]\delta_\rho(x-y)\delta_{\rho_0}(t-s)\,d\xi.$$

It is fairly easy to see that $\lim_{\rho_0\downarrow 0} R^t = 0$ and

$$\lim_{\alpha_0\downarrow 0} R^x = \int_{\mathbb{R}^d}\int_{\mathbb{R}^d}\Big(|v(x,\tau)-v(y,\tau)| - |v(x,\nu)-v(y,\nu)|\Big)\delta_\rho(x-y)\,dx\,dy, \\ \le \rho \sup_{t\in(\nu,\tau)} |v(\cdot,t)|_{BV(\mathbb{R}^d)},$$

$$\lim_{\alpha_0\downarrow 0} I = \|v(\cdot,\tau)-u(\cdot,\tau)\|_{L^1(\mathbb{R}^d)} - \|v(\cdot,\nu)-u(\cdot,\nu)\|_{L^1(\mathbb{R}^d)}.$$

We therefore get the following approximation inequality

$$\|v(\cdot,\tau)-u(\cdot,\tau)\|_{L^1(\mathbb{R}^d)} \\ \le \|v(\cdot,\nu)-u(\cdot,\nu)\|_{L^1(\mathbb{R}^d)} + \rho\sup_{t\in(\nu,\tau)}|v(\cdot,t)|_{BV(\mathbb{R}^d)} + \lim_{\alpha_0,\rho_0\downarrow 0}\Big(E_{\text{Conv}}+E_{\text{Diff}}\Big), \tag{16}$$

where

$$E_{\text{Conv}} = \iiiint_{\Pi_T\times\Pi_T} I_{\text{Conv}}\,d\xi,$$

$$E_{\text{Diff}} = -\iiiint_{\Pi_T\times\Pi_T} I^1_{\text{Diff}}\,d\xi - \lim_{\varepsilon\downarrow 0}\iiiint_{\Pi_T\times\Pi_T} I^2_{\text{Diff}}\,d\xi =: E^1_{\text{Diff}} - E^2_{\text{Diff}}.$$

Observe that

$$\begin{aligned}
E^2_{\text{Diff}} &= -\lim_{\varepsilon\downarrow 0} \iiiint\limits_{\Pi_T\times\Pi_T} 2\nabla_y \Big(\int_v^u \operatorname{sign}'_\varepsilon(v-\xi)\sqrt{A'(\xi)}\,d\xi\Big)\sqrt{B'(v)}\nabla_x v\phi\,d\xi \\
&= \lim_{\varepsilon\downarrow 0} \iiiint\limits_{\Pi_T\times\Pi_T} 2\Big(\int_v^u \operatorname{sign}'_\varepsilon(v-\xi)\sqrt{A'(\xi)}\,d\xi\Big)\sqrt{B'(v)}\nabla_x v\nabla_y\phi\,d\xi \\
&= -\lim_{\varepsilon\downarrow 0} \iiiint\limits_{\Pi_T\times\Pi_T} 2\Big(\int_v^u \operatorname{sign}'_\varepsilon(v-\xi)\sqrt{A'(\xi)}\,d\xi\Big)\sqrt{B'(v)}\nabla_x v\nabla_x\phi\,d\xi \\
&= \iiiint\limits_{\Pi_T\times\Pi_T} 2\operatorname{sign}(v-u)\sqrt{B'(v)}\sqrt{A'(v)}\nabla_x v\nabla_x\phi\,d\xi.
\end{aligned} \tag{17}$$

Writing $\operatorname{sign}(u-v)\nabla_y A(u) = \nabla_y\big|A(u)-A(v)\big|$ and using integration parts twice as well as the relation $\Delta_x\phi = \Delta_y\phi$, one can easily show that

$$\begin{aligned}
&\iiiint\limits_{\Pi_T\times\Pi_T} \operatorname{sign}(u-v)\nabla_y A(u)\cdot\nabla_y\phi\,d\xi \\
&\qquad = \iiiint\limits_{\Pi_T\times\Pi_T} \operatorname{sign}(v-u)\nabla_x A(v)\cdot\nabla_x\phi\,d\xi.
\end{aligned} \tag{18}$$

From (17), (18), and $\big\|\nabla_x\delta_\rho(x-y)\big\|_{L^1(\mathbb{R}^d)} = \frac{2d}{\rho}$, we get

$$\begin{aligned}
&\lim_{\alpha_0\downarrow 0} E_{\text{Diff}} \\
&\quad \le -\lim_{\alpha_0\downarrow 0} \iiiint\limits_{\Pi_T\times\Pi_T} \operatorname{sign}(v-u)\Big(B'(v)+A'(v)-2\sqrt{B'(v)}\sqrt{A'(v)}\Big)\nabla_x v\nabla_x\phi\,d\xi \\
&\quad \le \lim_{\alpha_0\downarrow 0} \iiiint\limits_{\Pi_T\times\Pi_T} \Big(\sqrt{B'(v)}-\sqrt{A'(v)}\Big)^2 |\nabla_x v|\,|\nabla_x\delta_\rho(x-y)|W_{\alpha_0}(t)\delta_{\rho_0}(t-s)\,d\xi \\
&\quad \le (\tau-\nu)\sup_{t\in(\nu,\tau)} |v(\cdot,t)|_{BV(\mathbb{R}^d)}\frac{2d}{\rho}\Big\|\sqrt{B'(v)}-\sqrt{A'(v)}\Big\|^2_{L^\infty(\mathbb{R}^d)}.
\end{aligned} \tag{19}$$

Arguing exactly as in [7], one can prove that

$$
\begin{aligned}
E_{\text{Conv}} &= \iiiint\limits_{\Pi_T\times\Pi_T} \Big(\operatorname{sign}(v-u)\,\big[\operatorname{div}_y k(y)\big(f(v)-g(v)\big) - \big(\operatorname{div}_y k(y) - \operatorname{div}_x l(x)\big)g(v)\big] \\
&\quad + \big(k(y)-l(x)\big)\cdot\nabla_x G(v,u) + k(y)\cdot\nabla_x\big(F(v,u)-G(v,u)\big)\Big)\phi\, d\xi \\
&= \iiiint\limits_{\Pi_T\times\Pi_T} \Big(\operatorname{sign}(v-u)\,\big[\operatorname{div}_y k(y)\big(f(v)-g(v)\big) - \big(\operatorname{div}_x k(x) - \operatorname{div}_x l(x)\big)g(v)\big] \\
&\quad + \big(k(x)-l(x)\big)\cdot\nabla_x G(v,u) + k(y)\cdot\nabla_x\big(F(v,u)-G(v,u)\big)\Big)\phi\, d\xi \\
&\quad + \iiiint\limits_{\Pi_T\times\Pi_T} \operatorname{sign}(v-u)\,\big(\operatorname{div}_x k(x) - \operatorname{div}_y k(y)\big)g(v)\phi\, d\xi \\
&\quad + \iiiint\limits_{\Pi_T\times\Pi_T} \operatorname{sign}(v-u)\,\big(k(y)-k(x)\big)\cdot\nabla_x G(v,u)\phi\, d\xi \\
&=: E^1_{\text{Conv}} + E^2_{\text{Conv}} + E^3_{\text{Conv}}.
\end{aligned}
\tag{20}
$$

Following [7], we derive the estimate

$$
\begin{aligned}
\lim_{\alpha_0\downarrow 0} E^1_{\text{Conv}}
&\le (\tau-\nu)\Big(\|g\|_{\text{Lip}} \sup_{t\in(\nu,\tau)} |v(\cdot,t)|_{BV(\mathbb{R}^d)}\,\|k-l\|_{L^\infty(\mathbb{R}^d)} + \|g\|_{L^\infty(I)}\,|k-l|_{BV(\mathbb{R}^d)} \\
&\quad + |k|_{BV(\mathbb{R}^d)}\,\|f-g\|_{L^\infty(I)} + \|k\|_{L^\infty(\mathbb{R}^d)} \sup_{t\in(\nu,\tau)} |v(\cdot,t)|_{BV(\mathbb{R}^d)}\|f-g\|_{\text{Lip}(I)}\Big).
\end{aligned}
\tag{21}
$$

Taking into account $\operatorname{div} k \in BV(\mathbb{R}^d)$ and $k \in \text{Lip}(\mathbb{R}^d)$, it is easy to show that

$$
\lim_{\alpha_0\downarrow 0} E^2_{\text{Conv}} \le |\operatorname{div} k|_{BV(\mathbb{R}^d)}\|g\|_{L^\infty(I)}(\tau-\nu)\rho, \tag{22}
$$

$$
\lim_{\alpha_0\downarrow 0} E^3_{\text{Conv}} \le \|k\|_{\text{Lip}(\mathbb{R}^d)}\|g\|_{\text{Lip}(I)} \sup_{t\in(\nu,\tau)} |v(\cdot,t)|_{BV(\mathbb{R}^d)}(\tau-\nu)\rho. \tag{23}
$$

Inserting (19), (21), (22), and (23) into (16), minimizing the result with respect to $\rho > 0$, and subsequently sending $\nu \downarrow 0$, we get (6). This concludes the proof of Theorem 1.1 when $B', A' > 0$. Note that (6) does not depend on the smoothness of v, u. Hence the proof in the general case $B', A' \ge 0$ can proceed via the L^1 convergence of the viscosity method (see Section 1).

References

[1] F. Bouchut and B. Perthame. Kružkov's estimates for scalar conservation laws revisited. *Trans. Amer. Math. Soc.*, 350(7):2847–2870, 1998.

[2] R. Bürger, S. Evje, and K. H. Karlsen. On strongly degenerate convection-diffusion problems modeling sedimentation-consolidation processes. *J. Math. Anal. Appl.*, 247(2):517–556, 2000.

[3] J. Carrillo. Entropy solutions for nonlinear degenerate problems. *Arch. Rational Mech. Anal.*, 147(4):269–361, 1999.

[4] B. Cockburn and G. Gripenberg. Continuous dependence on the nonlinearities of solutions of degenerate parabolic equations. *J. Differential Equations*, 151(2):231–251, 1999.

[5] M. S. Espedal and K. H. Karlsen. Numerical solution of reservoir flow models based on large time step operator splitting algorithms. In *Filtration in Porous Media and Industrial Applications (Cetraro, Italy, 1998)*, volume 1734 of *Lecture Notes in Mathematics*, pages 9–77. Springer, Berlin, 2000.

[6] K. H. Karlsen and N. H. Risebro. Convergence of finite difference schemes for viscous and inviscid conservation laws with rough coefficients. *Mathematical Modelling and Numerical Analysis*, 35(2):239–270, 2001.

[7] K. H. Karlsen and N. H. Risebro. On the uniqueness and stability of entropy solutions of nonlinear degenerate parabolic equations with rough coefficients. Preprint, Department of Mathematics, University of Bergen, 2000.

[8] R. A. Klausen and N. H. Risebro. Stability of conservation laws with discontinuous coefficients. *J. Differential Equations*, 157(1):41–60, 1999.

[9] S. N. Kružkov. First order quasi-linear equations in several independent variables. *Math. USSR Sbornik*, 10(2):217–243, 1970.

[10] N. N. Kuznetsov. Accuracy of some approximative methods for computing the weak solutions of a first-order quasi-linear equation. *USSR Comput. Math. and Math. Phys. Dokl.*, 16(6):105–119, 1976.

[11] B. J. Lucier. A moving mesh numerical method for hyperbolic conservation laws. *Math. Comp.*, 46(173):59–69, 1986.

[12] É. Rouvre and G. Gagneux. Solution forte entropique de lois scalaires hyperboliques-paraboliques dégénérées. *C. R. Acad. Sci. Paris Sér. I Math.*, 329(7):599–602, 1999.

[13] A. I. Vol'pert and S. I. Hudjaev. Cauchy's problem for degenerate second order quasilinear parabolic equations. *Math. USSR Sbornik*, 7(3):365–387, 1969.

S. Evje
RF-Rogaland Research, Thormøhlensgt. 55,
N–5008 Bergen, Norway
steinar.evje@rf.no

K.H. Karlsen
Department of Mathematics, University of Bergen,
Johs. Brunsgt. 12, N–5008 Bergen, Norway
kennethk@mi.uib.no

N.H. Risebro
Department of Mathematics, University of Oslo,
P.O. Box 1053, Blindern, N–0316 Oslo, Norway
nilshr@math.uio.no

International Series of Numerical Mathematics
Vol. 140,

A Lagrangian Central Scheme for Multi-Fluid Flows

Riccardo Fazio and Giovanni Russo

Abstract. We develop a central scheme for multi-fluid flows in Lagrangian coordinates. The main contribution is the derivation of a special equation of state to be imposed at the interface in order to avoid non-physical oscillations. The proposed scheme is validated by solving several tests concerning one-dimensional hyperbolic interface problems.

1. Introduction

We study a high-resolution central scheme for the system of conservation laws describing two gases separated by an interface. We consider a piston problem as described by Fazio and LeVeque [1]: a tube contains two different gases separated by an interface at some point $I(t) < L(t)$ where $I(t)$ and $L(t)$ are interface and piston position. The governing equations are Euler equations of gas dynamics,

$$\frac{\partial \mathbf{q}}{\partial t} + \frac{\partial}{\partial x}\left[\mathbf{f}(\mathbf{q})\right] = \mathbf{0} , \tag{1}$$

with

$$\mathbf{q} = [\rho, \rho u, E]^T ,$$
$$\mathbf{f}(\mathbf{q}) = \left[\rho u, \rho u^2 + p, (E+p)u\right]^T , \tag{2}$$

and with the constitutive law for ideal gases

$$p = (\gamma(x,t) - 1)\left(E - \frac{1}{2}\rho u^2\right) , \tag{3}$$

where ρ, u, E, and p denote density, velocity, total energy density per unit volume, and pressure of the gas. The polytropic constant $\gamma(x,t)$ takes the value γ_1 on $0 \le x < I(t)$, and γ_2 on $I(t) < x \le L(t)$. The motion of the piston is driven by Newton's equation

$$\frac{d^2 L}{dt^2} = \frac{A}{m}(p(L(t),t) - p_{out}(t)) , \tag{4}$$

where A is the area of the piston, m is its mass and $p_{out}(t)$ is the external pressure.

Note that by setting $A = 0$ we resume, from the piston model above, the classical tube problem with an interface and fixed boundaries.

2. Lagrangian formulation.

By introducing the Lagrangian coordinate ξ given by

$$\xi = \int_{x_0(t)}^{x} \rho(z,t)dz \ ,$$

where $x_0(t)$ denotes the Eulerian coordinate of the first fluid particle of the domain, the Euler equations (1)-(2) can be transformed in Lagrangian form

$$\frac{D\mathbf{q}}{Dt} + \frac{\partial}{\partial \xi}\left[\mathbf{f}(\mathbf{q})\right] = \mathbf{0} \ , \qquad (5)$$

which is also in conservation form with

$$\begin{aligned} \mathbf{q} &= [V, u, \mathcal{E}]^T \ , \\ \mathbf{f}(\mathbf{q}) &= [-u, p, up]^T \ , \end{aligned} \qquad (6)$$

here the time derivative is the Lagrangian derivative

$$D/Dt \ = \ \partial/\partial t + u\partial/\partial x \ ,$$

the new field variables are defined by $V(\xi,t) = \rho^{-1}, \mathcal{E} = E/\rho$, and the equation of state (3) becomes

$$p \ = \ (\gamma(\xi,t) - 1)\left(\mathcal{E} - \frac{1}{2}u^2\right)/V \ . \qquad (7)$$

The inverse transformation of coordinate is

$$x = x_0(t) + \int_0^{\xi} V(z,t)dz \ ,$$

and $x_0(t)$ satisfies the equation

$$\frac{d}{dt}x_0 = u(\xi = 0, t) \ .$$

Hence, $0 \le \xi \le \xi_{\max}$ will be our "computational domain" in which we have a fixed uniform grid with $\xi_i = (i - 1/2)\Delta\xi$ for $i = 1, 2, \ldots, N$ denoting the center of i-th cell, and $\Delta\xi = \xi_{\max}/N$.

3. The Nessyahu and Tadmor central scheme.

The Nessyahu and Tadmor central scheme [5] has the form of a predictor-corrector scheme

$$\begin{aligned} \mathbf{q}_j^{n+1/2} &= \mathbf{q}_j^n - \frac{\lambda}{2}\mathbf{f}_j' \ , \\ \mathbf{q}_{j+1/2}^{n+1} &= \frac{1}{2}(\mathbf{q}_j^n + \mathbf{q}_{j+1}^n) + \frac{1}{8}(\mathbf{q}_j' - \mathbf{q}_{j+1}') - \lambda\left(\mathbf{f}\left(\mathbf{q}_{j+1}^{n+1/2}\right) - \mathbf{f}\left(\mathbf{q}_j^{n+1/2}\right)\right) \ , \end{aligned}$$

where $\mathbf{q}_j^n$ denotes an approximation of the cell average of the field at time t_n

$$\mathbf{q}_j^n \approx \frac{1}{\Delta\xi} \int_{\xi_j-\Delta\xi/2}^{\xi_j+\Delta\xi/2} \mathbf{q}(\xi, t_n)\, d\xi$$

and $\lambda = \Delta t/\Delta\xi$. The time step Δt must satisfy the stability condition

$$\lambda \max_j \rho(A(\mathbf{q}_j^n)) < \frac{1}{2} \,,$$

where $\rho(A)$ denotes the spectral radius of the Jacobian matrix

$$A = \left[\frac{\partial f_i}{\partial q_k}\right] \,.$$

This condition ensures that the generalized Riemann problems with piece-wise smooth data at time t_n do not interact during the time step Δt.

$\mathbf{q}_j'/\Delta\xi$ and $\mathbf{f}_j'/\Delta\xi$ are a first order approximation of the space derivatives of the field and of the flux and can be computed in several ways. The simplest choice is

$$\begin{aligned} \mathbf{q}_j' &= \mathrm{MM}(\mathbf{q}_{j+1} - \mathbf{q}_j, \mathbf{q}_j - \mathbf{q}_{j-1}) \,, \\ \mathbf{f}_j' &= \mathrm{MM}(\mathbf{f}_{j+1} - \mathbf{f}_j, \mathbf{f}_j - \mathbf{f}_{j-1}) \,, \end{aligned}$$

where $\mathrm{MM}(v,w)$ is the min-mod limiter

$$\mathrm{MM}(v,w) = \begin{cases} \mathrm{sgn}(v) \cdot \min(|v|,|w|) & \text{if } \mathrm{sgn}(v) = \mathrm{sgn}(w) \\ 0 & \text{otherwise} \,. \end{cases}$$

Several other choices of the limiters are possible, as discussed in [5].

4. Balancing the pressure at the interface.

The development of this section starts from the observation that due to the presence of the interface several quantities (such as density and energy) may be discontinuous, but pressure and velocity have to be continuous across the interface. This condition is enforced in our scheme.

At the initial time we assume that the interface is located at the boundary between cell j_w and cell $j_w + 1$. Because of the use of Lagrangian coordinates, the interface will always separate cell j_w from cell $j_w + 1$ at even time steps, and it will be in the middle of cell $j_w + 1/2$ at odd time steps (see Figure 1).

Let us denote by subscript 1 and 2 the values of the field variables on the two sides of the interface cell at odd time steps. The balance of the pressure on the two sides of the interface, $p_1 = p_2$, gives

$$\frac{\gamma_1 - 1}{V_1}\left(\mathcal{E}_1 - \frac{1}{2}u^2\right) = \frac{\gamma_2 - 1}{V_2}\left(\mathcal{E}_2 - \frac{1}{2}u^2\right) \,. \tag{8}$$

We assume that on the two sides of the interface there are two different gases, whose ratio η of densities is equal to the ratio of molecular masses:

$$V_2 = \eta V_1 \,. \tag{9}$$

Note that this condition is physically correct if the temperature is continuous across the interface. The cell average of the specific volume V and energy $\mathcal{E}$ at the interface cell $j_w + 1/2$ are

$$V = \frac{V_1 + V_2}{2}, \quad \mathcal{E} = \frac{\mathcal{E}_1 + \mathcal{E}_2}{2} . \tag{10}$$

Making use of above relations (8, 9, 10) we get the following formula for the pressure $p = p_1 = p_2$ in terms of the field variables at the interface

$$p(V, u, \mathcal{E}) = \frac{(1+\eta)(\gamma_1 - 1)(\gamma_2 - 1)}{\eta\gamma_1 + \gamma_2 - 1 - \eta} \left(\mathcal{E} - \frac{1}{2}u^2\right) / V . \tag{11}$$

Note that by setting $\eta = 1$ and $\gamma_1 = \gamma_2 = \gamma$ we recover from (11) the classical equation of state for a polytropic gas. Once the interface quantities are known, the values of the field variables on the two sides can be computed

$$\begin{aligned} V_1 = \frac{2}{1+\eta} V , \quad \mathcal{E}_1 &= \frac{p(V,u,\mathcal{E})V_1}{\gamma_1 - 1} + \frac{1}{2}u^2 , \\ V_2 = \frac{2\eta}{1+\eta} V , \quad \mathcal{E}_2 &= \frac{p(V,u,\mathcal{E})V_2}{\gamma_2 - 1} + \frac{1}{2}u^2 . \end{aligned}$$

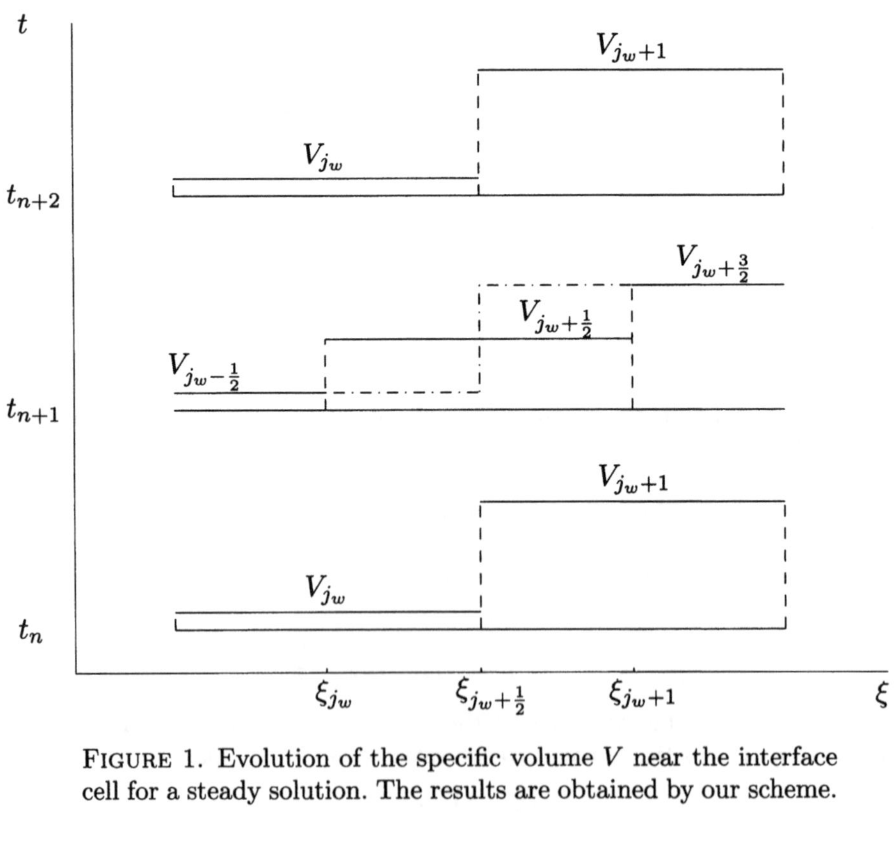

FIGURE 1. Evolution of the specific volume V near the interface cell for a steady solution. The results are obtained by our scheme.

Figure 1 shows the evolution of the specific volume obtained by our scheme on two time steps for a steady solution. A similar picture is obtained for the evolution of the energy. Note how the scheme maintains static solutions.

5. Numerical tests

In this section we report several numerical tests performed in order to validate the Lagrangian scheme.

5.1. Adiabatic approximation.

In this subsection we consider the adiabatic approximation for the piston problem of Section 1. In the adiabatic approximation the pressures of the two gases are equal and depend only on time. Furthermore, the entropies of both gases are constants, equal to their initial values. The above conditions alone are enough to describe the motion of the interface and of the piston by a single ordinary differential equation of second order. The above hypotheses are verified when the piston motion is slow compared with the sound speed and this is the case for a sufficiently high piston mass. In the case of polytropic gases, the ODE for the interface can be written explicitly

$$\frac{d^2I}{dt^2}\left(1+c_1\varrho I^{\varrho-1}\right)+c_1\varrho(\varrho-1)I^{\varrho-2}\left(\frac{dI}{dt}\right)^2=\frac{A}{m}\left(c_2I^{-\gamma_1}-p_{out}(t)\right), \quad (12)$$

with initial conditions $I(0)=I_0$ and $dI/dt(0)=v_0$ and

$$c_1=(L(0)-I(0))\left(\frac{1}{I(0)}\right)^{\varrho},$$
$$\varrho=\gamma_1/\gamma_2,$$
$$c_2=p_0[I(0)]^{\gamma_1}.$$

p_0 is the initial pressures on both sides of the interface. The piston position can be obtained from the interface position via the relation

$$L(t)=(L(0)-I(0))\left(\frac{I(t)}{I(0)}\right)^{\varrho}+I(t).$$

We solve Eq. (12) with the following data $v_0=0$, $p_0=1$, $A/m=0.01, p_{out}(t)=2, \gamma_1=1.4, \gamma_2=2.8$, and initial conditions

$$I(0)=0.3, \quad L(0)=1, \quad \frac{dL}{dt}(0)=0,$$

and then we solve Eq. (5) with the same data and initial conditions

$$(\rho,u,p)=\begin{cases}(\rho_1,u_1,p_1)=(1.0,0.0,1.0) & 0\le x\le I(0)\\ (\rho_2,u_2,p_2)=(1.0,0.0,1.0) & I(0)\le x\le L(0).\end{cases}$$

The position of the interface is computed from the numerical solution by solving the equation

$$\frac{dI}{dt}(t) = u_1(I(t),t) = u_2(I(t),t) \ .$$

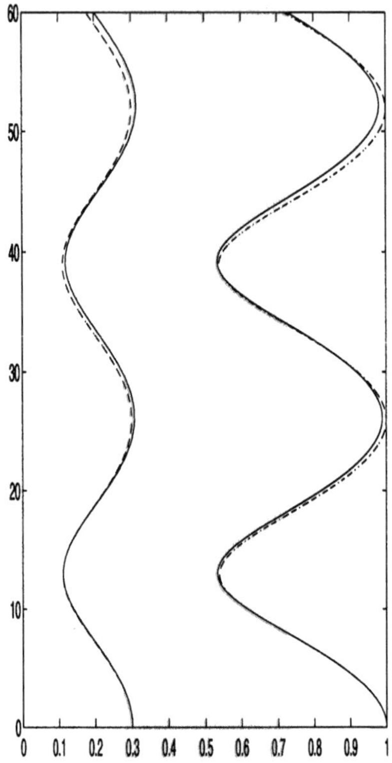

FIGURE 2. Adiabatic approximation: interface and piston positions in the $x-t$ plane, dashed line: adiabatic solution, solid line: numerical solution with 200 mesh-cells.

Figure 2 shows the good agreement between the adiabatic approximation and the numerical results obtained by setting $\Delta\xi = 0.005$. A Courant number $\lambda \max_j \rho(A(\mathbf{q}_j^n)) = 0.45$ has been used.

5.2. S. Karni test problems.

The five tests described here were proposed by S. Karni in order to test her hybrid approach to multi-fluid flows [4]. Moreover, variations of those tests were used by Jenny, Müller and Thomann [3] and by Fedkiw et al. [2].

The numerical tests in this section have been performed with a Courant number 0.4.

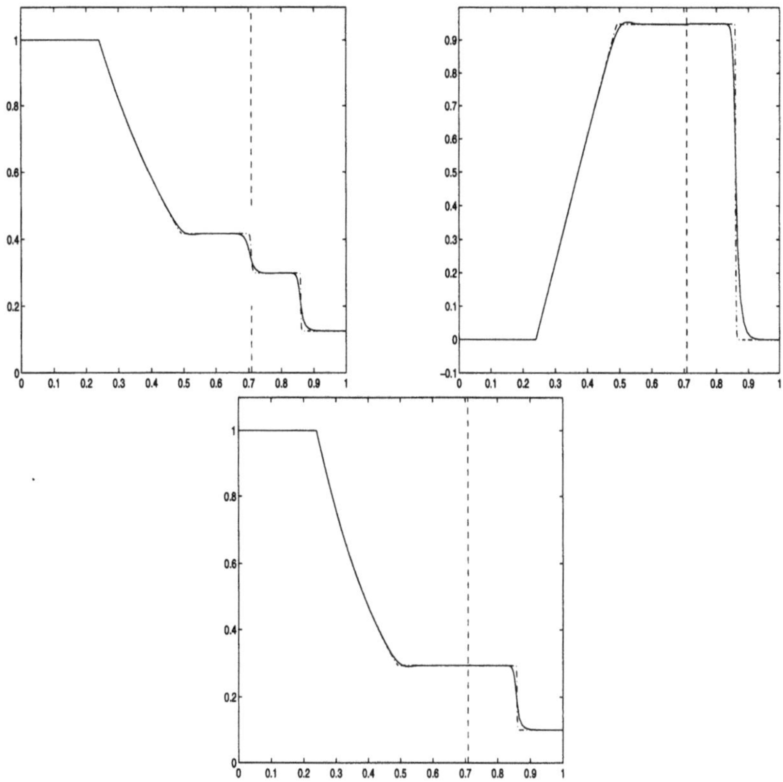

FIGURE 3. S. Karni's Test A. Top left: density. Top right: velocity. Bottom: pressure.

Test A is a classical tube problem with the interface initially at $x = 0.5$ and the following initial conditions

$$\begin{aligned} (\rho_1, u_1, p_1) &= (1.0, 0.0, 1.0) & 0 \le x < 0.5 \\ (\rho_2, u_2, p_2) &= (0.125, 0.0, 0.1) & 0.5 < x \le 1\ . \end{aligned}$$

with $\gamma_1 = 1.4$ and $\gamma_2 = 1.2$. Figure 3 illustrates the numerical results.

The physical setup for the other tests is to consider the interaction of a weak (strong) shock at Mach number of 1.1952 (3.6055) moving from left to right in air with an air/helium (Tests B and D, Case 1) or air/R22 (Tests C and D, Case 2) interface. Inflow (outflow) boundary conditions at the left (right) boundary were applied to all the remaining tests.

The initial conditions for Test B are as follows

$$\begin{aligned} (\rho_1, u_1, p_1) &= \begin{cases} (1.3333, 0.3535, 1.5) & \text{post-shock, air} \quad 0 \le x < 0.3 \\ (1.0, 0.0, 1.0) & \text{pre-shock, air} \quad 0.3 < x < 0.5 \end{cases} \\ (\rho_2, u_2, p_2) &= (0.1379, 0.0, 1.0) \qquad \text{pre-shock, helium} \quad 0.5 < x \le 1 \end{aligned}$$

with $\gamma_1 = 1.4$ and $\gamma_2 = 1.67$. The initial conditions for Test C are given by $\mathbf{q}_1$ as in Test B and

$$(\rho_2, u_2, p_2) = (3.1538, 0.0, 1.0) \qquad \text{pre-shock, R22} \qquad 0.5 < x \leq 1$$

with $\gamma_1 = 1.4$ and $\gamma_2 = 1.249$. The Figures 4 and 5 illustrate the numerical results.

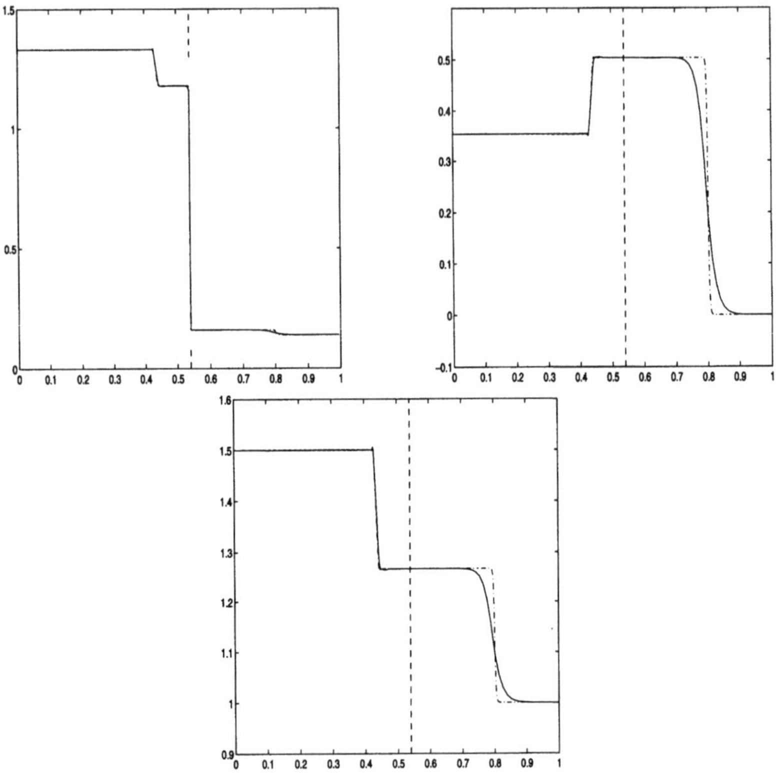

FIGURE 4. S. Karni's Test B. Top left: density. Top right: velocity. Bottom: pressure.

Within the Figures 3, 4 and 5 the solid lines and dashed-dotted lines are obtained with 400 and 4000 mesh-cells, respectively, and a dashed line indicates the position of the interface.

Tests D, Case 1 and Test D, Case 2 differ from Tests B and C, respectively, because the post-shock state is (in both cases) given by

$$(\rho_1, u_1, p_1) = (4.3333, 3.2817, 15.0) \qquad \text{post-shock, air} \qquad 0 \leq x < 0.3\ .$$

Figure 6 shows the numerical results obtained for the density with 400 mesh-cells for the Test D, Case 1 and Test D, Case 2. The spacing of grid points is uniform in Lagrangian coordinates, but in the physical domain it is determined

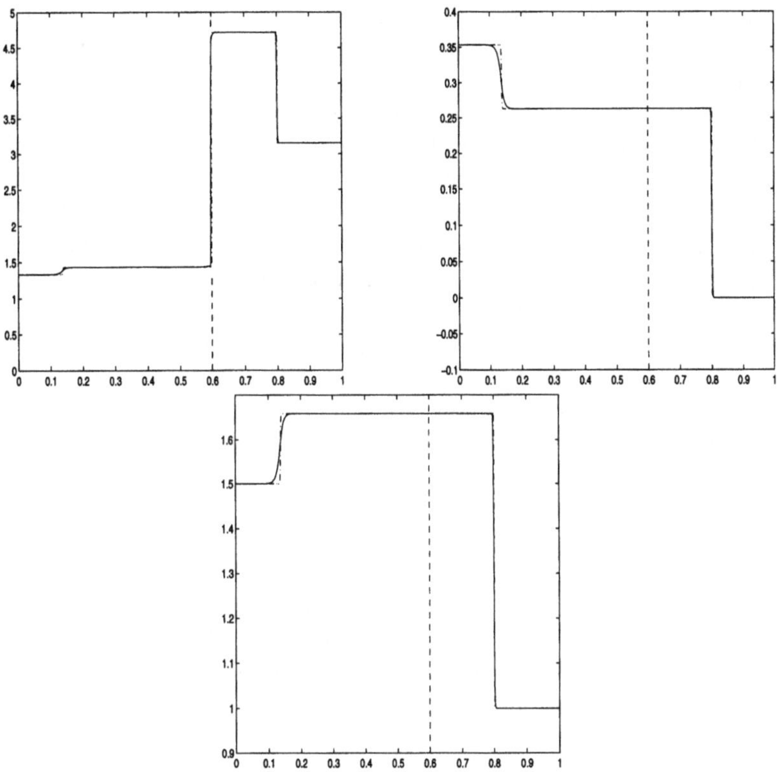

FIGURE 5. S. Karni's Test C. Top left: density. Top right: velocity. Bottom: pressure.

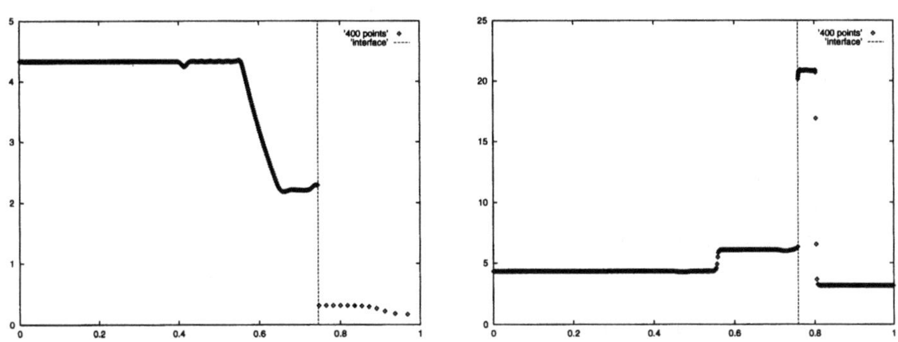

FIGURE 6. S. Karni's Density for Test D. Left: Case 1. Right: Case 2. The interface is marked by a dashed line.

by the density values. Note how, by setting a constant cell-width $\Delta\xi$, we get only 13 nodes on the right of the interface on the left frame of Figure 6. By contrast the number of cells in the case of the right frame of Figure 6 is much higher, and the discontinuities look sharper. Moreover, the over-shoot (under-shoot) on the left of the interface on the left frame of Figure 6 (right frame of the same Figure) is due to the following reason. The temperature across the interface is not necessarily continuous, and therefore assumption (9) with constant η is not appropriate. According to whether the correct value of the volume ratio V_2/V_1 (which is proportional to the temperature ratio) is greater or smaller than η, one observes an over-shoot (Figure 6, left) or an under-shoot (Figure 6, right), the latter being much less evident. A better condition is to impose that the interface is adiabatic, i.e. that no heat flows through it (and therefore the entropy is locally flat). The above possibility is presently under investigation.

Note that, in spite of the above mentioned difficulty and the lack of resolution on the low density regions, the scheme provides an accurate position of the waves.

References

[1] R. Fazio and R. J. LeVeque, *Moving-mesh methods for one-dimensional hyperbolic problems using CLAWPACK*, to appear on Comput. & Math. Appl. (2001).

[2] R. P. Fedkiw, T. Aslam, B. Merriman and S. Osher, *A non-oscillatory Eulerian approach to interfaces in multimaterial flows (the ghost fluid method)*, J. Comput. Phys., **152** (1999), 457–492.

[3] P. Jenny, B. Müller and H. Thomann, *Correction of conservative Euler solvers for gas mixture*, J. Comput. Phys. **132** (1997), 91–107.

[4] S. Karni, *Hybrid multifluid algorithms*, SIAM J. Sci. Comput. **17** (1996), 1019–1039.

[5] H. Nessyahu and E. Tadmor, *Nonoscillatory central differencing for hyperbolic conservation laws*, J. Comput. Phys. **87** (1990), 408–463.

R. Fazio
Department of Mathematics,
University of Messina,
Salita Sperone 31,
98166 Messina, Italy
`rfazio@dipmat.unime.it`

G. Russo
Department of Mathematics and Computer Science,
University of Catania,
Viale Andrea Doria 6,
95126 Catania, Italy
`russo@dmi.unict.it`

International Series of Numerical Mathematics
Vol. 140, © 2001 Birkhäuser Verlag Basel/Switzerland

Ultimate Boundedness, Propagation of Oscillations, and the Long-time Behaviour of Solutions to the Navier-Stokes Equations of Compressible Fluid Flows

Eduard Feireisl

Abstract. We review some recent results on *a priori* estimates, compactness, global existence, and the long-time behaviour of solutions to the Navier-Stokes equations of a compressible fluid flow.

1. Finite energy weak solutions

We shall study some properties of the *finite energy weak solutions* $\varrho = \varrho(t,x)$, $\mathbf{u} = \mathbf{u}(t,x)$ of the Navier-Stokes system of an isentropic compressible fluid:

$$\left\{ \begin{array}{c} \partial_t \varrho + \operatorname{div}(\varrho \mathbf{u}) = 0, \\ \\ \partial_t(\varrho \mathbf{u}) + \operatorname{div}(\varrho \mathbf{u} \otimes \mathbf{u}) - \mu \Delta \mathbf{u} - (\lambda + \mu) \nabla \operatorname{div} \mathbf{u} + a \nabla \varrho^\gamma = \varrho \mathbf{f}, \\ \\ \mu, a > 0, \ \lambda + \mu \geq 0, \ \gamma > 1 \end{array} \right\} \tag{1}$$

where $t \in I = (a,b)$, $x \in \Omega \subset R^3$, Ω a bounded domain with Lipschitz boundary. The system is complemented by the no-slip boundary conditions

$$\mathbf{u}|_{\partial\Omega} = 0 \tag{2}$$

and the energy inequality

$$\frac{d}{dt} E + \int_\Omega \mu |\nabla \mathbf{u}|^2 + (\lambda + \mu) |\operatorname{div} \mathbf{u}|^2 \, dx \leq \int_\Omega \varrho \mathbf{f} \cdot \mathbf{u} \, dx \tag{3}$$

where

$$E = E[\varrho, (\varrho \mathbf{u})] = \int_{\varrho > 0} \frac{1}{2} \frac{|\varrho \mathbf{u}|^2}{\varrho} + \frac{a}{\gamma - 1} \varrho^\gamma \, dx.$$

The finite energy weak solutions $\varrho \geq 0$, $\mathbf{u}$ belong to the classes

- $\varrho \in L^\infty_{loc}(I; L^\gamma(\Omega))$, $u^i \in L^2_{loc}(I; W^{1,2}_0(\Omega))$, $i = 1,2,3$,

---

Work supported by the Grant 201/98/1450 of GA ČR

- the energy E is locally integrable on I and the energy inequality (3) is satified in $\mathcal{D}'(I)$,
- the equations (1) hold in $\mathcal{D}'(I \times \Omega)$,
- the density is square integrable up to the boundary of Ω, i.e.,

$$\varrho \in L^2_{loc}(I; L^2(\Omega)).$$

Remark. The last condition may be replaced by a weaker hypothesis that ϱ is a renormalized solution of the continuity equation 1 (cf. [2]).

The existence of finite energy weak solutions on the time interval $I = R^+ = (0, \infty)$, $\Omega \subset R^N$ with prescribed initial density $\varrho(0)$ and momenta $(\varrho \mathbf{u})(0)$ was proved by LIONS [8] under the hypothesis $\gamma \geq \gamma(N)$.

In this note we shall address three topics: Ultimate boundedness or the existence of a bounded absorbing set, time propagation of oscillations in the density component, and the long-time behaviour of finite energy weak solutions.

It can be shown that for any finite energy weak solutions the continuity equation holds, in fact, on the whole space R^3,

$$\partial_t \varrho + \operatorname{div}(\varrho \mathbf{u}) = 0 \text{ in } \mathcal{D}'(I \times R^3)$$

provided ϱ, $\mathbf{u}$ were extended to be zero outside Ω (see [6, Lemma 3.1]). In particular, the total mass m is a constant of motion, i.e.,

$$m = \int_\Omega \varrho(t, x)\, dx \text{ is independent of } t \in I.$$

Rescaling the constants a, μ, and λ as the case may be we shall always assume $m = 1$.

The density ϱ and the momenta $\varrho \mathbf{u}$ enjoy certain additional continuity properties with respect to t (see LIONS [8] or [2, Section 1.1]):

$$\varrho \in C(I; L^\gamma_{weak}(\Omega)) \cap C(I; L^1(\Omega)), \ (\varrho \mathbf{u}) \in C(I; L^{\frac{2\gamma}{\gamma+1}}_{weak}(\Omega)).$$

The energy E is a lower semicontinuous function of t (see [2, Corollary 1.1]).

In the next two setions, we shall deal with a family $\{\varrho_n\}$, $\{\mathbf{u}_n\}$ of finite energy weak solutions having the following properties:

$$\left\{ \begin{array}{c} \varrho_n,\ \mathbf{u}_n \text{ are finite energy weak solutions of (1), (3) on } (a,b) \times \Omega \\ \\ \text{with the right-hand side } \mathbf{f}_n,\ \|\mathbf{f}_n\|_{L^\infty(I \times \Omega)} \leq F \text{ for all } n = 1, 2, ... \\ \\ \limsup_{t \to a+} E[\varrho_n, (\varrho_n \mathbf{u}_n)](t) \leq E_0 \text{ for all } n = 1, 2, ... \end{array} \right\} \tag{4}$$

2. Ultimate boundedness

Our first result, which is based on multipliers of the type "$\mathrm{div}^{-1} b(\varrho)$" for a suitable nonlinear function b, states that the energy of the sequence introduced in (4) is ultimately bounded (see [4, Theorem 1.1]):

Theorem 2.1. *Let $\gamma > \frac{5}{3}$ and let ϱ_n, $\mathbf{u}_n$ be a sequence of finite energy weak solutions satisfying (4). Then there exists a constant E_∞ depending solely on F and a time T which depends only on E_0 such that*

$$E[\varrho_n, (\varrho_n \mathbf{u}_n)](t) \leq E_\infty \text{ for all } t \in (a,b) \cap (T, \infty) \text{ and all } n = 1, 2, \dots$$

3. Propagation of oscillations

The values $\varrho_n(0)$ are only bounded in $L^\gamma(\Omega)$ and, consequently, they can form an oscillating sequence. Since the continuity equation is hyperbolic, these oscillations will propagate in time (cf. LIONS [8]). However, the amplitude of oscillations will decay at uniform time rate. To show this, let us introduce the defect measure

$$D(t) = \int_\Omega \overline{\varrho \log(\varrho)} - \overline{\varrho} \log(\overline{\varrho}) \, dx, \ t \in (a,b)$$

where $\overline{\varrho}$ denotes a weak limit of $\{\varrho_n\}$ and $\overline{\varrho \log(\varrho)}$ stands for a weak limit of $\{\varrho_n \log(\varrho_n)\}$. The following result was proved in [3]:

Theorem 3.1. *Let $\gamma \geq \frac{9}{5}$. There is a function Φ depending solely on $|\Omega|$,*

$$\Phi : R \mapsto R \text{ is continuous, strictly increasing, } \Phi(0) = 0,$$

such that

$$D(t_2) \leq \chi(t_2 - t_1) \text{ where } \chi' + \Phi(\chi) = 0, \ \chi(0) = D(t_1) \text{ for any } a < t_1 \leq t_2 < b.$$

Remark. The hypothesis $\gamma \geq 9/5$ can be relaxed to $\gamma > 5/3$ (see [2]).

4. Trajectory attractors

The preceding results show that the finite energy weak solutions emenating from bounded energy sets are asymptotically compact. Such a property can be used to obtain several results on the existence of attracting sets.

We introduce the quantity $U(t_0, t)$ playing the role of the evolution operator related to the problem (1), (2). Let $\mathcal{F}$ be a bounded subset of $L^\infty_{loc}(R; L^\infty(\Omega))$. We denote

$$U[E_0, \mathcal{F}](t_0, t) =$$

$$\Big\{ [\varrho, \mathbf{q}] \ \Big| \ \varrho = \varrho(t), \ \mathbf{q} = \varrho\mathbf{u}(t) \text{ where } \varrho, \mathbf{u} \text{ is a finite energy weak solution}$$
$$\text{of the problemx (1), (2) on an open interval } I, \ (t_0, t] \subset I, \text{ with } \mathbf{f} \in \mathcal{F},$$
$$\text{and such that } \limsup_{t \to t_0+} E[\varrho(t), (\varrho\mathbf{u})(t)] \leq E_0 \Big\}.$$

Analogously, we define an equivalent of the so-called short trajectory in the spirit of MÁLEK and NEČAS [9]:

$$U^s[E_0,\mathcal{F}](t_0,t) =$$

$$\Big\{[\varrho(\tau),\mathbf{q}(\tau)],\ \tau\in[0,1]\ \Big|\ \varrho(\tau)=\varrho(t+\tau),\ \mathbf{q}(\tau)=(\varrho\mathbf{u})(t+\tau)$$

where $\varrho, \mathbf{u}$ is a finite energy weak solution of the problem (1), (2)

on an open interval I, $(t_0, t+1]\subset I$, with $\mathbf{f}\in\mathcal{F}$,

$$\text{and such that } \limsup_{t\to t_0} E(t)\le E_0\Big\}.$$

Next, we set

$$\mathcal{F}^+ = \Big\{\mathbf{f}\ \Big|\ \mathbf{f} = \lim_{\tau_n\to\infty}\mathbf{h}_n(.+\tau_n) \text{ weak star in } L^\infty(R\times\Omega)$$

$$\text{for a certain } \mathbf{h}_n\in\mathcal{F} \text{ and } \tau_n\to\infty\Big\}$$

and, finally, let

$$\mathcal{A}^s[\mathcal{F}] =$$

$$\Big\{[\varrho(\tau),\mathbf{q}(\tau),\ \tau\in[0,1]\ |\ \varrho,\ \mathbf{q}=(\varrho\mathbf{u}) \text{ is a finite energy weak solution}$$

$$\text{of the problem (1), (2) on } I=R, \text{ with } \mathbf{f}\in\mathcal{F}^+ \text{ and } E\in L^\infty(R)\Big\}.$$

We report the following result on the existence of trajectory attractors (see [2, Theorem 3.1]):

Theorem 4.1. *Assume $\gamma>\frac{5}{3}$. Let $\mathcal{F}$ be a bounded subset of $L^\infty(R\times\Omega)$.*
Then the set $\mathcal{A}^s[\mathcal{F}]$ is compact in $L^\gamma((0,1)\times\Omega)\times[L^p((0,1)\times\Omega)]^3$ and

$$\sup_{[\varrho,\mathbf{q}]\in U[E_0,\mathcal{F}](t_0,t)}\Big[\inf_{[\overline{\varrho},\overline{\mathbf{q}}]\in\mathcal{A}^s[\mathcal{F}]}\Big(\|\varrho-\overline{\varrho}\|_{L^\gamma((0,1)\times\Omega)}+\|\mathbf{q}-\overline{\mathbf{q}}\|_{L^p((0,1)\times\Omega)}\Big)\Big]\to 0$$

$$\text{as } t\to\infty$$

for any $1\le p<\frac{2\gamma}{\gamma+1}$.

5. Attractors and the long-time convergence

Let us introduce the set

$$\mathcal{A}[\mathcal{F}] =$$

$$\Big\{[\varrho,\mathbf{q}]\ \Big|\ \varrho=\varrho(0),\ \mathbf{q}=(\varrho\mathbf{u})(0) \text{ where } \varrho,\mathbf{u} \text{ is a finite energy weak solution}$$

$$\text{of the problem (1), (2) on } I=R \text{ with } \mathbf{f}\in\mathcal{F}^+ \text{ and } E\in L^\infty(R)\Big\}.$$

The next statement shows that $\mathcal{A}[\mathcal{F}]$ is a global attractor in the sense of FOIAS and TEMAM [7] (see [2, Theorem 4.1]).

Theorem 5.1. *Suppose $\gamma > \frac{5}{3}$. Let $\mathcal{F}$ be a bounded subset of $L^\infty(R \times \Omega)$.*

Then $\mathcal{A}[\mathcal{F}]$ is compact in $L^\alpha(\Omega) \times L^{\frac{2\gamma}{\gamma+1}}_{weak}(\Omega)$ and

$$\sup_{[\varrho,\mathbf{q}] \in U[E_0,\mathcal{F}](t_0,t)} \left[\inf_{[\overline{\varrho},\overline{\mathbf{q}}] \in \mathcal{A}[\mathcal{F}]} \left(\|\varrho - \overline{\varrho}\|_{L^\alpha(\Omega)} + \left| \int_\Omega (\mathbf{q} - \overline{\mathbf{q}}) \cdot \phi \, dx \right| \right) \right] \to 0 \text{ as } t \to \infty$$

for any $1 \le \alpha < \gamma$ and any $\phi \in [L^{\frac{2\gamma}{\gamma-1}}(\Omega)]^3$.

The more delicate issue seems to be the strong convergence of the density in $L^\gamma(\Omega)$ and the momenta in $L^1(\Omega)$ in Theorem 5.1. The following assertion was proved in [2, Theorem 4.2]:

Theorem 5.2. *In addition to the hypotheses of Theorem 5.1, assume that the energy $E = E[\varrho, \mathbf{q}]$ is (sequentially) continuous on $\mathcal{A}[\mathcal{F}]$, specifically, for any sequence*

$$\{[\overline{\varrho_n}, \overline{\mathbf{q}}_n]\} \subset \mathcal{A}[\mathcal{F}] \text{ such that } \overline{\varrho}_n \to \overline{\varrho} \text{ in } L^1(\Omega), \; \mathbf{q}_n \to \mathbf{q} \text{ weakly in } L^1(\Omega)$$

one requires

$$E[\overline{\varrho}_n, \overline{\mathbf{q}}_n] \to E[\overline{\varrho}, \overline{\mathbf{q}}].$$

Then

$$\sup_{[\varrho,\mathbf{q}] \in U[E_0,\mathcal{F}](t_0,t)} \left[\inf_{[\overline{\varrho},\overline{\mathbf{q}}] \in \mathcal{A}[\mathcal{F}]} \left(\|\varrho - \overline{\varrho}\|_{L^\gamma(\Omega)} + \|\mathbf{q} - \overline{\mathbf{q}}\|_{L^1(\Omega)} \right) \right] \to 0 \text{ as } t \to \infty.$$

There is a very important particular case when the hypotheses of the preceding theorem are satisfied, namely, if $\mathcal{A}[\mathcal{F}]$ consists of a single stationary solution. We have the following generalization of the convergence results of [5] and NOVOTNÝ, STRAŠKRABA [10] (for the proof see [2, Theorem 4.3]:

Theorem 5.3. *Assume $\gamma > \frac{5}{3}$ and $\mathbf{f}$ is a measurable function bounded uniformly on $(0, \infty)$. Let F be globally Lipschitz continuous on $\overline{\Omega}$ such that the upper level sets*

$$[F > k] = \{x \in \Omega \mid F(x) > k\}$$

are connected for any $k \in R$. Moreover, let

$$(\mathbf{f}(. + \tau) - \nabla F) \to 0 \text{ weakly in } L^1((0,1) \times \Omega) \text{ as } \tau \to \infty.$$

Then any finite energy weak solution ϱ, $\mathbf{u}$ of the problem (1) , (2) on $I = (0, \infty)$ satisfies

$$\varrho(t) \to \varrho_s \text{ in } L^\gamma(\Omega), \; (\varrho\mathbf{u})(t) \to 0 \text{ in } L^1(\Omega) \text{ as } t \to \infty$$

where ϱ_s is the (unique) solution of the stationary problem

$$a\nabla\varrho_s^\gamma = \varrho_s \nabla F.$$

References

[1] E. Feireisl. Global attractors for the Navier-Stokes equations of three-dimensional compressible flow. *C. R. Acad. Sci. Paris, Sér. I*, **331**: 35–39, 2000.

[2] E. Feireisl. Propagation of oscillations, complete trajectories and attractors for compressible flows. *NoDEA*, 2000. To appear.

[3] E. Feireisl and H. Petzeltová. Asymptotic compactness of global trajectories generated by the Navier-Stokes equations of compressible fluid. *J. Differential Equations*, 1999. To appear.

[4] E. Feireisl and H. Petzeltová. Bounded absorbing sets for the Navier-Stokes equations of compressible fluid. *Commun. Partial Differential Equations*, 1999. To appear.

[5] E. Feireisl and H. Petzeltová. Large-time behaviour of solutions to the Navier-Stokes equations of compressible flow. *Arch. Rational Mech. Anal.*, **150**:77–96, 1999.

[6] E. Feireisl and H. Petzeltová. On integrability up to the boundary of the weak solutions of the Navier-Stokes equations of compressible flow. *Commun. Partial Differential Equations*, **25**(3-4), 2000. To appear.

[7] C. Foias and R. Temam. The connection between the Navier-Stokes equations, dynamical systems and turbulence. *In Directions in Partial Differential Equations, Academic Press, New York*, pages 55–73, 1987.

[8] P.-L. Lions. *Mathematical topics in fluid dynamics, Vol.2, Compressible models.* Oxford Science Publication, Oxford, 1998.

[9] J. Málek and J. Nečas. A finite-dimensional attractor for the three dimensional flow of incompressible fluid. *J. Differential. Equations*, **127**:498–518, 1996.

[10] A. Novotný and I. Straškraba. Convergence to equilibria for compressible Navier-Stokes equations with large data. *Annali Mat. Pura Appl.*, 2000. To appear.

E. Feireisl
Institute of Mathematics,
Czech Academy of Sciences,
Žitná 25,
115 67 Praha 1,
Czech Republic
feireisl@math.cas.cz

International Series of Numerical Mathematics
Vol. 140, © 2001 Birkhäuser Verlag Basel/Switzerland

Adaptive Methods for the Solution of Compressible Flow

M. Feistauer and A. Kliková

Abstract. The numerical calculations show that the local fineness of the mesh has an important influence on the precision of the numerical results. We will present in our communication two types of error indicators, which are developed for Euler and Navier-Stokes equations. These indicators are used for adaptive refinement of the computational mesh. The local adaptive refinement leads to the improving the precision of the approximate solution and minimizing the time of the computation. Details can be found in [8].

1. Formulation of the continuous and discrete problem

The complete system consisting of the continuity equation, the Navier-Stokes equations, the energy equation and the state equation, describing viscous compressible flow in a plane bounded domain Ω and time interval $(0, T)$ is written in the form

$$\frac{\partial w}{\partial t} + \sum_{s=1}^{2} \frac{\partial \boldsymbol{f}_s(w)}{\partial x_s} = \sum_{s=1}^{2} \frac{\partial \boldsymbol{R}_s(w, \nabla w)}{\partial x_s} \quad \text{in } Q_T = \Omega \times (0, T), \tag{1}$$

where

$$\begin{aligned}
&\boldsymbol{w} = (\rho, \rho v_1, \rho v_2, e)^{\mathrm{T}}, \quad \boldsymbol{w} = \boldsymbol{w}(\boldsymbol{x}, t), \quad \boldsymbol{x} \in \Omega,\ t \in (0, T), \\
&\boldsymbol{f}_s(\boldsymbol{w}) = (\rho v_s, \rho v_s v_1 + \delta_{s1} p, \rho v_s v_2 + \delta_{s2} p, (e + p)\, v_s)^{\mathrm{T}}, \\
&\boldsymbol{R}_s(\boldsymbol{w}) = (0, \tau_{s1}, \tau_{s2}, \tau_{s1} v_1 + \tau_{s2} v_2 + k \partial\theta/\partial x_s)^{\mathrm{T}}, \\
&\tau_{ij} = \lambda \operatorname{div} \boldsymbol{v}\, \delta_{ij} + \mu(\partial v_i/\partial x_j + \partial v_j/\partial x_i), \\
&p = (\gamma - 1)\,(e - \rho|\boldsymbol{v}|^2/2), \quad e = \rho(c_v\theta + |\boldsymbol{v}|^2/2).
\end{aligned} \tag{2}$$

We use the following notation: t – time, $x = (x_1, x_2)$, x_s $(s = 1, 2)$ – Cartesian coordinates, ρ – density, p – pressure, θ – absolute temperature, e – total energy, $\boldsymbol{v} = (v_1, v_2)$ – velocity, δ_{ij} – Kronecker delta, $\gamma > 1$, c_v, k, $\mu > 0$ – given constants, $\lambda = -2\mu/3$.

The above system is equipped with *initial conditions* prescribing the state $\boldsymbol{w}$ on Ω at time $t = 0$ and *boundary conditions*: At inlet we prescribe ρ, v_1, v_2 and set $\partial\theta/\partial\boldsymbol{n} = 0$, on fixed walls we assume that $v_1 = v_2 = 0$, $\partial\theta/\partial\boldsymbol{n} = 0$ and at outlet we set $-pn_j + \sum_{i=1}^{2} \tau_{ij} n_i = 0$, $j = 1, 2$, and $\partial\theta/\partial n = 0$. (Here $\partial/\partial \boldsymbol{n}$ is the derivative in the direction of the unit outer normal $\boldsymbol{n} = (n_1, n_2)$ to the boundary $\partial\Omega$).

The numerical method for obtaining the approximate solution, which has been applied with success, is the *inviscid–viscous operator splitting* ([3]). In this case we split (1) into the inviscid system

$$\frac{\partial \boldsymbol{w}}{\partial t} + \sum_{s=1}^{2} \frac{\partial \boldsymbol{f}_s(\boldsymbol{w})}{\partial x_s} = 0, \tag{3}$$

and the purely viscous system

$$\frac{\partial \boldsymbol{w}}{\partial t} = \sum_{s=1}^{2} \frac{\partial \boldsymbol{R}_s(\boldsymbol{w}, \nabla \boldsymbol{w})}{\partial x_s}, \tag{4}$$

and discretize these separately.

We denote the basic finite volume triangulation as $\mathcal{D}_h = \{D_i\}_{i\in J}$ (J is an index set of positive integers). After that we construct a new triangulation $\mathcal{T}_h$. The vertices of $\mathcal{T}_h$ are the barycenters of triangles of the fundamental grid $\mathcal{D}_h$ together with the vertices of $\mathcal{D}_h$ which lie on the boundary $\partial\Omega_h$. The triangulation $\mathcal{T}_h$ is called *adjoint triangulation.* The inviscid system (3) is discretized by the FV method on the mesh $\mathcal{D}_h$, whereas the purely viscous system (4) is discretized using conforming piecewise linear finite elements on a triangulation $\mathcal{T}_h$ of Ω. One time step $t_k \to t_{k+1}$ is split into two *fractional steps*:

1.0.1. STEP I (*inviscid FV step on the mesh $\mathcal{D}_h$*): Assume that the values $\boldsymbol{w}_i^k$, $i \in J$, approximating the solution on the finite volumes D_i at time t_k are known. Compute the values $\boldsymbol{w}_i^{k+1/2}$, $i \in J$, from the FV formula

$$\boldsymbol{w}_i^{k+1/2} = \boldsymbol{w}_i^k - \frac{\tau_k}{|D_i|} \sum_{j\in S(i)} \boldsymbol{H}(\boldsymbol{w}_i^k, \boldsymbol{w}_j^k, \boldsymbol{n}_{ij})\,|\Gamma_{ij}| \tag{5}$$

equipped with inviscid boundary conditions. (Cf. [1, Par. 7.3], [2], [5]). Here $S(i)$ is the set of neighbours of D_i, $|\,D_i\,|$ is the area of D_i, $\boldsymbol{n}_{ij} = (n_{ij}^1, n_{ij}^2)$ is the unit outer normal pointing from the triangle D_i to its neighbour D_j and $|\Gamma_{ij}|$ is the length of the edge between D_i and D_j. $\boldsymbol{H}$ is the so-called *numerical flux.* In the practical computations the 2D version of the well-known Osher-Solomon numerical flux has been used (see, e.g.,[10]).

1.0.2. STEP II (*viscous FE step on the mesh $\mathcal{T}_h$*): Define the FE approximation $\tilde{\boldsymbol{w}}_h^{k+1/2}$ with values $\tilde{\boldsymbol{w}}_h^{k+1/2}(P_i) = \boldsymbol{w}_i^{k+1/2}$ at the nodes P_i, $i \in J$ (= vertices of triangles $T \in \mathcal{T}_h$). On $\partial\Omega$, the viscous boundary conditions are used. Compute the FE approximation $\tilde{\boldsymbol{w}}_h^{k+1}$ as the solution of the following problem:

$$\text{(i) } \tilde{\boldsymbol{w}}_h^{k+1} \text{ satisfies the viscous Dirichlet boundary conditions,} \tag{6}$$

$$\text{(ii)} (\tilde{\boldsymbol{w}}_h^{k+1}, \varphi)_h = (\tilde{\boldsymbol{w}}_h^{k+1/2}, \varphi)_h - \tau_k\, a_h(\tilde{\boldsymbol{w}}_h^{k+1/2}, \varphi)$$

for all test functions $\varphi = (\varphi_1, \ldots, \varphi_4)$ such that φ_j $(j = 1, \ldots, 4)$ is piecewise linear and vanishes at the nodes on the part of $\partial\Omega$ where the j-th component $\boldsymbol{w}_j$ of the state vector $\boldsymbol{w}$ satisfies the Dirichlet boundary condition.
Now set $\boldsymbol{w}_i^{k+1} := \tilde{\boldsymbol{w}}_h^{k+1}(P_i)$ for $i \in J,\ k := k+1$ and go to Step I.

In (6), $(\boldsymbol{w}, \varphi)_h$ and $a_h(\boldsymbol{w}, \varphi)$ denote the approximation of

$$\int_\Omega \boldsymbol{w}\, \varphi\, dx \qquad \text{and} \qquad \int_\Omega \sum_{s=1}^{2} \boldsymbol{R}_s(\boldsymbol{w}, \nabla \boldsymbol{w})\, \partial\varphi/\partial x_s\, dx,$$

respectively, with the aid of the numerical quadrature using the vertices of triangles as integration points.

The following *stability conditions* are used:

$$\frac{\tau_k}{|D_i|}\, |\partial D_i| \max_{j \in S(i)} \rho(\mathbb{P}(\boldsymbol{w}_i^k, \boldsymbol{n}_{ij}) \leq \text{CFL} \approx 0.85, \quad i \in J, \tag{7}$$

$$\frac{3}{4}\, \frac{h(T)}{\sigma(T)}\, \frac{\tau_k}{|T|} \max(\mu, k) \leq \text{CFL}, \quad T \in \mathcal{T}_h, \tag{8}$$

where $\mathbb{P}(\boldsymbol{w}, \boldsymbol{n}) = \sum_{s=1}^{2} (D\, \boldsymbol{f}_s(\boldsymbol{w})/D\boldsymbol{w})\, n_s$, $\rho(\mathbb{P})$ = spectral radius of the matrix $\mathbb{P}$ and $\sigma(T)$ is the radius of largest circle inscribed in T. Conditions (7) and (8) are obtained on the basis of linearization and in analogy with a scalar problem. In case of the Euler equations, which describe the inviscid compressible flow, we solve the system

$$\frac{\partial \boldsymbol{w}}{\partial t} + \sum_{i=1}^{2} \frac{\partial \boldsymbol{f}_i(\boldsymbol{w})}{\partial x_i} = 0\,, \text{in } Q_T. \tag{9}$$

Here we use the finite volume method on a triangular unstructured mesh for gaining the approximate solution. This method is mentioned in Step I, where the explicite numerical scheme has the form

$$\boldsymbol{w}_i^{k+1} = \boldsymbol{w}_i^k - \frac{(t_{k+1} - t_k)}{|\, D_i\, |} \sum_{j \in S(i)} \boldsymbol{H}(\boldsymbol{w}_i^k, \boldsymbol{w}_j^k, \boldsymbol{n}_{ij})\, |\Gamma_{ij}|, \tag{10}$$

for $i \in J$.

The accuracy of the solution of transonic flow is increased with the aid of *automatic adaptive mesh refinement* in the vicinity of shock waves and in boundary layers, based on a suitable shock indicator or error indicator ([4], [5], [6], [7], [9]).

2. Adaptive mesh refinement

The adaptive mesh refinement methods are based on detecting such areas of the computational mesh where the error of the numerical solution highly arises. The regions where the error is high, are mainly regions containing discontinuities (slightly smeared by viscosity) or boundary layers. Our goal is to develop an adaptive iterative strategy allowing the construction of successively imbedded mesh refinements. To this end, we proceed in the following way. In advance the value of the tolerance

of the refinement is prescribed. It belongs to the interval [0,1]. >From the piecewise constant approximate solution the error indicator is computed. It is piecewise constant and normalized into the interval [0,1]. We refine all volumes on which the value of indicator is greater than a prescribed tolerance. To do it, we use the so called red-green refinement. For triangles with one of their angle greater than 110° we replace the red-green refinement by red*-green one. After the refinement, the solution process is repeated. In the case, when no coarsening is carried out, the method is, of course, suitable for steady flow.
The numerical tests have been carried out for the 2D GAMM channel, for a cascade of profiles and for the profile NACA 0012.
In what follows we describe two error indicators applied according to the described strategy.

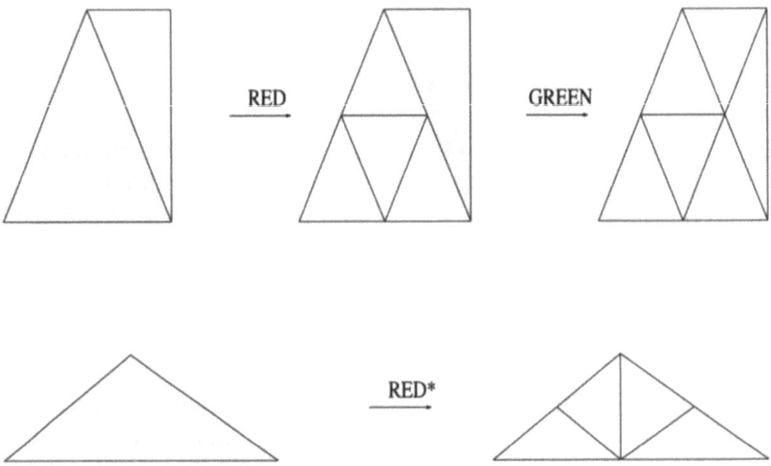

FIGURE 1. Refinement technique.

3. Superconvergence error indicator for the Euler equations

The idea of the superconvergence indication of the error is the following:
1. From $\boldsymbol{w}_h$ its piecewise linear reconstruction $\boldsymbol{w}_h^*$ is computed. This function is determined uniquely by values calculated at each vertex of the computational mesh. These values are weighted averages of the values $\boldsymbol{w}_i^k$ given on triangles which have the corresponding vertex in common. As weights the areas of these triangles are used.
2. We construct a function $\boldsymbol{g} = (g^1, g^2, g^3, g^4)$, $\boldsymbol{g} := \boldsymbol{w}_h^* - \boldsymbol{w}_h$, which is piecewise linear and discontinuous in the domain Ω. The array of values of the norm $\| \boldsymbol{g} \|_{L^2(T_i)}$ is further denoted as *superconvergence error indicator.* Numerical results obtained with the aid of this indicator show that the adaptive refinement based on it has a

remarkable influence on the accuracy of the numerical solution and on the length of the computational time. The indicator is able to focus the shock wave very sharply.

4. $\mathbf{H}^{-1}$-norm error indicator for the Navier-Stokes equations

The present approach is based on the weak formulation of the system of compressible Navier-Stokes equations, written in the form (1). Multiplying (1) by a test function $\boldsymbol{\varphi} = (\varphi^1, \varphi^2, \varphi^3, \varphi^4)^{\mathrm{T}} \in \mathbf{H}_0^1(\Omega) = [H_0^1(\Omega)]^4$, integrating it over Ω and the time interval $I = (t_k, t_{k+1})$ and using the Green's theorem, we obtain the identity

$$0 = \int_\Omega (\boldsymbol{w}(x, t_{k+1}) - \boldsymbol{w}(x, t_k))\boldsymbol{\varphi} dx - \int_I \int_\Omega \sum_{s=1}^{2} \boldsymbol{f}_s(\boldsymbol{w}) \frac{\partial \boldsymbol{\varphi}}{\partial x_s} dx dt$$

$$+ \int_I \int_\Omega \sum_{s=1}^{2} \boldsymbol{R}_s(\boldsymbol{w}, \nabla \boldsymbol{w}) \frac{\partial \boldsymbol{\varphi}}{\partial x_s} dx dt =: \boldsymbol{a}_k(\boldsymbol{w}, \nabla \boldsymbol{w}, \boldsymbol{\varphi}) \tag{11}$$

There exists an element $\boldsymbol{\mathcal{A}}_k(\boldsymbol{w}, \nabla \boldsymbol{w}) \in \mathbf{H}^{-1}(\Omega)$ such that the equation (11) is equivalent to the operator equation $\boldsymbol{\mathcal{A}}_k(\boldsymbol{w}, \nabla \boldsymbol{w}) = 0$. The used FV-FE operator splitting method yields on the time level t_k a piecewise constant solution $\boldsymbol{w}_h^k$. By $\tilde{\boldsymbol{w}}_h^k$ we denote the piecewise linear reconstruction of $\boldsymbol{w}_h^k$ and set $\boldsymbol{r}_h^k = \boldsymbol{\mathcal{A}}_k(\boldsymbol{w}_h^k, \nabla \tilde{\boldsymbol{w}}_h^k)$ (= the approximation of the residual). We approximate the $\boldsymbol{H}^{-1}$ norm of this residual in the form $\|\boldsymbol{r}_h^k\|_{\boldsymbol{H}^{-1}(\Omega)} \approx \left(\sum_{D_i \in \mathcal{D}_h} \eta^2(i)\right)^{1/2}$, where η is called the $\mathbf{H}^{-1}$-norm error indicator and is expressed according to the operator splitting method in two parts corresponding to the inviscid and viscous fluxes:

$$\eta^2(i) = \beta^2(i) + \gamma^2(i),$$

$$\beta^2(i) = \sum_{\ell=1}^{4} \left(\beta_k^\ell(i)\right)^2,$$

$$\gamma^2(i) = \sum_{\ell=1}^{4} \left(\gamma_k^\ell(i)\right)^2,$$

where for $\ell = 1, \dots, 4$

$$\beta_k^\ell(i) = \max_{j \in S(i)} \frac{h_{ij}}{|\varphi_{ij}|_{H_0^1(\Omega)}} \left| \sum_{s=1}^{2} n_{ij}^s \left(f_s^\ell(\boldsymbol{w}_i^k) - f_s^\ell(\boldsymbol{w}_j^k)\right) \right|, \tag{12}$$

$$\gamma_k^\ell(i) = \max_{j \in S(i)} \frac{h_{ij}}{|\varphi_{ij}|_{H_0^1(\Omega)}} \left| \sum_{s=1}^{2} n_{ij}^s \left(R_s^\ell(\boldsymbol{w}_i^k, \nabla \tilde{\boldsymbol{w}}_i^k) - R_s^\ell(\boldsymbol{w}_j^k, \nabla \tilde{\boldsymbol{w}}_j^k)\right) \right|. \tag{13}$$

Here $S(i)$ is the set of neighbours D_j of D_i, n_{ij}^s the components of unit outer normal pointing form D_i to its neighbour D_j, h_{ij} is the length of the edge Γ_{ij} between D_i and D_j. φ_{ij} are test functions chosen from a particular set of functions defined in

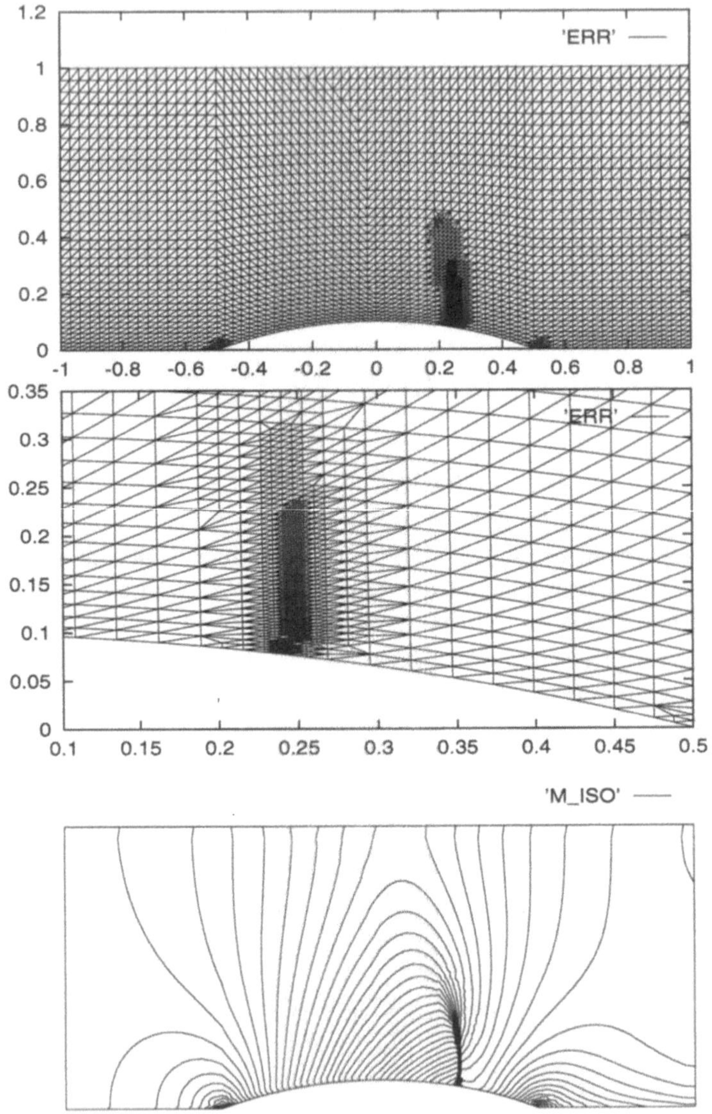

FIGURE 2. GAMM channel: Computational mesh (whole and zoom) obtained after three adaptive refinements on the basis of the superconvergence error indicator and corresponding Mach number isolines (inviscid flow with the inlet Mach number at the inlet $M_{inlet} = 0.67$).

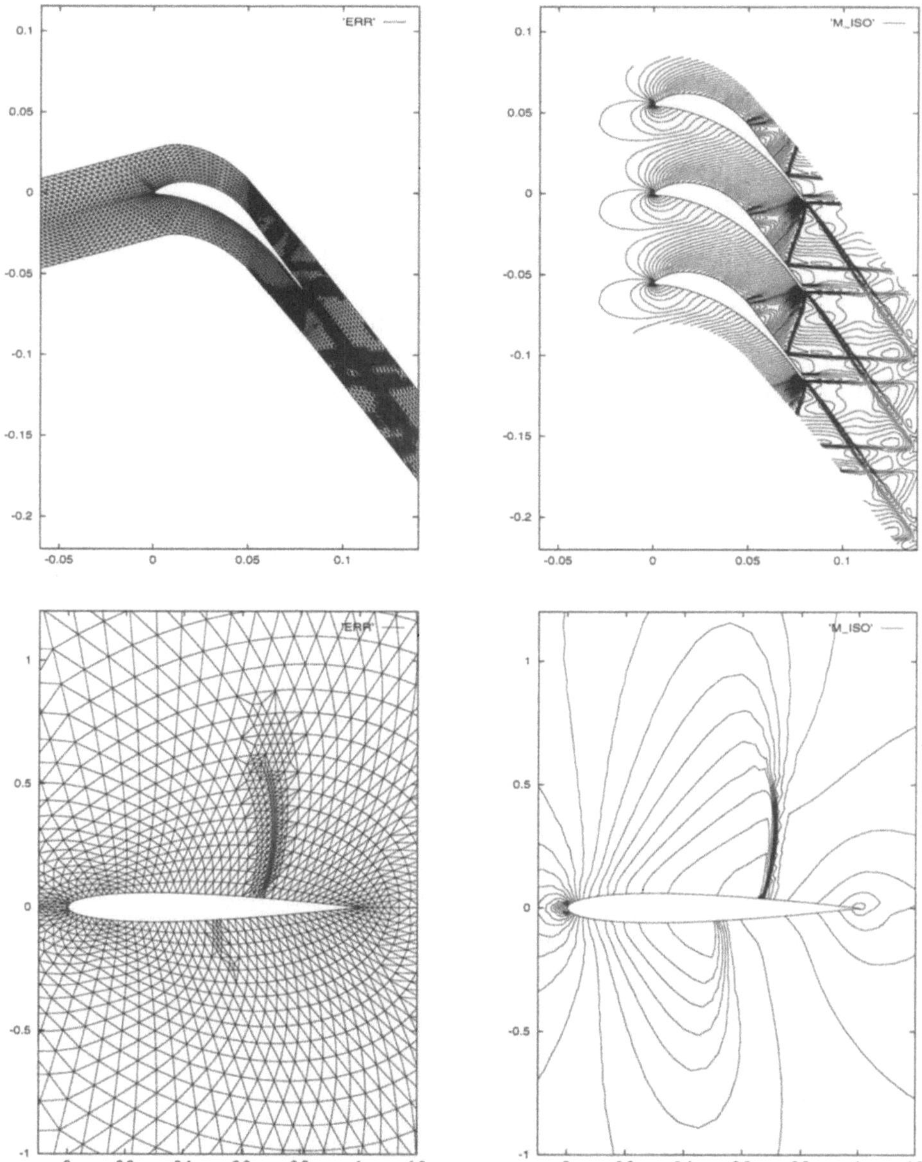

FIGURE 3. Cascade of profiles and profile NACA0012: Computational mesh obtained after the first (second) adaptive refinement on the basis of the superconvergence error indicator and corresponding Mach number isolines. Cascade of profiles: $M_{inlet} = 0.3223$; NACA profile: $M_\infty = 0.8$, $\alpha = 1.25^o$.

the following way:
Using this notation we define the set $\mathcal{E}_{\mathcal{D}_h}$ of test functions as follows

$$\begin{aligned}\mathcal{E}_{\mathcal{D}_h} = \{ \quad & \varphi_{ij} \in C(\overline{\Omega}); \\ & \varphi_{ij} \text{ is linear on } D_i^{jk}, D_j^{ik}, k = 1, \ldots, 3, \\ & \varphi_{ij} = 1 \text{ in the middle of the } \Gamma_{ij}, \\ & \operatorname{supp} \varphi_{ij} = \bigcup_{k=1}^{3} D_i^{jk} \cup \bigcup_{k=1}^{3} D_j^{ik}, \\ & D_i, D_j \text{ are neighbours}\}.\end{aligned}$$

The subtriangles D_i^{jk}, D_j^{ik} are depicted in Figure 4.

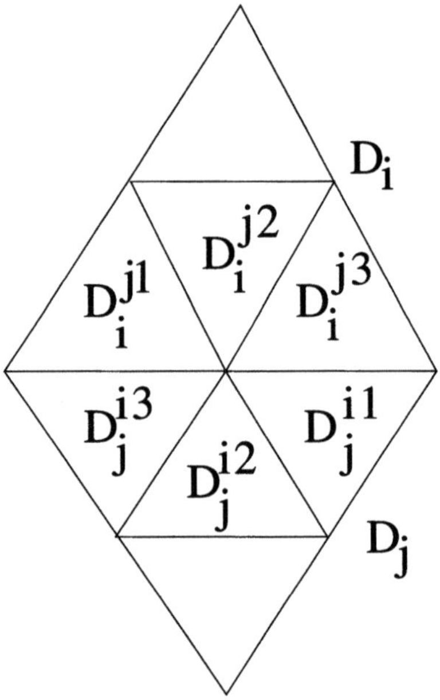

FIGURE 4. Support of the function φ_{ij}

The $\mathbf{H}^{-1}$-norm error indicator allows us to detect regions with shock waves as well as boundary layers quite accurately.

Acknowledgments
This work has been supported under the Grant No. 201/99/0267 of the Czech Grant Agency and under the Grant No. MSM 113200007.

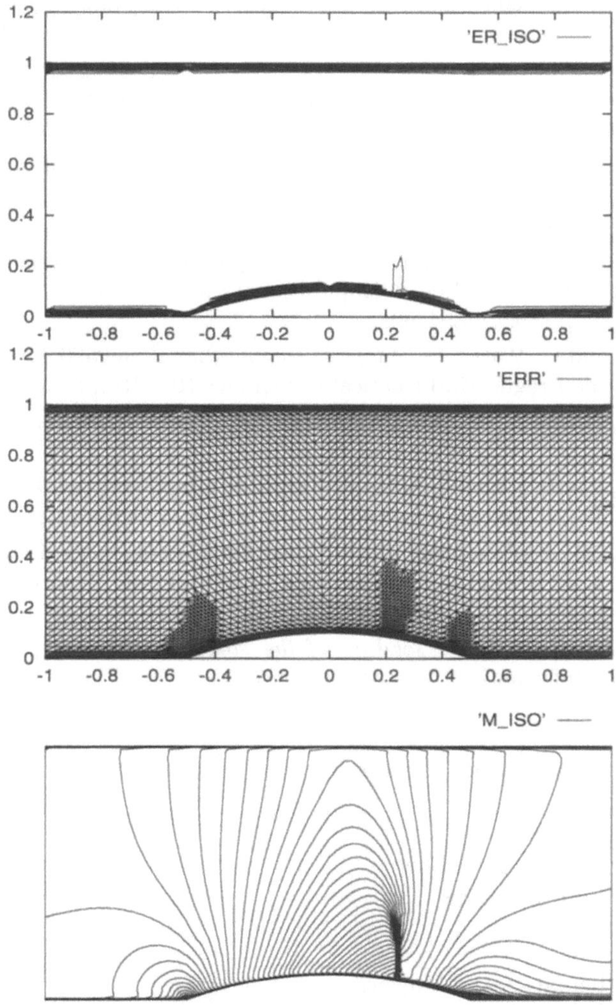

FIGURE 5. GAMM chanel: $\mathbf{H}^{-1}$-norm error indicator isolines. Computational mesh refined on the basis of this indicator. Mach number isolines computed on the refined mesh for the viscous flow with $Re = 10^5$ and $M_{inlet} = 0.671$.

References

[1] M. Feistauer, *Mathematical methods in fluid dynamics.* Longman Scientific & Technical, Monographs and Surveys in Pure and Applied Mathematics **67**, Harlow, 1993.

[2] M. Feistauer, J. Felcman and M. Lukáčová, *Combined finite elements–finite volume solution of compressible flow*, Journal of Comput. and Applied Math., **63** (1995), 179–199.

[3] M. Feistauer, J. Felcman and V. Dolejší, *Numerical simulation of compresssible viscous flow through cascades of profiles*, ZAMM, **76** (1996), 297–300.

[4] J. Felcman and V. Dolejší, *Adaptive methods for the solution of the Euler equations in elements of the blade machines*, ZAMM, **76** (1996), 301–304.

[5] J. Felcman, V. Dolejší and M. Feistauer, *Adaptive finite volume method for the numerical solution of the compressible Euler equations*, in: J. Périaux S. Wagner, E. H. Hirschel and R. Piva, Eds., Computational Fluid Dynamics '94 (John Wiley and Sons, Stuttgart) (1994), 894–901.

[6] J. Felcman and G. Warnecke, *Adaptive computational methods for gas flow*, in: Proceedings of the Prague Mathematical Conference (ICARIS, Prague) (1996), 99–104.

[7] A. Kliková, *Adaptive methods for the Euler and Navier-Stokes equations*, ZAMM **99** (1999).

[8] A. Kliková, *Finite volume – finite element solution of compressible flow*, PhD Disertation, Faculty of Mathematics and Physics, Charles University, Prague, 2000.

[9] A. Kliková, M. Feistauer and J. Felcman, *Adaptive methods for problems of fluid dynamics*, in: J. Holenda, I. Marek, Eds., Software and Algorithms of Numerical Mathematics'99 (Univ.of Pilsen) (1999).

[10] S. P. Spekreijse, *Multigrid solution of the steady Euler equations*, Centrum voor Wiskunde en Informatica, Amsterdam, 1987.

M. Feistauer
Faculty of Mathematics and Physics,
Charles University,
Sokolovská 83,
186 75 Praha 8, Czech Republic
`feist@karlin.mff.cuni.cz`

A. Kliková
Faculty of Mathematics and Physics,
Charles University,
Sokolovská 83,
186 75 Praha 8, Czech Republic
`klikova@cts.cuni.cz`

International Series of Numerical Mathematics
Vol. 140,

The MoT-ICE: A New Multi-dimensional Wave-propagation-algorithm Based on Fey's Method of Transport. With Application to the Euler- and MHD-equations

Michael Fey, Sebastian Noelle, and Christian von Törne

Abstract. We present a new multi-dimensional wave-propagation algorithm for magneto-hydrodynamical flows based on Fey's MoT (Method of Transport) and Noelle's MoT-ICE (MoT with interfaced-centered evolution). Numerical experiments show the potential of the method, but also the need of further analysis and algorithmic development.

1. Multidimensional Schemes

Since the work of Godunov, Van Leer, Harten-Lax and Roe, the numerical solution of systems of hyperbolic conservation laws is dominated by Riemann-solver based schemes. These one-dimensional schemes are usually extended to several space-dimensions either by using dimensional-splitting on cartesian grids or by the finite-volume approach on unstructured grids. The first systematic criticism of using one-dimensional Riemann-solvers for multi-dimensional gas-dynamics goes back to Roe himself: the Riemann-solver is applied in the grid- rather than the flow-direction, which may lead to a misinterpretation of the local wave-structure of the solution.

Since the mid-eighties, Roe, Deconinck, Van Leer and many others developed the so-called fluctuation-splitting schemes for the equations of gas-dynamics (see [4] for references). Other multi-dimensional approaches include Colella's Corner-Transport-Upwind (CTU) scheme [3], LeVeque's CLAWPACK [11], the Weighted-Average-Flux (WAF) scheme of Billet and Toro [1] and the Evolution-Galerkin-Method [12].

In this contribution we focus on Fey's Method of Transport (**MoT**) [6, 7], which belongs to the family of flux-vector-splitting schemes. The starting point of Fey's algorithm is a multi-dimensional wave-model, which leads to a reformulation of the system of conservation laws as a finite set of coupled nonlinear advection equations. At the beginning of each timestep, these coupled nonlinear equations are decomposed into a set of linear scalar advection equations, which are then solved numerically using characteristic schemes.

2. Sonic Points

One-dimensional first-order flux-vector-splitting schemes split the flux-vector in the center of a cell into components which are transported across the left and right boundaries of that cell. At sonic points, where eigenvalues change their sign, this procedure (which we would like to call **C**ell-**C**entered-**E**volution) may lead to inconsistencies [18]. Indeed, we have constructed a linear advection equation with smooth variable coefficients and smooth solutions for which Fey's first-order scalar characteristic scheme diverges at the sonic points.

3. The MoT-ICE

Motivated by this discovery we have developed a new version of the MoT based on **I**nterface-**C**entered-**E**volution, the **MoT-ICE**. We refer to [15] for the full details of the new method. The multi-dimensional wave-models are inherited from Fey's Method of Transport or other flux-vector-splitting schemes, while a Lax-Friedrichs-type predictor-step provides auxiliary transport-velocities on the cell-interfaces.

For the new method, we have proved uniform first- resp. second-order consistency, including at sonic points. Numerical experiments confirm second-order-accuracy for smooth solutions and high-resolution nonoscillatory shock-capturing properties for discontinuous solutions. For example, the new scheme produces excellent radially symmetric solutions for two-dimensional flows, see Figure 1.

The second-order version of the new MoT-ICE is several times faster than the second-order MoT-CCE and seems to be as fast as standard second-order algorithms. This gain of efficiency is partly due to an improved linearisation and decomposition of the nonlinear system into advection equations, and to our particularly simple characteristic transport algorithm for the resulting linear advection equations.

The method has been implemented in a general framework. Main applications are multi-d Euler- and magnetohydrodynamic (MHD) simulations, but other systems of hyperbolic equations can be implemented as well.

4. Parallelization

Multi-dimensional high-resolution computations are extremely time-consuming. This is especially true for wave models with comparatively many waves (such as MHD) as well as any 3d application. Parallelization of the code becomes essential. In [19] the first author has developed a block-oriented approach where macro

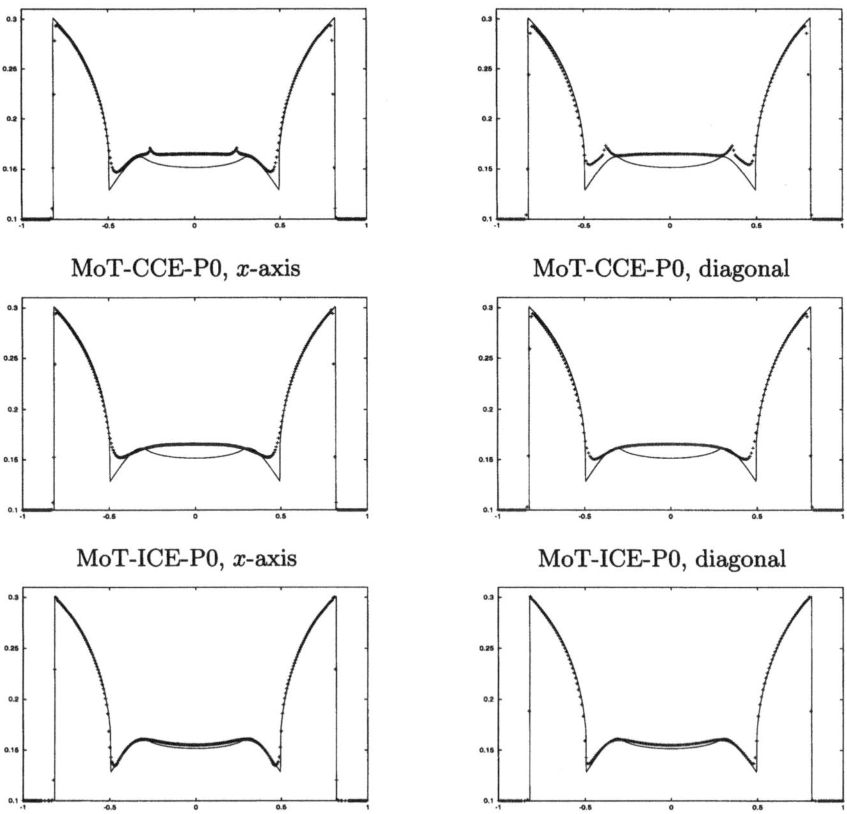

MoT-CCE-P0, x-axis — MoT-CCE-P0, diagonal

MoT-ICE-P0, x-axis — MoT-ICE-P0, diagonal

MoT-ICE-P1, x-axis — MoT-ICE-P1, diagonal

FIGURE 1. Explosion problem for the shallow water-equations. Plot of water-height for grids of 320×320 points. First-order MoT-CCE (top), the new first-order MoT-ICE (middle) and the new second-order MoT-ICE (bottom). Left column: cuts along the x-axis. Right column: cuts along the diagonal. Solid line: one-dimensional solution with 3200 points. Note the kinks at the sonic points for the MoT-CCE.

elements (blocks) that contain an equal amount of computational work are distributed to the different processors. Load balancing is done using a space filling curve based on work by Zumbusch [8]. This approach is well-suited for two and three dimensions and adaptive grids.

Results of high resolution computations are shown in figure 2 (see [10], example 3, for the setup).

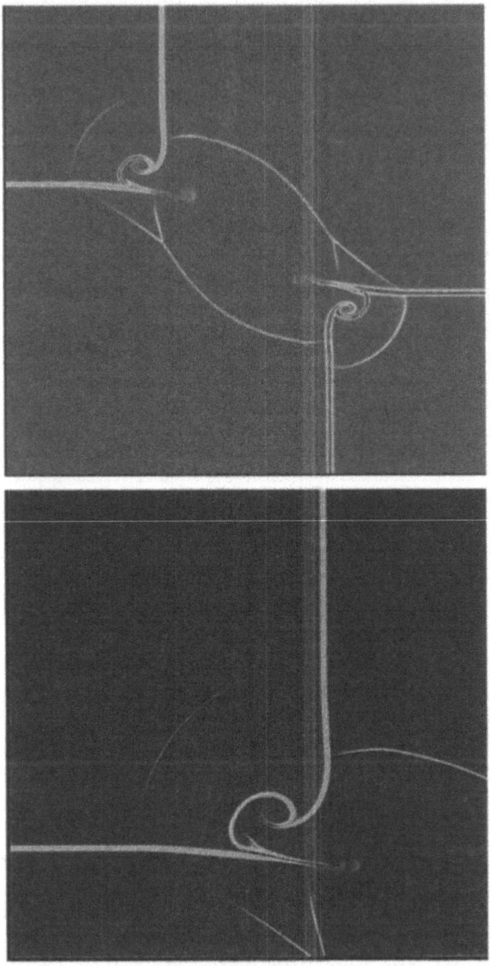

FIGURE 2. Two-dimensional Riemann problem for the Euler equations, schlieren-type plots. $\Delta x = 1/1024$ (top) and $\Delta x = 1/2048$ (bottom). The lower picture is a detail of the NW region of the top picutre.

5. Ideal Magnetohydrodynamics (MHD)

A second application is the system of ideal MHD equations which describes the motion of conducting fluid in a magnetic field. In [19] the first author has developed a second order version of the MoT-ICE for the two-dimensional MHD equations based on a wave model of M. Fey.

One very important issue in this system is the so-called "div $\underline{\mathbf{B}}$ constraint": The set of equations for the magnetic field $\underline{\mathbf{B}}$ implicates that its divergence is a constant in time, such that if initial data yield div $\underline{\mathbf{B}} = 0$, analytically this will remain so for all time.

Unfortunately, this is not necessarily true for the numerical solution. Due to small numerical errors and numerical viscosity, small amounts of divergence will be created. This of course violates the constraint, and a non-vanishing divergence raises the question of the physical relevance of the numerical solution [17].

There are some approaches to overcome this problem: First, one can use a Helmholtz-decomposition of the numerically computed $\underline{\mathbf{B}}$-field, splitting it up in a divergence-free part and a gradient field. This decomposition can be computed by solving a Poisson's problem, with the potential as the unknown and the divergence as the right hand side. A similar problem is known in the incompressible Navier-Stokes equations, the constraint here is on the velocity field $\underline{\mathbf{u}}$.

Second, a source term proportional to div $\underline{\mathbf{B}}$ can be introduced to the MHD equations. This idea is due to Powell [16]. A possible interpretation of this approach is that any divergence created by the numerical scheme is advected away with the speed of the fluid.

Both these approaches can lead to problems. The Helmholtz-decomposition needs some boundary values for the divergence. This problem is not so severe as it can be overcome [5]. But since the Poisson equation is of elliptic type, information travels with infinite speed – in other words, divergence created at one end of the domain may change the value of the magnetic field at some very distant point of the domain. This is in contrast with the finite characteristic velocities of the hyperbolic system under consideration and so not a desirable feature.

In Powell's idea, a problem arises when there is a stagnation point of the velocity field in the computational domain. Following the interpretation of the "advected divergence", non-vanishing divergence will accumulate at the stagnation point which will lead to physically questionable solutions.

Another interesting approach is by Fey and Limacher [13]. In addition to the wave model, they implement a different explicit solver for the evolution of the magnetic field in such a way that the divergence due to a given discrete operator is not changed. The solver is entirely local, so that no information is passed on at infinite speed.

A recent way of dealing with the divergence problem is by Kemm, Munz et. al. ([9, 14]). The idea is to modify the equation for the magnetic field and to solve an additional equation for a quantity that my be interpreted as a generalized Lagrange multiplier incorporating the constraint on $\underline{\mathbf{B}}$.

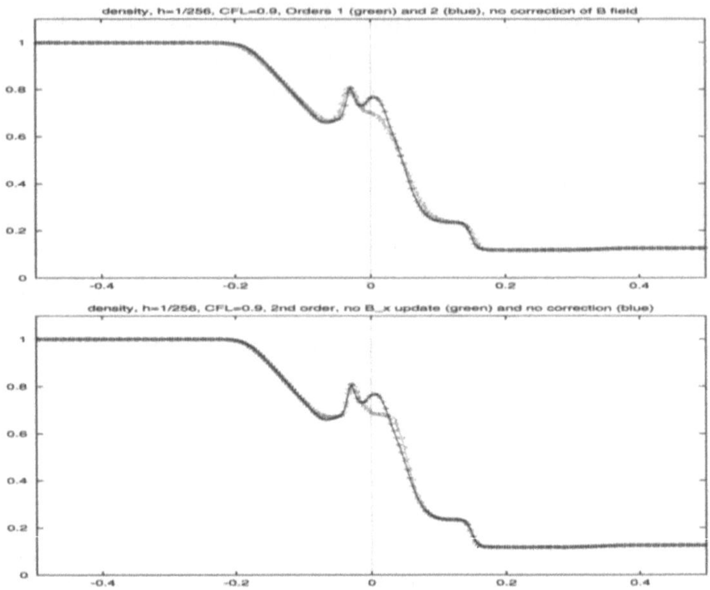

FIGURE 3. MHD Riemann Problem of Brio and Wu. Top: 1st and 2nd order MoT-ICE without correction of magnetic field. Bottom: 2nd order MoT-ICE with and without magnetic-field correction.

6. Numerical experiments for MHD

First results for the 1d MHD Riemann problem by Brio and Wu [2] with different correction techniques for the magnetic field are shown in figures 3 and 4. The top left figure shows a comparison of a first and a second order MoT-ICE without any treatment for the divergence constraint; we can see that while the first order method does not have any obvious problems after the compound wave, the second order method does. However, if we do not update the B_x compoment at all, the undesirable effect disappears (top right). The bottom two graphics show a comparison between second order MoT-ICE with no B_x update and a second order MoT-ICE with Powell's source term (left) and with the use of an projection scheme (right). The use of a projection scheme seems to be a bit sharper than Powell's source term, however, Powell's source term is much cheaper in terms of computation time. Numerical experiments for twodimensional MHD-flows are reported in [19]. This is work in progress, and there are still some stability problems which need further analysis.

7. Future Extensions

In ongoing work we are developing the MoT-ICE into a fully adaptive, distributed parallel code, capable of handling general geometries.

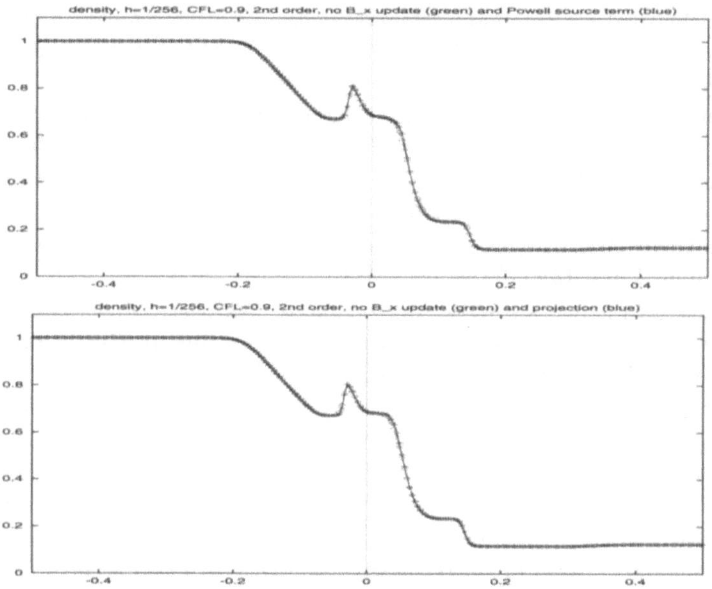

FIGURE 4. MHD Riemann Problem of Brio and Wu. Different corrections of magnetic field. Top: Powell source term versus exact correction. Bottom: projection method versus exact correction.

8. Acknowledgement

This work was financially supported by the DFG priority research program "Analysis und Numerik von Erhaltungssätzen" (ANumE).

References

[1] S. Billet and E. Toro, *On WAF-type schemes for multidimensional hyperbolic conservation laws,* J. Comput. Phys. **130** (1997), 1 – 24.

[2] M. Brio and C.C. Wu, *An upwind differencing scheme for the equations of ideal magnetohydrodynamics,* J. Comp. Phys. **75** (1988), 236–243.

[3] P. Colella, *Multidimensional upwind methods for hyperbolic conservation laws,* J. Comput. Phys. **87** (1990), 171 – 200.

[4] H. Deconinck, H. Paillère, R. Struijs and P. Roe, *Multidimensional upwind schemes based on fluctuation-splitting for systems of conservation laws,* Comput. Mech. **11** (1993), 323 – 340.

[5] H. DeSterck, private communication, 1998.

[6] M. Fey, *Multidimensional upwinding. I. The method of transport for solving the Euler equations,* J. Comput. Phys. **143** (1998), 159 – 180.

[7] M. Fey, *Multidimensional upwinding. II. Decomposition of the Euler equations into advection equations,* J. Comput. Phys. **143** (1998), 181 – 199.

[8] M. Griebel and G. Zumbusch, *Parallel multigrid in an adaptive PDE solver based on hashing,* Proceedings of ParCo '97, editors: E. D'Hollander, G.R. Joubert, F.J. Peters, U. Trottenberg, Elsevier, 589–599, 1998.

[9] F. Kemm, C.-D. Munz, A. Dedner, D. Kröner and M. Wesenberg, *Hyperbolic divergence cleaning for MHD equations,* Preprint (in preparation), March 2000.

[10] P.D. Lax and X.-D. Liu, *Solution of two-dimensional Riemann problems of gas dynamics by positive schemes.* SIAM J. Sc. Comp. **19** (1998), 319–340.

[11] R.J. LeVeque, *Wave propagation algorithms for multidimensional hyperbolic systems,* J. Comput. Phys. **131** (1997), 327 – 353.

[12] M. Lukacova-Medvidova, K.W. Morton and G. Warnecke, *Evolution Galerkin methods for hyperbolic systems in two space dimensions,* Report 97-44, Univ. Magdeburg, Germany (1997). Math. Comp. (2000).

[13] R. Limacher, *Simulation der MHD-Gleichungen mit der Transportmethode,* Diploma thesis, ETH Zurich, 2000 (unpublished).

[14] C.-D. Munz, R. Schneider, E. Sonnendrücker and U. Voss, *Maxwell's equations when the charge conservation is not satisfied,* to appear in C. R. Acad. Sci. Paris, 1999.

[15] S. Noelle, *The MoT-ICE: a new high-resolution wave-propagation algorithm for multidimensional systems of conservation laws based on Fey's Method of Transport,* J. Comput. Phys **164** (2000), 283 – 334.

[16] K.G. Powell, *An approximate Riemann solver for magnetohydrodynamics (that works in more than one dimension),* ICASE Report No. 94-24, Langley, VA, 1994.

[17] J.J. Quirk, *A contribution to the great Riemann solver debate,* Intern. J. Numer. Meth. Fluids **18** (1994), 555–574.

[18] J. Steger and R. Warming, *Flux vector splitting of the inviscid gas-dynamic equations with applications to finite difference methods,* J. Comput. Phys. **40** (1981), 263 – 293.

[19] C. v. Törne, *MOTICE: adaptive, parallel numerical solution of hyperbolic conservation laws.* Dissertation, Bonner Math. Schriften **334** (2000).

M. Fey
Seminar für Angewandte Mathematik, ETH Zürich,
CH 8092 Zürich, Switzerland
`fey@sam.math.ethz.ch`

S. Noelle
Institut für Geometrie und Praktische Mathematik, RWTH Aachen,
Templergraben 55, 52056 Aachen, Germany
`noelle@igpm.rwth-aachen.de`

C. v. Toerne
Computational Fluid Dynamics, Bayer AG,
51368 Leverkusen, Germany
`christian.vontoerne.cv@bayer-ag.de`

International Series of Numerical Mathematics
Vol. 140, © 2001 Birkhäuser Verlag Basel/Switzerland

Positive Decompositions of the Euler Equations into Advection Equations

Michael Fey and Susanne A. Zimmermann

Abstract. The Method of Transport is a genuinely multi-dimensional finite volume scheme to solve the Euler equations. It is based on decomposing the Euler equations into a finite number of advection equations, and solving the resulting equations with some advection solver. In this paper we investigate how the decomposition and the advection solver must be chosen such that the resulting scheme preserves positivity of density and pressure.

1. Introduction

In technical applications there often is a need to simulate the flow of a gas. This can be done by analysing the macroscopic physics of a gas, which leads to the Euler equations, or by considering the microscopic physics of a gas, which leads to the Boltzmann equation. The Euler equations are closely related to the Boltzmann equations; accordingly, there are two approaches to solve the Euler equations numerically: one based on the Euler equations, the other based on the Boltzmann equation. This distinction was first made by Harten, Lax, and van Leer [7], who called the corresponding finite volume schemes Godunov, resp. Boltzmann type schemes.

No matter how a numerical scheme for the Euler equations is derived we expect it to have certain properties apart from being consistent with the equations. Properties needed to ascertain good quality of the numerical solution are for example: robust handling of discontinuities, no grid dependencies of the numerical solution in multi dimensions, conservation of mass, momentum, and energy, positivity of density and pressure, discrete entropy inequalities.

Godunov type schemes are constructed in the framework of conservation laws and are in general based on the (exact or approximate) solution of local Riemann problems. Roughly speaking they have the following properties: they are robust in handling discontinuities, in particular they are exact on stationary contact discontinuities. However, negative densities and pressure occur (cf. [3]) and entropy may be produced. In addition, the step from one to two space dimensions is not straightforward; most Godunov type schemes are extended to multi dimensions by the use of operator splitting techniques.

Boltzmann (or kinetic) schemes on the other hand extend naturally to multi dimensions because they trace the path of individual gas particles. The first kinetic scheme was introduced by Sanders and Prendergast [11]; Deshpande [2] and Perthame [9, 10] derived similar schemes. All of these schemes preserve positivity of density and pressure (cf. [10]), some of them lead to a discrete entropy inequality ([2],[9]). While this group of schemes is robust in handling discontinuities it is not exact on stationary contact discontinuities, which is the reason why these schemes are quite diffusive.

There is a group of schemes that - although constructed in the framework of conservation laws - do not solve any Riemann problems: flux-vector splitting schemes. Gressier et al. [6] show that if a flux-vector splitting scheme preserves positivity of density and pressure it cannot be exact on stationary contact discontinuities. This indicates that flux-vector splitting schemes have properties similar to those of kinetic schemes; in fact their relation was already pointed out in [7].

In this paper we investigate properties of Fey's Method of Transport [4, 5], a genuinely multi-dimensional scheme derived in the framework of conservation laws. It is based on decomposing the Euler equations into advection equations and solving the advection equations separately by a genuinely multi-dimensional scheme. In one space dimension it can be viewed as a flux-vector splitting scheme; we will therefore investigate whether it preserves positivity of density and pressure. This will be done by writing the Method of Transport as a kinetic scheme (cf. [12, 13]).

This paper is organized as follows. In section 2 we present the main ideas of the Method of Transport; in section 3 we write it as a kinetic scheme. In section 4 we present admissible decompositions and advection solvers. Finally, in section 5, we present some numerical results and conclusions.

2. The Method of Transport

The Method of Transport is derived in the framework of hyperbolic conservation laws:

$$\frac{\partial}{\partial t}\mathbf{U}(\vec{\boldsymbol{x}},t) + \nabla \cdot \underline{\boldsymbol{F}}(\mathbf{U}(\vec{\boldsymbol{x}},t)) = 0. \tag{1}$$

Here $\mathbf{U} : \Omega \subset \mathbb{R}^N \times \mathbb{R}^+ \to \mathbb{R}^M$ is the state vector, and $\underline{\boldsymbol{F}} : \mathbb{R}^M \to \mathbb{R}^{M\times N}$ is the flux matrix. The divergence acts on rows of $\underline{\boldsymbol{F}}$. In the case of the Euler equations we have

$$\mathbf{U} = \begin{pmatrix} \rho \\ \rho\vec{\boldsymbol{u}} \\ E \end{pmatrix}, \quad \underline{\boldsymbol{F}}(\mathbf{U}) = \begin{pmatrix} \rho \\ \rho\vec{\boldsymbol{u}} \\ E \end{pmatrix} \vec{\boldsymbol{u}}^T + p \begin{pmatrix} \vec{\mathbf{0}} \\ \underline{\boldsymbol{I}} \\ \vec{\boldsymbol{u}}^T \end{pmatrix};$$

ρ denotes the density, $\rho\vec{\boldsymbol{u}} \in \mathbb{R}^N$ the momentum, E the energy, and p the pressure of the gas. $\underline{\boldsymbol{I}}$ is the identity in $\mathbb{R}^{N\times N}$, and $\vec{\mathbf{0}}$ the zero vector in $\mathbb{R}^N$. The equation of state relates the pressure with the conserved quantities:

$$p = (\gamma - 1)\left(E - 1/2\rho\vec{u}^T\vec{u}\right).$$

γ is the ratio of the heat capacities.

The Method of Transport is based on two ideas: first decompose the Euler equations into a finite number of advection equations, then solve each of the advection equations by a genuinely multi-dimensional scheme.

2.1. Decomposition into advection equations.

Assume that $\mathbf{U}$ and $\mathbf{F}(\mathbf{U})$ can be decomposed in the following way:

$$\mathbf{U} = \sum_{i=0}^{I} \mathbf{S}_i, \quad \mathbf{F}(\mathbf{U}) = \sum_{i=0}^{I} \mathbf{S}_i \vec{a}_i^T, \tag{2}$$

where $\mathbf{S}_i = \mathbf{S}_i(\mathbf{U}) \in \mathbb{R}^M$, $\vec{a}_i = \vec{a}_i(\mathbf{U}) \in \mathbb{R}^N$, and $I \in \mathbb{N}$. Instead of considering the original equations (1) we attempt to solve the equivalent equations

$$\sum_{i=0}^{I} (\mathbf{S}_i)_t + \sum_{i=0}^{I} \nabla \cdot (\mathbf{S}_i \vec{a}_i^T) = 0.$$

We decouple the above equations, and solve each of the advection equations

$$(\mathbf{S}_i)_t + \nabla \cdot (\mathbf{S}_i \vec{a}_i^T) = 0, \quad i = 0, \dots I$$

separately.

2.2. Solving the advection equations.

Consider advection equations

$$\mathbf{S}_t + \nabla \cdot (\mathbf{S}\vec{a}^T) = 0, \tag{3}$$

with initial data $\mathbf{S}(\vec{x}, t_n) := \mathbf{S}^n(\vec{x})$ and $\vec{a}(\vec{x}, t_n) := \vec{a}^n(\vec{x})$. We will now derive genuinely multidimensional finite volume schemes to solve this equation.

First divide the computational domain Ω into cells Ω_j. The unknowns in a cell Ω_j at time t_n are cell averages denoted by $\mathbf{S}_j^n$. They are updated to time $t_n + \Delta t$ by the subtraction of all fluxes leaving the cell Ω_j and the addition of all fluxes entering from neighbouring cells:

$$\mathbf{S}_j^{n+1} := \mathbf{S}_j^n - \frac{1}{|\Omega_j|} \sum_{k \neq j} (\mathbf{F}_{\Omega_j \Omega_k} - \mathbf{F}_{\Omega_k \Omega_j}). \tag{4}$$

Note that we are considering fluxes to and from all neighbouring cells, including those neighbours having only one point in common with Ω_j.

The contributions $\mathbf{F}_{\Omega_j \Omega_k}$ approximate the physical flux from the cell Ω_j into the cell Ω_k. They are defined as follows:

$$\mathbf{F}_{\Omega_j \Omega_k} := \int_{\Omega_k} \int_{\Omega_j} \mathbf{S}(\vec{y}, t_n) \delta[\vec{z}(t_n + \Delta t, \vec{y}) - \vec{x}] d\vec{y}\, d\vec{x}, \tag{5}$$

where $\vec{z}(\tau, \vec{y})$ is the characteristic curve solving

$$\dot{\vec{z}}(\tau) = \vec{a}(\vec{z}(\tau), \tau), \quad \vec{z}(t_n) = \vec{y}. \tag{6}$$

Note that so far the time development of $\vec{a}$ is not known.

Let $\vec{z}^{-1}(\tau, \vec{x})$ denote the inverse function of $\vec{z}(\tau, \vec{y})$: if $\vec{x} = \vec{z}(\tau, \vec{y})$ then $\vec{z}^{-1}(\tau, \vec{x}) := \vec{y}$. With this notation we have the following lemma:

Lemma 2.1. *The flux (5) has the form*

$$F_{\Omega_j \Omega_k} = \int_{\widetilde{\Omega}_{MoT}} \mathbf{S}(\xi, t_n)\, d\xi, \quad \text{where } \widetilde{\Omega}_{MoT} := \vec{z}^{-1}(t_n + \Delta t, \Omega_k) \cap \Omega_j. \tag{7}$$

The order of the scheme depends on the decomposition (2), and on the order of the scheme to solve (3). For more details see [4, 5].

In the next section we investigate whether the Method of Transport preserves positivity of density and pressure.

3. The Method of Transport written as a kinetic scheme

Kinetic schemes are based on the relation between the Euler equations and the Boltzmann equation. Its unknown is a distribution function $f(\vec{x}, \vec{v}, t) : \mathbb{R}^N \times \mathbb{R}^N \times \mathbb{R}^+ \to \mathbb{R}^+$. In the case $\gamma = 1 + 2/N$ macroscopic quantities of a gas are recovered by calculating the following moments of the distribution function:

$$\mathbf{U}(\vec{x}, t) := \int_{\mathbb{R}^N} \mathbf{D}(\vec{v})\, f(\vec{x}, \vec{v}, t)\, d\vec{v}, \quad \mathbf{D}(\vec{v}) := \begin{pmatrix} 1 \\ \vec{v} \\ \vec{v}^T \vec{v}/2 \end{pmatrix}.$$

The group of kinetic schemes proposed in [11, 2, 10] consists in performing the following steps: Discretize the computational domain Ω and calculate the initial cell averages. Given cell averages $\mathbf{U}_j^n$ at time t_n calculate values $\mathbf{U}_j^{n+1}$ at time $t_n + \Delta t$ as follows:

1. Let $\mathbf{U}^n(\vec{x})$ be a representation of $\mathbf{U}_j^n$. Construct an initial distribution function $f_0(\mathbf{U}^n(\vec{x}), \vec{v}) \geq 0$ representing the initial state $\mathbf{U}^n(\vec{x})$.
2. Solve the collision-less Boltzmann equation $f_t + \vec{v} \cdot \nabla f = 0$ exactly for a time-step Δt with the initial data $f(\vec{x}, \vec{v}, t_n) = f_0(\mathbf{U}^n(\vec{x}), \vec{v})$.
3. Set $\mathbf{U}_j^{n+1} := \dfrac{1}{|\Omega_j|} \displaystyle\int_{\Omega_j} \int_{\mathbb{R}^N} \mathbf{D}(\vec{v}) f_0(\mathbf{U}^n(\vec{x} - \vec{v}\Delta t), \vec{v})\, d\vec{v}\, d\vec{x}.$

The order of the scheme depends on the choice of the initial distribution function $f_0(\mathbf{U}, \vec{v})$, and on the representation $\mathbf{U}^n(\vec{x})$. Modifications need to be made in the case $\gamma \neq 1 + 2/N$.

The following theorem was proved by Perthame [10]:

Theorem 3.1. *If the initial distribution function $f_0(\mathbf{U}^n(\vec{x}), \vec{v})$ is nonnegative then the kinetic scheme 1.-3. preserves positivity of density and pressure.*

Now we construct a positive initial distribution function such that the corresponding kinetic scheme is the same as the Method of Transport. Consider distribution functions of the following type (cf. [12]):

$$f_0^i(\mathbf{U},\vec{v}) := A_i(\mathbf{U})\,\delta[\vec{v}-\vec{a}_i(\mathbf{U})], \quad A_i(\mathbf{U}) \geq 0, \quad i=0,\dots I.$$

The moments of these distribution functions are:

$$\mathbf{Z}_i(\mathbf{U}) := A_i(\mathbf{U})\,\mathbf{D}(\vec{a}_i(\mathbf{U})), \quad i=0,\dots I. \tag{8}$$

Applying the kinetic scheme 1.-3. leads to a numerical scheme for the advection equation $(\mathbf{Z}_i)_t + \nabla\cdot(\mathbf{Z}_i\vec{a}_i^T) = 0$:

$$(\mathbf{Z}_i)_j^{n+1} := \frac{1}{|\Omega_j|}\int_{\Omega_j} \mathbf{Z}_i(\mathbf{U}^n(\underbrace{\vec{x}-\vec{v}^*(\vec{x})\Delta t}_{=:\vec{z}_{kin}^{-1}(\vec{x},\Delta t)}))\,d\vec{x}, \tag{9}$$

where $\vec{v}^*(\vec{x})$ is the solution of

$$\vec{v} = \vec{a}_i(\vec{x}-\vec{v}\Delta t). \tag{10}$$

Rewritten as in (4) we have the following contributions:

$$\mathbf{F}_{\Omega_j\Omega_k} := \int_{\widetilde{\Omega}_{kin}} \mathbf{Z}_i(\mathbf{U}^n(\xi))\,d\xi \quad \text{where} \quad \widetilde{\Omega}_{kin} := \vec{z}_{kin}^{-1}(t_n+\Delta t,\Omega_k)\cap\Omega_j.$$

The above implies that there is a strong analogy between the Method of Transport and the kinetic scheme based on distribution functions f_0^i: the moments $\mathbf{Z}_i$ correspond to the decomposition (2), and the integration (9) corresponds to the advection solver (4),(7). Therefore, our first positivity result is:

Theorem 3.2. *A Method of Transport based on a kinetic decomposition (2) ($\mathbf{S}_i$ have the form (8)), and on a kinetic advection solver (4),(7) ($\vec{z}^{-1}(\tau,\vec{x}) = \vec{z}_{kin}^{-1}(\tau,\vec{x})$) is a kinetic scheme. As a consequence it preserves positivity of density and pressure.*

This first result is quite restrictive: the number of kinetic decompositions is limited, in particular higher order kinetic decompositions in multi dimensions might not exist at all (depending on the advecting velocities $\vec{a}_i$). The problem is that the decompositions of the density, momentum and energy are strongly related through (8).

Therefore, we use a separate distribution function for each conserved quantity:

$$\mathbf{f}_0(\mathbf{U},\vec{v}) := \left(f_0^0,\cdots,f_0^{N+1}\right)^T \quad \text{with} \quad \int_{\mathbb{R}^N} \mathbf{f}_0(\mathbf{U},\vec{v})\,d\vec{v} = \mathbf{U}.$$

However, we have to make sure that the kinetic scheme based on this generalized distribution function preserves positivity and density. Bouchut [1] proved a similar lemma:

Lemma 3.3. *Consider the kinetic scheme 1.-3. with the initial distribution function* $\mathbf{f}_0(\mathbf{U},\vec{\boldsymbol{v}})$. *If the following conditions are satisfied at time* t_n *then the scheme preserves positivity of density and pressure.*

$$f_0^0(\mathbf{U},\vec{\boldsymbol{v}}) \geq 0, \qquad f_0^{N+1}(\mathbf{U},\vec{\boldsymbol{v}}) \geq 0, \tag{11}$$

$$\sum_{j=1}^{N}(f_0^j(\mathbf{U},\vec{\boldsymbol{v}}))^2 \leq 2f_0^0(\mathbf{U},\vec{\boldsymbol{v}})f_0^{N+1}(\mathbf{U},\vec{\boldsymbol{v}}). \tag{12}$$

For the Method of Transport we consider the distribution function

$$\mathbf{f}_0(\mathbf{U},\vec{\boldsymbol{v}}) := \sum_{i=0}^{I}\mathbf{S}_i(\mathbf{U})\,\delta[\vec{\boldsymbol{v}}-\vec{\boldsymbol{a}}_i(\mathbf{U})].$$

The conditions of lemma 3.3 are satisfied if each $\mathbf{S}_i$ has positive density and pressure. Therefore our second, more general, positivity result is:

Theorem 3.4. *A Method of Transport based on a positive decomposition (2) (each* $\mathbf{S}_i$ *has positive density and pressure), and on a kinetic advection solver (cf. theorem 3.2) is a generalized kinetic scheme satisfying (11),(12). As a consequence it preserves positivity of density and pressure.*

In the next section we give some examples for positive decompositions and kinetic advection solvers.

4. Positive decompositions and kinetic advection solvers

4.1. Kinetic advection solvers.

For a first order advection solver we assume that all quantities are piecewise constantly represented. Then, (10) becomes

$$\vec{\boldsymbol{v}} = \vec{\boldsymbol{a}}_j \quad \text{if} \quad \vec{\boldsymbol{x}} - \vec{\boldsymbol{a}}_j\Delta t \in \Omega_j.$$

We have the following first order transformation $(\vec{\boldsymbol{z}}_{kin}^{-1})^{1o}$:

$$(\vec{\boldsymbol{z}}_{kin}^{-1}(\vec{\boldsymbol{x}},\Delta t))^{1o} = \vec{\boldsymbol{x}} - \vec{\boldsymbol{a}}_j\Delta t \quad \text{if} \quad \vec{\boldsymbol{x}} - \vec{\boldsymbol{a}}_j\Delta t \in \Omega_j.$$

For a second order advection solver we assume that $\vec{\boldsymbol{a}}$ is linearly represented in the cell Ω_j, i. e. $\vec{\boldsymbol{a}}(\vec{\boldsymbol{x}}) = \vec{\boldsymbol{a}}_0 + \underline{\boldsymbol{A}}_1\vec{\boldsymbol{x}}$. Then, (10) reduces to

$$\vec{\boldsymbol{v}} = \vec{\boldsymbol{a}}_0 + \underline{\boldsymbol{A}}_1(\vec{\boldsymbol{x}} - \vec{\boldsymbol{v}}\Delta t).$$

The solution $\vec{\boldsymbol{v}}^*$ is $\vec{\boldsymbol{v}}^*(\vec{\boldsymbol{x}}) = (\underline{\boldsymbol{I}} + \underline{\boldsymbol{A}}_1\Delta t)^{-1}(\vec{\boldsymbol{a}}_0 + \underline{\boldsymbol{A}}_1\vec{\boldsymbol{x}})$ and therefore the second order kinetic transformation $(\vec{\boldsymbol{z}}_{kin}^{-1})^{2o}$ is

$$(\vec{\boldsymbol{z}}_{kin}^{-1}(\vec{\boldsymbol{x}},\Delta t))^{2o} := \vec{\boldsymbol{x}} - (\underline{\boldsymbol{I}} + \underline{\boldsymbol{A}}_1\Delta t)^{-1}(\vec{\boldsymbol{a}}_0 + \underline{\boldsymbol{A}}_1\vec{\boldsymbol{x}})\Delta t.$$

This transformation is obtained in the framework of section 2 if (6) is solved under the assumption that $\vec{\boldsymbol{a}}_t = -(\nabla\vec{\boldsymbol{a}})\vec{\boldsymbol{a}}$.

4.2. Positive decompositions.

The following first order decomposition was proposed in [4, 5] and is motivated by symmetry properties of the Euler equations, in particular by the fact that the characteristics lie on the Mach cone. It is related to the eigenvalues and right eigenvectors of the Jacobian of the flux matrix.

As advecting velocities we choose $\vec{a}_0(\mathbf{U}) := \vec{u}$, and $\vec{a}_i(\mathbf{U}) := \vec{u} + c\vec{n}_i$ for $i = 1 \dots I$ (c is the speed of sound). The vectors $\vec{n}_i$ satisfy certain properties. It is easy to check that the following decomposition is positive:

$$\begin{aligned} \mathbf{S}_0^{1o}(\mathbf{U}) &:= \frac{\gamma - 1}{\gamma}\rho \begin{pmatrix} 1 \\ \vec{u} \\ \vec{u}^T\vec{u}/2 \end{pmatrix}, \\ \mathbf{S}_i^{1o}(\mathbf{U}) &:= \frac{1}{\gamma I}\rho \begin{pmatrix} 1 \\ \vec{u} \\ H \end{pmatrix} + \frac{\rho c}{\gamma I} \begin{pmatrix} \vec{0} \\ \underline{\boldsymbol{I}} \\ \vec{u}^T \end{pmatrix} \vec{n}_i, \quad i = 1 \dots I \end{aligned}$$

H is the total enthalpy $(E + p)/\rho$. Note that in the case $\gamma 1 + 2/N$ the above decomposition is kinetic.

First order decompositions are extended to second order by the addition of correction terms:

$$\mathbf{S}_i^{2o}(\mathbf{U}, \nabla\mathbf{U}) := \mathbf{S}_i^{1o}(\mathbf{U}) + \mathbf{K}_i(\mathbf{U}, \nabla\mathbf{U})\Delta t, \quad i = 0, \dots I,$$

where $\mathbf{K}_i(\mathbf{U}, \nabla\mathbf{U}) \in \mathbb{R}^{N+2}$, $i = 0, \dots I$. These correction terms must add up to zero so that first order accuracy is not destroyed; a simple way to fulfill this condition is to define $\mathbf{K}_i := \underline{\boldsymbol{K}}\vec{n}_i$, $i = 0, \dots I$, where $\underline{\boldsymbol{K}}$ is a $(N + 2) \times N$-matrix, and $\vec{n}_0 := \vec{0}$. The components of $\underline{\boldsymbol{K}}$ are determined by the comparison of the Taylor expansions of the exact and numerical solutions. In one space dimension they are:

$$\mathbf{K} := (k^\rho, k^{\rho u}, k^E)^T := \left(0, \frac{1}{2\gamma}(3 - \gamma)\rho c u_x, u k^{\rho u} + \frac{p}{\gamma - 1}c_x\right)^T.$$

The resulting decomposition is not necessarily positive, therefore it might be necessary to limit the gradients.

5. Numerical results and conclusions

As a test problem we choose the 1-2-0-3 Riemann problem proposed in [3]. Both the first and the second order Method of Transport (based on the decompositions and advection solvers from the previous section) do not produce any negative density or pressure. Note that for the second order calculation a flux limiter is used.

We have seen that if a Method of Transport is based on a positive decomposition and on a kinetic advection solver it will preserve positivity of density and

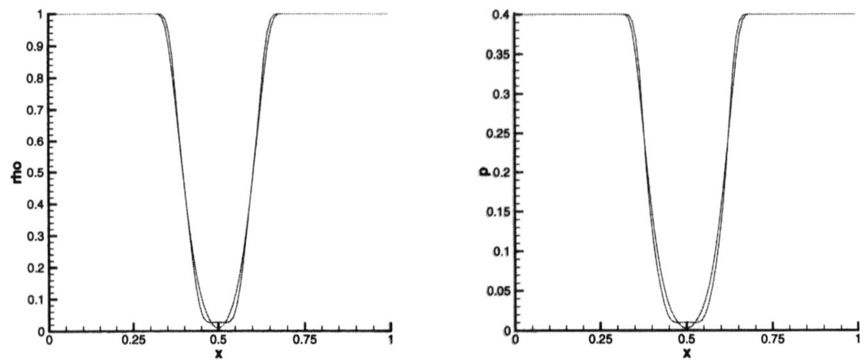

FIGURE 1. 1-2-0-3 Riemann problem: first and second order calculations of density and pressure

pressure. In [13] this will be explained in more detail; in addition, we will investigate whether other advection solvers (e. g. those proposed in [5] or [8]) have the same property.

References

[1] F. Bouchut. *Construction of BGK models with a family of kinetic entropies for a given system of conservation laws.* J. Stat. Phys. **95(1-2)** (1999).

[2] S. M. Deshpande. *A second order accurate, kinetic-theory based method for inviscid compressible flows.* NASA Langley Tech. paper No. 2613, 1986.

[3] B. Einfeldt, C. D. Munz, P. L. Roe, and B. Sjögreen. *On Godunov-type Methods near low Densities.* J. Comp. Phys. **92** (1991).

[4] M. Fey. *Multidimensional upwinding part I: the method of transport for solving the Euler equations.* J. Comp. Phys. **143(1)** (1998).

[5] M. Fey. *Multidimensional upwinding part II: decomposition of the Euler equations into advection equations.* J. Comp. Phys. **143(1)** (1998).

[6] J. Gressier, P. Villedieu, and J.-M. Moschetta. *Positivity of flux vector splitting schemes.* J. Comp. Phys. **155** (1999).

[7] A. Harten, P.D. Lax, and B. Van Leer. *On upstream differencing and Godunov-type schemes for hyperbolic conservation laws.* SIAM Review **25** (1983).

[8] S. Noelle. *The MoT-ICE: a new high-resolution wave-propagation algorithm for multi-dimensional systems of conservation laws based on Fey's Method of Transport.* J. Comp. Phys. **164** (2000).

[9] B. Perthame. *Boltzmann type schemes for gas dynamics and the entropy property.* SIAM J. Numer. Anal. **27** (1990).

[10] B. Perthame. *Second order Boltzmann schemes for compressible Euler equations in one and two space variables.* SIAM J. Numer. Anal. **29** (1992).

[11] H. Sanders, and K.H. Prendergast. *On the origin of the 3 kiloparsec arm.* Astrophys. J. **188** (1974).

[12] S. A. Zimmermann. *The Method of Transport for the Euler Equations written as a kinetic scheme.* Proc. 7th Intern. Conf. Hyperbolic Problems, Zürich 1998. Birkhäuser, **1999**.

[13] S. A. Zimmermann. *Positivity of the Method of Transport for the Euler equations.* In preparation.

M. Fey
Seminar for Applied Mathematics,
ETH Zürich,
Rämistr. 101,
CH-8092 Zürich
fey@sam.math.ethz.ch

S. A. Zimmermann
Seminar for Applied Mathematics,
ETH Zürich,
Rämistr. 101,
CH-8092 Zürich
szimmer@sam.math.ethz.ch

International Series of Numerical Mathematics
Vol. 140, © 2001 Birkhäuser Verlag Basel/Switzerland

The Einstein-Dirac-Yang/Mills Equations: Black Holes

Felix Finster

Abstract. We consider a static, spherically symmetric system of a Dirac particle in a classical gravitational and $SU(2)$ Yang-Mills field. We prove that the only black-hole solutions of the corresponding Einstein-Dirac-Yang/Mills equations are the Bartnik-McKinnon black-hole solutions of the $SU(2)$ Einstein-Yang/Mills equations; thus the spinors must vanish identically. This indicates that the Dirac particles must either disappear into the black-hole or escape to infinity.

1. Introduction

In General Relativity, gravity is described geometrically through Einstein's field equations. The understanding of these equations has been driven by the discovery of exact solutions. We here restrict attention to solutions with an event horizon, usually referred to as black-hole solutions. The most prominent black-hole solution is the Kerr-Newman solution, which includes the Schwarzschild and Reissner-Nordström solutions as limiting cases. It describes the gravitational field of a charged rotating point mass. According to the well-known no-hair conjecture, the Kerr-Newman black hole should be even universal in the sense that every stable, stationary black hole should be completely characterized by its mass, its angular momentum, and its electric charge. Other interesting black-hole solutions arise when gravity is coupled to an $SU(2)$ Yang-Mills field [1, 2, 3]. Unfortunately, these so-called Bartnik-McKinnon black-hole solutions are unstable [4], and thus do not serve as a proper counterexample to the no-hair conjecture.

More recently, black-hole solutions were studied for the coupling of gravity with Dirac particles. If one thinks of the limit of classical point particles, this should correspond e.g. to the situation where particles rotate on stable orbits around a black hole. Surprisingly, this classical picture breaks down as a consequence of the quantum mechanical character of Dirac particles. Namely, it is shown in [5] that the only black-hole solutions of a spherically symmetric, static Einstein-Dirac-Maxwell (EDM) system are the Reissner-Nordström solutions, even when the Dirac particles are allowed to have angular momentum. Subsequently, we considered systems where gravity is coupled to both Dirac particles and an $SU(2)$ Yang-Mills field. The inclusion of the Yang-Mills field changes the equations considerably, and we therefore studied the cases with and without angular momentum of the Dirac particles separately. Without angular momentum, one gets a consistent system of equations involving two real spinor functions [6, Section 2]. It is proved in [7] under rather weak regularity assumptions on the form of the horizon that the only black-hole solutions of these Einstein-Dirac-Yang/Mills (EDYM) equations are the Bartnik-McKinnon black-hole solutions. In the case with angular momentum [8], one gets equations with four real spinor functions. This makes the analysis much harder, and we could show the absence of black-hole solutions only under stronger assumptions on the regularity of the horizon. The above non-existence results for black holes involving Dirac particles indicate that the no-hair conjecture might indeed be true if matter is taken into account on a quantum mechanical level.

In this short report, we will state our non-existence result for the 2-spinor EDYM equations [7]. Furthermore, we give an outline of the proof, omitting some of the technical details.

2. Non-Existence of Black-Hole Solutions

We consider a Lorentzian metric in polar coordinates $(t, r, \vartheta, \varphi)$ of the form

$$ds^2 = \frac{1}{T(r)^2}\, dt^2 \;-\; \frac{1}{A(r)}\, dr^2 \;-\; r^2(d\vartheta^2 + sin^2\vartheta\, d\varphi^2)$$

with positive functions A and T. The Dirac wave function is described by two real functions $(\alpha(r), \beta(r))$, and the potential $w(r)$ corresponds to the magnetic component of an $SU(2)$ YM field. Then the EDYM equations are (cf. [6, Section 2] for the derivation)

$$\sqrt{A}\,\alpha' = \frac{w}{r}\,\alpha \;-\; (m+\omega T)\,\beta \tag{1}$$

$$\sqrt{A}\,\beta' = (-m+\omega T)\,\alpha \;-\; \frac{w}{r}\,\beta \tag{2}$$

$$r\,A' = 1-A \;-\; \frac{1}{e^2}\,\frac{(1-w^2)^2}{r^2} \;-\; 2\omega T^2(\alpha^2+\beta^2) \;-\; \frac{2}{e^2}\,Aw'^2 \tag{3}$$

$$\begin{aligned} 2rA\,\frac{T'}{T} &= -1+A+\frac{1}{e^2}\,\frac{(1-w^2)^2}{r^2} + 2mT(\alpha^2-\beta^2) \;-\; 2\omega\, T^2(\alpha^2+\beta^2) \\ &\quad +4\,\frac{T}{r}\,w\,\alpha\beta \;-\; \frac{2}{e^2}\,Aw'^2 \end{aligned} \tag{4}$$

$$r^2 A\, w'' = -(1-w^2)\,w \,+\, e^2\, rT\alpha\beta \;-\; r^2\,\frac{A'\,T-2A\,T'}{2T}\,w'\,. \tag{5}$$

Here (1) and (2) are the Dirac equations, (3) and (4) are the Einstein equations, and (5) is the Yang-Mills equation. The constants m, ω, and e are the rest mass of the Dirac particle, its energy, and the YM coupling constant, respectively.

By a black-hole solution of this EDYM system we mean a solution defined in the region $r > \rho > 0$ for which the surface $r = \rho$ is an event horizon, i.e.

$$\lim_{r \searrow \rho} A(r) = 0 \quad \text{and} \quad A(r) > 0 \quad \text{if} \quad r > \rho\,. \tag{6}$$

In addition, we assume the following conditions:

$$\int_{r_0}^{\infty} (\alpha^2 + \beta^2)\, \frac{\sqrt{T}}{A}\, dr \;<\; \infty \qquad \text{for every } r_0 > \rho \tag{7}$$

(the spinors are normalizable),

$$\lim_{r \to \infty} \frac{r}{2}\,(1 - A(r)) \;<\; \infty \tag{8}$$

(finite ADM mass),

$$\lim_{r \to \infty} T(r) \;=\; 1$$

(the gravitational field is asymptotically flat Minkowskian),

$$\lim_{r \to \infty} \left(w(r), w'(r)\right) \qquad \text{is finite} \tag{9}$$

(well-behavedness of the Yang-Mills field).

Concerning the event horizon $r = \rho$, we make the following three regularity assumptions (cf. [5]):

(I): The volume element $\sqrt{|\det g_{ij}|} = |\sin \vartheta|\, r^2 A^{-1}\, T^{-2}$ is smooth and non-zero on the horizon; i.e.

$$T^{-2} A^{-1},\; T^2 A \in C^1([\rho, \infty))\,.$$

(II): The strength of the Yang-Mills field F_{ij} is given by

$$\mathrm{Tr}(F_{ij} F^{ij}) \;=\; \frac{2Aw'^2}{r^2} + \frac{(1 - w^2)^2}{r^4}\,.$$

We assume that this scalar is bounded near the horizon; i.e. we assume that outside the event horizon and near $r = \rho$,

$$w \text{ and } Aw'^2 \text{ are bounded.} \tag{10}$$

(III): The function $A(r)$ is monotone increasing outside of and near the event horizon.

If assumptions (I) or (II) were violated, an observer freely falling into the black hole would feel strong forces when crossing the horizon. Assumption (III) is considerably weaker than the corresponding assumption in [5]; indeed, in [5] we assumed that the function $A(r)$ obeyed a power law near the event horizon.

We can now state our main result.

Theorem 2.1. *Every black-hole solution of the EDYM equations (1)–(5) satisfying the regularity conditions (I), (II), and (III) coincides with a BM black-hole of the corresponding Einstein-Yang/Mills (EYM) equations; that is, the spinors α and β must vanish identically outside the event horizon.*

Thus amending the EYM equations by taking quantum mechanical Dirac particles into account, does not yield any new types of black-hole solutions.

3. Non-Existence of Black-Hole Solutions.

Assume that we have a black-hole solution of the EDYM equations (1)–(9), with event horizon at $r = \rho$, satisfying the regularity assumptions (I)–(III), where the spinors $(\alpha(r), \beta(r))$ do not vanish identically in the region $r > \rho$. We will show that this leads to a contradiction. We split up the proof into two cases, depending on whether the function $A^{-\frac{1}{2}}$ is or is not integrable near the event horizon.

Case 1. $A^{-\frac{1}{2}}$ is integrable near $r = \rho$.

We first use a standard Gronwall-type estimate to show that the spinors are bounded near the event horizon, both from above and from below.

Lemma 3.1. *If $A^{-\frac{1}{2}}$ is integrable near $r = \rho$, then there exist positive constants c, ε such that*

$$c \le \alpha^2(r) + \beta^2(r) \le \frac{1}{c} \qquad \text{if } \rho < r < \rho + \varepsilon \,. \tag{1}$$

Proof. If we multiply (1) by α and (2) by β and add, we obtain

$$\begin{aligned} \sqrt{A}\, \frac{d}{dr}(\alpha^2 + \beta^2) &= 2 \begin{pmatrix} \alpha \\ \beta \end{pmatrix}^T \begin{pmatrix} \frac{w}{r} & -m \\ -m & -\frac{w}{r} \end{pmatrix} \begin{pmatrix} \alpha \\ \beta \end{pmatrix} \\ &\le 2 \sqrt{m^2 + \frac{w^2}{r^2}}\, (\alpha^2 + \beta^2) \,, \end{aligned} \tag{2}$$

where in the last estimate we have computed the eigenvalues of the above matrix. Since (α, β) is a non-trivial solution, the uniqueness theorem for ODEs implies that $(\alpha^2 + \beta^2)(r) \neq 0$ on all intervals of the form $(\rho, \rho + \varepsilon)$. Thus dividing (2) by $\sqrt{A}\,(\alpha^2 + \beta^2)$ and integrating from $r > \rho$ to $\rho + \varepsilon$ gives

$$|\log(\alpha^2 + \beta^2)(\rho + \varepsilon) - \log(\alpha^2 + \beta^2)(r)| \le 2 \int_r^{\rho+\varepsilon} A^{-\frac{1}{2}}(s) \sqrt{m^2 + \frac{w^2(s)}{s^2}}\, ds \,. \tag{3}$$

Since $w(r)$ is bounded near the event horizon (by Assumption (II)), and $A^{-\frac{1}{2}}$ is integrable near the event horizon, we can take the limit $r \searrow \rho$ in (3) to get the desired result. □

Now the equation for AT^2 yields that ω must vanish.

Corollary 3.2. *If $A^{-\frac{1}{2}}$ is integrable near $r = \rho$, then $\omega = 0$.*

Proof. Combining (3) and (4), we have

$$r\,(AT^2)' = -4\omega\, T^4\, (\alpha^2 + \beta^2) + T^3 \left[2m\, (\alpha^2 - \beta^2) + \frac{4w}{r}\, \alpha\beta \right] - \frac{4}{e^2}\, (Aw'^2)\, T^2 \,. \tag{4}$$

From Assumption (II), together with the last result, we see that the coefficients of T^4, T^3, and T^2 on the right-hand-side of (4) are bounded. From Assumption (I) we see that the left-hand side of (4) is bounded near the event horizon. Since $T(r) \to \infty$ as $r \searrow \rho$ (again by (I)), we conclude from (4) and Lemma 3.2 that $\omega = 0$. □

In view of this last result, we can write the Dirac equations (1) and (2) as

$$\sqrt{A}\,\alpha' = \frac{w}{r}\,\alpha - m\,\beta \tag{5}$$

$$\sqrt{A}\,\beta' = -m\,\alpha - \frac{w}{r}\,\beta\,. \tag{6}$$

If we multiply (5) by β and (6) by α, and add, we obtain

$$\sqrt{A}\,(\alpha\beta)' = -m\,(\alpha^2+\beta^2) < 0\,, \tag{7}$$

so that $\alpha\beta$ is monotone decreasing. Thus $(\alpha\beta)(r)$ has a limit for $r \to \infty$ (which might be $-\infty$). Since $\alpha^2+\beta^2 \geq 2\,|\alpha\beta|$, we see that the normalization condition (7) will be satisfied only if this limit is zero. We thus have proved the following lemma.

Lemma 3.3. *If $\omega = 0$, then $(\alpha\beta)(r)$ is a positive decreasing function tending to 0 as $r \to \infty$.*

Remark. We did not use the fact that $A^{-\frac{1}{2}}$ is integrable to prove this lemma.

Next we consider the YM equation (5), which can be written equivalently as an equation for Aw',

$$r^2(Aw')' = -w\,(1-w^2) + e^2\,\frac{r(\sqrt{A}\,T)\alpha\beta}{\sqrt{A}} + r^2\,\frac{(AT^2)'}{2AT^2}(Aw')\,. \tag{8}$$

Since Aw'^2 is bounded, we see that $A^2w'^2 \to 0$ as $r \searrow \rho$ and thus $Aw' \to 0$ as $r \searrow \rho$. In view of the last lemma, together with Assumption (I), we see that for r near ρ, we can write (8) in the form

$$(Aw')' = d(r) + \frac{c(r)}{\sqrt{A(r)}} \tag{9}$$

where $d(r)$ is bounded, and $c(r)$ is a positive function which is bounded away from zero near the event horizon. It thus follows from (9) that we have

$$(Aw')'(r) \geq d + \frac{c}{\sqrt{A(r)}}\,, \tag{10}$$

for $r > \rho$ (r near ρ), where d and c are constants, and $c > 0$. As is shown in [7, Lemma 3.5], this inequality implies that A is transversal on the event horizon, i.e.

$$A(\rho) = 0 \quad\text{and}\quad 0 < A'(\rho) < \infty\,. \tag{11}$$

We omit the technical details of the proof of transversality here and merely explain in words how (11) is derived. First of all, the right side of the A-equation (3) (for

$\omega = 0$) is bounded near the event horizon according to the regularity assumption (II) and has a finite limit for $r \searrow \rho$. Hence $A'(\rho)$ exists and is finite. To see that $A'(\rho)$ cannot be zero, one proceeds indirectly. Under the assumption that $A'(\rho) = 0$, A decays near the event horizon faster than linearly. As a consequence, the right side of (10) diverges for $r \searrow \rho$ at a rate faster than $\sim (r-\rho)^{-\frac{1}{2}}$. On the left of (10), on the other hand, we write Aw' as $\sqrt{A}\,(\sqrt{A}w')$ and use that $\sqrt{A}w'$ is bounded according to assumption (II) to conclude that the left side of (10) diverges for $r \searrow \rho$ at a certain rate. A rather delicate comparison of the two growth rates yields a contradiction.

According to (11), our Einstein metric has near the event horizon the same qualitative features as the Schwarzschild metric. For the Schwarzschild metric, the non-existence of normalizable solutions of the Dirac equation was shown in [9] using so-called matching conditions and a radial flux argument. These arguments also go through here, and we conclude that the spinors must vanish identically outside the horizon. This completes the proof of Theorem 2.1 in the case that $A^{-\frac{1}{2}}$ is integrable near the event horizon.

Case 2. $A(r)^{-\frac{1}{2}}$ is not integrable near $r = \rho$.

We break the proof up into two sub-cases:

$$\text{(i)} \quad \omega \neq 0\,, \qquad \text{(ii)} \quad \omega = 0\,.$$

Suppose first that we are in Case (i), $\omega \neq 0$. Then we can use the following "oscillation argument" to show that the spinors are bounded near the event horizon.

Proposition 3.4. *Assume that $\omega \neq 0$; then there exist constants $c_1 > 0$, $\varepsilon_1 > 0$ such that*

$$c_1 \;\leq\; (\alpha^2+\beta^2)(r) \;\leq\; \frac{1}{c_1}\,, \qquad \rho < r < \rho + \varepsilon_1. \tag{12}$$

Proof. We define the matrix $B(r)$ by

$$B(r) \;=\; \begin{pmatrix} 1 - \dfrac{m}{\omega T} & -\dfrac{w}{r\omega T} \\ -\dfrac{w}{r\omega T} & 1 + \dfrac{m}{\omega T} \end{pmatrix}. \tag{13}$$

Our proof is based on an extension of [9, Lemma 5.1]. In this earlier paper we used the fact that the matrix elements in (13) are monotone near the horizon. In the case considered here, we do not have these hypotheses, and we must therefore work harder.

Since $T(r) \to \infty$ as $r \searrow \rho$ and $w(r)$ is bounded (by Hypothesis (II)), we see that both $\frac{m}{\omega T}$ and $\frac{w}{r\omega T}$ tend to zero as $r \searrow \rho$. Thus $B(r)$ is close to the identity matrix when r is near ρ. If we define $F(r)$ by

$$F(r) = \langle \Phi(r),\; B(r)\,\Phi(r) \rangle$$

(where $<,>$ denotes the usual Euclidean inner product), then we can find constants $c > 0$ and $\varepsilon_1 > 0$ such that

$$\frac{1}{c}\,|\Phi(r)|^2 \;\le\; F(r) \;\le\; c\,|\Phi(r)|^2\,, \qquad \rho < r < \rho + \varepsilon_1. \tag{14}$$

Furthermore, an easy calculation using the Dirac equations (1) and (2) yields that

$$F'(r) = <\Phi(r),\; B'(r)\,\Phi(r)>\,,$$

and thus if $\rho < r < \rho + \varepsilon_1$,

$$-\sqrt{2}\left[\left|\left(\frac{m}{\omega T}\right)'\right| + \left|\left(\frac{w}{r\omega T}\right)'\right|\right] |\Phi|^2 \;\le\; F'(r) \;\le\; \sqrt{2}\left[\left|\left(\frac{m}{\omega T}\right)'\right| + \left|\left(\frac{w}{r\omega T}\right)'\right|\right] |\Phi|^2\,,$$

because the sup-norm of the matrix B' is bounded by

$$|B'| \;\le\; \sqrt{2}\left[\left|\left(\frac{m}{\omega T}\right)'\right| + \left|\left(\frac{w}{r\omega T}\right)'\right|\right] \;:=\; a(r)\,.$$

Thus from (14), we obtain

$$-c\,a(r)\,F(r) \;\le\; F'(r) \;\le\; c\,a(r)\,F(r) \tag{15}$$

on $\rho < r < \rho + \varepsilon_1$. Dividing this inequality by $F(r)$ and applying Lemma 3.6, we obtain

$$-c\,a(r) \;\le\; \frac{F'(r)}{F(r)} \;\le\; c\,a(r)\,. \tag{16}$$

The regularity assumptions (I), (II), and the monotonicity (III) imply that the functions $a(r)$ is integrable near the event horizon (see [7, Lemma 3.6] for details). Thus if we integrate (16) from ρ to $\rho + \varepsilon_1$, we see that $\log F(r)$ is bounded from above and below near $r = \rho$. Upon exponentiating and substituting into (14), we see (as in the proof of Lemma 3.2) that (12) holds. □

Now consider the differential equation for (AT^2), (4). As $r \searrow \rho$, the right-hand-side tends to $-\infty$, while the left-hand side is bounded. This contradiction implies that $\omega \neq 0$ cannot hold.

The remaining case in the proof of Theorem 2.1 is when we assume that $A^{-\frac{1}{2}}$ is not integrable near the event horizon, and $\omega = 0$. In this case, we note that Lemma 3.4 holds (cf. the remark after the statement of this lemma). As in Case 1 above, we find that (10) holds near $r = \rho$. Integrating this inequality from $r > \rho$ to $\rho + \varepsilon$, we see that for r near ρ, the left-hand side is bounded while the right-hand side can be made arbitrarily large. This contradiction completes the proof of Theorem 2.1.

This is joint work with Joel Smoller and Shing-Tung Yau.

References

[1] Bartnik. R., and McKinnon, J., *Particlelike solutions of the Einstein-Yang-Mills equations,* Phys. Rev. Lett. **61** (1988) 141-144

[2] Künzle, H.P., and Masood-ul-Alam, A.K.M., *Spherically symmetric static* $SU(2)$ *Einstein-Yang-Mills fields,* J. Math. Phys. **31** (1990) 928-935

[3] Smoller, J., Wasserman, A., and Yau, S.-T., *Existence of black hole solutions for the Einstein-Yang/Mills equations,* Commun. Math. Phys. **154** (1993) 377-401

[4] Wald, R.M., *On the instability of the* $n = 1$ *Einstein-Yang-Mills black holes and mathematically related systems,* J. M. Phys. **33** (1992) 248-255

[5] Finster, F., Smoller, J., and Yau, S.-T., *Non-existence of black hole solutions for a spherically symmetric, static Einstein-Dirac-Maxwell system,* gr-qc/9810048, Commun. Math. Phys. **205** (1999) 249-262

[6] Finster, F., Smoller, J., and Yau, S.-T., *The interaction of Dirac particles with non-Abelian gauge fields and gravity – Bound States,* gr-qc/0001067, Nucl. Phys. B **584** (2000) 387-414

[7] F. Finster, J. Smoller, and S.-T. Yau, *The interaction of Dirac particles with non-Abelian gauge fields and gravity – Black Holes,* gr-qc/9910047, Mich. Math. J. **47** (2000) 199-208

[8] F. Finster, J. Smoller, and S.-T. Yau, *Absence of stationary, spherically symmetric black hole solutions for Einstein-Dirac-Yang/Mills equations with angular momentum,* gr-qc/0005028

[9] F. Finster, J. Smoller, and S.-T. Yau, *Non-Existence of Time-Periodic Solutions of the Dirac Equation in a Reissner-Nordström Black Hole Background,* gr-qc/9805050, J. Math. Phys. **41** (2000) 2173-2194

F. Finster
Max Planck Institute for Mathematics in the Sciences,
Leipzig, Germany
`felix.finster@mis.mpg.de`

International Series of Numerical Mathematics
Vol. 140, © 2001 Birkhäuser Verlag Basel/Switzerland

A Numerical Study on Viscous Profiles of MHD Shock Waves

Heinrich Freistühler and Christian Rohde

Abstract. The equations of ideal magnetohydrodynamics(MHD) admit a variety of different types of shock waves. In this paper it is shown that viscous profiles for all these types, including, in particular, the so-called intermediate waves, can be computed in a reliable manner. Relying on a refined bifurcation analysis for the Rankine-Hugoniot equations we present numerical results for several situations of interest. Thereby we extend known analytical results.

1. Introduction

Plane waves in classical magnetohydrodynamics (MHD) are governed by the hyperbolic-parabolic system of differential equations

$$\begin{aligned} \rho_t + (\rho v)_x &= 0, \\ (\rho v)_t + (\rho v^2 + p + \tfrac{1}{2} \mid \vec{b} \mid^2)_x &= \lambda v_{xx}, \\ (\rho \vec{w})_t + (\rho v \vec{w} - \vec{b})_x &= \mu \vec{w}_{xx} \\ \vec{b}_t + (v \vec{b} - \vec{w})_x &= \nu \vec{b}_{xx}, \\ \mathcal{E}_t + (v(\mathcal{E} + p + \tfrac{1}{2} \mid \vec{b} \mid^2) - \vec{w} \cdot \vec{b})_x &= (\lambda v v_x + \mu \vec{w} \cdot \vec{w}_x + \nu \vec{b} \cdot \vec{b}_x + \kappa \theta_x)_x. \end{aligned}$$

Here $\rho, \theta, p > 0$ denote density, temperature, and pressure of the fluid, while $v \in \mathbb{R}$ stands for the longitudinal velocity, $\vec{w} = (w_1, w_2) \in \mathbb{R}^2$ for the transverse velocity, $\vec{b} = (b_1, b_2) \in \mathbb{R}^2$ for the magnetic field, and $\mathcal{E} \in \mathbb{R}$ for the total energy. With the internal energy ϵ, the total energy splits up as

$$\mathcal{E} = \rho(\epsilon + \frac{1}{2}(v^1 + |\vec{w}|^2)) + \frac{1}{2}|\vec{b}|^2.$$

For definiteness, we consider an ideal fluid, i.e. $\epsilon = c_v \theta$ and $p = \rho R \theta$ for $c_v, R > 0$. The vector $\delta \in [0, \infty)^4$ of dissipation parameters consists of the fluid viscosities λ, μ, the resistivity ν, and the heat conductivity κ.
The MHD-system (1) belongs to the class of conservation laws of the form

$$u_t + f(u)_x = (D(u, \delta) u_x)_x, \qquad \delta \in [0, \infty)^4. \tag{1}$$

We recover the equations (1) from (1) by setting $u = (\rho, \rho v, \rho\vec{w}, \vec{b}, \mathcal{E}) \in \mathbb{R}^7$, and defining an appropriate flux $f : \mathbb{R}^7 \to \mathbb{R}^7$, respectively dissipation $D : \mathbb{R} \times [0,\infty)^4 \to \mathbb{R}^{7\times 7}$. If all components of δ vanish (1) becomes a purely hyperbolic conservation law. Let $q \in \mathbb{R}^7$ be given. If $u^\pm$ and s satisfy the Rankine–Hugoniot relation

$$f(u^\pm) - su^\pm = q, \tag{2}$$

the function

$$U(x,t) = \begin{cases} u^- : x < st, \\ u^+ : x > st, \end{cases}$$

is a weak solution of the hyperbolic conservation law, with other words: a shock wave.

A function $\phi \in C^1(\mathbb{R}, \mathbb{R}^7)$ is called a viscous profile of the shock wave U if it satisfies

$$D(\phi, \delta)\phi' = f(\phi) - s\phi - q \tag{3}$$

and

$$\phi(\pm\infty) = u^\pm. \tag{4}$$

Viscous profiles of shock waves played a prominent rôle for the analysis of (general) conservation laws since many years. The question of the (non)linear stability of viscous profiles as solutions of the PDE with respect to perturbations of the viscous profile itself has attracted a lot of people in the last decade. Many of these stability problems –at least up to now– cannot be treated analytically but computationally. The numerical algorithms used rely on the viscous profile[4, 14], that most often has to be determined numerically too, for instance by the algorithm we presented and tested in [8, 9].

The purpose of this paper is twofold. First we want to show that our numerical scheme is reliable, even in complicated situations as represented by the ODE-system (3) in the MHD case. Note that in contrast to earlier work where our algorithm was tested for some subsystems of (3) we will deal here with the full system (3). Although our algorithm is designed to cover the general case (1) we will focus in this paper (exclusively) on the MHD–system (1). The MHD–system is particularly interesting since it provides an example of a physically relevant system that admits as weak solutions non–Laxian shock waves of overcompressive and undercompressive type, usually called intermediate in this context. Concerning the (non-)existence of viscous profiles for intermediate shock waves the analysis is rather subtle and not complete. For an overview we refer to [7] and references therein. As a second task therefore we want to present numerical evidence of the (non-)existence of viscous profiles for situations which are not (completely) covered by analytical results. In particular, our results support the conjecture that intermediate shock waves, which are characteristic on one or both sides, can have structure.

The plan of the paper is as follows. In Section 2 we will report on some new results of a bifurcation analysis for (2). These results are necessary to state the specific

MHD–examples to be discussed in the following Section 3. They will deal with one global and two local bifurcation scenarios.

2. Brief Discussion of the Rankine-Hugoniot-Locus

In this section we collect some results of a bifurcation analysis for the Rankine–Hugoniot relation (2) in the MHD case. For this example we can assume without loss of generality that we have

$$s = 0 \text{ and } q = (1, j, 0, 0, c, 0, e), \quad c \geq 0, j, e \in \mathbb{R}, \tag{5}$$

which is a consequence of Galilean invariance, directional equivariance, and rescaling arguments[10]. Using (5) simplifies the bifurcation analysis greatly; we obtain for (2):

$$\begin{aligned} \rho v &= 1, \\ v + R\theta/v + |\vec{b}|^2/2 &= j, \\ \vec{w} - \vec{b} &= \mathbf{0}, \\ vb_1 - w_1 &= c, \\ vb_2 - w_2 &= 0, \\ \tfrac{v^2+|\vec{w}|^2}{2} + (c_v + R)\theta + v|\vec{b}|^2 - \vec{w}\cdot\vec{b} &= e. \end{aligned}$$

At first consider the case $c > 0$. It is well known that there can be up to four distinct states that solve (6). The two fast states u_0, u_1 satisfy $v_0 > v_1 > 1$ while the slow states u_2, u_3 satisfy $v_3 < v_2 < 1$. The typical configuration in the b_1v–plane is displayed in Figure 1. The subsequent lemma gives a more precise statement on the existence of *physical* solutions $u_0, \ldots, u_3$, i.e. states with positive pressure.

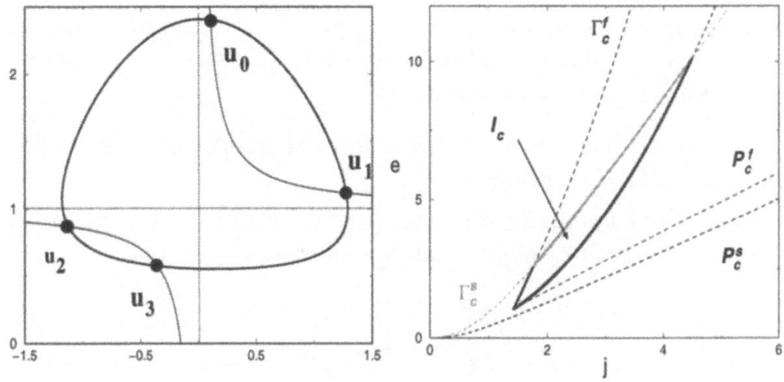

FIGURE 1. Null clines of (6) in b_1v–plane and the set $\mathcal{I}_c$.

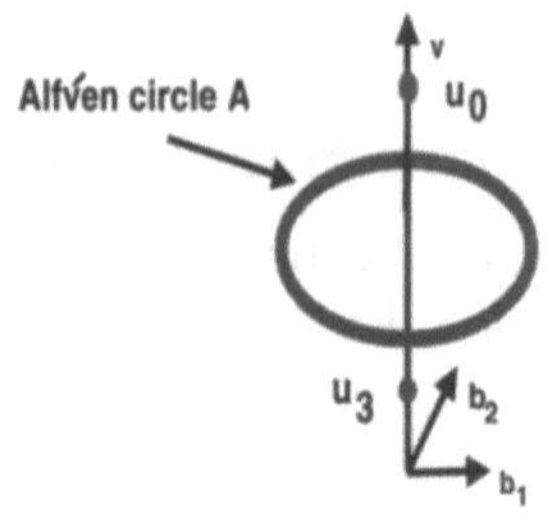

FIGURE 2. Physical states for $c = 0$.

Lemma 2.1. *For the adiabatic coefficient $\gamma = 1 + R/c_v$, let $\bar{c}$ be the smallest (positive) solution of*

$$c^{2/3} - \sqrt{\gamma+1}\frac{2-\gamma}{3\gamma}c = \frac{2}{3\gamma}.$$

For each $c \in (0, \bar{c})$ there is an open bounded set $\mathcal{I}_c \subset (0,\infty)^2$ such that $(j, e) \in \mathcal{I}_c$ if and only if there are four distinct physical states $u_0, \ldots, u_3 \in \mathbb{R}^7$ satisfying (6).

Lemma 2.1 is illustrated in the right picture of Figure 1. The curve Γ_c^s denotes the set of all points $(j, e) \in \mathbb{R}^2$ such that (6) has exactly one slow solution denoted by $u_{2=3}$. For points $(j, e) \in \mathbb{R}^2$ to the left of Γ_c^s there are no slow solutions, for $(j, e) \in \mathbb{R}^2$ to the right of Γ_c^s there are two slow solutions u_2, u_3. Γ_c^f marks the analogous partition of the je–plane for the fast solutions. The curve P_c^f consists of an upper and a lower part, ending in a cusp-type singularity for $(j, e) = (1 + \frac{3}{2}c^{2/3}, \frac{1}{2} + \frac{3}{2}c^{2/3} + \frac{3}{2}c^{4/3})$. It identifies, for the upper (lower) part, the loci where the pressure $p_0 = \rho_0 R\theta_0$ ($p_1 = \rho_1 R\theta_1$) vanishes and changes sign. Parts of these three curves –marked with solid lines in Figure 1–form the boundary of $\mathcal{I}_c$.
The singular case $c = 0$ is of particular interest. The states u_1, u_2 degenerate to a onedimensional curve of states solving (6).

Lemma 2.2. *Let $c = 0$ and $\gamma = 1 + R/c_v$. There is an open bounded set $\mathcal{I}_0 \subset (0,\infty)^2$ such that $(j, e) \in \mathcal{I}_0$ if and only if*

(i) *there are two physical states $u_0, u_3 \in \mathbb{R}^7$ with $v_0 > 1 > v_3$ solving (6), and*
(ii) *there is a set A of physical states solving (6) given by*

$$A = \left\{(\rho, v, \vec{w}, \vec{b}, \theta) \,|\, \rho = v = 1, |\vec{w}| = |\vec{b}| = r, \theta = \tfrac{1}{R}(j - \tfrac{r^2}{2} - 1))\right\},$$
$$r = \sqrt{2\gamma j - 2(\gamma-1)e - \gamma - 1}.$$

A is called the Alfvén circle.

Figure 2 shows the geometry of the physical states in the $b_1 b_2 v$–space as given in Lemma 2.2.

Remark 2.3. *Although some results presented in this section –up to our knowledge– cannot be found in the literature we stress that they just refine and complement findings that trace back to the early work of Germain or Kulikovskij& Liubimov*[12, 13]. *For a proof of (a more detailed statement of) Lemma 2.1 and similar results we refer to* [10]. *Lemma 2.2 and more information on the case c = 0 in connection with the nonstrict hyperbolicity of the MHD equations can be found in* [6, 7].

3. Viscous Profiles in MHD

The corresponding ODE (3) in the MHD case becomes

$$\begin{aligned} \lambda\dot{v} &= v + p + \tfrac{1}{2} \mid \vec{b} \mid^2 - j, \\ \mu\dot{\vec{w}} &= \vec{w} - \vec{b}, \\ \nu\dot{\vec{b}} &= v\vec{b} - \vec{w} + (c,0)^T, \\ \kappa\dot{\theta} &= c_v\theta - \tfrac{1}{2}(|\vec{w}|^2 - 2\vec{b}\cdot\vec{w} + v|\vec{b}|^2) - \tfrac{v^2}{2} + jv + \vec{b}\cdot(c,0)^T - e. \end{aligned} \qquad (\Sigma^6)$$

Obviously solutions of (6) are rest points of Σ^6. Before reporting on numerical results for Σ^6 let es make some remarks on the computational method we use. Following the pioneering work of Beyn[2], we use the direct approach of computing the viscous profiles as heteroclinic orbits of the ODE–system rather than considering them as possible large–time limits of the PDE–system (cf. [9] for details). The latter would require stability properties of the traveling waves which we are far from knowing at this stage. Such properties are rather the objective of further investigations *after* the profiles have been found.

Our algorithm leads to ordinary boundary value problems on finite intervals which are solved with the help of the code COLNEW[1].

3.1. A Global Bifurcation

Conley and Smoller showed that the (Lax-type) shock waves $u_0 \to u_1$ and $u_2 \to u_3$ admit a viscous profile[16] for all $\delta \in (0,\infty)^4$ and all q such that the associated rest points exist. The situation for the intermediate waves is more complicated.

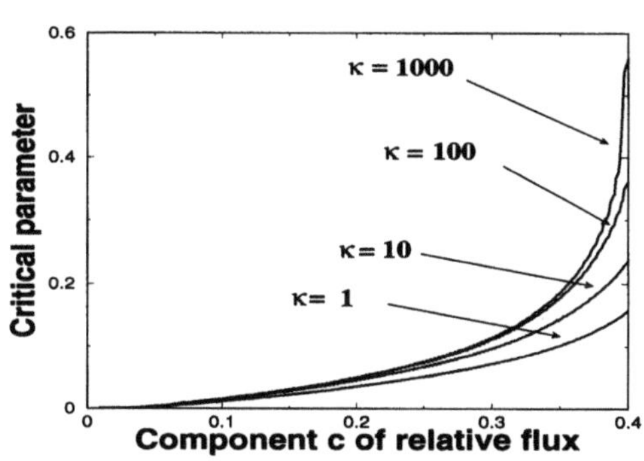

FIGURE 3. Threshold ω^* versus c.

The known (analytical and numerical) results from literature support the following conjecture:

There exists a threshold $\omega^* = \omega^*(q, \mu/\lambda, \kappa/\lambda) > 0$ *such that the following holds for all* $c \in (0, \bar{c})$, $(\jmath, e) \in \mathcal{I}_c$, *and* $\delta = (\nu, \lambda, \mu, \kappa) \in (0, \infty)^4$: *If* $\nu/\lambda > \omega^*$, *then all intermediate shocks (for the given q) have viscous profiles (for the given* δ*). Conversely, if* $\nu/\lambda < \omega^*$, *then no intermediate shock wave has a profile.*

A proof of this conjecture for small μ and κ can be found in [11]. Numerical work we have done suggests that the conjecture holds also true globally, that means, in particular for big values of μ and κ. The numerical algorithm that we use to decide the validity of the conjecture is described in [10]. It also lets us compute the threshold ω^* directly. Critical values ω^* are displayed in Figure 3 for $\mu = 0.01$, different values for κ, and a sequence $(\jmath_c, e_c) \in \mathcal{I}_c$ with $c \to 0$.

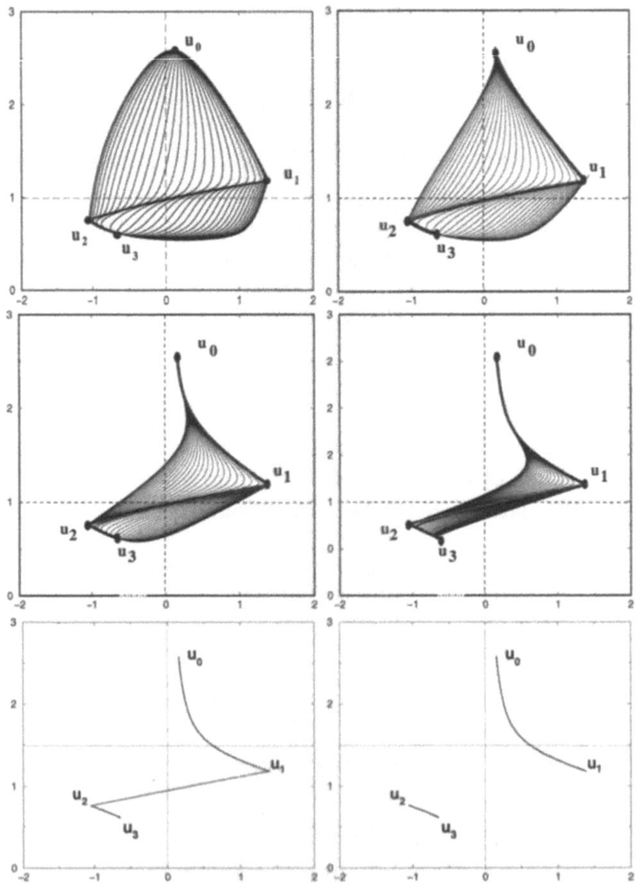

FIGURE 4. Projection to b_1v–plane for $\nu = 7.5, 1.0, 0.25, 0.06, \nu^*, 0.02$.

In a set of pictures in Figure 4 we display viscous profiles of all types except $u_0 \to u_3$ projected to the b_1v–plane. For the set $c = 0.15$, $(j, e) = (2.68, 4.23) \in \mathcal{I}_c$, $\lambda = \kappa = 1$, $\mu = 0.01$ are fixed and ν varies. the threshold value in this case computes as $\omega^* = \nu^*/1 \approx 0.0492$.
Note that an analysis of the restpoints $u_0, \ldots, u_3$ shows that the viscous profiles of type $u_0 \to u_2$, $u_1 \to u_2$, $u_2 \to u_3$ ($u_0 \to u_2$, $u_1 \to u_3$) $\{u_0 \to u_3\}$ form a submanifold of dimension 1 (2) {3} if they exist.

3.2. The Alfvén Circle

For the case $c = 0$ (cf. Figure 2) it is known that there are always viscous profiles connecting u_0 with each point on the Alfvén circle A and viscous profiles connecting each point of A with u_3. The associated shock waves are called switch–on, respectively switch–off shocks. The existence of viscous profiles for any choice of δ matches with the observation from our numerical investigations that the threshold value ω^* vanishes as c tends to 0, cf. Figure 3. Recall that $\omega^* \to 0$ suggests that in the limit there remains no configuration for which only the Lax type shock waves $u_0 \to u_1$ and $u_2 \to u_3$ have profiles.
In this section we present a set of computations for fixed values of j, e and δ but for different c tending to 0. The results, projected into the b_1v–plane are displayed in Figure 5.

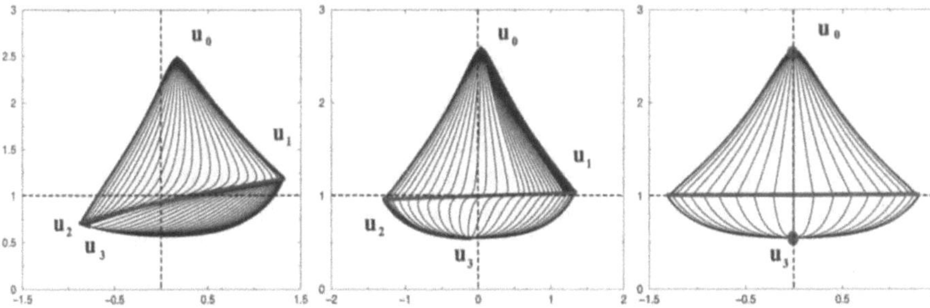

FIGURE 5. Viscous profiles of type $u_0 \to u_1$, $u_1 \to u_2$, $u_2 \to u_3$ and $u_1 \to u_3$ for $c = 0.25, 0.05, 0$.

For $c = 0$ the phase portrait, is rotationally symmetric with respect to $b_1 = b_2 = 0$ (cf. Figure 6).

3.3. Collapse of Two Rest Points

In this section we consider points $(j, e) \in \partial\mathcal{I}_c$, $c \in (0, \bar{c})$, such that we have two fast solutions u_0, u_1, and furthermore one slow solution $u_{2=3}$ as indicated in Figure 7. Although a codimension–1 phenomenen in the bifurcation picture of the profile ODE, situations, where either u_0 and u_1 or u_2 and u_3 coincide are important. These cases, in which intermediate shocks become characteristic on one side, occur generically in the composite waves(combinations of a shock wave with an adjacent rarefaction wave) solving Riemann problems for (1) with $\delta = 0$ [15, 3, 5].

Our computations show that the intermediate shock waves of type $u_0 \to u_{2=3}$ and $u_1 \to u_{2=3}$ indead can have structure. In Figure 8 the numerical results for the viscous profiles of types $u_0 \to u_{2=3}$ and $u_1 \to u_{2=3}$ are displayed. While the four rest points $u_0, \ldots, u_3$ for $(j, c) \in \mathcal{I}_c$ are hyperbolic, here we have zero eigenvalue for the linearized flux in $u_{2=3}$. This leads to a center manifold emerging from the rest point. The projection to the b_1b_2–plane and the blow up of the neighbourhood of $u_{2=3}$ shows the typical structure for the viscous orbits if a center manifold arises.

Acknowledgments: The authors were supported in part by DFG–Schwerpunktprogramm DANSE and the European TMR project Hyperbolic Conservation Laws.

References

[1] G. Bader and U. Ascher, *A New Basis Implementation for a Mixed Order Boundary Value ODE Solver.* SIAM J. Sci. Stat. Comput., **8** (1987), 483-500.

[2] W.–J. Beyn, *The numerical computation of connecting orbits in dynamical systems.* IMA J. Numer. Anal., **9** (1990), 379-405.

[3] M. Brio and C.C. Wu, *An Upwind Differencing Scheme for the Equations of Ideal Magnetohydrodynamics.* J. Comput. Phys., **75(2)** (1988), 400-422.

[4] L.Q. Brin, *Numerical Testing of the Stability of Viscous Shock Waves.* PhD–Thesis, Indiana University **(1998)**.

[5] H. De Sterck, B.C. Low, and S. Podts, *Complex Magnetohydrodynamic bow Shock Topology in Field–aligned low–β Flow around a Perfectly Conducting Cylinder,* Physics of Plasma, **5(11)** (1998), 4015-4027.

[6] H. Freistühler, *Anomale Schocks, strukturell labile Lösungen und die Geometrie der Rankine–Hugoniot Bedingungen.* Doctoral Thesis, Ruhr-Universität Bochum **(1987)**.

[7] H. Freistühler, *Contributions to the Mathematical Theory of Magnetohydrodynamic Shock Waves.* Stud. Adv. Math., **3** (1997), 175-187.

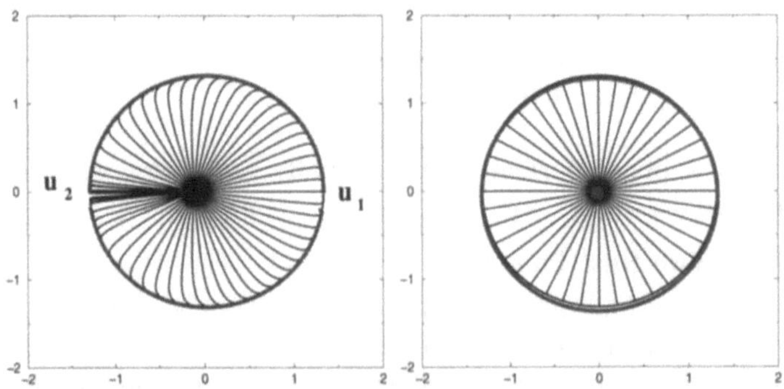

FIGURE 6. The orbit pair $u_1 \to u_2$ and the one–parameter family $u_1 \to u_3(c = 0.05$, left picture), viscous profiles for switch–off shocks($c = 0$, right picture).

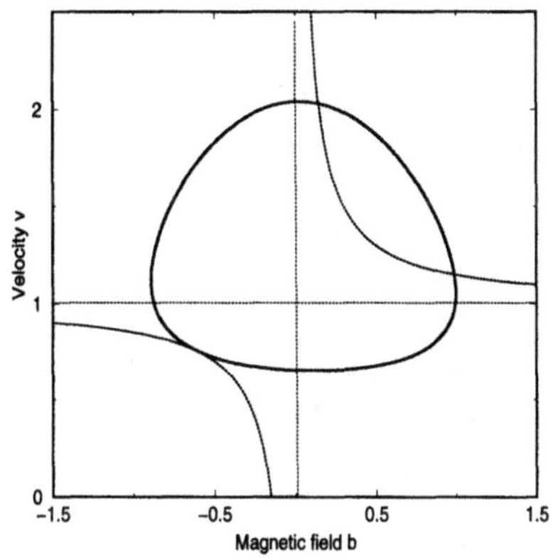

FIGURE 7. Null clines of (6) in b_1v–plane for one slow solution $u_{2=3}$.

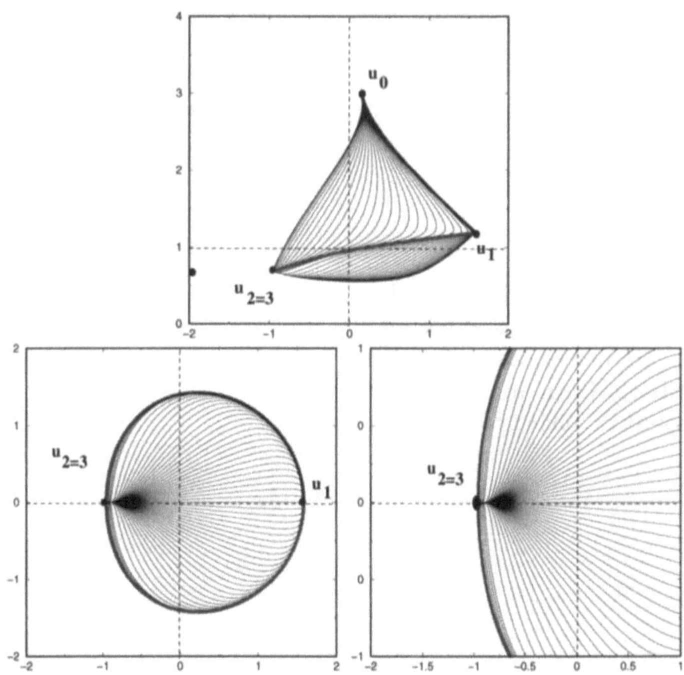

FIGURE 8. b_1v–plane (above) and projection to b_1b_2–plane with zoom (below).

[8] H. Freistühler and C. Rohde, *Numerical Methods for Viscous Profiles of Non-Classical Shock Waves.* In: Hyperbolic problems: Theory, Numerics, Applications:

Seventh International Conference in Zürich 1998/ ed. by Michael Fey; Rolf Jeltsch (1999).

[9] H. Freistühler and C. Rohde, *Numerical Computation of Viscous Profiles for Hyperbolic Conservation Laws.*, to appear in Math. Comput..

[10] H. Freistühler and C. Rohde, *A Numerical Study of Existence and Bifurcation of MHD Shock Profiles.*, in preparation.

[11] H. Freistühler and P. Szmolyan, *Existence and Bifurcation of Viscous Profiles for all Intermediate Magnetohydrodynamic Shock Waves.* SIAM J. Math. Anal., **26(1)** (1995), 112-128.

[12] P. Germain, *Contribution à la theorie des ondes de choc en magnétohydrodynamique de fluides.* O.N.E.R.A. Publ. No.97 (1995).

[13] A. G. Kulikovskij and G. A. Lyubimov, *On the Structure of an Inclined Magnetohydrodynamic Shock Wave.* J. Appl. Math. Mech.,**25** (1961), 171-179.

[14] M. Liefvendahl, *On the Stability of Viscous Shock Waves.* Licentiate's Thesis, Royal Institute of Technology Stockholm **(1999)**.

[15] T-P. Liu, *The Riemann Problem for General Systems of Conservation Laws.* J. Diff. Eqs. **18** (1975), 218-234.

[16] J. Smoller, *Shock Waves and Reaction-Diffusion Equations.* Springer **(1983)**.

H. Freistühler
Max-Planck-Institut für Mathematik in den Naturwissenschaften,
Inselstr. 22-26,
D-04103 Leipzig, Germany
`hfreist@mis.mpg.de`

C. Rohde
Institut für Angewandte Mathematik,
Albert–Ludwigs–Universität Freiburg,
10, Hermann–Herder–Straße,
D-79104 Freiburg, Germany
`chris@mathematik.uni-freiburg.de`

International Series of Numerical Mathematics
Vol. 140,

A Vanishing Debye Length Limit in a Hydrodynamic Model for Semiconductors

Ingenuin Gasser and Pierangelo Marcati

Abstract. A small Debye length limit combined with a relaxation limit in the hydrodynamic model for semiconductors is analysed. The limit problem is identified. Entropy methods and L^∞ type estimates originating from the existence proof are used in order to proof the result.

1. Introduction

We consider the bipolar isentropic hydrodynamic model for semiconductors (HD) in one space dimension. Denoting by n, p, j, g, $P(n)$, $Q(p)$ and E the charge densities, current densities and pressures of the electrons and the holes, respectively and the electric field, the scaled equations of the bipolar HD model for semiconductors are given by

$$\begin{aligned} &n_t + j_x = 0 \\ &j_t + (\frac{j^2}{n} + P(n))_x = nE - j/\tau \\ &p_t + g_x = 0 \\ &g_t + (\frac{g^2}{p} + Q(p))_x = -pE - g/\tau \\ &\lambda^2 E_x = n - p - C \end{aligned} \tag{1}$$

where $x \in \mathbb{R}$ and $t \geq 0$. We prescribe initial conditions

$$\begin{aligned} n(t=0,x) &= n_0(x), & j(t=0,x) &= j_0(x), \\ p(t=0,x) &= p_0(x), & g(t=0,x) &= g_0(x). \end{aligned}$$

The positive constants τ and λ denote the relaxation time and the Debye length, respectively. We consider the isentropic case, i.e. the pressures $P = P(n) = n^{\gamma_n}$, $Q = Q(p) = p^{\gamma_p}$ are functions of the densities only, where we assume $\gamma_n, \gamma_p > 1$. The function $C = C(x)$ is the so called doping profile describing the background charge in the semiconductor. For details on the model see [14]. Similar models are used in plasma physics. In the case of a plasma there is in general no fixed background charge, i.e. $C \equiv 0$. Also, the relaxation term in (1) is used only for so called weakly ionised plasmas consisting of a underlying neutral part of the fluid and the two positively and negatively charged parts of the fluid (see [7] and references therein).

The problem (1) is well known in the literature. The existence of a global in time entropy solution of the system (1) was shown in [16]. Since the proof in [16] uses the existence results for the isentropic gas dynamic equations as a given input, the condition $1 < \gamma_n, \gamma_p \leq \frac{5}{3}$ in [16] is not necessary any more. In the mean time global existence in time for the isentropic gasdynamic equation was shown for $1 < \gamma_n, \gamma_p \leq \infty$ (see [12, 11]) which cancels the above more restrictive condition.

The small Debye length limit $\lambda \to 0$ is known as challenging problem. On the hydrodynamic level the only rigorous results concern classical solutions (see [2]) or traveling wave solutions [3]. To the authors knowledge there is no result for general weak entropy solutions.

On the other hand the relaxation limit is well understood. The long time rescaling

$$s_1 = \tau t, \qquad E^\tau(s_1, x) = E(\frac{s_1}{\tau}, x)$$

$$J^\tau(s_1, x) = \tfrac{1}{\tau} j(\tfrac{s_1}{\tau}, x), \qquad G^\tau(s_1, x) = \frac{1}{\tau} g(\frac{s_1}{\tau}, x)$$

gives (the superscript τ indicates from now on the τ–dependence)

$$\begin{aligned}
&n^\tau_{s_1} + J^\tau_x = 0\\
&\tau^2 J^\tau_{s_1} + (\tau^2 \frac{J^{\tau 2}}{n^\tau} + P(n^\tau))_x = n^\tau E^\tau - J^\tau\\
&p^\tau_{s_1} + G^\tau_x = 0\\
&\tau^2 G^\tau_s + (\tau^2 \frac{G^{\tau 2}}{p^\tau} + \tau^2 Q(p^\tau))_x = -p^\tau E^\tau - G^\tau\\
&\lambda E^\tau_x = n^\tau - p^\tau - C.
\end{aligned}$$

As $\tau \to 0$ one obtains the Drift Diffusion model

$$\begin{aligned}
n_{s_1} &= (P(n))_{xx} + (nE)_x\\
p_{s_1} &= (Q(p))_{xx} - (pE)_x\\
\lambda^2 E_x &= n - p - C
\end{aligned} \tag{2}$$

for the limits n, p, E of n^τ, p^τ, E^τ, respectively. This result was shown first by [13] in the unipolar case and in [16] for the bipolar case. In [7] the unipolar limit is given based on a different method. In [10] and [9] a 3-dimensional version for the unipolar and bipolar case, respectively, is given. The papers [17, 8] concern the isothermal versions of the relaxation limit, i.e. the case $\gamma_n = \gamma_p = 1$.

Rescaling in the Drift Diffusion model

$$s = \frac{s_1}{\lambda^2}, \quad F^\lambda(s,x) = \lambda^2 E(\lambda^2 s, x), \quad n^\lambda(s,x) = n(\lambda^2 s, x), \quad p^\lambda(s,x) = p(\lambda^2 s, x)$$

we obtain

$$\begin{aligned} n_s^\lambda &= \lambda^2(P(n^\lambda))_{xx} + (n^\lambda F^\lambda)_x \\ p_s^\lambda &= \lambda^2(Q(p^\lambda))_{xx} - (p^\lambda F^\lambda)_x \\ F_x^\lambda &= n^\lambda - p^\lambda - C \end{aligned}$$

(here the superscript λ denotes the dependence on λ). In the limit $\lambda \to 0$ the pure drift model (PD)

$$\begin{aligned} n_s &= +(nF)_x \\ p_s &= -(pF)_x \\ F_x &= n - p - C \end{aligned} \tag{3}$$

for the limits n, p, F of n^λ, p^λ, E^λ, respectively, is obtained. This limit is well understood (see [1, 4, 5, 15, 18]). However, the question arises if the PD model can be obtained directly from the HD model. If this is possible it is not clear what kind of approximation the PD model - long or short time - represents.

The idea of this paper is to combine the relaxation limit $\tau \to 0$ and the vanishing Debye length limit $\lambda \to 0$. Assuming in (1) the relation $\lambda^2 = \tau^{1+\alpha}$ with $-1 < \alpha < 1$ and rescaling

$$\begin{aligned} s &= \tfrac{1}{\tau^\alpha} t, & J^\tau(s,x) &= \tau^\alpha j(\tau^\alpha s, x), \\ F^\tau(s,x) &= \tau^{1+\alpha} E(\tau^\alpha s, x), & G^\tau(s,x) &= \tau^\alpha g(\tau^\alpha s, x) \end{aligned}$$

gives

$$\begin{aligned} & n_s^\tau + J_x^\tau = 0 \\ & \tau^{1-\alpha} J_s^\tau + (\frac{\tau^{1-\alpha} J^{\tau 2}}{n^\tau} + \tau^{1+\alpha} P(n^\tau))_x = n^\tau F^\tau - J^\tau \\ & p_s^\tau + G_x^\tau = 0 \\ & \tau^{1-\alpha} G_s^\tau + (\tau^{1-\alpha} \frac{G^{\tau 2}}{p^\tau} + \tau^{1+\alpha} Q(p^\tau))_x = -p^\tau F^\tau - G^\tau \\ & F_x^\tau = n^\tau - p^\tau - C. \end{aligned} \tag{4}$$

Formally, for $-1 < \alpha < 1$ we obtain the pure drift (PD) model as $\tau \to 0$

$$\begin{aligned} n_s &= -(nF)_x \\ p_s &= (pF)_x \\ F_x &= n - p - C \end{aligned} \tag{5}$$

for the limit quantities n, p, F of n^τ, p^τ, F^τ, respectively, with initial data $n(s=0,x) = n_0(x)$, $p(s=0,x) = p_0(x)$. In this scaling we see that the PD model represents a short time approximation if $\alpha > 0$ (small Debye length limit dominates) and a long time approximation if $\alpha < 0$ (relaxation limit dominates). In the case $\alpha = 0$ the two competing limits balance.

This paper concerns the rigorous proof of the above formal considerations. In the next section we formulate the main result and comment on the proof.

2. The main result

The main result of this article reads

Theorem 2.1. *Let* n^τ, p^τ, J^τ G^τ *and* F^τ *be an entropy solution of (1), (2). Let* $C \in L^1(\mathbb{R}) \cap L^\infty(\mathbb{R})$. *Let* $S > 0$. *Let* $\gamma_n = \gamma_p = \gamma > 2$ *and* $-1 < \alpha < \frac{\gamma-5}{3(\gamma-1)}$. *Then, as* $\tau \to 0$ *the following convergences (after extracting subsequences) hold:*

- $n^\tau \rightharpoonup n$, $p^\tau \rightharpoonup p$ *weakly in* $L^\gamma((0,S) \times \mathbb{R})$,
- $F^\tau \to F$ *strongly in* $C([0,S]; L^s_{loc}(\mathbb{R}))$ *with* $1 < s < \infty$,
- $J^\tau \rightharpoonup J = nF$, $G^\tau \rightharpoonup G = pF$ *weakly in* $L^q_{loc}((0,S) \times \mathbb{R})$ *with* $q = 2\frac{\gamma}{\gamma+1}$,

and the limits n, p *and* F *satisfy the system (5) in* $D'([0,S) \times \mathbb{R})$ *with the initial data* $n(s=0,x) = n_0(x)$ *and* $p(s=0,x) = p_0(x)$.

The proof of the above combined limit result can be found in [6]. A difficulty in the proof of the above result lies in the fact that the standard entropy (in rescaled form)

$$e^\tau(s) = \int_{\mathbb{R}} [\tau^{-2\alpha}\frac{J^{\tau 2}}{2n^\tau} + \tau^{-2\alpha}\frac{G^{\tau 2}}{2p^\tau} + \frac{1}{\gamma_n - 1}P(n^\tau) + \frac{1}{\gamma_p - 1}Q(p^\tau)]dx.$$

and the related entropy inequality does not give strong enough uniform estimates in τ as is it the case for fixed λ. The idea of the proof is to introduce a "modified entropy"

$$\begin{aligned} e^\tau(s) &= \int_{\mathbb{R}} \Big\{ \tau^{-2\alpha}\frac{(J^\tau - n^\tau F^\tau)^2}{2n^\tau} + \tau^{-2\alpha}\frac{(G^{\tau 2} + p^\tau F^\tau)^2}{2p^\tau} \\ &\quad + \frac{1}{\gamma_n - 1}P(n^\tau) + \frac{1}{\gamma_p - 1}Q(p^\tau) \Big\}\, dx. \end{aligned}$$

involving the electric field. Mathematically speaking, this is nothing but a rearrangement of the standard entropy inequality. However, this rearrangement combined with L^∞-type estimates originating in the existence proofs (see [13]) gives the necessary uniform bounds. Finally, the Poisson equation is used in order to provide strong convergence of the electric field needed for the passage to the limit in the nonlinear terms.

A final comment on the restrictions $-1 < \alpha < \frac{\gamma-5}{3(\gamma-1)}$ requiring $\gamma > 2$ should be made. The formal limit works for $-1 < \alpha < 1$. In the proof the additional restriction is needed. We believe that it is a technical problem only. In [6] the limit is performed also in the unipolar case, where this γ dependent restriction does not appear.

Acknowledgements. The first author acknowledge partial support from the european TMR network *Asymptotic Methods in Kinetic Theory* (contract number ERB FMRX CT97 0157). The second author acknowledges partial support from the the TMR Network "Hyperbolic Systems of Conservation Laws" (contract number ERB FMRX CT96 0033).

References

[1] Brezzi, F. and Markowich, P.A., 'A convection-diffusion problem with small diffusion coefficient arising in semiconductor physics', *Bollettino UMI*, 7 (2B), 903-930 (1988).

[2] Cordier, S. and Grenier, E., 'Quasineutral limit of Euler-Poisson systems arising from plasma physics', *Comm. PDE*, 25, 1099-1113 (2000).

[3] Cordier, S., Degond, P., Markowich, P.A., Schmeiser, C., 'Quasineutral limit of travelling waves for the Euler-Poisson model', in Gary (ed.), Mathematical and numerical aspects of wave propagation. *Proceedings of the third international conference, Mandelieu-La Napoule, France, April 24-28*, 724-733 (1995).

[4] Gasser, I., 'The quasineutral limit in the drift diffusion model with nonlinear diffusion', *Hamburger Beiträge zur Angewandten Mathematik, Reihe A*, 143, (1999), *NoDEA*, vol. 8 (3), 237-249 (2001).

[5] Gasser, I., Levermore, C.D., Markowich, P.A., Schmeiser, C., 'The initial time layer problem and the quasineutral limit in the drift diffusion model', *preprint 859 of the Erwin Schrödinger Institut Vienna* (2000), to appear in Europ. J. Appl. Math..

[6] Gasser, I., Marcati, P., 'The Combined Relaxation and Vanishing Debye Length Limit in the Hydrodynamic Model for Semiconductors', *Math. Meth. in the Appl. Sci.* M^2AS, vol. 24 (2), 81-92 (2001).

[7] Jüngel, A. and Peng, Y.J., 'A hierarchy of hydrodynamic models for plasmas. Part II: zero-relaxation-time limits', *Comm. Partial Differential Equations*, 24 (5-6), 1007-1033 (1999).

[8] Junca, S. and Rascle, M., 'Relaxation du système d'Euler-Poisson isotherme vers les équations de dérive-diffusion', *Quart. Appl. Math.*, vol. 58 (3), 511-521 (2000).

[9] Lattanzio, C., 'On the 3-D bipolar isentropic Euler-Poisson model for semiconductors and the drift-diffusion limit', *Math. Models Methods Appl. Sci.*, vol. 10 (3), 351-360 (2000).

[10] Lattanzio, C. and Marcati, P., 'The relaxation to the drift-diffusion system for the 3-D isentropic Euler-Poisson model for semiconductors', *Discrete Contin. Dynam. Systems* 5, 449-455 (1999).

[11] Lions, P.L., Perthame, B., Souganidis, P.E., 'Existence and stability of entropy solutions for the hyperbolic systems of isentropic gas dynamics in Eulerian and Lagrangian coordinates', *Commun. Pure Appl. Math.*, 49 (6), 599-638 (1996).

[12] Lions, P.L., Perthame, B., Tadmor, E., 'Kinetic formulation of the isentropic gas dynamics and p-systems', *Commun. Math. Phys.*, 163 (2), 415-431 (1994).

[13] Marcati, P. and Natalini, R., 'Weak solutions to a hydrodynamic model for semiconductors and relaxation to the drift-diffusion equation', *Arch. Rational Mech. Anal.*, 129, 129-145 (1995).

[14] Markowich, P.A., Ringhofer, C. and Schmeiser, C., Semiconductors equations, Springer Verlag, Wien, New York, 1990.

[15] Markowich, P.A. and Szmolyan, P., 'A system with convection-diffusion equations with small diffusion coefficient arising in semiconductor physics', *JDE*, 81, 234-254 (1989).

[16] Natalini, R., 'The bipolar hydrodynamic model for semiconductors and the drift-diffusion equations', *J. of Math. Anal. and Appl.*, 198, 262-281 (1996).

[17] Poupaud, F., Raslce, M. and Vila, J.P., 'Global solutions to the isothermal Euler-Poisson system with arbitrarily large data', *J. Differ. Equations*, 123 (1), 93-121 (1995).

[18] Ringhofer, C., 'An asymptotic analysis of a transient p-n-junction model', *SIAM J. Appl. Math.*, 47, 624-642 (1987).

I. Gasser
Fachbereich Mathematik,
Universität Hamburg,
Bundesstrasse 55,
D-20146 Hamburg, Germany
gasser@math.uni-hamburg.de

P. Marcati
Dipartimento di Matematica Pura ed Applicata,
Universitá delgli Studi dell' Aquila,
I-67100 L'Aquila, Italy
marcati@univaq.it

International Series of Numerical Mathematics
Vol. 140, © 2001 Birkhäuser Verlag Basel/Switzerland

Dynamic Mesh Adaption for Supersonic Reactive Flow

Thomas Geßner and Dietmar Kröner

Abstract. Dynamic mesh adaption lets us resolve the disparate physical scales of reactive flow problems. This powerful tool gives us the possibility to simulate demanding problems with detailed reaction mechanisms on a workstation. We used a finite volume space discretization based on unstructured triangular meshes. For time–dependent problems, an explicit time discretization with time–scale splitting and local time stepping is used. To validate the whole numerical scheme, we simulate test problems as well as realistic hydrogen–oxygen reactions.

1. Introduction

For reactive flow problems adaptive mesh refinement (AMR) algorithms are used to resolve the disparate physical scales occurring e.g. in reaction fronts and shocks. Starting with Bell et al. [2] several researchers used this technique to reduce the computational cost of reactive flow simulations, see e.g. [14], [9], [17], [11]. In contrast to most of these researchers we prefer unstructured triangular meshes to ensure a maximum of flexibility for the representation of complex geometries.

In general, differential equations presenting a wide range of scales are called stiff (see e.g. [16]). The stiffness of the reactive flow equations is a problem associated with the short time scales. To cope with this problem, our scheme includes special integration methods (see Section 3). Beyond this problem associated with the short time scales, there is a further numerical phenomenon associated with the spatial resolution used to solve the system of equations: In the stiff case, spurious unphysical solutions that only occur in numerics can be observed, such as discontinuities moving with wrong speed and bifurcations instead of a single reaction front. This problem of "spurious numerical solutions" is systematically analyzed by LeVeque and Yee in [16] and can be solved by effectively resolving the scales of the stiff source term via methods such as adaptive mesh refinement (cf. references above), front tracking, or subcell resolution. Another approach, pursued in [3], [10], is to prevent spurious numerical solutions without resolving the scales of the stiff source term.

2. Physical and Chemical Model

The physical/chemical model bases on the compressible Euler equations of gas dynamics in two space dimensions describing the inviscid ideal gas flow. Additionally we have the conservation laws with source terms for an arbitrary number of different species. For details concerning the model beyond the following brief description, we refer e.g. to [23].

The basic quantities in the system of equations are the density ρ, the velocities $v_{1,2}$ in $x-$ and y–direction, the specific total energy E, the pressure p, and the mass fractions z_i of the different gas components $i \in \{1, 2, \ldots, N\}$. Then the complete system of conservation laws can be written in the following form

$$\partial_t u + \partial_x f_1(u) + \partial_y f_2(u) = s(u), \tag{1}$$

where

$$u = (\rho, \rho v_1, \rho v_2, \rho E, \rho z_1, \ldots, \rho z_N)^t \tag{2}$$

is the vector of the conservative variables. The nonlinear fluxes are

$$f_1(u) = \begin{pmatrix} \rho v_1 \\ \rho {v_1}^2 + p \\ \rho v_1 v_2 \\ (\rho E + p) v_1 \\ \rho v_1 z_1 \\ \vdots \\ \rho v_1 z_N \end{pmatrix}, \quad f_2(u) = \begin{pmatrix} \rho v_2 \\ \rho v_1 v_2 \\ \rho {v_2}^2 + p \\ (\rho E + p) v_2 \\ \rho v_2 z_1 \\ \vdots \\ \rho v_2 z_N \end{pmatrix}$$

and the right hand side (RHS) is

$$s(u) = (0, 0, 0, 0, \dot{m}_1, \ldots, \dot{m}_N)^t,$$

the source term including the chemical production rates $\dot{m}_i$ of the different gas components. The chemical production rate of species i for an arbitrary number R of chemical reactions is

$$\dot{m}_i = M_i \sum_{k=1}^{R} (\nu''_{ik} - \nu'_{ik})\, \omega_k, \tag{3}$$

where M_i is the molecular weight of species i, ω_k is the reaction rate of elementary reaction $k \in \{1, 2, \ldots, R\}$, and ν'_{ik}, ν''_{ik} are the stoichiometric coefficients. The reaction rates ω_k are expressed as follows

$$\omega_k := k_k \prod_{i=1}^{N} \left(\frac{\rho z_i}{M_i} \right)^{\nu'_{ik}} \quad \text{with} \quad k_k(T) := A_k T^{\beta_k} \exp\left(\frac{-E_{act_k}}{\mathbf{R} T} \right).$$

The rate coefficients $k_k(T)$ are modeled by a modified Arrhenius law, where A_k is the frequency factor, β_k is the pre–exponential temperature exponent, E_{act_k} is the activation or threshold energy of reaction k, and $\mathbf{R}$ is the universal gas constant. The heats of reaction are modeled as follows: The specific enthalpy h for an ideal–gas mixture is

$$h := \sum_{i=1}^{N} z_i h_i(T) \quad \text{where} \quad h_i(T) := h_i^0 + \int_{T^0}^{T} c_{pi}(T')dT'$$

is the partial specific enthalpy of species i which depends only on T. Thereby c_{pi} is the specific heat capacity at constant pressure, and h_i^0 is the reference enthalpy at the reference temperature of species i. To close the system the equation of state (EOS) for a mixture of thermally ideal gases

$$p = \frac{\rho \mathbf{R} T}{\overline{M}}, \quad \text{with} \quad \overline{M} := \left(\sum_{i=1}^{N} \frac{z_i}{M_i} \right)^{-1},$$

and the caloric EOS

$$p = \rho h - \rho E + \frac{1}{2}\rho \left|\vec{v}\right|^2 \tag{4}$$

are necessary. As a simplification of (4), the equation of state for a polytropic ideal gas with the specific heat ratio γ can be used

$$h := c_p T + h^0, \quad c_p = \frac{\gamma}{\gamma - 1} \frac{\mathbf{R}}{\overline{M}}. \tag{5}$$

An interpreter of the popular CHEMKIN^{TM}–II format (see [13]) allows the application of any desired reaction mechanism. To handle the thermo–chemical data of any desired species, we implemented a JANAF interpreter (see [22]).

3. Numerical Scheme

The important ingredients besides the dynamic mesh adaption, which is specified in the next section, are briefly described as follows. The numerical scheme bases on an unstructured conforming triangular discretization $\mathbf{T}$ of the considered domain with elements T_j. The convective terms in (1) are discretized using an upwind finite volume scheme (see e.g. [15]). Different methods to integrate the chemical source terms (3) have been implemented. Among them are the integrator for stiff systems of ordinary differential equations of Kaps and Rentrop [12] (A–stable of fourth order) and the hybrid second order scheme for stiff ODE's associated with chemical kinetics by Young and Boris [24]. We restrict ourselves to the description of the explicit time stepping. Furthermore an implicit time stepping can be used (see [8]). The adaptive step size control of the explicit schemes takes the CFL condition and an a posteriori error estimate of the source term integration into account. This step size control is combined with a time–scale splitting and a local time stepping for the source term integration. The resulting explicit numerical scheme including these features is

$$\begin{aligned} u_j^0 &:= u_0(x_j), \quad x_j \in T_j \in \mathbf{T} \\ u_j^{n,k+1} &:= \mathbf{S}\left(u_j^n, u_j^{n,1}, \ldots, u_j^{n,k}, \Delta t_{\mathrm{ST}}^{n,k}, s\right), \quad u_j^{n,0} := u_j^n \\ u_j^{n+1} &:= u_j^{n,K_n} - \frac{\Delta t_{\mathrm{FV}}^n}{|T_j|} \sum_{l=1}^{3} g_{jl}(u_j^n, u_{jl}^n), \end{aligned}$$

with a source term integration method

$$\mathbf{S} : \left(\mathrm{I\!R}^K, \mathrm{I\!R}^{K \cdot K_n}, \mathrm{I\!R}, C^2(\Omega, \mathrm{I\!R}^K)\right) \to \mathrm{I\!R}^K$$

and a numerical flux $g_{jl}(\cdot,\cdot)$. There the step sizes of the source term integration $\Delta t_{\mathrm{ST}}^{n,k}$ and the finite volume scheme Δt_{FV}^n have to fulfill the condition

$$\Delta t_{\mathrm{FV}}^n = \sum_{k=1}^{K_n} \Delta t_{\mathrm{ST}}^{n,k}.$$

Concerning the efficiency of this numerical method it is very important to note that the source term integration can be performed in each element T_j of the triangulation $\mathbf{T}$ independently with a local time step. A synchronization is only necessary before calculating the numerical fluxes in the new time step.

4. Dynamic Mesh Adaption

The aim of local mesh adaption is an equally distributed numerical error over the discretization domain to reach a provided accuracy with minimal numerical cost. Therefore the fineness of the grid is locally increased or decreased. For time–dependent problems a dynamic mesh adaption is necessary. The present theoretical shortcomings concerning rigorous error estimation do not prevent the successful (as demonstrated by the literature mentioned at the Section 1) use of mesh adaption for complex compressible (reactive) flow problems.

The criterion we use to control this dynamic mesh adaption bases on heuristic considerations founded on experiences in non–reactive flow simulations [6], theoretical results [18], and numerical experiments (see Section 5). It evaluates three different quantities and takes the space and time derivities of the flow and the local chemical production rate into consideration. As quantity controlling the mesh adaption in pure flow simulations we define the following improved (in comparison with [6]) weighted space–time–gradient

Definition 4.1. (local weighted flow quantity) *Consider the triangle $T_j \in \mathbf{T}$ with the constant vector valued u_j^n at the time t^n. Let v_j^n be a scalar component of the conservative flow variables (see (2)) of u_j^n. The values in the neighbors T_{jl} of T_j*

($l \in \{1,2,3\}$) and at time t^{n-1} have the corresponding indices. Then

$$\eta_j^n := \left(|T_j| \left(\left(\frac{v_j^n - v_j^{n-1}}{\Delta t^n} \right)^2 + \sum_{l=1}^{3} \left(\frac{v_j^n - v_{jl}^n}{w_j - w_{jl}} \right)^2 \right) \right)^{\frac{1}{2}}$$

defines the local weighted flow quantity, where w_j is the center of gravity of T_j. (In boundary elements of $\mathbf{T}$ the sum is calculated using the reduced set of neighboring triangles).

To characterize the local intensity of the chemical reaction we define

Definition 4.2. (local reaction quantity) *In the triangle $T_j \in \mathbf{T}$ with the constant vector–valued u_j^n at the time t^n*

$$\mu_j^n := \max_{i=1\ldots N} \left\{ \left| \dot{m}_i(u_j^n) \right| \right\}$$

defines the local reaction quantity, where $\dot{m}_i$ as defined in (3).

The rigorous error estimates in [18] demonstrated, that the source term has to be taken into consideration. Our numerical studies in [8] imply the same conclusion.

Now our mesh adaption criterion is the worst case combination of the two local quantities defined above:

Definition 4.3. (local reactive flow mesh adaption quantity) *Let*

$$\overline{\eta}^n := \frac{1}{\sharp I} \sum_{j \in I} \eta_j^n, \quad \overline{\mu}^n := \frac{1}{\sharp I} \sum_{j \in I} \mu_j^n$$

be the averages of η_j^n and μ_j^n on the triangulation $\mathbf{T}$, where I is the set of indices of $\mathbf{T}$. Then

$$\zeta_j^n := \max \left(\frac{\eta_j^n}{\overline{\eta}^n}, \frac{\mu_j^n}{\overline{\mu}^n} \right)$$

is the local relative reactive flow mesh adaption quantity on triangle T_j at time t^n.

To simplify the control of the AMR for different problems, the flow and the reaction quantities participate relatively to their mean values on $\mathbf{T}$ in the reactive flow quantity ζ_j^n. In the following criterion controlling the dynamic mesh adaption the parameters are almost independent of the reactive flow simulation performed.

Definition 4.4. (reactive flow mesh adaption criterion) *Consider the triangle $T_j \in \mathbf{T}$ at the time t^n with ζ_j^n according Definition 4.3. Then we mark T_j for refinement or coarsening as follows:*

$$\begin{array}{ll} \text{if } \zeta_j^n > \mathrm{A_{FINE}} & \rightarrow \text{mark } T_j \text{ for refinement} \\ \text{if } \zeta_j^n < \mathrm{A_{COARSE}} & \rightarrow \text{mark } T_j \text{ for coarsening} \end{array}$$

with positive constants $\mathrm{A_{FINE}} > \mathrm{A_{COARSE}}$.

Good choices for A_{COARSE}/A_{FINE} are e.g. 0.1/0.5 for a very sensitive mesh refinement up to 0.7/1.3 if only the most intense structures in the numerical solution should force a mesh refinement. Altogether the controlling of the dynamic mesh adaption requires some additional parameters to increase the efficiency of the algorithm and to make it more comfortable to handle. For details concerning the comfortable user control of the dynamic mesh adaption we refer to [8].

We use the conforming mesh adaption algorithm of Bänsch [1]. His approach bases on local bisection and the inverse coarsening process.

5. Numerical Results

The numerical scheme described in the previous sections is now applied to approximate the solutions of different reactive flow problems modeled analogously Section 2. All numerical experiments pursue the object to validate all parts of the numerical schemes and to show the flexibility (and limitations) of the numerical method concerning different problems and setups. We compare the results of our numerical simulations with ZND solutions (cf. e.g. [23]) and determine ignition time, velocity of reaction front, detonation cell size in comparison with analytical results, other numerical results, and experiments. Additionally, we want to determine the best and fastest numerical method by comparison of the respective components, e.g. numerical fluxes, time discretization, or source term integration. The results of these comparisons can be found in [8]. To receive an impression of the dynamic of the numerical results briefly presented in Subsections 5.2, 5.3, 5.4, we refer to the video tape/DVD [7].

5.1. ZND Detonation Waves

At first, we consider the following simplified problem: The model described in Section 2 for the single irreversible reaction R $\rightarrow$ P from reactant to product (corresponding to unburnt and burnt gas) together with the equation of state for a polytropic gas (5). For this setup, the theory of Zeldovich, von Neumann, and Döring (ZND; see e.g. [23]) makes an "exact" solution in one space dimension available. An exact solution with a very small half reaction length $L_{1/2}$ (to classify the length scale of the reaction; see [5]) is chosen. Therefore spurious solutions are very likely in the absence of a fine spatial discretization.

Figure 1 shows the pressure for different numerical simulations with and without AMR. The fine parts of the adaptive mesh are enlightened. In each part of Figure 1, the thick line (with the small peak) in front of each part of it shows the "exact" ZND solution. Without AMR, there is only 1/1000 of a point in the $L_{1/2}$. A typical spurious solution traveling with wrong speed instead of a single reaction wave can be observed. With dynamic mesh adaption, the solution is quite right. In this example, 1/8 point in the $L_{1/2}$ is necessary to get an impression of the reaction zone.

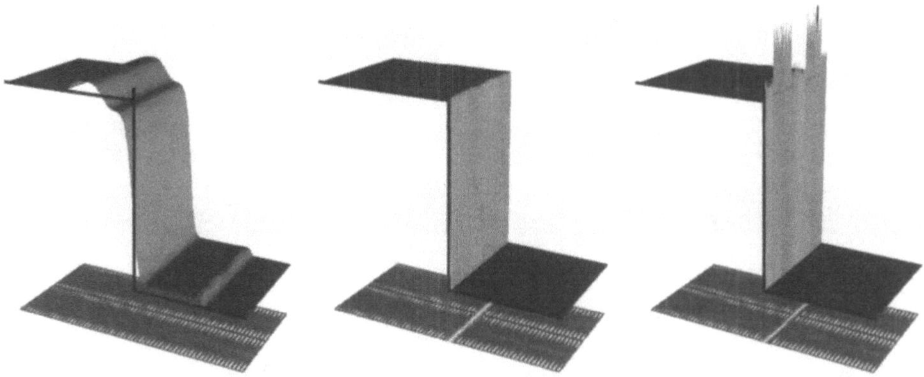

FIGURE 1. Numerical Simulation of a ZND Detonation in 2–D

5.2. Unstable Detonation Waves in Two Space Dimensions

The unstable behavior of detonation waves in two space dimensions is extensively studied in [4]. Using the same simplified setup as in the previous simulation, we consider an unstable detonation in the shock frame (transformation in Galilean coordinates). Figure 2 compares the results of numerical simulations with and without dynamic mesh adaption at two different times. The reaction front com-

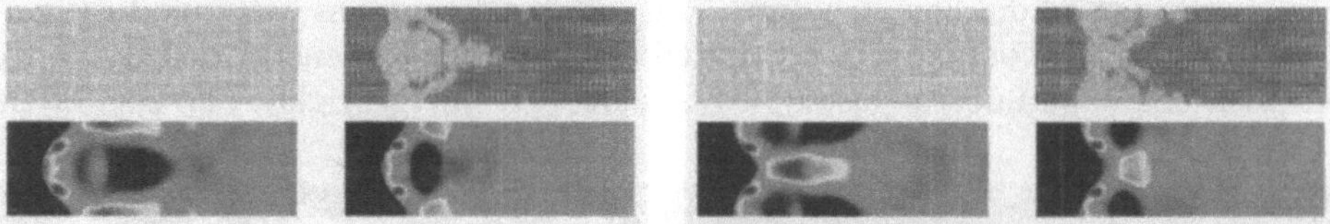

FIGURE 2. Unstable detonation in 2–D: with/without AMR

puted with AMR is as well resolved as on the uniformly refined mesh with distinct minor computational cost. Solely the subsonic effects behind the reaction front are better resolved on the uniformly refined mesh.

5.3. Detonation in a Channel with Barriers: Transition to a Deflagration

Everything else being the same as in the previous simulation, the travelling detonation wave (in laboratory frame) now meets a cascade of barriers. The resulting complex reactive flow severely challenges the mesh adaptor. Behind the first barrier, the transition of the detonation to a deflagration can be observed. Figure 3 shows the adaptive mesh, the density, and reactant respectively product at subsequent times. During the whole simulation all structures in the numerical solution

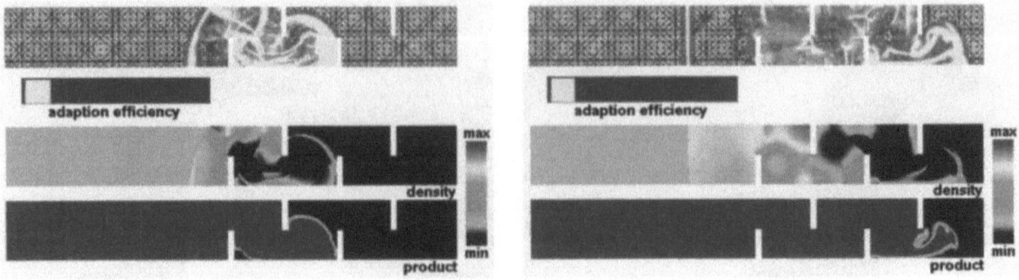

FIGURE 3. Dynamic mesh adaption in channel with cascade of barriers

are resolved. For example the reaction front is captured by the dynamic mesh adaption as well as the less intense discontinuities in the density. The bar chart (in the middle of each snapshot) compares the numerical cost of the displayed simulation with the cost of a ficticious simulation on a uniformly refined mesh. In this complex geometry this ratio is less then 15%, which is an indication for the efficiency of the adaptive mesh refinement.

5.4. Two Dimensional Unstable Hydrogen–Oxygen Detonation

Now we consider an unstable low-pressure hydrogen–oxygen detonation with high diluent argon. A detailed reaction mechanism with 9 species and 48 elementary reactions (see [19]) is used in the model. Low–pressure detonations have been extensively studied (e.g. [20]), because such mixtures are known to produce extremely regular detonation structures and cellular detonation patterns. In Figure 4, we qualitatively compare the computed released chemical energy density on the left with the experimental result of Strehlow [21] on the right. The size of the

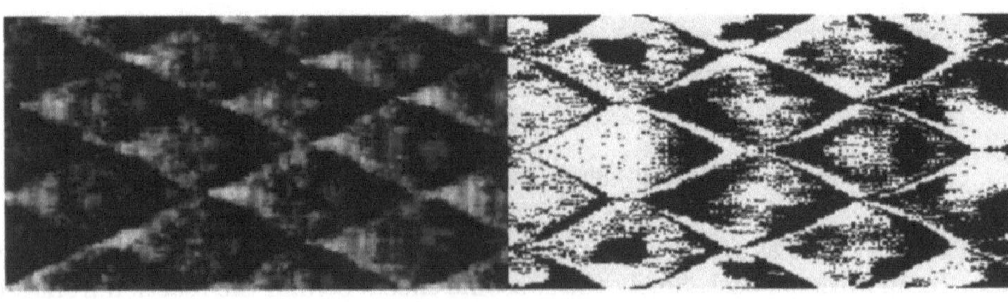

FIGURE 4. Regular detonation cells: Simulation – Experiment of Strehlow

detonation cells (displayed in Figure 4) is approximately $5.7 \times 3.0\,cm$, and the reaction front is propagating with $1600\,\frac{m}{s}$. Oran et. al. [20] simulated the same experiment with a different numerical approach on a powerful parallel computer. Their computed cell size is $5.5 \times 3.0\,cm$ with a reaction front velocity of $1625\,\frac{m}{s}$. The detonation cell size measured in experiments is larger, but the aspect ratio of our computed and the experimental cells is pretty much the same (see Figure 4).

The reason for the larger experimental cell sizes is probably the energy loss at the walls, which cause the detonation to propagate less rapidly (see [20]).

The AMR is very efficient for the simulation in Figure 4. Less than 2% of computing time is necessary comparing the displayed simulation with a fictitious one based on a uniformly refined mesh.

6. Conclusion

The dynamic mesh adaption makes the resolution of the disparate physical scales of reactive flow problems possible. Also the other components of the numerical method we did not describe in detail are very important to increase its efficiency. Among these ingredients are the local time stepping, and the integrator for stiff systems of ODE. In summary, we obtain the possibility to simulate challenging reactive flow problems with detailed reaction mechanisms on a workstation with moderate cost.

Acknowledgment: This work has been partially supported by the German Research Association (DFG; Deutsche Forschungsgemeinschaft; Grant DFG Kr 795/5-1).

References

[1] E. Bänsch, *Local mesh refinement in 2 and 3 dimensions*, Report 6, SFB 256, Bonn, 1989.

[2] J. B. Bell, P. Colella, J. A. Trangenstein, and M. Welcome, *Adaptive methods for high Mach number reacting flow*, AIAA Paper 87–1168–CP, 1987.

[3] A. C. Berkenbosch, E. F. Kaasschieter, and R. Klein, *Detonation capturing for stiff combustion chemistry*, Combust. Theory Modeling **2** (1998), 313–348.

[4] A. Bourlioux and A. Majda, *Theoretical and numerical structures for unstable two–dimensional detonations*, Comb. and Flame **90** (1992), 211–229.

[5] A. Bourlioux, A. Majda, and V. Roydburd, *Theoretical and numerical structures for unstable one–dimensional detonations*, SIAM J. Appl. Math. **51** (1991), 303–343.

[6] T. Geßner, *Zeitabhängige Adaption für Finite Volumen Verfahren höherer Ordnung am Beispiel der Euler–Gleichungen der Gasdynamik*, Institut für Angewandte Mathematik, Universität Bonn, 1994, Diplomarbeit.

[7] T. Geßner, *Numerical simulation of unstable detonations*, Video and Multimedia at 3ecm (S. Xambo-Descamps and S. Zarzuela, eds.), Springer, 2000.

[8] T. Geßner, *Timedependent adaption for supersonic combustion waves modeled with detailed reaction mechanisms*, Mathematische Fakultät, Universität Freiburg i. Br., 2001, Dissertation.

[9] F.-K. Hebeker, *An adaptive finite element method for detonation waves: Toward time limit zero*, Preprint 98-27 (SFB 359), Heidelberg, 1998.

[10] C. Helzel, R. J. LeVeque, and G. Warnecke, *A modified fractional step method for the accurate approximation of detonation waves*, Technical Report 99-04, Department of Applied Mathematics, University of Washington, 1999.

[11] R. Jeltsch and P. Klingenstein, *Error estimates for the position of discontinuities in hyperbolic conservation laws with source terms which are solved using operator splitting*, Comput. Visual. Sci. **1** (1999), 231–249.

[12] P. Kaps and P. Rentrop, *Generalized Runge–Kutta methods of order four with step-size control for stiff ordinary equations*, Numer. Math. **33** (1979), 55–68.

[13] R. J. Kee, F. M. Rumpley, and J. A. Miller, CHEMKIN–II*: A FORTRAN chemical kinetics package for the analysis of gas phase chemical kinetics*, SAND89 8009, Sandia National Laboratories, Livermore, 1989.

[14] P. Klingenstein, *Nonlinear hyperbolic conservation laws with source term: Errors of the shock location*, Dissertation 12019, ETH–Zürich, 1997.

[15] D. Kröner, *Numerical schemes for conservation laws*, Wiley, Teubner, Stuttgart, 1997.

[16] R. J. LeVeque and H. C. Yee, *A study of numerical methods for hyperbolic conservation laws with stiff source terms*, J. Comput. Phys. **86** (1990), 187–210.

[17] E. Loth, S. Sivier, and J. Baum, *Adaptive unstructured finite element method for two–dimensional detonation simulations*, Shock Waves **8** (1998), 47–53.

[18] M. Ohlberger, *A posteriori error estimates for vertex centered finite volume approximations of convection–diffusion reaction equations*, Mathematical Modelling and Numerical Analysis **35** (2001), no. 2, 355–387.

[19] E. S. Oran and J. P. Boris, *Theoretical and computational approach to modeling flame ignition*, Prog. Aeronaut. Astronautics **76** (1981), 154–171.

[20] E. S. Oran, J. W. Weber Jr., E. I. Stefaniw, M. H. Lefebvre, and J. D. Anderson Jr., *A numerical study of a two–dimensional* H_2–O_2–Ar *detonation using a detailed chemical reaction model*, Comb. and Flame **113** (1998), 147–163.

[21] R. A. Strehlow, *The nature of traverse waves in detonations*, Astronautica Acta. **14** (1969), 539–548.

[22] D. R. Stull and H. Prophet, *JANAF thermochemical tables*, Tech. report, U. S. Department of Commerce, 1971.

[23] F. A. Williams, *Combustion theory*, The Benjamin/Cummings Publishing Company, Inc., 1985.

[24] T. R. Young and J. P. Boris, *A numerical technique for solving stiff ordinary differential equations associated with the chemical kinetics in reactive–flow problems*, J. Phys. Chem. **81** (1977), 2424–2427.

T. Geßner
D. Kröner
Institut für Angewandte Mathematik,
Albert–Ludwigs–Universität Freiburg,
Hermann–Herder–Str. 10
D-79104 Freiburg, Germany
`gessner@mathematik.uni-freiburg.de`
`dietmar@mathematik.uni-freiburg.de`

International Series of Numerical Mathematics
Vol. 140, © 2001 Birkhäuser Verlag Basel/Switzerland

A High-Resolution Scheme for the Elastic-Plastic Wave Equation

Guido Giese and Rolf Jeltsch

Abstract. In this paper we present a numerical method of high order for solving the multidimensional elastic-plastic wave equation. The basic idea is to decompose the conservation law into advection equations which can be solved numerically. Furthermore, the occurrence of plasticity makes it necessary to compute numerical approximations of the stress-strain relationship.

1. Introduction

The wave equation of an elastic material represents a hyperbolic conservation law for the momentum and the strain variables and can be solved with standard methods for conservation laws. However, for solids which undergo plastic deformation, only the momentum is conserved. In this work, we follow the ansatz of Fey [1], [2] who developed a high order scheme called Method of Transport for solving the multidimensional Euler equations. The basic idea is to decompose the partial differential equation into scalar advection equations which can be solved numerically.

Although the elastic-plastic wave equation is not a pure conservation law, we were able to use the ansatz of Fey to construct a genuinely multidimensional numerical scheme of high order for plastic waves. In addition to the flux calculation which allows the update of strain and velocity variables in each time step, we have to construct an appropriate update for the stress variables.
In this paper we first give a brief introduction into the governing equations of elasto-plasticity. Afterwards, we present our numerical scheme and finally we give some numerical results in 2-D and 3-D.

2. Governing Equations

We use a formulation of the elastic-plastic wave equation as a first order system, which means we have to use three physical variables: The symmetric stress tensor $\underline{\sigma}$, the symmetric strain tensor $\underline{\epsilon}$ and the velocity vector $\vec{v}$. Furthermore, we need the deviatoric stress tensor $\underline{s}$ which is defined as $s_{ij} = \sigma_{ij} - \frac{1}{3}\delta_{ij}\sigma_{kk}$.

With these variables, we get three equations describing the conservation of momentum ($\rho =$ density):

$$\rho \frac{\partial}{\partial t} v_i = \sum_{j=1}^{3} \frac{\partial}{\partial x_j} \sigma_{ij} \qquad i = 1, \ldots, 3$$

and six compatibility relations between velocity and strain variables

$$\frac{\partial}{\partial t} \epsilon_{ij} = \frac{1}{2} \left(\frac{\partial}{\partial x_j} v_i + \frac{\partial}{\partial x_i} v_j \right)$$

which are due to the fact that both strain and velocity variables are derivatives of the displacement vector $\vec{u}$. Altogether we obtain nine equations describing wave phenomena in solids. However, we still need an equation of state connecting stress and strain variables. For plasticity there exists a relationship between infinitesimal changes of stress and strain only:

$$\frac{d}{dt} \epsilon_{ij} = \frac{1+\nu}{E} \frac{d}{dt} \sigma_{ij} - \frac{\nu}{E} \delta_{ij} \frac{d}{dt} \sigma_{kk} + \sigma_{ij} \frac{d}{dt} \chi \tag{1}$$

with $\nu =$ Poisson's ratio, $E =$ Young's modulus. For elasticity equation (1) reduces with $\frac{d}{dt}\chi \equiv 0$ to Hooke's law.

In order to distinguish between elastic and plastic deformation we use the so called von Mises yield function:

$$f(\underline{s}) = \frac{1}{2} s_{ij} s_{ij} =: \kappa^2 \tag{2}$$

Basically, plasticity occurs if the current function $\kappa(t)$ attains the value of $\kappa_0(t)$ which is the maximal value of $\kappa(t)$ in the past:

$$\kappa_0(t) = \max_{t_0 \le t' \le t} \kappa(t')$$

Furthermore, for this yield-criterion the function $\frac{d}{dt}\chi$ can be written in the form

$$\dot{\chi} = \frac{1}{2\kappa} \left(\frac{1}{\mu_p(\kappa)} - \frac{1}{\mu} \right) \dot{\kappa} \tag{3}$$

with a measured function $\mu_p(\kappa) \le \mu$ and the shear modulus μ. Hence, we have a system of the form

$$\boldsymbol{V}_t + \nabla \cdot c\underline{\boldsymbol{L}}(\vec{v}, \underline{\boldsymbol{\sigma}}) = 0 \qquad (c = \text{wave-speed}) \tag{4}$$

$$\underline{\dot{\boldsymbol{\sigma}}} = \underline{\boldsymbol{C}}(\underline{\boldsymbol{\sigma}}) : \underline{\dot{\boldsymbol{\epsilon}}} \tag{5}$$

Thus the governing equations consists of a flux equation (4) and an ODE (5) in every point.

The most important difference between the elastic wave equation and the standard wave equation $\phi_{tt} - c^2 \Delta \phi = 0$ is the existence of compression waves with wave-speed c_1 and shear waves with wave-speed $c_2 < c_1$. For the elastic-plastic wave equation even more wave-speeds occur.

3. Numerical Scheme

Due to the structure of the equations (4) and (5) our numerical scheme consists of two steps:

1. Solve the flux equation (4) with the Method of Transport.
2. Integrate the stress-strain relationship (5) in time for every cell of the grid.

We will now discuss both steps in detail.

3.1. The Method of Transport

The basic idea of the Method of Transport is to rewrite the flux equation (4) equivalently as a coupled system of advection equations. Therefore, we introduce a set of direction vectors $\vec{\boldsymbol{n}}_i$, $i = 1, \ldots, k$ which have to fulfill the following two consistency relations:

$$\sum_{i=1}^{k} \vec{\boldsymbol{n}}_i = 0, \qquad \sum_{i=1}^{k} \vec{\boldsymbol{n}}_i \vec{\boldsymbol{n}}_i^T = \underline{\boldsymbol{I}} \tag{6}$$

With these definitions we can rewrite the flux equation as follows:

$$\frac{\partial}{\partial t}\boldsymbol{V} \quad + \qquad \nabla \cdot c\underline{\boldsymbol{L}} = 0 \tag{7}$$

$$\Longleftrightarrow \qquad \frac{1}{k}\sum_{i=1}^{k}\{\frac{\partial}{\partial t}(\boldsymbol{V} + \underline{\boldsymbol{L}}\vec{\boldsymbol{n}}_i) \quad + \quad \nabla \cdot c(\boldsymbol{V} + \underline{\boldsymbol{L}}\vec{\boldsymbol{n}}_i)\vec{\boldsymbol{n}}_i^T\} = 0 \tag{8}$$

Each of these advection terms transports the quantity $\boldsymbol{V} + \underline{\boldsymbol{L}}\vec{\boldsymbol{n}}_i$ at the velocity $c\vec{\boldsymbol{n}}_i$.

Note that equation (8) is strictly equivalent to the original equation (7). Our approximation consists of decoupling the system, i.e. at a certain time-step t^n we define the independent quantities

$$\boldsymbol{R}_i(\vec{\boldsymbol{x}}, t^n) := \boldsymbol{V}(\vec{\boldsymbol{x}}, t^n) + \underline{\boldsymbol{L}}(\vec{\boldsymbol{x}}, t^n)\vec{\boldsymbol{n}}_i$$

and then solve the advection equations

$$(\boldsymbol{R}_i)_t + \nabla \cdot \left(\boldsymbol{R}_i c\vec{\boldsymbol{n}}_i^T\right) = 0 \quad \forall i \tag{9}$$

independently on the time interval $I = [t^n, t^{n+1}]$. At the next time-step t^{n+1} the update for the vector $\boldsymbol{V}$ reads:

$$\boldsymbol{V}(\vec{\boldsymbol{x}}, t^{n+1}) = \frac{1}{k}\sum_{i=1}^{k} \boldsymbol{R}_i(\vec{\boldsymbol{x}}, t^{n+1}) \tag{10}$$

This yields a first order approximation for the exact solution in time if the consistency relations (6) hold, since

$$\boldsymbol{V}_t = \frac{1}{k}\sum_{i=1}^{k}(\boldsymbol{R}_i)_t = -\frac{1}{k}\sum_{i=1}^{k}\nabla \cdot (c\boldsymbol{R}_i\vec{\boldsymbol{n}}_i^T).$$

However, the approximation is not of second order since

$$\boldsymbol{V}_{tt} \neq \frac{1}{k}\sum_{i=1}^{k}(\boldsymbol{R}_i)_{tt} = \frac{1}{k}\sum_{i=1}^{k}\nabla\cdot\left(\nabla\cdot(c\boldsymbol{R}_i\vec{\boldsymbol{n}}_i^T)c\vec{\boldsymbol{n}}_i^T\right)$$

To obtain approximations of higher order in time one has to add correction terms into the numerical fluxes, i.e. one uses slightly modified quantities in the advection equations

$$\boldsymbol{R}_i := \boldsymbol{V} + (\underline{\boldsymbol{L}} + \underline{\boldsymbol{K}})\,\vec{\boldsymbol{n}}_i$$

where the correction matrix $\underline{\boldsymbol{K}}$ can be found by comparing the Taylor expansion of the exact solution to the expansion of the numerical scheme.

So far our discussion has been semi-discrete only since we haven't discretized the space variables yet. The space disretization can be done as for all types of finite volume methods, i.e. one has to add the following two steps:

- At a certain time-level one computes cell-averages for each cell I_{ij} [1] for the quantity $\boldsymbol{V}$ according to equation (10):

$$\bar{\boldsymbol{V}}_{ij}^{n+1} = \frac{1}{|I_{ij}|}\sum_{l=1}^{k}\int_{I_{ij}}\boldsymbol{R}_l(\vec{\boldsymbol{y}}, t^{n+1})d\vec{\boldsymbol{y}}$$

- Before each time-step one has to reconstruct the function $\boldsymbol{V}(\vec{\boldsymbol{x}}, t)$ by polynomials from the previously updated cell-averages.

It is important to mention that any wave-model fulfilling the conditions (6) will lead to a consistent numerical scheme. The simplest model which can be found are the four diagonal propagation directions:

$$\vec{\boldsymbol{n}}_1 = \begin{pmatrix}1\\1\end{pmatrix}, \quad \vec{\boldsymbol{n}}_2 = \begin{pmatrix}1\\-1\end{pmatrix}, \quad \vec{\boldsymbol{n}}_3 = \begin{pmatrix}-1\\1\end{pmatrix}, \quad \vec{\boldsymbol{n}}_4 = \begin{pmatrix}-1\\-1\end{pmatrix}.$$

However, to obtain good numerical results it is important to take into account more physical information to choose an appropriate wave-model. For this reason, we choose a wave-model for the decomposition of our 2-D or 3-D wave equation which reduces automatically to the decomposition into right eigenvectors for an elastic 1-D problem which we know is exact in 1-D, which means that if we simulate a 1-D problem with our multidimensional scheme it will be equivalent to the transport of right eigenvectors at the speed of the eigenvalues of the Jacobian matrix in 1-D. An example for such a wave-model in 2-D can be found in [3].

To conclude, the Method of Transport yields an update for the vector $\boldsymbol{V}$ which is high order in time due to the use of correction terms and high order in space by using a polynomial reconstruction in each cell.

[1]We assumed a grid in 2-D.

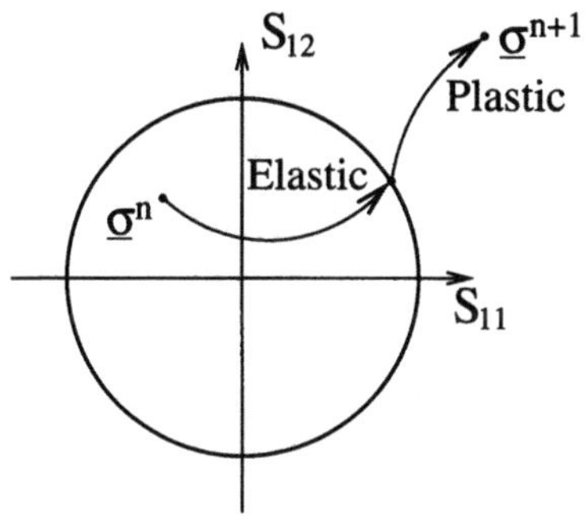

FIGURE 1. Integration of the stress-strain relationship in the stress space from $\underline{\sigma}^n$ to the new state $\underline{\sigma}^{n+1}$. The ODE-solver has to be restarted on the yield surface since the ODE can be discontinuous at the transition from elasticity to plasticity.

3.2. Stress Update

The Method of Transport we described in the previous section yields an update for the strain and velocity vector contained in the vector $\boldsymbol{V}$ in each time-step. For the update of the stress variables $\underline{\sigma}$ one has to integrate the ODE (5) in time. The problem is that the strain variables $\underline{\epsilon}$ are only known at the discrete time-levels t^n and t^{n+1}. However, to integrate equation (5) in time we have to know the derivative $\dot{\underline{\epsilon}}$ on the whole time-interval $[t^n, t^{n+1}]$. Therefore, we reconstruct the strain-path in time by polynomials:

$$\underline{\epsilon}(\vec{x}, t) = \underline{\epsilon}(\vec{x}, t^n) + \underline{a}(\vec{x})(t - t^n) + \underline{b}(\vec{x})(t - t^n)^2 + \dots \quad \forall t \in [t^n, t^{n+1}]$$

With the help of this reconstruction one can integrate equation (5). It is important to note that during the integration a transition from elasticity to plasticity might occur where the right hand side of the ODE is discontinuous. In such a case one has to find the exact transition point (e.g. by bisection or with Newton's method) and then restart the ODE-solver at this point (cf. Figure 4).

4. Numerical Example for Plane-Strain

We consider a problem in 2-D under so-called plane-strain, i.e. we assume

$$\epsilon_{33} = \sigma_{23} = \sigma_{13} = v_3 \equiv 0.$$

We consider two compression waves hitting a crack in a plate (cf. Figure 4).
The initial conditions for the compression waves are chosen as

$$c_1 = 1, \qquad c_2 = \frac{1}{\sqrt{3}}.$$

The initial conditions for the compression waves are chosen as

$$v_2^0 = 0.55, \quad \sigma_{22}^0 = \frac{v_2^0}{c_1}$$

and the initial value for the yield parameter κ_0 is

$$\kappa_0 = 1 \quad \text{and} \quad \frac{\mu_p}{\mu} = \frac{3}{16}.$$

In addition to the physical behavior observed for an elastic solid a zone of plastic

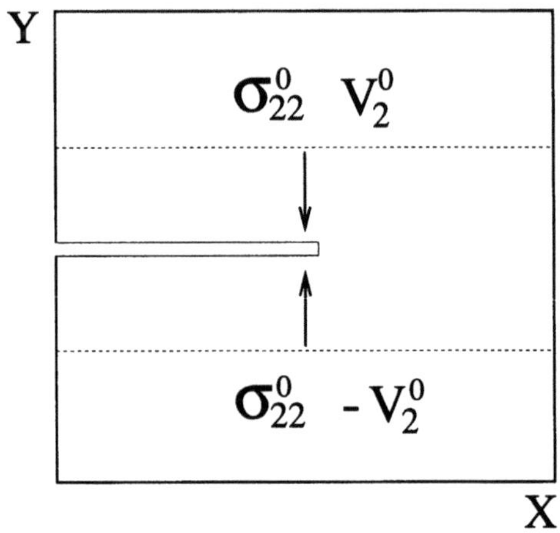

FIGURE 2. Compression waves carrying the velocity $\pm v_2^0$ and the stress component σ_{22}^0 approaching a crack.

deformation is created at the crack-tip, as we can see in Figure 4.

5. Numerical Results in 3-D

We consider a body in three space dimensions with a crack along the x-direction, i.e. all points in the set $M = \{(x,y,z)^T | y = 0, \quad z = 1 \quad \text{and} \quad 0 \leq x \leq 1\}$ are assumed to belong to the crack. Analogous to the 2-D-case a plane shear wave propagates from both sides along the y-axis towards the crack (cf. Figure 5).

Both shear waves carry the stress and velocity component $\sigma_{12} = 0.5$ and $v_1 = -(\pm\frac{\sigma_{12}}{c_2})$, with the positive and negative sign indicating the positive and negative propagation direction along the y-axis. The wave-speeds and the yield-condition are chosen as in our previous 2-D example.

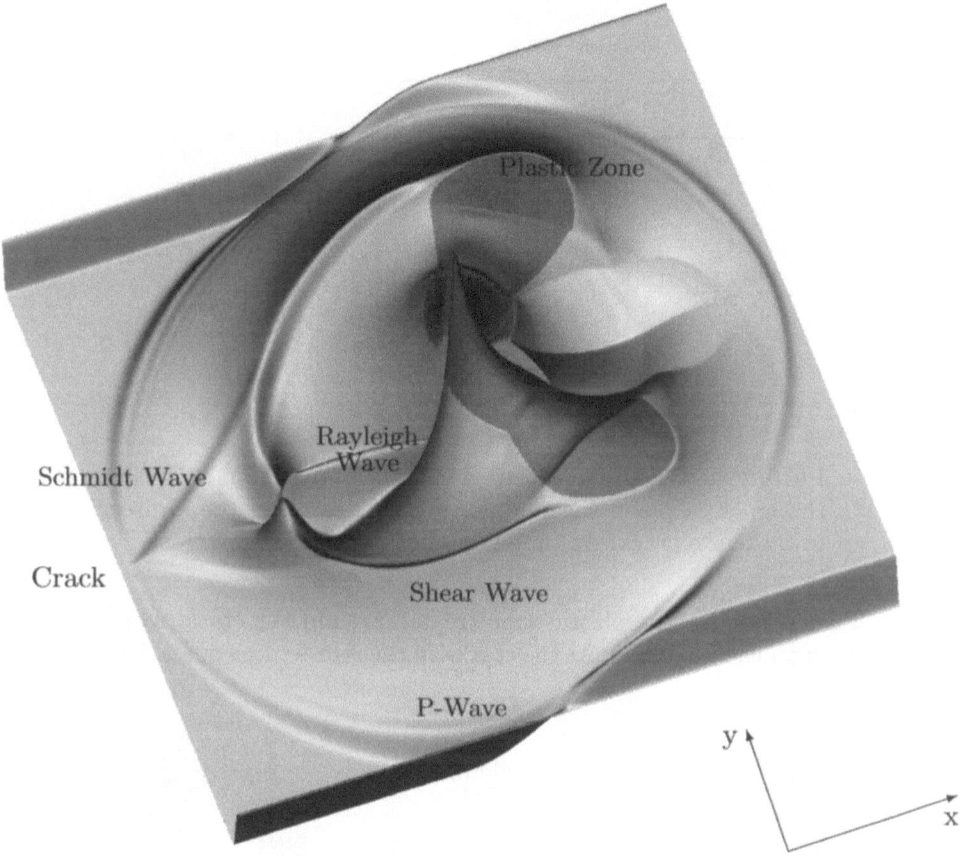

FIGURE 3. Numerical computation of the maximal shear stress around the crack-tip in the elastic case. The singularity, the compression wave, the shear wave, the Rayleigh wave, the von Schmidt wave and the plastic zone are well resolved (grid: 500 × 500 cells).

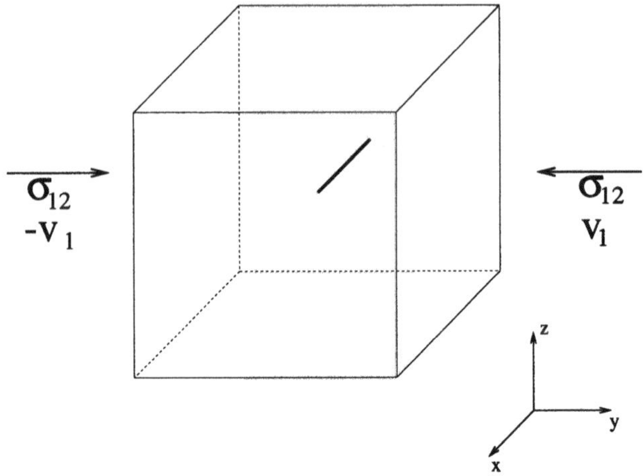

FIGURE 4. One dimensional crack in a body. From both sides of the crack, a plane shear waves propagates along the y-direction towards the crack. At time $t = 0$ both waves hit the crack.

Figure 5 shows the plastic zone created around the crack, which is a rather flat region on the x-z-plane. At the crack-tip the stress turns out to be very intense.

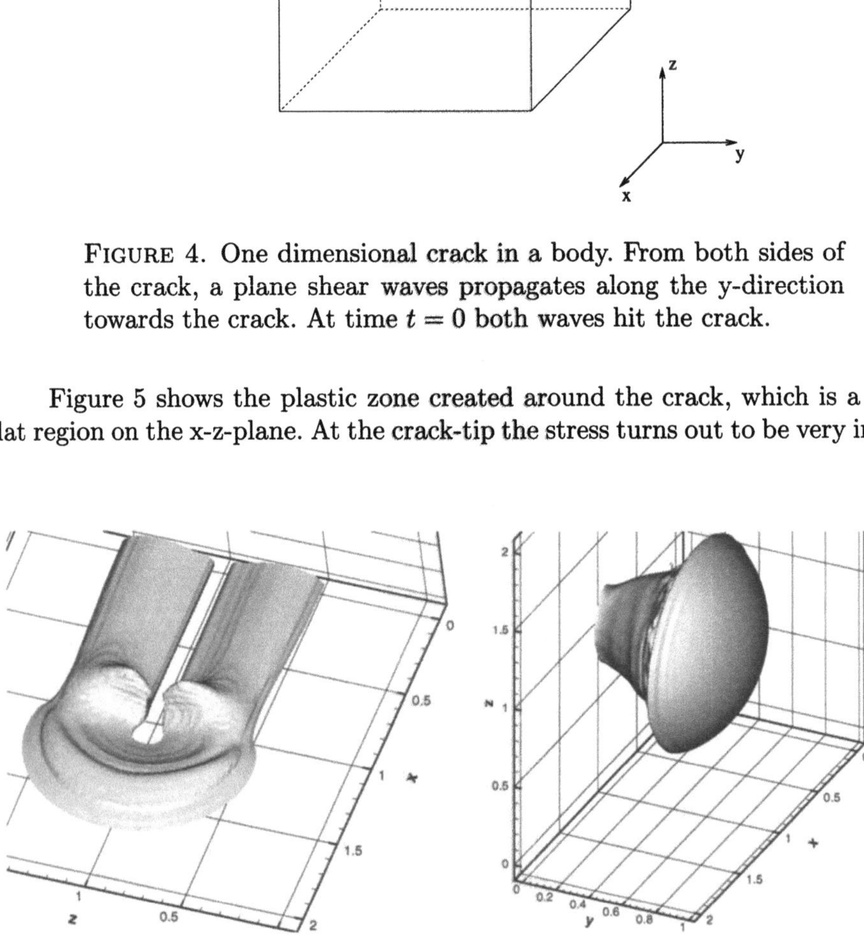

FIGURE 5. Plastic zone created around the crack (left Figure), which is concentrated on the x–z-plane. Plot of the hyper-surface $\sigma_{22}(x, y, z, t = 0.885) \equiv -0.002$ (right Figure). As in 2-D a spherical wave is produced at the crack-tip. Due to symmetry only the half-space $y \geq 0$ is plotted (grid: 128 $\times$ 64 $\times$ 128 cells).

References

[1] M. Fey. *Multidimensional Upwinding. Part I. The Method of Transport for Solving the Euler Equations. Journal of Computational Physics 143, 159-180 (1998).*

[2] M. Fey. *Multidimensional Upwinding. Part II. Decomposition of the Euler Equations into Advection Equations. Journal of Computational Physics 143, 181-199 (1998).*

[3] G. Giese *Decomposition of the Elastic-Plastic Wave Equation into Advection Equations. Proceedings of the Seventh International Conference on Hyperbolic Problems in Zürich, February 1998. Birkhäuser Verlag 1999.*

[4] G. Giese *A Genuinely Multi-Dimensional High-Resolution Scheme for the Elastic-Plastic Wave Equation. PhD-thesis Nr. 13472 ETH-Zürich.*

[5] X. Lin. *Numerical Computation of stress waves in solids.* Akademie Verlag, 1996

G. Giese
`guido.giese@db.com`

R. Jeltsch
Seminar for Applied Mathematics,
ETH Zentrum,
Switzerland
`jeltsch@sam.math.ethz.ch`

International Series of Numerical Mathematics
Vol. 140, © 2001 Birkhäuser Verlag Basel/Switzerland

Stability for Temple Class Systems with L^∞ Boundary Data

Paola Goatin

Abstract. We construct a continuous semigroup of solutions to the initial-boundary value problem for Temple class systems, on a domain of $\mathbf{L}^\infty$ functions. Trajectories of the semigroup depend $\mathbf{L}^1$ Lipschitz continuously on the initial and boundary data. We apply the result to the study of topological properties of sets of attainable profiles.

1. Introduction

We consider the initial-boundary value problem for a strictly hyperbolic system of conservation laws in one space dimension,

$$u_t + f(u)_x = 0, \tag{1}$$
$$u(0,x) = \overline{u}(x), \tag{2}$$
$$u(t,0) = \widetilde{u}(t), \tag{3}$$

on the domain $\Omega = \{(t,x) \in I\!R^2 : t \geq 0, x \geq 0\}$. Here $f : U \mapsto I\!R^N$ is a smooth vector field defined on some open set $U \subseteq I\!R^N$. The problem (1.1)-(1.3) is usually not well-posed when the boundary data are required to be assumed pointwise in the (strong) sense (1.3), even when (1.1) is a linear system (see [13]). In fact, different notions of the boundary condition have been considered in the literature, see [1, 14, 15] for definitions and references. Here, following [10, 11], we will deal with the (weak) form

$$f(u(0+,t)) \in f\big(\mathcal{V}(\widetilde{u}(t))\big), \qquad t > 0, \tag{4}$$

where $\mathcal{V}(\widetilde{u}(t)) \subset U$ is a time-dipendent set (the set of *admissible boundary values*) to be defined from the boundary data, and $f(u(0+,t))$ is the weak trace (whose existence is treated later) of the flux $f(u(t,x))$ at the boundary $x = 0$.

Most of the study on the boundary condition for (1.1) has been restricted to the scalar equation. We recall that, for scalar conservation laws, the existence and continuous dependence on the initial and boundary data of global solutions to the mixed problem (1.1)-(1.3) was proved within domains of $\mathbf{L}^\infty$-functions [12, 14]. On the other hand, in the case of $N \times N$ systems, global existence and stability of weak solutions has been established only for data with small total variation [1, 16, ...], [2, 3].

Here, having in mind to study the initial-boundary value problem from the point of view of control theory (where it is natural to regard the boundary data as varying into a prescribed set of $\mathbf{L}^\infty$-controls), we extend the existence and the stability results in [12, 14] to the case of Temple class systems with genuinely nonlinear characteristic fields.

We rely on the stability result obtained in [9] for the Cauchy problem for Temple class systems with $\mathbf{L}^\infty$-initial data, and we apply the same technique to prove $\mathbf{L}^1$-stability with respect to initial and boundary data in a domanin of $\mathbf{L}^\infty$ functions with possibly unbounded total variation.

In the case of scalar equations, the mixed problem with integrable (possibly unbounded) boundary data was first studied from the point of view of control by F. Ancona and A. Marson [5, 6]. On the same line, we fix a set $\mathcal{U} \subset \mathbf{L}^\infty(I\!R^+)$ of admissible controls, and we study the set of attainable profiles at a fixed time T

$$\mathcal{A}(T,\mathcal{U}) \doteq \left\{u(T,\cdot) : \ u \text{ is a solution to (1.1)-(1.3) with } \widetilde{u} \in \mathcal{U}\right\},$$

and at a fixed point in space $\overline{x} > 0$

$$\mathcal{A}(\overline{x},\mathcal{U}) \doteq \left\{u(\cdot,\overline{x}) : \ u \text{ is a solution to (1.1)-(1.3) with } \widetilde{u} \in \mathcal{U}\right\}.$$

Applications to calculus of variations and problems of optimization motivate the study of topological properties of $\mathcal{A}(T,\mathcal{U})$, $\mathcal{A}(\overline{x},\mathcal{U})$. We establish closure and compactness of the attainable sets in connection with a class of $\mathbf{L}^\infty$-boundary controls satisfying convex constraints.

2. Preliminaries and statement of the main result

Call $A(u) \doteq Df(u)$ the Jacobian matrix of f and denote by $\lambda_1(u) < \ldots < \lambda_N(u)$ its eigenvalues. In the sequel, we will assume that

$$\lambda_{N-p}(u) \leq 0 < \lambda_{N-p+1}(u) \qquad \text{for every } u \in U, \tag{5}$$

for some fixed $p \in \{1,\ldots,N\}$. The right and left eigenvectors of $A(u)$ will be written as $r_i(u)$, $l_i(u)$ respectively, and normalized so that

$$\big|r_i(u)\big| = 1, \qquad l_i(u)\cdot r_j(u) = \begin{cases} 1 & \text{if } i = j, \\ 0 & \text{if } i \neq j. \end{cases} \tag{6}$$

We recall that the i-th field is *genuinely nonlinear* if, by a suitable orientation of the eigenvectors r_i, at every point $u \in \Omega$ one has $D\lambda_i \cdot r_i > 0$.

Definition 2.1. *The system* (1.1) *is of Temple class if there exists a system of coordinates* $w = (w_1,\ldots,w_N)$ *consisting of Riemann invariants, and such that the level sets* $\{u \in \Omega;\ w_i(u) = \text{constant}\}$ *are hyperplanes (see* [15]*).*

Throughout the following, $w_i(t,x) \doteq w_i\big(u(t,x)\big)$ denotes the i-th Riemann coordinate of u. For a Temple class system, the integral curve of the vector field r_i through a point u_0 is the straight line described by the $N-1$ linear equations

$$w_j(u) = w_j(u_0) \qquad\qquad j \neq i. \tag{7}$$

In particular, shock and rarefaction curves coincide.

In connection with (2.3), we consider a convex, compact polytope $K \subset U$ having the form

$$K = \Big\{u \in U;\ w_i(u) \in [a_i,\ b_i] \qquad i = 1,\ldots,N\Big\}, \tag{8}$$

and assume that, as u varies in K, a strengthened version of the strict hyperbolicity condition holds, namely

(SH) For any given $u_1,\ldots,u_N \in K$, the characteristic speeds at these points satisfy $\lambda_1(u_1) < \ldots < \lambda_N(u_N)$. Moreover, the eigenvectors $r_1(u_1),\ldots,$ $r_N(u_N)$ are linearly independent.

Observe that the above assumption is certainly satisfied if the system is strictly hyperbolic and K is contained in a small neighborhood of a given point. Moreover, by genuine nonlinearity and assumption (2.1) one can deduce the injectivity of the flux function f.

We introduce here a definition of weak solution to (1.1)-(1.3) which includes an entropy admissibility condition of Oleinik type on the decay of positive waves. In the following, for any given pair of left and right states u_L, $u_R \in U$, we shall denote by $u(t,x) = W(\xi = x/t; u_L, u_R)$ the self-similar solution of the standard Riemann problem for (1.1) with initial data

$$u(0,x) = \begin{cases} u_L & \text{if } x < 0, \\ u_R & \text{if } x > 0, \end{cases}$$

while, for any given boundary state $\widetilde{u}$, we shall define the set $\mathcal{V}(\widetilde{u})$ of *admissible states at the boundary* as

$$\mathcal{V}(\widetilde{u}) \doteq \{W(0+; \widetilde{u}, u_R)|\ u_R \in U\}.$$

Definition 2.2. *A function $u : \Omega \mapsto U$ is an entropy weak solution of* (1.1) – (1.3) *on $\Omega_T := [0,T] \times I\!R^+$, if it is continuous as a function from $]0,T]$ into $\mathbf{L^1}_{loc}$, and the following properties hold:*

(i) u is a weak solution to (1.1) – (1.2), *in the sense that*

$$\int_0^T\!\!\int_0^{+\infty} \big(u(t,x)\cdot\phi_t(t,x) + f(u(t,x))\cdot\phi_x(t,x)\big)dx\ dt + \int_0^{+\infty} \overline{u}(x)\cdot\phi(0,x)dx = 0$$

for every C^1 function ϕ with compact support contained in the set $\{(t,x) \in I\!R^2; x > 0\}$.

(ii) the flux $f(u)$ admits a weak trace at the boundary $x = 0$, i.e. there exists a measurable function $\Psi : [0,\ T] \mapsto I\!R^N$ such that*

$$f(u(\cdot,x)) \ \underset{x\to 0+}{\overset{*}{\rightharpoonup}}\ \Psi \qquad in\ \mathbf{L}^\infty([0,\ T]),$$

and the boundary condition (1.3) *is satisfied in the following sense*

$$\Psi(t) \in f\big(\mathcal{V}(\widetilde{u}(t))\big) \qquad \textit{for a.e.} \quad 0 \le t \le T. \tag{9}$$

(iii) u satisfies the following entropy conditions on the decay of positive waves in time and in space. There exists some constant $C > 0$*, depending only on the system* (1.1)*, so that*

(a) For any $0 < t \le T$*, and for a.e.* $0 < x < y$*, there holds*

$$w_i(t,y) - w_i(t,x) \le C\,\frac{y-x}{t} \qquad \textit{if} \qquad i \in \{1,\dots,N-p\}, \tag{10}$$

$$w_i(t,y) - w_i(t,x) \le C\left\{\frac{y-x}{t} + \log\left(\frac{y}{x}\right)\right\} \; \textit{if } i \in \{N-p+1,\dots,N\}. \tag{11}$$

(b) For a.e. $x > 0$*, and for a.e.* $0 < \tau_1 < \tau_2 \le T$*, there holds*

$$w_i(\tau_2,x) - w_i(\tau_1,x) \le C\,\log\left(\frac{\tau_2}{\tau_1}\right) \qquad \textit{if} \qquad i \in \{1,\dots,N-p\}, \tag{12}$$

$$w_i(\tau_2,x) - w_i(\tau_1,x) \le C\left\{\frac{\tau_2-\tau_1}{x} + \log\left(\frac{\tau_2}{\tau_1}\right)\right\} \; \textit{if } i \in \{N-p+1,\dots,N\}. \tag{13}$$

Remark 2.3. The set of admissible flux values at the boundary $f\big(\mathcal{V}(\widetilde{u})\big)$ can be expressed in Riemann coordinates as

$$f\big(\mathcal{V}(\widetilde{u})\big) = \Big\{f(u)\ ;\ w_j(u) = w_j(\widetilde{u}) \qquad \text{for all } \ j = N-p+1,\dots,N\Big\}.$$

Hence, by the invertibiliy of the map $f : U \mapsto f(U)$, the above boundary condition (2.5) is equivalent to the set of equalities

$$w_j\big(f^{-1}(\Psi(t))\big) = w_j\big(\widetilde{u}(t)\big) \ \text{ for a.e. } \ 0 \le t \le T,\ j = N-p+1,\dots,N. \tag{14}$$

This means that the boundary condition (2.5) guarantees that, at almost every time $t \in [0,T]$, the solution to the Riemann problem for (1.1), having left and right initial states $u^L = \widetilde{u}(t)$, $u^R = f^{-1}(\Psi(t))$, contains only waves with non positive speeds and, in particular, its restriction to the region $[t, +\infty]\times]0,+\infty[$ takes costant value $f^{-1}(\Psi(t))$.

Remark 2.4. The notion of boundary condition considered here avoids the presence of reflected waves, that can not be treated with our technique, which moreover relies on the degenerate quality of Temple systems, and therefore seems unlikely to generalize to other systems. See [15] for other boundary conditions appearing in realistic models.

Due to the presence of the boundary data, the flow map $u(0,\cdot) \mapsto u(t,\cdot)$ is not time homogeneus. To recast the problem in a semigroup framework, it is convenient to incorporate the function $\widetilde{u}$ in the domain of the semigroup. More precisely, consider the positively invariant domain $\mathcal{D}$ of couples $\mathbf{p} = (\overline{u},\widetilde{u})$, where $\overline{u} \in \mathbf{L}^1(I\!R^+,K)$, $\widetilde{u} \in \mathbf{L}^1(I\!R^+,K)$. We define

$$\text{Tot.Var.}(\mathbf{p}) \doteq \text{Tot.Var.}(\overline{u}) + \text{Tot.Var.}(\widetilde{u}) \tag{15}$$

and we introduce the distance

$$d(\mathbf{p}_1, \mathbf{p}_2) \doteq \|\overline{u}_1 - \overline{u}_2\|_{\mathbf{L}^1} + \|f(\widetilde{u}_1) - f(\widetilde{u}_2)\|_{\mathbf{L}^1}. \tag{16}$$

With the above notations, we construct a semigroup S acting on $\mathcal{D}$, in the sense that

$$\begin{array}{rcccccc} S & : & I\!\!R^+ & \times & \mathcal{D} & \mapsto & \mathcal{D} \\ & & t & , & \mathbf{p} & \mapsto & S_t\mathbf{p}, \end{array} \tag{2.13}$$

where, if $\mathbf{p} = (\overline{u}, \widetilde{u})$, $S_t\mathbf{p} = (E_t\mathbf{p}, \mathcal{T}_t\widetilde{u})$. Here $\mathcal{T}_t$ is the translation operator, i.e. $(\mathcal{T}_t\widetilde{u})(s) \doteq \widetilde{u}(t+s)$, while the evolution operator $E : I\!\!R^+ \times \mathcal{D} \mapsto \mathbf{L}^1(I\!\!R^+, K)$ is such that $E_t\mathbf{p} = u(t, \cdot)$, u being the solution to (1.1)-(1.3).

Our main result is concerned with the existence of a semigroup generated by the system (1.1) on the domain $\mathcal{D}$.

Theorem 2.5. *Let the system* (1.1) *be of Temple class, let it satisfy the strict hyperbolicity condition* (SH) *and* (2.1) *and assume that all characteristic fields are genuinely nonlinear. Then there exist a continuous semigroup S of the form* (2.13) *and some constant $C > 0$, depending only on the system and on the domain K, so that, for every fixed $\delta > 0$, letting $L := L_t(\delta) = C(1 + \log(t/\delta))$, one has*

$$\|E_t(\overline{u}, \widetilde{u}) - E_t(\overline{v}, \widetilde{v})\|_{\mathbf{L}^1([\delta,+\infty[)} \leq L_t \cdot \left(\|\overline{u} - \overline{v}\|_{\mathbf{L}^1(I\!\!R^+)} + \|f(\widetilde{u}) - f(\widetilde{v})\|_{\mathbf{L}^1([0,t])}\right) \tag{2.14}$$

for all $(\overline{u}, \widetilde{u}), (\overline{v}, \widetilde{v}) \in \mathcal{D}$, $t \geq \delta$. Moreover, the map $(t, x) \mapsto E_t(\overline{u}, \widetilde{u})(x)$ yields an entropy weak solution to the initial-boundary problem $(1.1) - (1.3)$ *that admits a strong $\mathbf{L}^1$ trace at the boundary $x = 0$, i.e. there exists a measurable map $\psi : I\!\!R^+ \mapsto U$ such that*

$$\lim_{x \to 0+} \int_0^\tau \left|E_t(\overline{u}, \widetilde{u})(x) - \psi(t)\right| dt = 0 \qquad \text{for all } \tau \geq 0. \tag{2.15}$$

Remark 2.6. The uniqueness (up to the domain) of a Lipschitz continuous map as E_t having the property that each trajectory provides an entropy weak solution to $(1.1) - (1.3)$ that admits a strong $\mathbf{L}^1$ at the boundary $x = 0$, and at the initial time $t = 0$, is proved in [4].

3. Outline of the proof

We follow closely the procedure used for the Cauchy problem (see [4, 9] for the details).

By a front tracking algorithm [7, 8], we construct a sequence of semigroups $S^\nu = (E^\nu, \mathcal{T})$ on a domain $\mathcal{D}^\nu$ of piecewise constant functions, whose trajectories are approximate solutions of (1.1).

Next we proceed towards an estimate of the Lipschitz constant for S^ν independent of ν and the total vatiation, that is, for any $\delta > 0$ we look for an estimate of the type

$$\|E_t^\nu \mathbf{p}_1 - E_t^\nu \mathbf{p}_2\|_{\mathbf{L}^1([\delta,+\infty[)} \leq L \cdot d(\mathbf{p}_1, \mathbf{p}_2) \tag{3.1}$$

for some constant L and any $\mathbf{p}_1, \mathbf{p}_2 \in \mathcal{D}^\nu$, $\nu \geq 1$. To this aim, we introduce a class of suitable continuous paths (*pseudopolygonals*), such that any two couples $\mathbf{p}_1, \mathbf{p}_2$ in $\mathcal{D}^\nu$ can be joined by a psudopolygonal γ. By deriving a priori estimates on the $\mathbf{L}^1$-norm of γ, we prove (3.1) for some constant L independent of total variation.

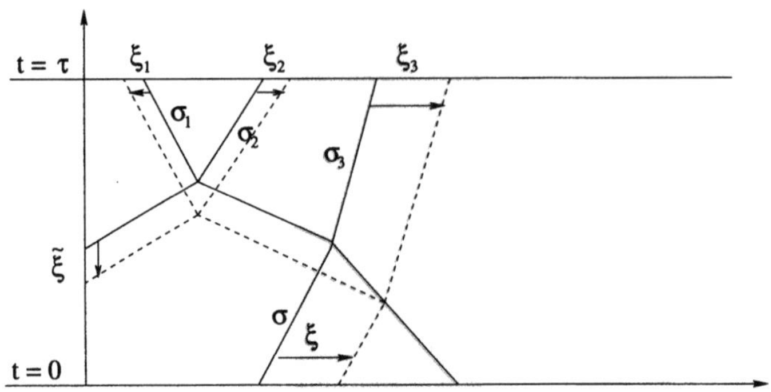

FIGURE 1

Let's fix $\nu \geq 1$ and consider a piecewise constant solution $u(t,\cdot) := E_t^\nu(\overline{u}, \widetilde{u})$ constructed by the front tracking algorithm, for some $(\overline{u}, \widetilde{u}) \in \mathcal{D}^\nu$. We then perturb this solution, shifting the (space) locations x_α of the jumps in $\overline{u}$ at rates ξ_α, and the (time) locations t_α of the jumps in $\widetilde{u}$ at rates $\widetilde{\xi}_\alpha$ (fig. 1). In other words, for θ suitably close to zero, the perturbation $u^\theta(t,\cdot)$ will be a function with jumps at the points $x_\beta^\theta = x_\beta + \theta\xi_\beta$. In the same way, $u^\theta(\cdot, x)$ will have jumps at times $t_\beta^\theta = t_\beta - \theta\widetilde{\xi}_\beta$ with $\widetilde{\xi}_\beta = \xi_\beta/\lambda_{k_\beta}$, where $t_\beta = t_\beta(x)$ denote the locations of all jumps in $u(\cdot,x)$ having nonzero slope λ_{k_β}. As long as the wave-front configuration of the functions u, u^θ is the same, the *space-shifts* $\xi_\beta(t)$ and the *time-shifts* $\widetilde{\xi}_\beta(x)$ are uniquely determined as linear functions of the shifts ξ_α, $\widetilde{\xi}_\alpha$. We collect here some properties of these shift differentials (for the proofs see [4, 9]).

Lemma 3.1. *Consider a bounded, open region Γ in Ω. Call σ_α, $\alpha = 1,\ldots,M$ the fronts entering Γ and let ξ_α be their space-shifts. Assume that the fronts leaving Γ, say σ_β, $\beta = 1,\ldots,M'$, are linearly independent. Then the shifts ξ_β are uniquely determined by the linear relation*

$$\sum_{\alpha=1}^{M} \xi_\alpha \sigma_\alpha = \sum_{\beta=1}^{M'} \xi_\beta \sigma_\beta \,. \tag{3.2}$$

This means that the shift rates of the outgoing fronts depend only on the shift rates of the incoming ones, and not on the order in which these wave-fronts interact inside Γ. More precisely, one can perform the following two operations, without changing the shift rates of the outgoing fronts:

(O1) Switch the order in which three fronts interact (fig. 2).

FIGURE 2

(O2) Invert the order of two fronts at $t = 0$ or at $x = 0$, provided that both fronts have zero shift rate (fig 3).

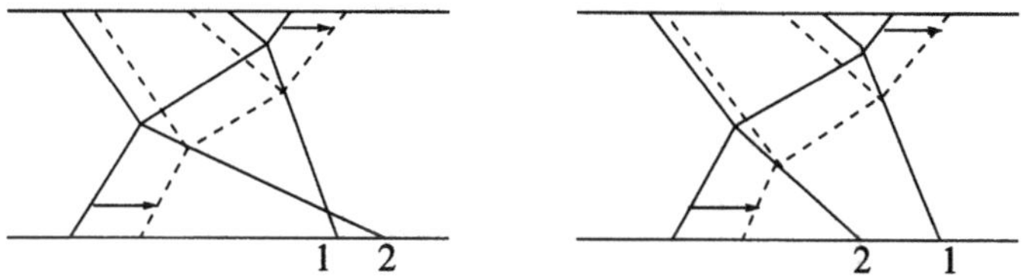

FIGURE 3

These properties will be repeatedly used to get the needed estimates. Indeed, in the calculation of a shift rate, we can suitably alter the order of wave interactions and thus reduce the problem to a case where the wave-front configuration is particularly simple.

Lemma 3.2. *Assume that a front tracking solution u contains two wave-fronts of the same k-th family, originating at distinct points and located at $x_{\alpha'}(t) \leq x_{\alpha''}(t)$, $t \in [t_0,\ t_1]$, $0 \leq t_0 < t_1 \leq T$. Then it is possible to assign shift rates ξ_α, $\widetilde{\xi}_{\alpha'}$ to all fronts in $\overline{u}$ and $\widetilde{u}$ so that $\xi_{\alpha'} = 1$ and moreover, in the corresponding solution u, all fronts $x_\beta(t)$ outside the strip $\Gamma \doteq \{(t,x) \in \Omega;\ t \in [t_0'',\ t_1'],\ x_{\alpha'}(t) \leq x \leq x_{\alpha''}(t)\}$ have zero shift rate.*

In other words, the perturbation of the initial data can be chosen so that one particular front shifts at unit rate, but the corresponding solution remains unaffected outside the region Γ.

Lemma 3.3. *Let u be a front-tracking solution of* (1.1) – (1.3), *and consider two wave-fronts, say $t \mapsto x(t)$, $t \in [\tau_x, T]$, and $t \mapsto y(t)$, $t \in [\tau_y, T]$, originating at two points $(\tau_x, \overline{x})$, $(\tau_y, \overline{y})$ of the boundary of Ω. Then there exists a second front tracking solution $\widehat{u}$ with two fronts $\widehat{x}(\cdot)$, $\widehat{y}(\cdot)$ such that the following holds.*

(i) $\widehat{x}(\tau_x) = \overline{x}$, $\widehat{y}(\tau_y) = \overline{y}$, $\widehat{x}(T) = x(T)$, $\widehat{y}(T) = y(T)$.

(ii) $\widehat{u} = u$ *in a neighborhood of* $(\tau_x, \overline{x})$, $(\tau_y, \overline{y})$, $(T, x(T))$, $(T, y(T))$.

(iii) *Tot. Var.*$\{(\widehat{u}(0,\cdot),\ \widehat{u}(\cdot,0))\} \leq C_1$, *for some constant C_1 depending only on the system* (1.1) *and on the set K.*

Due to genuine nonlinearity, the amount of positive waves in $u(t,\cdot)$ contained in a bounded interval $[a,\ b]$, measured in Riemann coordinates, dacays in time. We have the following result:

Lemma 3.4. *Consider a front tracking solution $u(t,\cdot) = E_t^\nu(\overline{u},\ \widetilde{u})$, with $\overline{u}$, $\widetilde{u}$ containing together at most M shock fronts of the k-th family. Then, there exists some* *constant C_1 depending only on the system* (1.1) *such that, for each $\tau > 0$, and for every bounded interval $[a,\ b]$, $a > 0$, one has:*

$$\textit{Tot. Var.}\big\{w_k(\tau,\cdot);\ [a,\ b]\big\} \le 2C_1\,\frac{b-a}{\tau} + \|w_k\|_{L^\infty} + (M+1)2^{1-\nu}\,, \tag{3.3}$$

for $k = 1,\dots,N-p$;

$$\textit{Tot. Var.}\big\{w_k(\tau,\cdot);\ [a,\ b]\big\} \le 2C_1\left\{\frac{b-a}{\tau} + \log\left(\frac{b}{a}\right)\right\} + \|w_k\|_{L^\infty} + (M+1)2^{1-\nu}\,, \tag{3.4}$$

for $k = N-p+1,\dots,N$.

We can now recover the key estimate on shift differentials.

Lemma 3.5. *Let $u(t,\cdot) = E_t^\nu(\overline{u},\widetilde{u})$ be a front tracking solution, with $\overline{u}$, $\widetilde{u}$ containing together M shocks. Assume that the fronts of $\overline{u}$ located at x_β (respectively, the fronts of $f(\widetilde{u})$ starting at t_β) are shifted with shift rate ξ_β ($\widetilde{\xi}_\beta = \xi_\beta/\lambda_{k_\beta}$ respectively) and amplitude σ_β (or $\widetilde{\sigma}_\beta$). Then there exists a constant C_2 depending only on the system* (1.1) *and on the domain K, such that for any $\delta > 0$, and for every $\tau \ge \delta$, calling $\xi_\alpha(\tau)$, $\sigma_\alpha(\tau)$ the shift rates and the amplitudes of the fronts in $u(\tau,\cdot)$, we have*

$$\sum_{x_\alpha(\tau)>\delta} |\xi_\alpha(\tau)\,\sigma_\alpha(\tau)| \le C_2(1+M2^{-\nu})(1+\log(\tau/\delta))\Big(\sum_\beta |\xi_\beta\,\sigma_\beta| + \sum_\beta |\widetilde{\xi}_\beta\,\widetilde{\sigma}_\beta|\Big). \tag{3.5}$$

4. Trace at the boundary

Since we are dealing with functions of possibly unbounded total variation, the limit of the solution $u(t,x)$ does not exist in general for $x \to 0+$. In order to show that the trajectories of the flow map E_t constructed as limit of front tracking approximations provide solutions with a strong $\mathbf{L}^1$ trace at the boundary $x = 0$, we will make use of the following estimates on shift-differentials of paths of approximate solutions along vertical segments of the domain Ω.

Lemma 4.1. *Under the same assumptions as in Lemma 3.5, there exists some constant C_3 depending only on the system* (1.1) *and on the domain K such that, for any $t_2 > t_1 > 0$, and for every $x > 0$, denoting with $\widetilde{\xi}_\alpha(x)$, $\widetilde{\sigma}_\alpha(x)$ the shift rates and the amplitudes of the fronts in $f(u(\cdot,x))$ having nonzero speed, and crossing the line $\{(t,x)\ ;\ t \ge 0\}$ at time $t_\alpha(x)$, there holds*

$$\sum_{t_\alpha(x)\in[t_1,\ t_2]} |\widetilde{\xi}_\alpha(x)\,\widetilde{\sigma}_\alpha(x)| \le C_3(1+M2^{-\nu})(1+\log(t_2/t_1))\Big(\sum_\beta |\xi_\beta\,\sigma_\beta| + \sum_\beta |\widetilde{\xi}_\beta\,\widetilde{\sigma}_\beta|\Big). \tag{4.1}$$

Let now $u(t,x) \doteq E_t\mathbf{p}(x)$, $\mathbf{p} = (\overline{u},\ \widetilde{u}) \in D$, be the weak solution obtained as limit of front tracking approximations $u^\nu(t,x) \doteq E_t\mathbf{p}^\nu(x)$, where $\mathbf{p}^\nu = (\overline{u}^\nu,\ \widetilde{u}^\nu)$ converges to $\mathbf{p}$ in $\mathbf{L}^1$ as $\nu \to \infty$. Since every $\mathbf{p}^\nu$ is piecewise constant, one can easily verify that any function $u^\nu(t,x)$, $\nu \geq 1$, has bounded total variation and pointwise satisfies the boundary condition (2.10)

$$\lim_{x\to 0+} w_j\big(u^\nu(t,x)\big) = w_j\big(\widetilde{u}^\nu(t)\big) \text{ for a.e. } t \geq 0,\ j = N-p+1,\dots,N. \tag{4.2}$$

Now, fix $\tau_2 > \tau_1 > 0$. By Lemma 4.1 and because of the invertibility property of the flux function f, there will be a countable set $\mathcal{N}$, and some constant $C_4 = C_4(\tau_1,\tau_2) > 0$ (depending only on τ_1,τ_2) such that

$$\Big\| w_j\big(u^\nu(\cdot,x)\big) - w_j\big(u(\cdot,x)\big)\Big\|_{\mathbf{L}^1([\tau_1,\ \tau_2])} \leq C_4 \cdot d(\mathbf{p}^\nu,\mathbf{p}) \tag{4.3}$$

for all $x \in I\!R^+ \setminus \mathcal{N}$, $\nu \geq 1$. Then, (4.2), (4.3) together imply that, for any $j = N-p+1,\dots,N$, the functions $w_j\big(u(\cdot,x)\big)$, $w_j\big(f(u(\cdot,x))\big)$ have a strong limit as $x \to 0$ and

$$\lim_{x\to 0+} \int_{\tau_1}^{\tau_2} \big|w_j\big(u(t,x)\big) - w_j\big(\widetilde{u}(t)\big)\big| dt = 0\,,$$

thus showing that $u(t,x)$ fulfills the boundary condition (2.10). On the other hand, because of the Oleinik type conditions (2.6) on the decay of the positive waves, also $w_j\big(u(\cdot,x)\big)$, $j = 1,\dots,N-p$, have a strong limit as $x \to 0$, which completes the proof of the existence of the strong $\mathbf{L}^1$ trace of $u(t,x)$ at $x = 0$, and hence concludes the proof of the theorem.

5. Properties of the attainable sets for Temple class systems with integrable boundary control

Following [5, 6], we turn now to study the mixed initial-boundary value problem (1.1)-(1.3) from the point of view of control theory, taking a fixed initial data $\overline{u} \in \mathbf{L}^1(I\!R^+,K)$ and considering, in connection with a prescribed set $\mathcal{U} \subseteq \mathbf{L}^1(I\!R^+,K)$ of boundary data regarded as admissible controls, the *attainable sets* for (1.1)-(1.3):

$$\mathcal{A}(T,\mathcal{U}) \doteq \Big\{E_T(\overline{u},\widetilde{u})(\cdot)\ ;\ \widetilde{u}\in\mathcal{U}\Big\},\quad \mathcal{A}(\overline{x},\mathcal{U}) \doteq \Big\{E_{(\cdot)}(\overline{u},\widetilde{u})(\overline{x})\ ;\ \widetilde{u}\in\mathcal{U}\Big\},$$

i.e. the sets of all profiles that can be attained at a fixed time $T > 0$, or at a fixed point in space $\overline{x} > 0$, by entropy weak solutions of (1.1)-(1.3) with initial data $\overline{u}$, and boundary data $\widetilde{u}$ that vary in $\mathcal{U}$. Relying on the well-posedness theory provided by the above results, we establish the compactness of $\mathcal{A}(T,\mathcal{U})$, $\mathcal{A}(\overline{x},\mathcal{U})$ for a class $\mathcal{U}$ of admissible boundary controls that satisfy convex constraints.

Theorem 5.1. *Let K be a set of the form (2.4), and $J \subseteq \{1,\dots,N\}$ a set of indices such that $J \supseteq \{N-p,\dots,N\}$. Define*

$$\mathcal{U} \doteq \Big\{\widetilde{u}\in\mathbf{L}^1(I\!R^+,K)\ ;\ w_j(\widetilde{u}(t)) \in [c_j,\ d_j],\quad \textit{for a.e. } t\geq 0,\quad \textit{for all } j\in J\Big\},$$

for some $-\infty < c_j \leq d_j < +\infty$, $j \in J$. Then $\mathcal{A}(T,\mathcal{U})$, $T > 0$, and $\mathcal{A}(\overline{x},\mathcal{U})$, $\overline{x} > 0$, are compact subsets of $\mathbf{L}^1{}_{loc}(I\!R^+,K)$.

Acknowledgments

The results presented here have been completed in all the details in collaboration with Fabio Ancona in [4].

References

[1] D. Amadori, Initial-boundary value problems for nonlinear systems of conservation laws, *NoDEA* **4** (1997), 1-42.

[2] D. Amadori and R. M. Colombo, Continuous dependence for 2×2 conservation laws with boundary, *J. Differential Equations* **138** (1997), no. 2, 229-266.

[3] D. Amadori and R. M. Colombo, Characterization of viscosity solutions for conservation laws with boundary, *Rend. Sem. Mat. Univ. Padova* **99** (1998), 219-245.

[4] F. Ancona and P. Goatin, Uniqueness and stability of $\mathbf{L}^\infty$ solutions for Temple class systems with boundary and properties of the attainable sets, submitted.

[5] F. Ancona and A. Marson, On the attainable set for scalar non-linear conservation laws with boundary control, *SIAM Journal on Control and Optimization* **36** (1998), no. 1, 290-312.

[6] F. Ancona and A. Marson, Scalar non-linear conservation laws with integrable boundary data, *Nonlinear Anal.* **35** (1999), 687-710.

[7] P. Baiti and A. Bressan, The semigroup generated by a Temple class system with large data, *Differ. Integ. Equat.* **10** (1997), 401-418.

[8] A. Bressan, Global solutions of systems of conservation laws by wave-front tracking, *J. Math. Anal. Appl.* **170** (1992), 414-432.

[9] A. Bressan and P. Goatin, Stability of $\mathbf{L}^\infty$ solutions of Temple class systems, *Differ. Integ. Equat.*, **13** (10-12) (2000), 1503-1528.

[10] F. Dubois and P.G. LeFloch, Boundary conditions for non-linear hyperbolic systems of conservation laws, *J. Differential Equations* **71** (1988), 93-122.

[11] K.T. Joseph and P.G. LeFloch, Boundary layers in weak solutions of hyperbolic conservation laws, *Arch. Rational Mech. Anal.* **147** (1999), 47-88.

[12] B.L.. Keyfitz, Solutions with shocks, *Comm. Pure Appl. Math.* **24** (1971), 125-132.

[13] H.O. Kreiss, Initial-boundary value problems for hyperbolic systems, *Comm. Pure Appl. Math.* **23** (1970), 277-298.

[14] P.G. LeFloch, Explicit formula for scalar non-linear conservation laws with boundary condition, *Math. Methods Appl. Sci.* **10** (1988), 265-287.

[15] D. Serre, *Systemes de Lois de Conservation*, Diderot Editeur, 1996.

[16] M. Sablé-Tougeron, Méthode de Glimm et probléme mixte, *Ann. Inst. Henri Poincaré* **10**, no. 4, (1993), 423-443.

P. Goatin
Centre de Mathématiques Appliquées
& CNRS, U.M.R. 7641, Ecole Polytechnique,
91128 Palaiseau Cedex, France
goatin@cmap.polytechnique.fr

International Series of Numerical Mathematics
Vol. 140, © 2001 Birkhäuser Verlag Basel/Switzerland

Linear Stability of Shock Profiles for Systems of Conservation Laws with Semi-linear Relaxation

Pauline Godillon

Abstract. The Evans function theory, which has recently been applied to the study of linear stability of viscous shock profiles, is developed below for semi-linear relaxation. We study the linear stability of shock profiles in the case of an extreme Lax shock. The results we obtain are similar to those found for viscous approximations by Gardner and Zumbrun [5].

1. Introduction

Let $n \geq 2$ and consider a $n \times n$ hyperbolic system of conservation laws

$$u_t + f(u)_x = 0, \;\; x \in \mathbf{R}\,, t \geq 0, \tag{1}$$

where f is a smooth flux. The semi-linear relaxation approximation is obtained by introducing a stiff source term to (1) and closure laws :

$$\left\{ \begin{array}{rcl} u_t + v_x & = & 0, \\ v_t + \mathsf{a}^2 u_x & = & \dfrac{1}{\tau}(f(u) - v), \end{array} \right. \tag{2}$$

where τ is the (positive) relaxation parameter and a is a positive constant. Taking the formal limit of (2) as τ tends to 0, one obtains the *local equilibrium*, $v = f(u)$, and the *equilibrium system*, which is in fact the original system (1). Performing an asymptotic Chapman-Enskog type expansion [9] on (2), one finds formally

$$u_t + f(u)_x = \tau((\mathsf{a}^2 - \mathrm{d}f(u)^2)u_x)_x + O(\tau^2). \tag{3}$$

To ensure the stability, (3) must be dissipative, that is the viscosity must be positive. So the classical assumption, known as the *subcharacteristic condition* [8] is

$$\mathsf{a} > \rho(\mathrm{d}f(u)), \; \forall\, u, \tag{4}$$

where ρ denotes the spectral radius. This condition is equivalent to the fact that the characteristics of the equilibrium system (1) are subcharacteristic to the characteristics $\pm\mathsf{a}$ of the system (2). This model was introduced by Jin and Xin [8] in order to obtain non-oscillatory schemes for systems of conservation laws. Serre [11] proved that the solutions of (2) converge, as τ tends to 0, towards an entropy solution of (1).

We consider a non-characteristic discontinuity $(u^-, u^+; s)$ of System (1) of *arbitrary strength* satisfying the Rankine-Hugoniot condition. We assume that the speed s is subcharacteristic in order to ensure, along with further conditions on the nature of the discontinuity, the existence of a shock profile $\bar{U}_\tau = (\bar{u}_\tau, \bar{v}_\tau)^T$ for (2) asymptotically connecting the end states $(u^-, f(u^-))^T$ and $(u^+, f(u^+))^T$ with speed s [9]. Our subcharacteristic condition thus reads

$$\mathsf{a} > \max(\rho(\mathrm{d}f(u^-)), \rho(\mathrm{d}f(u^+)), |s|).$$

We also assume that (1) is strictly hyperbolic at the end states $u^\pm$.

We aim here to obtain (computable) necessary conditions of linear stability of extreme Lax shock profiles by means of an Evans function [3] in the same way as it has been used for the viscous approximation [5, 2] and for semi-discrete shock profiles [1].

2. Background

Assumptions Let $(u^-, u^+; s)$ be a discontinuity for (1) satisfying the Rankine-Hugoniot condition

$$f(u^+) - f(u^-) = s(u^+ - u^-). \tag{5}$$

We make the following assumptions :

H 1. *the subcharacteristic condition holds :* $\mathsf{a} > \max(\rho(\mathrm{d}f(u^-)), \rho(\mathrm{d}f(u^+)), |s|)$,

H 2. *System (1) is strictly hyperbolic at* $u^\pm$*, that is,* $\mathrm{d}f(u^\pm)$ *has* distinct *real eigenvalues, that we denote by* $\alpha_1^\pm < \ldots < \alpha_n^\pm$,

H 3. *the discontinuity is non-characteristic :* $s \notin \mathrm{Sp}(\mathrm{d}f(u^\pm))$,

H 4. *the discontinuity* $(u^-, u^+; s)$ *is a Lax n-shock :*

$$\begin{array}{ccccc} \alpha_n^+ & < & s, & & \\ \alpha_{n-1}^- & < & s & < & \alpha_n^-. \end{array}$$

This last condition means that there are $n+1$ characteristics entering the shock and $n-1$ outgoing ones. For further details, see [10].

Shock profiles A shock profile $\bar{U}_\tau = (\bar{u}_\tau, \bar{v}_\tau)^T : \xi = x - st \mapsto (\bar{u}_\tau(\xi), \bar{v}_\tau(\xi))^T$ for $(u^-, u^+; s)$ is a traveling wave of (2) asymptotically connecting $U^- := (u^-, v^-)^T$ and $U^+ := (u^+, v^+)^T$ with speed s. This profile obeys

$$\begin{cases} -s\bar{u}_\tau' + \bar{v}_\tau' = 0, \\ -s\bar{v}_\tau' + \mathsf{a}^2\bar{u}_\tau' = \dfrac{1}{\tau}(f(\bar{u}_\tau) - \bar{v}_\tau), \\ \lim\limits_{x\to\pm\infty} \bar{u}_\tau(\xi) = u^\pm, \quad \lim\limits_{x\to\pm\infty} \bar{v}_\tau(\xi) = v^\pm. \end{cases} \tag{6}$$

that is, integrating and combining the equations of (6),

$$\begin{cases} \bar{u}_\tau' = \dfrac{1}{\tau\zeta^2}(f(\bar{u}_\tau) - \bar{f} - s\bar{u}_\tau), \\ \bar{v}_\tau = s\bar{u}_\tau + \bar{f}, \\ \lim\limits_{\xi\to\pm\infty} \bar{u}_\tau(\xi) = u^\pm, \quad \lim\limits_{\xi\to\pm\infty} \bar{v}_\tau(\xi) = f(u^\pm). \end{cases} \tag{7}$$

The end states $u^\pm$ are hyperbolic in the ODE sense. Indeed, $\mathsf{a} > |s|$ because of **(H1)** and the eigenvalues of $A^\pm := \mathrm{d}f(u^\pm) - sI_n$ are non-zero because of **(H2)**. Consequently, $\zeta^{-2}A^\pm$ has non-zero, distinct, real eigenvalues.

Remark 2.1.

- *Note that $\bar{U} : \xi \mapsto \bar{U}_\tau(\xi/\tau)$ does not depend on τ since $\bar{U} := (\bar{u}, \bar{v})^T$ satisfies*

$$\begin{cases} \zeta^2\bar{u}' = f(\bar{u}) - \bar{f} - s\bar{u}, \\ \bar{v} = s\bar{u} + \bar{f}, \\ \lim\limits_{\xi\to\pm\infty} \bar{u}(\xi) = u^\pm, \quad \lim\limits_{\xi\to\pm\infty} \bar{v}(\xi) = f(u^\pm). \end{cases} \tag{8}$$

  Moreover, taking $\tau = 1$ in System (7), we find System (8).
- *The existence of a semi-linear relaxation profile is equivalent to the existence of a viscous shock profile with scalar viscosity $B = \zeta^2 I_n$. For* weak Lax shocks, *there actually are such profiles (see* [4, 10]*).*

Since we can derive a profile $\bar{U}_\tau$ of (7) quite easily from a profile $\bar{U}$ of (8) (Remarks 2.1), we now make the additional assumption

H 5. *there exists a profile $\bar{U} = (\bar{u}, \bar{v})^T$ for (8) asymptotically connecting $(u^-, f(u^-))^T$ and $(u^+, f(u^+))^T$ with speed s.*

As far as the asymptotic behavior of $\bar{U}$ at $x = +\infty$ is concerned, there is only one negative eigenvalue of A^+, that we denote a_n^+, such that

$$\lim_{x\to+\infty} e^{-a_n^+ x/\zeta^2}\bar{u}'(x) \neq 0,$$

that is, at $x = +\infty$, a_n^+/ζ^2 is the decay rate of $\bar{u}$, which is tangent to an eigenvector r_n^+ of A^+ associated with a_n^+. Furthermore, let $(r_1^+, \dots, r_n^+)$ a basis of eigenvectors of A^+ associated with $\{a_k^+\}_{k\in\{1,\dots,n\}}$. We stress the fact that the eigenvalues $(a_k^+)_{k\in\{1,\dots,n\}}$ are not ordered in increasing order to cover all the cases of decaying rate of $\bar{u}$. At $x = -\infty$, $\bar{u}$ is tangent to an eigenvector r_n^- of A^- associated with $a_n^- := \alpha_n^-$ and the decay rate of $\bar{u}$ is a_n^-/ζ^2.

3. Linear stability

Let us linearize System (2) about $\bar{U}_\tau$:

$$\begin{cases} u_t + v_x = 0, \\ v_t + \mathsf{a}^2 u_x = \dfrac{1}{\tau}(\mathrm{d}f(\bar{u}_\tau)u - v), \end{cases}$$

that is

$$\begin{pmatrix} u \\ v \end{pmatrix}_t = -\begin{pmatrix} 0_n & I_n \\ a^2 I_n & 0_n \end{pmatrix}\begin{pmatrix} u \\ v \end{pmatrix}_x + \frac{1}{\tau}\begin{pmatrix} 0_n & 0_n \\ df(\bar{u}_\tau) & -I_n \end{pmatrix}\begin{pmatrix} u \\ v \end{pmatrix}$$

$$=: L_\tau(x, \partial_x, \bar{U}_\tau(x))\begin{pmatrix} u \\ v \end{pmatrix}.$$

The profile $\bar{U}_\tau$ of (2) is **linearly stable** if the differential operator L_τ has no spectrum in the right half-plane. Let P_τ be such that $P_\tau(U) = U(\cdot/\tau)$ for all functions U. Note that $L_\tau \circ P_\tau = \tau^{-1} P_\tau \circ L_1$ (see Remarks 2.1). As we are only interested in the real part of the spectrum of L_τ, we set $\tau = 1$ and we study the spectrum of L_1.

Theorem 3.1. *For a Lax shock, given* **(H1)-(H5)**, *a necessary condition of linear stability of $\bar{U}$ is*

$$\det(r_1^-, \dots, r_n^-) \cdot \det(r_1^-, \dots r_{n-1}^-, [u]) > 0. \tag{9}$$

Remark 3.2.
- *Note that correctly orienting the eigenvectors that do not appear an even number of times in the formula (9) is crucial ! Here, these orientations are given in our study of the asymptotic behavior of $\bar{U}$.*
- *Gardner and Zumbrun [5] obtained a similar condition for viscous Lax shock profiles of 2×2 systems and they conjectured the condition that Benzoni-Gavage, Serre and Zumbrun [2] proved in the $n \times n$ case.*
- *Condition (9) can be interpreted as follows :* *let $\Pi = \mathrm{span}\{r_1^-, \dots, r_{n-1}^-\}$. The hyperplane Π divides $\mathbf{R}^n$ into two halves. If the profile $\bar{U}$ is linearly stable, then the segment $[u^-, u^+]$ and r_n^-, which is tangent to $\bar{u}$ at $-\infty$, are in the same half-space of $\mathbf{R}^n$.*

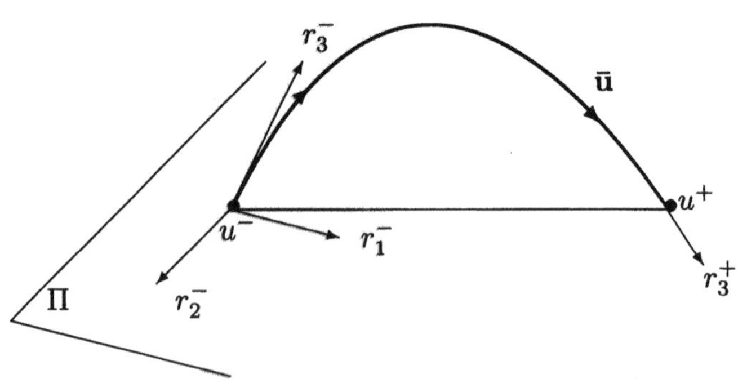

FIGURE 1. A Lax 3-shock satisfying Condition (9) for $n = 3$

4. Proof

4.1. Stationary shocks

Recall that we have set $\tau = 1$. Appropriately choosing the coordinate frame, we change to a stationary shock. Let $\xi = x - st$, $\tilde{u}(\xi, t) := u(x,t)$ and $\tilde{v}(\xi,t) := v(x,t)$. System (2) is equivalent to

$$\begin{cases} \tilde{u}_t - s\tilde{u}_\xi + \tilde{v}_\xi = 0, \\ \tilde{v}_t - s\tilde{v}_\xi + \mathsf{a}^2\tilde{u}_\xi = f(\tilde{u}) - \tilde{v}, \end{cases}$$

that is,

$$\begin{cases} \tilde{u}_t + \tilde{\tilde{v}}_\xi = 0, \\ \tilde{\tilde{v}}_t + s\tilde{u}_t + \zeta^2\tilde{u}_\xi - s\tilde{\tilde{v}}_\xi = \tilde{f}(\tilde{u}) - \tilde{\tilde{v}}, \end{cases} \tag{10}$$

where $\tilde{\tilde{v}} = \tilde{v} - s\tilde{u} - \bar{f}$ and $\tilde{f}(\tilde{u}) = f(\tilde{u}) - s\tilde{u} - \bar{f}$.

Remark 4.1. 1. *Since $(a_k^{\pm})_{k\in\{1,\dots,n\}}$ are the eigenvalues of $\mathrm{d}\tilde{f}(u^{\pm})$,* **(H1)** *equivalently reads $\forall k \in \{1,\dots,n\}$, $|a_k^{\pm} + s| < \mathsf{a}$ and $|s| < \mathsf{a}$.*
2. *The change in v induces a change for the profile : $\bar{\tilde{v}}(\xi) = 0$, $\forall \xi$.*

Combining the two equations of System (10), dropping tildes and renaming ξ in x, we shall now consider the following system

$$\begin{cases} u_t + v_x = 0, \\ v_t + \zeta^2 u_x - 2sv_x = f(u) - v. \end{cases} \tag{11}$$

Linearizing (11) about the profile $\bar{U} = (\bar{u}, 0)^T$, we obtain

$$\begin{cases} u_t + v_x = 0, \\ v_t + \zeta^2 u_x - 2sv_x = \mathrm{d}f(\bar{u})u - v, \end{cases}$$

that is

$$\begin{pmatrix} u \\ v \end{pmatrix}_t = \left(-\begin{pmatrix} 0_n & I_n \\ \zeta^2 I_n & -2sI_n \end{pmatrix}\partial_x + \begin{pmatrix} 0_n & 0_n \\ \mathrm{d}f(\bar{u}) & -I_n \end{pmatrix}\right)\begin{pmatrix} u \\ v \end{pmatrix} =: M\begin{pmatrix} u \\ v \end{pmatrix}.$$

The (unbounded) operator $M : D(M) \subset L^2(\mathbf{R}, \mathbf{R}^{2n}) \to L^2(\mathbf{R}, \mathbf{R}^{2n})$ is closed with dense domain $D(M) = H^1(\mathbf{R}, \mathbf{R}^{2n})$.

4.2. The essential spectrum of M

In order to be able to construct an analytic Evans function in the open right half-plane, we check that the spectrum of M in the right half-plane consists only of eigenvalues. Following [5], the essential spectrum of M, $\sigma_{ess}(M)$, is the complementary set of the union of the resolvent set of M and of the set of the isolated eigenvalues of M of finite multiplicity. Using Fourier techniques and a key theorem by Henry [7], it is straightforward [6] to prove that the essential spectrum of M lies in $\{\mathrm{Re}(\lambda) < 0\} \cup \{0\}$.

4.3. Definition of an Evans function

The eigenvalue equation associated with the differential operator M is $M\phi = \lambda\phi$, that can be rewritten as a first-order ODE with variable coefficients

$$\phi' = \frac{1}{\zeta^2}\begin{pmatrix} df(\bar u) - 2s\lambda I_n & -(\lambda+1)I_n \\ -\lambda\zeta^2 I_n & 0_n \end{pmatrix}\phi =: \mathbf{A}(x,\lambda)\phi. \tag{12}$$

The matrix $\mathbf{A}$ is clearly analytic in λ and $\mathbf{C}^\infty$ in x because f is smooth and admits limits $\mathbf{A}^\pm(\lambda)$ at $x = \pm\infty$. An Evans function D associated with the differential operator M is a Wronskian of solutions of (12), chosen so as to ensure that the zeroes of D in the open right half-plane correspond to unstable eigenvalues of M. Moreover, since we proved that M has no essential spectrum in the open right half-plane, $D : \lambda \mapsto D(\lambda)$ is analytic in $\mathrm{Re}(\lambda) > 0$. We then apply R. Gardner and K. Zumbrun's *gap lemma* [5] that allows to extend a Wronskian of solutions of (ODE) System (12), analytic in the open right-half plane, to a neighborhood of 0. For real λ, System (12) is real, so we can set on D to be real. Since M has only eigenvalues in the open right half-plane, we derive a necessary condition of linear stability of $\bar U$ from the fact that D must have same signs in a neighborhood of $\lambda = 0$ and at $\lambda = +\infty$ for positive real λ. But, at $\lambda = 0$, $\bar U'$ is a solution of $\phi' = \mathbf{A}(x,0)\phi$ which decays at $x = \pm\infty$, so D vanishes at $\lambda = 0$. Since the gap lemma allows us to compute $D'(0)$ which has the same sign as D about $\lambda = 0$, our necessary condition reads

$$D'(0).D(+\infty) > 0.$$

4.4. Sign of D about $\lambda = 0$

About $\lambda = 0$, the eigenvalues and eigenvectors associated with the stable (respectively unstable) subspace of $\mathbf{A}^+(\lambda)$ (resp. $\mathbf{A}^-(\lambda)$) are

$$\forall k \in \{1,\dots,n\},\quad \mu_k^+(0) = a_k^+/\zeta^2,\quad V_k^+(0) = (r_k^+, 0)$$

$$\left(\text{resp.}\quad \begin{array}{lll} \forall k \in \{1,\dots,n-1\}, & \mu_{n+k}^- \sim -\lambda/a_k^-, & V_{n+k}^-(0) = (r_k^-, a_k^- r_k^-)^T, \\ & \mu_{2n}^+(0) = a_n^-/\zeta^2, & V_{2n}^-(0) = (r_n^+, 0). \end{array}\right)$$

We then apply the gap lemma : there are $2n$ functions $\phi_1, \dots, \phi_{2n}$ of (x,λ), depending analytically on λ, which satisfy the equation $\phi' = \mathbf{A}(x,\lambda)\phi$ and whose asymptotic behavior in x is

$$\begin{aligned} \phi_k(x,\lambda) &\underset{x\to+\infty}{=} e^{\mu_k^+(\lambda)x}\left(V_k^+(\lambda) + O\left(e^{-\gamma|x|}\right)\right), \\ \phi_{n+k}(x,\lambda) &\underset{x\to-\infty}{=} e^{\mu_{n+k}^-(\lambda)x}\left(V_{n+k}^-(\lambda) + O\left(e^{-\gamma|x|}\right)\right). \end{aligned} \tag{13}$$

The Evans function about $\lambda = 0$ reads :

$$D(\lambda) := e^{-\int_0^x \mathrm{tr}\mathbf{A}(\sigma,\lambda)d\sigma}\det(\phi_1(x,\lambda), \dots, \phi_{2n}(x,\lambda)).$$

The derivative $\bar U'$ satisfies $\bar U'' = \mathbf{A}(\cdot,0)\bar U'$, so we choose, at $\lambda = 0$, $\phi_n(\cdot,0) = \phi_{2n}(\cdot,0) := \bar U'$, referring to the asymptotic behavior of the eigenfunctions ϕ given by (13). Since $\mathbf{A}(\sigma,0)$ is real-valued, $\exp\left(-\int_0^x \mathrm{tr}(\mathbf{A}(\sigma,0))d\sigma\right)$ is positive. So

$$\mathrm{sgn}(D'(0)) = \mathrm{sgn}\left(\det(\phi_1(\cdot,0), \dots, \phi_{2n-1}(\cdot,0), (\partial_\lambda\phi_{2n} - \partial_\lambda\phi_n)(\cdot,0))\right).$$

Denoting $\phi_k(\cdot,0) = (u_k, v_k)^T$, $k \in \{1,\dots,n\}$, we have $v_k' = 0$ and, consequently, $v_k = 0$ for $k \in \{1,\dots,n\} \cup \{2n\}$, and $v_k = a_{k-n}^- r_{k-n}^-$ for $k \in \{n+1,\dots,2n-1\}$ because of the asymptotic behavior of v_k (13).
Let $z_k = \partial_\lambda \phi_k(\cdot,0) = (w_k, t_k)$, $k \in \{n, 2n\}$. Since ϕ_k and System (12) depend analytically on λ, z_k is a solution of

$$z' = \mathbf{A}(\cdot,0)z - \frac{1}{\zeta^2}\begin{pmatrix} 2sI_n & I_n \\ \zeta^2 I_n & 0_n \end{pmatrix}\phi(\cdot,0).$$

Consequently, $t_k' = -\bar{u}'$. Integrating from x to $+\infty$ (respectively from $-\infty$ to x), we find $t_j = u^+ - \bar{u}$ (resp. $t_{n+j} = u^- - \bar{u}$). Consequently, $t_{n+j} - t_j = -[u]$. Thus

$$\begin{aligned} \operatorname{sgn}(D'(0)) &= \operatorname{sgn}\left(\det\begin{pmatrix} u_1 & \dots & u_n & u_{n+1} & \dots & u_{2n-1} & w_{2n} - w_n \\ 0 & \dots & 0 & a_1^- r_1^- & \dots & a_{n-1}^- r_{n-1}^- & -[u] \end{pmatrix}\right) \\ &= (-1)^n \operatorname{sgn}\left(\det(u_1, \dots, u_{2n}) \cdot \det(r_1^-, \dots, r_{n-1}^-, [u])\right). \end{aligned}$$

Since $\operatorname{sgn}(D'(0))$ does not depend on x, we find that

$$\operatorname{sgn}(D'(0)) = (-1)^n \cdot \operatorname{sgn}(\det(r_1^+, \dots, r_{n-1}^+, \bar{u}') \cdot \det(r_1^-, \dots, r_{n-1}^-, [u])).$$

4.5. Large $|\lambda|$ behavior

Lemma 4.2. *There is a positive real Λ such that*

$$\forall \lambda \in (\Lambda + \infty),\ \operatorname{sgn}(D(\lambda)) = (-1)^n \cdot \operatorname{sgn}(\det(r_1^+, \dots, r_n^+) \cdot \det(r_1^-, \dots, r_n^-)).$$

Proof : We use a homotopy method [2] to join M and M_0, defined as

$$M_0\begin{pmatrix} u \\ v \end{pmatrix} := \begin{pmatrix} 0_n & -I_n \\ -\zeta^2 I_n & -2sI_n \end{pmatrix}\begin{pmatrix} u \\ v \end{pmatrix}_x.$$

Let θ be in $[0,1]$ and M_θ be $M_\theta := \theta M + (1-\theta)M_0$. Using the same argument as for M, one proves that there exists $\Lambda > 0$ such that

$$\forall \lambda \in (\Lambda, +\infty), \forall \theta \in [0,1],\ M_\theta - \lambda \text{ is invertible.}$$

So, defining a continuation of the Evans function $\tilde{D}$ to the variables (θ, λ) associated with $\mathbf{A}(x,\theta,\lambda)$ that we derive from M_θ in the same way as we derived $\mathbf{A}$ from M, we find that $\tilde{D}$ does not vanish in $\mathbf{S} = [0,1] \times (\Lambda, +\infty)$, so its sign is constant on $\mathbf{S}$ and

$$\operatorname{sgn}(D(+\infty)) = \operatorname{sgn}(\tilde{D}(1,+\infty)) = \operatorname{sgn}(\tilde{D}(0,+\infty)).$$

The eigenvalues of $\mathbf{A}(x,0,\lambda)$ are $\nu_- = \lambda/(\mathsf{a}+s)$ with multiplicity n and $\nu_+ = \lambda/(s-\mathsf{a})$ with multiplicity n too. Thanks to Assumption **(H3)**, we know that $\nu_- > 0$ and $\nu_+ < 0$. So, for $k \in \{1,\dots,n\}$,

$$\phi_k(x,0,\lambda) = e^{(\nu_+)x}\begin{pmatrix} r_k^+ \\ -\frac{\lambda}{\nu_+} r_k^+ \end{pmatrix} \text{ and } \phi_{n+k}(x,0,\lambda) = e^{(\nu_-)x}\begin{pmatrix} r_k^- \\ -\frac{\lambda}{\nu_-} r_k^- \end{pmatrix}.$$

But $\{(r_1^+, -\lambda r_1^+/\nu_+)^T, \ldots, (r_n^+, -\lambda r_n^+/\nu_+)^T\}$ and $\{(r_1^-, -\lambda r_1^-/\nu_-)^T, \ldots, (r_n^-, -\lambda r_n^-/\nu_-)^T\}$ have the same orientation as $\{V_1^+(\lambda), \ldots, V_n^+(\lambda)\}$ and $\{V_{n+1}^-(\lambda), \ldots, V_{2n}^-(\lambda)\}$ respectively, because these bases have been chosen continuously with respect to λ. We thus obtain, for some $\lambda_0 > \Lambda$,

$$\operatorname{sgn}(\tilde{D}(0, +\infty)) = \operatorname{sgn}\left(\lambda_0^n \left(\frac{\nu_- - \nu_+}{\nu_-\nu_+}\right)^n \cdot \det(r_1^+, \ldots, r_n^+) \cdot \det(r_1^-, \ldots, r_n^-)\right).$$

So, $\operatorname{sgn}(\tilde{D}(0, +\infty)) = (-1)^n \operatorname{sgn}(\det(r_1^+, \ldots, r_n^+) \cdot \det(r_1^-, \ldots, r_n^-))$. As the bases do not twist along the real axis from $\lambda = 0$ to $\lambda = \lambda_0$ at $\theta = 1$,

$$\operatorname{sgn}(D(+\infty)) = (-1)^n \cdot \operatorname{sgn}(\det(r_1^+, \ldots, r_n^+) \cdot \det(r_1^-, \ldots, r_n^-)).$$

References

[1] S. Benzoni-Gavage, *Stability of semi-discrete shock profiles by means of an Evans function in infinite dimensions,* Preprint ENS Lyon, (1999).

[2] S. Benzoni-Gavage, D. Serre, and K. Zumbrun, *Alternate Evans functions and viscous shock waves,* SIAM J. Math. Anal., **32**(2001), 929–962.

[3] J. W. Evans, *Nerve axon equations I. Linear approximations,* Indiana Univ. Math. J., **21** (1971/72), 877–885.

[4] L. R. Foy, *Steady state solutions of hyperbolic systems of conservation laws with viscosity terms,* Comm. Pure Appl. Math., **17** (1964), 177–188.

[5] R. A. Gardner and K. Zumbrun, *The gap lemma and geometric criteria for instability of viscous shock profiles,* Comm. Pure Appl. Math., **51(7)** (1998), 797–855.

[6] P. Godillon, *Linear stability of shock profiles for systems of conservation laws with semi-linear relaxation,* Physica D, **148** (2001), 289–316.

[7] D. Henry, *Geometric theory of semilinear parabolic equations. Lecture Notes in Mathematics* **840** 1981, Springer-Verlag, Berlin.

[8] S. Jin and Z. P. Xin, *The relaxation schemes for systems of conservation laws in arbitrary space dimensions*, Comm. Pure Appl. Math., **48(3)** (1995), 235–276.

[9] T.-P. Liu, *Hyperbolic conservation laws with relaxation*, Comm. Math. Phys., **108(1)** (1987), 153–175.

[10] D. Serre, *Systems of conservation laws. 1,* 1999, Cambridge University Press, Cambridge. Hyperbolicity, entropies, shock waves, Translated from the 1996 French original by I. N. Sneddon.

[11] D. Serre, *Relaxations semi-linéaire et cinétique des systèmes de lois de conservation,* Ann. IHP, Anal. non-linéaire, **17** (2000), 169–192.

P. Godillon
Unité de Mathématiques Pures et Appliquées (CNRS UMR # 5669),
ENS Lyon, 46 Allée d'Italie, 69364 Lyon Cédex 07
`Pauline.Godillon@umpa.ens-lyon.fr`

International Series of Numerical Mathematics
Vol. 140, © 2001 Birkhäuser Verlag Basel/Switzerland

A Nonconservative Numerical Approach for Hyperbolic Systems with Source Terms: The Well-Balanced Schemes

Laurent Gosse

Abstract. We present an introduction to the Well-Balanced schemes in both cases of scalar and systems of nonhomogeneous balance laws. Our construction is based on a reformulation of the source terms as nonconservative products. Some simple self-similar Riemann solvers and thus most Godunov-type schemes can be deduced. We provide also a few relevant computational results.

1. Introduction

The purpose of these notes is to provide an easy introduction to the field of Well-Balanced numerical schemes in the sense of [13] for systems of nonhomogeneous balance laws. This class of algorithms has been introduced first for 1D scalar equations in order to fix two problems inherent to classical approaches (method of lines, time-splitting ...). The degradation of performances experienced by these conventional numerical processes in presence of source terms comes mostly from the fact that they do not meet the following requirements:

(i) a stability criterion which does not depend on the *stiffness* of the source term, see *e.g.* [8],

(ii) a good ability to preserve static time-asymptotic solutions with possibly very high spatial gradients.

We propose here a construction based on a nonconservative reformulation of the nonhomogeneous problems which permits to derive a wide class of Riemann-based numerical schemes relying on simple *self-similar* solvers. The structure of these solvers ensures that both properties (i), (ii) should hold on reasonably coarse computational grids. From now on, we assume working on a cartesian space-time discretization which is determined by the parameters $h, \Delta t$. For related works, we refer *e.g.* to [2, 3, 4, 15, 17, 18, 22, 23, 25] and the quotations therein; see also [12] for a first study in the special context of a relaxation system.

Partially supported by the European Union TMR project HCL #ERBFMRXCT960033.

2. General source terms in 1D scalar balance laws

We consider the following Cauchy problem:

$$\begin{cases} \partial_t u + \partial_x f(u) = g(x,u), \qquad x \in \mathbb{R}, t > 0 \\ f \in C^2(\mathbb{R}), f'' > 0,\ u \mapsto g(x,u) \in C^1(\mathbb{R}) \text{ and } x \mapsto g(x,u) \in C^1_c(\mathbb{R}) \\ g(.,0) = 0 \text{ and } \exists M \in \mathbb{R}^+ \text{ such that } u.g(.,u) \leq 0 \text{ for } |u| \geq M \\ u(.,0) = u_0 \in L^1 \cap BV(\mathbb{R}). \end{cases} \tag{1}$$

Here, $BV(\mathbb{R})$ denotes the space of functions with bounded variation. The existence, uniqueness and stability theory of the entropy solutions of (1) has been established in [16]. Because of the presence of the source term on the right-hand-side, the derivation of a reliable and robust numerical scheme is still a difficult task, [17].

We introduce a function $a \in C^1(\mathbb{R})$ given by $a(x) = x$; (1) reads:

$$\partial_t u + \partial_x f(u) - g(a,u)\partial_x a = 0, \qquad x \in \mathbb{R}, t > 0. \tag{2}$$

It does not seem possible to transform this last equation into an homogeneous one. But we can see it as a very elementary Temple system, [14], if we add the static constraint on a: $\partial_t a = 0$. This system can be solved within the framework of Kružkov as long as a remains at least Lipschitz continuous, in particular, the Riemann problem is highly ambiguous since a so-called *nonconservative product*, [5], appears in the equations.

Because of the structure of the equation (2), it is fortunately possible to give a unique self-similar solution to the Riemann problem for the following singular *homogeneous* system:

$$\begin{cases} \partial_t u + \partial_x f(u) - \left[g(a^h,u)\partial_x a^h\right]_\Phi = 0, \\ \partial_t a^h = 0, \end{cases} \tag{3}$$

where a^h, $h > 0$ being given, is some piecewise constant approximation of a:

$$a^h(x) = j.h \text{ if } x \in C_j \stackrel{def}{=} \Big[(j-1/2)h, (j+1/2)h\Big[.$$

The reaction term becomes a Dirac comb and induces some stationary contact discontinuities in the solution of (3). According to the formalism of the *families of paths*, [5], the nonconservative term rendering the localized action of the source term reads:

$$\left[g(a^h,u)\partial_x a^h\right]_\Phi = \sum_{j\in\mathbb{Z}} h\left(\int_0^1 g(\Phi(s;V_j,V_{j+1})).ds\right)\delta(x-(j+1/2)h), \tag{4}$$

where $V = (u, a^h)$ and δ stands for the Dirac measure located at $x = 0$. The path Φ can be derived following different methodologies, see *e.g.* [6, 10, 11]; it reads:

$$[0,1] \ni s \mapsto \Phi\left(s; \begin{pmatrix} u \\ a^h \end{pmatrix}(x-0), \begin{pmatrix} u \\ a^h \end{pmatrix}(x+0)\right) = \begin{pmatrix} \tilde{u}(s) \\ \tilde{a}(s) \end{pmatrix}, \tag{5}$$

where

$$\partial_s f(\tilde{u})(s) = h.g(\tilde{a},\tilde{u})(s) \text{ and } \tilde{a}(s) = a((j+s)h). \tag{6}$$

This provides a way to solve in a unique manner the Riemann problem for (3) as long as f' does not vanish inside the domain. This pathological situation has been investigated by several authors, see *e.g.* [14, 20], and goes beyond the scope of these notes.

Therefore, under the assumption $f' \neq 0$, we can construct a sequence of piecewise constant approximations u^h to u, the entropy solution of (1) by means of any Riemann-based numerical scheme. This class contains the ones of Glimm, Godunov, Murman-Roe *etc* ... One advantage is that the reaction term is going to be treated directly inside the Riemann solver and the error coming from the projection step will be reduced. Then, it is well-known that the Godunov scheme is **exact** at steady-state for *homogeneous* problems. This entails also very interesting stability properties for the schemes since the source creates no time-step restriction (see [6, 11] for the case of the Godunov scheme). The last and very good point is that all these schemes will preserve the piecewise constant approximations of the static solutions derived in [21].

To be more precise, the Godunov scheme for (1) relying on (3), (5) consists in solving Riemann problems at each endpoint of the cells C_j, $j \in \mathbb{Z}$. Assuming that the condition $f' > 0$ holds true, it reads in standard notations:

$$u_j^{n+1} = u_j^n - \frac{\Delta t}{h}\Big[f(u_j^n) - f(u_{j-\frac{1}{2}}^n)\Big]. \tag{7}$$

And as a consequence of (4), (6), the following jump relation is of use:

$$u_{j-\frac{1}{2}}^n = \tilde{u}(1),\ \tilde{u}(0) = u_{j-1}^n. \tag{8}$$

Taking advantage of (6), one can rewrite the scheme (7) in an integral form

$$u_j^{n+1} = u_j^n - \frac{\Delta t}{h}\Big[f(u_j^n) - f(u_{j-1}^n)\Big] + \Delta t \int_0^1 g(\tilde{a},\tilde{u})(s).ds. \tag{9}$$

which emphasizes its ability to preserve **exactly** static regimes satisfying (6). In constrast, the "interface method" , [15], corresponds to a mid-point rule approximation of the integral term in (2.9) in the case $f'(u)$ does not change sign (see also [3, 22]). We can define u^h, a sequence of approximate solutions of (1) depending on h

$$u^h(x,t) = u_j^n \text{ for } x \in C_j,\ t \in [n\Delta t, (n+1)\Delta t[, \tag{10}$$

which is uniformly stable in $L^1 \cap BV(\mathbb{R})$ and $L^1(\mathbb{R})$-equicontinuous in time under the homogeneous CFL condition *only*, [6, 7, 11]. More precisely, we have the result:

Theorem 2.1. *Let u be the entropy solution in the sense of Kružkov of (1). Under the further assumptions $f'(u) > 0$ and $\sup_{j,n} |f'(u_j^n)|\Delta t \le h$, the following error estimate holds for all $t \in \mathbb{R}^+$:*

$$\|u^h(.,t) - u(.,t)\|_{L^1(\mathbb{R})} \le \exp(Nt)\mathbf{O}\left(TV(u_0)\sqrt{h} + h\right),$$

where $N = \sup_{x\in\mathbb{R};u,u^h}\left\{\partial_u g(x,u)\right\}$.

Another nice feature of this construction lies in its ability to be applied also to rather general strictly hyperbolic *systems* of balance laws.

3. Extension to hyperbolic systems of balance laws

We consider now the more difficult problem of finding a good numerical process to approximate the solutions of the following $n \times n$ strictly hyperbolic system:

$$\left\{\begin{array}{c} \partial_t U + \partial_x f(U) = g(x,U), \qquad U \in B(0,\varrho)_{\mathbb{R}^n},\ x \in \mathbb{R},\ t > 0, \\ U(0,.) = U_0 \in \left(L^1 \cap BV(\mathbb{R})\right)^n, \\ |\lambda_i(U)| \geq c > 0 \text{ for all } i \in \{1,\dots,n\}, \\ \sup_U \left\{\|g(.,U)\|_{\mathbb{R}^n} + \|\nabla_U g(.,U)\|_{M_n(\mathbb{R})}\right\} \leq \omega \in L^1 \cap L^\infty(\mathbb{R}), \end{array}\right. \tag{11}$$

each characteristic field being genuinely nonlinear or linearly degenerate. Of course, f and g are assumed the same smoothness than in the preceding section. This meets essentially the hypotheses of the existence theorem in [19] provided $TV(U_0)$ and $\|\omega\|_{L^1(\mathbb{R})}$ are small enough.

It is thus completely straightforward to follow the same canvas in order to derive a Well-Balanced scheme. We introduce a function $a(x) = x$ at the level of the continuous problem (11) and the equations read:

$$\partial_t U + \partial_x f(U) - g(a,U)\partial_x a = 0, \qquad \partial_t a = 0.$$

We go one step beyond and replace a by a piecewise constant approximation a^h as it is of use in any formally first-order scheme for (11); we are once again led to tackle a singular nonconservative system:

$$\left\{\begin{array}{c} \partial_t U + \partial_x f(U) - \left[g(a^h,U)\partial_x a^h\right]_\Phi = 0 \\ \partial_t a^h = 0. \end{array}\right. \tag{12}$$

The nonconservative products still have the form (4) and the family of paths extends very naturally, [6, 9, 10]:

$$[0,1] \ni s \mapsto \Phi\left(s; \begin{pmatrix} U \\ a^h \end{pmatrix}(x-0), \begin{pmatrix} U \\ a^h \end{pmatrix}(x+0)\right) = \begin{pmatrix} \tilde{U}(s) \\ \tilde{a}(s) \end{pmatrix}, \tag{13}$$

where

$$\partial_s f(\tilde{U})(s) = h.g(\tilde{a},\tilde{U})(s) \text{ and } \tilde{a}(s) = a((j+s)h). \tag{14}$$

The theorem of Lax generalized in [5] or a direct construction, [1], provide a unique self-similar solution to the Riemann problem for (12), (13), (14). Therefore, we are in position to use any Riemann-based numerical scheme as long as no eigenvalue of the Jacobian matrix of f vanishes. This last hypothesis is called the "non-resonance assumption" and appears also *e.g.* in [1, 19]. At this point, it is theoretically possible to derive a Godunov scheme, but a severe drawback will be the computational cost of the elementary Riemann problems resolution. A robust alternative based on flux-vector splitting has been presented for one- or two-dimensional problems in [9]. Here, we propose to derive a very simple linearized Well-Balanced Roe-type scheme for (11) following the ideas of [10, 24].

We want to avoid solving the Riemann problem (12) exactly for initial states (W_L, W_R). One idea is to introduce a local linearized problem whose simple self-similar solution shall be a reasonable approximation. Let us denote $A_\Phi(W_L, W_R)$ such a generalized Roe matrix and consider:

$$\partial_t W + A_\Phi(W_L, W_R)\partial_x W = 0, \tag{15}$$

where $W = (U, a^h)$ is somehow an "augmented unknowns vector". Then, according to the Roe conditions, a matrix A_Φ has the following form:

$$A_\Phi(W_L, W_R) = \begin{pmatrix} A(U_L, U_R) & -G_\Phi(W_L, W_R) \\ 0 & 0 \end{pmatrix}$$

where $A(U_L, U_R)$ is a usual Roe matrix for the homogeneous problem, *i.e.* satisfies the jump relation $A(U_L, U_R).(U_R - U_L) = f(U_R) - f(U_L)$. Therefore, we just have to find a reliable average value for the source terms; according to [10, 24], it reads:

$$G_\Phi(W_L, W_R) = \int_0^1 G(\Phi(s; W_L^*, W_R^*)).ds = \int_0^1 G(\tilde{a}, \tilde{U})(s).ds$$

where Φ is of course the path (13), (14). This choice still requires the complete solution of the Riemann problem in order to know the states (W_L^*, W_R^*) located at the endpoints of the steady nonconservative contact discontinuity. One simple possibility is thus to decide to fix $W_L^* = W_L$ and this is enough to ensure the Well-Balanced property.

A last problem could be the resolution of the steady differential system (14). Despite the fact that powerful O.D.Es solvers exist in the literature, one can decide to use very simple average values coming from piecewise-constant or -linear approximations of $\tilde{U}$. In the simplest case, one recovers the well-known and widely used value, [22]:

$$G_\Phi(W_L, W_R) = \frac{G(\tilde{a}(0), U_L) + G(\tilde{a}(1), U_R)}{2}.$$

We close this section mentioning that it is possible to use this approach also for theoretical purposes. Indeed, in [1], a ε-wave-front tracking algorithm relying on the Riemann solver (12), (13), (14) is studied. Some compactness properties are established under the same assumptions than in [19]. Moreover, the limit solutions as $\varepsilon, h \to 0$ determine a uniformly Lipschitz semigroup defined on a closed domain of integrable functions with small total variation. At last, its trajectories can be characterized by means of local integral estimates.

4. Numerical results

a. An initial boundary value problem.

As an illustrative test-case, we consider the following equation:

$$\begin{cases} \partial_t u + \partial_x \left(\frac{u^2}{2}\right) = k(x)u(1-u) \\ x \in [0,1] \text{ and } 0 < t \leq 3.5 \\ u(0,t) = 0.5; \; u(x,0) = 0 \end{cases} \tag{16}$$

As a consequence of the time-stabilization results proven in [21], the steady-state for (16), independently of the initial datum, is given by:

$$v(x) = 1 - \frac{1}{2}.\exp\left(-\int_0^x k(s).ds\right). \tag{17}$$

We compare the accuracy of the numerical stationary curves for the proposed nonconservative scheme (7), (8) and a classical time-splitting algorithm which corresponds approximately to a Godunov scheme for the "time-localized" equation, [12]:

$$\partial_t u + \partial_x \left(\frac{u^2}{2}\right) = k(x) \sum_{n \in \mathbb{N}^*} \Delta t.u(1-u).\delta(t - n\Delta t),$$

and reads (compare with (9)):

$$u_j^{n+1} = u_j^n - \frac{\Delta t}{2h}\left[(u_j^n)^2 - (u_{j-1}^n)^2\right] + \Delta t.k(jh).u_j^n(1-u_j^n). \tag{18}$$

Thus, when reaching the numerical steady-state, the discretization (18) achieves only an upwind first-order approximation of the static relation (6) for (16). We selected the following "delta-like" function:

$$k(x) = 20 \exp\left(-\frac{(x-0.5)^2}{\varepsilon}\right), \qquad \varepsilon = 0.01. \tag{19}$$

At time $T = 3.5$, all the waves are out of the domain $x \in [0,1]$. We display the numerical solutions in Figure 1. The advantages of this nonconservative approach are clearly noticeable. We took for both schemes (7), (18) $h = 0.03$ and $\Delta t = 0.02$.

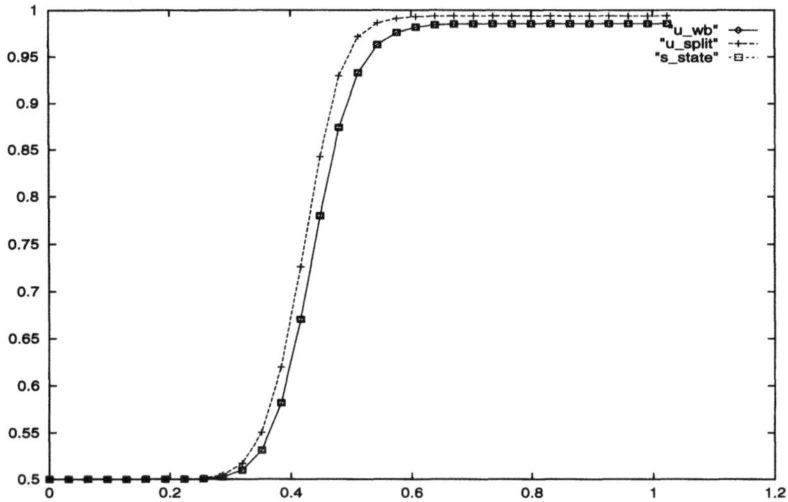

FIGURE 1. Exact and numerical solutions for (16), (19) at $T = 3.5$.

b. A supersonic flow in a divergent nozzle.

Let us consider a perfect gas flow in a cylindric smooth and slowly-varying variable cross-section duct. The variation of section is denoted by $x \mapsto S(x)$ with $S \in C^1([0,1];\mathbb{R})$. We get a one-dimensional system written in conservative form with a source term:

$$\left\{ \begin{array}{c} \partial_t(\rho S) + \partial_x(\rho u S) = 0 \\ \partial_t(\rho u S) + \partial_x(S(\rho u^2 + p)) = pS'(x) \\ \partial_t(\rho e S) + \partial_x(Su(\rho e + p)) = 0 \end{array} \right. \tag{20}$$

As usual, ρ, u, i are respectively the density, velocity, internal energy of the fluid; $e = i + \frac{u^2}{2}$, $p = (\gamma - 1)\rho i$ being the total energy and the pressure. It is convenient to introduce new variables for (20): $R = \rho S$, $P = pS$, $h = i + \frac{P}{R}$, $H = e + \frac{P}{R} = h + \frac{u^2}{2}$ and we denote W the augmented unknown vector: $W = (R, Ru, Re, S)$. The test is carried out with a supersonic divergent nozzle introduced in [26]. The cross-section of the nozzle is given by $S(x) = 1.398 + 0.347\tanh(8x - 4)$, $x \in [0,1]$. The initial data are those of a gas at rest: $\rho_0 = 0.125$, $u = 0$, $e_0 = i_0 = 2.5$, $\gamma = 1.4$, and we put a supersonic left boundary condition: $\rho_l = 0.502$, $u_l = 1.299$, $p_l = 0.381$. The flow is highly transient, but a stationary regime appears for both methods around the time $t \simeq 3.6$: see Figure 2 for a comparison between the present approach and a more conventional algorithm of the type (18). The stopping criterion was simply given by the decay of the residues in the L^∞ norm as in [9]. The parameters are fixed initially to $h = 0.02$ and $\Delta t = 0.007$; this corresponds to a maximal CFL number close to 1.

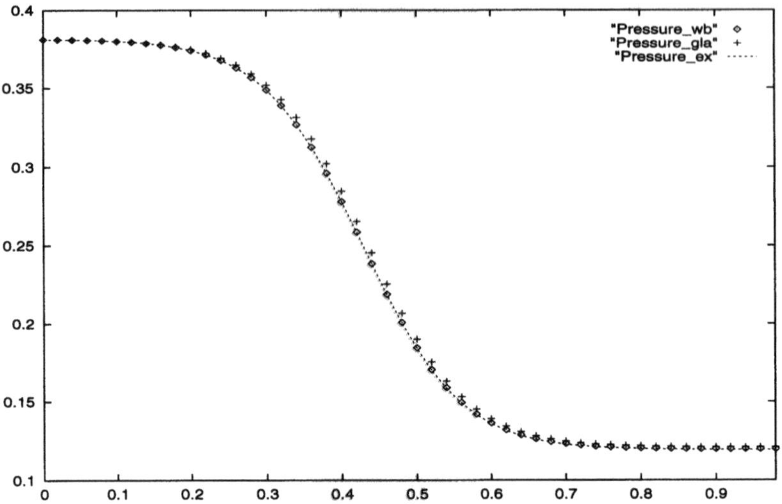

FIGURE 2. Theoretical and numerical values of the pressure for (20) at steady state.

References

[1] D. AMADORI, L. GOSSE AND G. GUERRA, *Global BV entropy solutions and uniqueness for hyperbolic systems of balance laws*, Quaderno Univ. di Milano 30/2000. Available at: http://www.math.ntnu.no/conservation/2000/050.html

[2] R. BOTCHORISHVILI, *Implicit kinetic schemes for scalar conservation laws*, Tech. report INRIA (2000).

[3] R. BOTCHORISHVILI, B. PERTHAME, A. VASSEUR, *Equilibrium schemes for scalar conservation laws with stiff sources*, Tech. report INRIA (2000).

[4] S. CLERC, Ph.D. Thesis, University Paris VI Jussieu (1997).

[5] G. DAL MASO, P.G. LEFLOCH AND F. MURAT, *Definition and weak stability of a nonconservative product*, J. Math. Pures Appl. **74** (1995) 483 – 548.

[6] L. GOSSE, Ph.D. Thesis, University Paris IX Dauphine (1997).

[7] L. GOSSE, *A priori error estimate for a well-balanced scheme for inhomogeneous scalar conservation laws*, C.R. Acad. Sc. Paris Série I **327** (1998) 467 – 472.

[8] L. GOSSE, *Sur la stabilité des approximations implicites des lois de conservation scalaires non-homogènes*, C.R. Acad. Sc. Paris Série I **329** (1999) 79 – 84.

[9] L. GOSSE, *A well-balanced flux splitting scheme designed for hyperbolic systems of conservation laws with source terms*, Comp. Math. Applic. **39** (2000) 135 – 159.

[10] L. GOSSE, *A well-balanced scheme using non-conservative products designed for hyperbolic systems of conservation laws with source terms*, Math. Mod. Meth. Appl. Sci. **11** (2001) 339 – 365.

[11] L. GOSSE, *Localization effects and measure source terms in numerical schemes for balance laws*, to appear in Math. Comp.

[12] L. GOSSE, *Localization effects and measure source terms for a quasilinear relaxing system*, preprint 2000-51, Università dell'Aquila.

[13] J. GREENBERG, A.Y. LEROUX, *A well balanced scheme for the numerical processing of source terms in hyperbolic equations*, SIAM J. Numer. Anal. **33** (1996) 1 – 16.

[14] E. ISAACSON AND B. TEMPLE, *Convergence of the 2×2 Godunov method for a general resonant nonlinear balance law*, SIAM J. Appl. Math. **55** (1995) 625 – 640.

[15] S. JIN, *A steady-state capturing method for hyperbolic systems with geometrical source terms* , to appear in Math. Mod. Numer. Anal.

[16] S.N. KRUŽKOV, *First order quasilinear equations in several independant space variables*, Math. USSR Sbornik, **10** (1970) 217 – 243.

[17] A.Y. LEROUX, *Riemann solvers for some hyperbolic problems with a source term*, Actes du CANum'98 (Arles, 1998), 75–90 (electronic), ESAIM Proc. **6** SMAI 1999. Available at: http://www.emath.fr/Maths/Proc/Vol.6/leroux/leroux.htm

[18] R.J. LEVEQUE, *Balancing source terms and flux gradients in high resolution Godunov methods: the quasi steady wave propagation algorithm*, J. Comp. Phys. **146** (1998) 346 – 365.

[19] T.P. LIU, *Quasilinear hyperbolic systems*, Comm. Math. Phys. **68** (1979) 141 – 172.

[20] T.P. LIU, *Nonlinear resonance for quasilinear hyperbolic equation*, J. Math. Phys. **28** (1987) 2593 – 2602.

[21] C. MASCIA AND A. TERRACINA, *Long-time behavior for conservation laws with source in a bounded domain*, J. Diff Eq. **159** (1999) 485 – 514.

[22] P.L. ROE, *Upwind differencing schemes for hyperbolic conservation laws with source terms* in **Nonlinear Hyperbolic Problems**, C. Carasso, P.-A. Raviart, D. Serre Eds, Lecture Notes in Mathematics **1270** Springer-Verlag (1986) 41–51

[23] G. RUSSO, *Central schemes for balance laws*, in this volume.

[24] I. TOUMI, *A weak formulation of Roe's approximate Riemann solver*, J. Comp. Phys. **102** (1992) 360–373.

[25] B. VANLEER, *On the relation between the upwind differencing schemes of Godunov, Engquist-Osher and Roe*, SIAM J. Sci. Statist. Comp. **5** (1984) 1 – 20.

[26] H.C. YEE, R.M. BEAM AND R.F. WARMING, *Boundary approximations for implicit schemes for one-dimensional inviscid equations of gas dynamics*, AIAA Journal **20** (1982) 1203–1211

L. Gosse
Dipartimento di Matematica Pura e Applicata,
Università degli Studi di L'Aquila,
Via Vetoio, Località Coppito,
I-67100 L'Aquila, Italy

International Series of Numerical Mathematics
Vol. 140, © 2001 Birkhäuser Verlag Basel/Switzerland

Multidimensional Artificial Dissipation for the Numerical Approximation of Conservation Laws

Thorsten Grahs and Thomas Sonar

Abstract. We adapt ideas from image processing, namely the concept of anisotropic diffusion and integrate them in the concept of stabilizing numerical schemes for conservation laws by adding some nonlinear artificial dissipation as introduced by Harten [5]. We resume this concept and supplement it with a multidimensional nonlinear anisotropic diffusion filters, to allow different dissipation directions. We combine this with investigations concerning the production of numerical entropy inside the scheme, taking entropy production as a measure for the dose of artificial dissipation necessary to stabilize the algorithm.

1. Governing equations

We consider hyperbolic conservation laws in two space dimensions

$$\partial_t u + \partial_x f(u) + \partial_y g(u) = 0 \tag{1}$$

where the fluxes f and g are assumed to be differentiable.

Independently of the smoothness of the initial conditions discontinuities develop in general within a finite time so that weak solutions u defined by

$$\int_\omega (u\partial_t \Phi + f(u)\partial_x \Phi + g(u)\partial_y \Phi)\; dx\, dy\, dt = 0$$

have to be considered. Here, $\omega \subset \mathbb{R}_0^+ \times \mathbb{R}^2$, $\Phi \in C_0^1(\omega)$ and $u \in L^1 \cap L^\infty(\omega)$. The additional entropy equality for the conservation law is

$$\partial_t \eta(u) + \partial_x \psi(u) + \partial_y \varphi(u) \leq 0 \tag{2}$$

with the entropy $\eta(u)$ and the corresponding entropy fluxes $\psi'(u) = \eta'(u)f'(u)$, $\varphi'(u) = \eta'(u)g'(u)$, (see [7, 14] for details).

2. Anisotropic diffusion

One interest in image processing is to remove small scale oscillations from the picture while simultaneously enhance the crucial structures of the image. To this purpose one tool in this framework is anisotropic diffusion. Since this is not an everyday tool in the framework of the numerical treatment of conservation laws we give a short overview.

Here one takes local information about the structure of the data and dissipates parallel to important features like edges or corners. A short description of the idea is given below, for details see [16].

One may start with a pre-smoothing of the data which means convolution with a Gaussian kernel to get smoothed data

$$U_\delta := K_\delta * U, \qquad K_\delta(x,y) := \frac{1}{2\pi\delta^2}\exp\left(-\frac{x^2+y^2}{2\delta^2}\right).$$

The parameter δ denotes the width of the Gaussian. The pre-smoothing is used to remove the small scale oscillations (noise) from the data in order to analyze only the crucial structures.

Since we are concerned with discrete numerical data pre-smoothing in our context means solving the discrete heat equation with initial data $U_{i,j}$ for $t \in [0,T]$ with $T = \frac{1}{2}\delta^2$ which is the numerical equivalent to Gaussian smoothing. We get discrete pre-smoothed data $U_{i,j;\delta}$. From this information we compute the structure tensor

$$J_0(\nabla U_{i,j;\delta}) := \nabla U_{i,j;\delta} \otimes \nabla U_{i,j;\delta}$$

which should give us some information about the local structure of the data. For the discrete nabla operator we use simple finite differences, i.e.

$$\nabla U_{i,j;\delta} := \left(\frac{U_{i+1,j;\delta} - U_{i-1,j;\delta}}{2\Delta x}, \frac{U_{i,j+1;\delta} - U_{i,j-1;\delta}}{2\Delta y}\right)^T.$$

The structure tensor is symmetric positive semidefinite and possesses eigenvalues

$$\lambda_1 = |\nabla U_{i,j,\delta}|^2, \quad \lambda_2 = 0$$

corresponding to the eigenvectors

$$\nu_1 = \nabla U_{i,j,\delta}, \quad \nu_2 = \nabla^\perp U_{i,j,\delta},$$

where

$$\nabla^\perp U_{i,j,\delta} := \left(-\frac{U_{i,j+1;\delta} - U_{i,j-1;\delta}}{2\Delta y}, \frac{U_{i+1,j;\delta} - U_{i-1,j;\delta}}{2\Delta x}\right)^T.$$

The eigenvalues and eigenvectors contain the information gained from the local analysis of the data, i.e. the eigenvectors define the direction across and parallel an edge, respectively, while the eigenvalues give the contrast in the eigendirections in the language of image processing.

In order to average the structure tensor data in the vicinity of each grid point component-wise convolution with the Gaussian Kernel of width ρ may be applied. This leads to

$$J_\rho(U_{i,j;\delta}) := K_\rho * \left(\nabla U_{i,j,\delta} \otimes \nabla^\perp U_{i,j,\delta}\right),$$

where ρ reflects the width of the averaging region. Hence we get the matrix

$$J_\rho =: V_\rho \Lambda_\rho V_\rho^{-1} =: \begin{pmatrix} j_{11} & j_{12} \\ j_{12} & j_{22} \end{pmatrix} \tag{3}$$

with eigenvalues

$$\lambda_{1,2;\rho} = \frac{1}{2}\left(j_{11} + j_{22} \pm \sqrt{(j_{11} - j_{22})^2 + 4j_{12}^2}\right)$$

and eigenvectors

$$\begin{aligned} \nu_{1;\rho} &= \begin{pmatrix} 2j_{12} \\ j_{22} - j_{11} + \sqrt{(j_{11} - j_{22})^2 + 4j_{12}^2} \end{pmatrix} \\ \nu_{2;\rho} &= \begin{pmatrix} j_{11} - j_{22} - \sqrt{(j_{11} - j_{22})^2 + 4j_{12}^2} \\ 2j_{12} \end{pmatrix} \end{aligned}$$

which again are orthogonal.

With this information one is able to construct a suitable diffusion matrix

$$D := V_\rho L V_\rho^{-1}$$

where V_ρ, according to (3), is the matrix of the eigenvectors. The diagonal matrix L will not, like Λ_ρ, contain the eigenvalues of J_ρ, but gives some freedom to steer the amount of diffusion required for the corresponding diffusion direction. This can be achieved by an anisotropic regularization of the Perona-Malik model [10] which can be found in Weickert's book [16]:

$$\begin{aligned} l_1 &= \vartheta(\lambda_{1;\rho}) \\ l_2 &= 1 \\ \vartheta(s) &= \begin{cases} 1 & ; \quad s \leq 0 \\ 1 - \exp\left(\frac{-C_m}{(s/\lambda)^m}\right) & ; \quad s > 0 \end{cases} \end{aligned} \tag{4}$$

The values of m and C_m are chosen in such a way, that the so-called flux $\Phi(s) := s\vartheta(s)$ is increasing in an interval $s \in [0, \lambda]$ and decreasing in $s \in]\lambda, \infty[$.

With this approach one is able to construct a discrete anisotropic nonlinear diffusion model for

$$\partial_t u = \nabla \cdot [D(u)\nabla u].$$

A detailed report on constructing such dissipation tensors for scalar conservation laws can be found in [2]. We point out that in our opinion the different smoothing procedures are not desirable in the framework of conservation laws. Since we are working with highly nonlinear approximations we require information from all scales of the data. Moreover, the computational cost for solving the heat equation is very high.

3. The underlying scheme

We consider a splitting scheme of the form

$$U_{i,j}^{n+1} = \mathcal{D}(\Delta t)\mathcal{C}(\Delta t)U_{i,j}^n$$

where $\mathcal{C}$ is the operator associated with the convective part (in our framework the Lax-Wendroff method) while $\mathcal{D}$ denotes the operator representing the nonlinear anisotropic diffusion filter. The above splitting is known to be of first order in time only but our considerations also work for more sophisticated splittings like Strang's. Note that time accuracy in our steady test case is by no means mandatory. The splitting means that in each time step we apply the discrete convection (i.e. the Lax-Wendroff scheme) first and the nonlinear diffusion part afterwards.

The anisotropic nonlinear diffusion equation was weighted with a factor $\Delta x \Delta y$ in order to guarantee consistency, i.e. we compute

$$\begin{aligned} w_{i,j}^0 &:= \mathcal{C}(\Delta t) U_{i,j}^n \\ w_{i,j}^1 &= w_{i,j}^0 - \Delta t \Delta x \Delta y \nabla \cdot \left(D(w_{i,j}^0) \nabla w_{i,j}^0 \right) \\ U_{i,j}^{n+1} &= w_{i,j}^1 . \end{aligned}$$

For the discretization of the Lax-Wendroff scheme (see [11, 15]) we use the form

$$\begin{aligned} &\frac{U_{i,j}^{n+1} - U_{i,j}^n}{\Delta t} + \frac{F_{i+1,j}^n - F_{i-1,j}^n}{2\Delta x} + \frac{G_{i,j+1}^n - G_{i,j-1}^n}{2\Delta y} = \qquad (5) \\ &\quad \frac{\kappa_x}{2} \left[A_{i+1/2,j}^n \left(\frac{F_{i+1,j}^n - F_{i,j}^n}{\Delta x} + \frac{G_{i+1/2,j+1/2}^n - G_{i+1/2,j-1/2}^n}{\Delta y} \right) \right. \\ &\quad \left. - A_{i-1/2,j}^n \left(\frac{F_{i,j}^n - F_{i-1,j}^n}{\Delta x} + \frac{G_{i-1/2,j+1/2}^n - G_{i-1/2,j-1/2}^n}{\Delta y} \right) \right] \\ &+ \frac{\kappa_y}{2} \left[B_{i,j+1/2}^n \left(\frac{F_{i+1/2,j+1/2}^n - F_{i-1/2,j+1/2}^n}{\Delta x} + \frac{G_{i,j+1}^n - G_{i,j}^n}{\Delta y} \right) \right. \\ &\quad \left. - B_{i,j-1/2}^n \left(\frac{F_{i+1/2,j-1/2}^n - F_{i-1/2,j-1/2}^n}{\Delta x} + \frac{G_{i,j}^n - G_{i,j-1}^n}{\Delta y} \right) \right] . \end{aligned}$$

Here we have used $F_{i,j}^n := f(U_{i,j}^n)$ and $A_{i\pm1/2,j}^n := (f'(U_{i\pm1,j}^n) + f'(U_{i,j}^n))/2$, $B_{i,j\pm1/2}^n := (g'(U_{i,j\pm1}^n) + g'(U_{i,j}^n))/2$, $G_{i\pm1/2,j\pm1/2}^n := (G_{i\pm1,j\pm1}^n + G_{i,j})/2$, et cetera. The grid coefficients are $\kappa_x := \Delta t/\Delta x$, $\kappa_y := \Delta t/\Delta y$.

If σ denotes the maximum value of A and B and if we assume $\Delta x = \Delta y =: h$, then the Lax-Wendroff scheme can be shown to be linearly stable under the somehow pessimistic CFL-condition $\frac{\Delta t}{h}\sigma \le \frac{1}{2\sqrt{2}}$ (see [15]).

4. The entropy based diffusion tensor

Related to the foregoing section we are interested in constructing a nonlinear diffusion tensor with data dependent diffusion directions. In this section we describe a diffusion matrix based on the entropy production of the underlying scheme. This means, we distinguish between regions where the entropy inequality is violated, associated with oscillations and smooth regions .

As a discrete analogue of (2) we introduce the discrete entropy inequality

$$\partial_t \eta_{i,j} + \frac{1}{\Delta x}\left(\Psi_{i+1/2,j}^n - \Psi_{i-1/2,j}^n \right) - \frac{1}{\Delta y}\left(\Phi_{i,j+1/2}^n - \Phi_{i,j-1/2}^n \right) \le 0,$$

where $\eta_{i,j}(t)$ is the entropy located at a grid point and $\Psi^n_{i+1/2,j}$ and $\Phi^n_{i,j+1/2}$ are numerical entropy fluxes consistent with ψ and φ in just the same sense as the numerical fluxes are with the fluxes, see [8, 9, 12, 13]. Using the chain rule we get

$$\partial_t \eta(u) = \partial_u \eta \partial_t u$$

and write

$$\begin{aligned} E_{i,j} := \quad & -\eta'_{i,j}\left(\tfrac{1}{\Delta x}\left[\hat{f}_{i+1/2,j} - \hat{f}_{i-1/2,j}\right] + \tfrac{1}{\Delta y}\left[\hat{g}_{i,j+1/2} - \hat{g}_{i,j-1/2}\right]\right) \\ & + \left(\tfrac{1}{\Delta x}\left[\Psi^n_{i+1/2,j} - \Psi^n_{i-1/2,j}\right] + \tfrac{1}{\Delta y}\left[\Phi^n_{i,j+1/2} - \Phi^n_{i,j-1/2}\right]\right) \end{aligned} \tag{6}$$

for the numerical entropy inequality for the cell (i,j). Here $\hat{f}, \hat{g}$ denotes the corresponding numerical fluxes of the associated scheme, i.e. the Lax-Wendroff fluxes.

In smooth regions we are looking for a scheme which introduces the necessary stabilizing diffusion in the direction of the characteristic curves. This seems reasonable since this is the direction where the information goes.

In the vicinity of shocks the characteristic curves run into the shock and smoothing along the characteristic lines means smoothing across the front and hence smearing of the shock and loss of accuracy. Hence in the vicinity of shock fronts oscillations occure due to the underlying second order scheme. This is related to regions where the entropy inequality is violated, (i.e. $E_{ij} > 0$) even if at the shock self (6) is fulfilled. Thus we construct a diffusion tensor with diffusion direction parallel to this front in order to dissipate parallel to the shock.

With (6) we are able to state an indicator for the numerical entropy production

$$e^+_{i,j} := \begin{cases} E_{i,j} & ; E_{i,j} > 0 \\ 0 & \text{else,} \end{cases} \tag{7}$$

used to specify regions where the discrete entropy inequality is violated.

Note that numerical rarefaction waves also violates the entropy inequality. Nevertheless, the gradients there are too small, so that the scheme behaves quite well in this regions.

4.1. Construction of the diffusion tensor in smooth regions

Since a simple central scheme is unstable we need a kind of background diffusion that guarantees a stable behavior in the flow computation. In smooth regions, i.e. $E_{i,j} = 0$, the flow follows the characteristics and we are interested to design a diffusion tensor which reveals this property and gives all dissipation parallel to the characteristic curves, while avoiding dissipation perpendicular to them.

Since the characteristic equations for (1) are given by $dx/ds = f' =: A(u)$ and $dy/ds = g' =: B(u)$, we end up with $dy/dx = B/A$. The characteristic curves are described locally by the vector $v := (A(u),\, B(u))^T$.

To construct the diffusion tensor D_C with diffusion direction parallel and perpendicular to the characteristic curves, we take the vectors $V := (v_1, v_2) := (v, v^{\perp})$ as the eigenvectors of the tensor and define (compare (3)):

$$D_C \quad := \quad V \Lambda V^{-1} \tag{8}$$

This reads as

$$\begin{aligned} D_{\mathcal{C}} &:= \begin{pmatrix} A & -B \\ B & A \end{pmatrix} \begin{pmatrix} \lambda_1 & 0 \\ 0 & \lambda_2 \end{pmatrix} \frac{1}{A^2+B^2} \begin{pmatrix} A & B \\ -B & A \end{pmatrix} \\ &= \frac{1}{A^2+B^2} \begin{pmatrix} \lambda_1 A^2 + \lambda_2 B^2 & (\lambda_1 - \lambda_2)AB \\ (\lambda_1 - \lambda_2)AB & \lambda_1 B^2 + \lambda_2 A^2 \end{pmatrix}. \end{aligned}$$

Since we are interested in restricting diffusion perpendicular to the characteristics we set the eigenvalue corresponding to the vector v_1 to one, i.e. $\lambda_1 = 1$ and the eigenvalue corresponding to v_2 to null, i.e. $\lambda_2 = 0$. Hence we get

$$D_{\mathcal{C}} := \frac{1}{A^2+B^2} \begin{pmatrix} A^2 & AB \\ AB & B^2 \end{pmatrix}.$$

For a numerical analyst this looks very familiar and we are able to state the following

Lemma:

> The Lax-Wendroff scheme in the case of constant coefficients is the scheme with dissipation parallel to the characteristic curves.

Proof.
If we linearize (1), i.e. $u_t + Au_x + Bu_y = 0$ the Taylor series expansion reads

$$u(x, t+h) = u(x,t) + hu_t(x,t) + \tfrac{1}{2}h^2 u_{tt} + \mathcal{O}(h^3).$$

If we plug in the time derivate $u_t = -Au_x - Bu_y$, we get

$$\begin{aligned} u(x,t+h) &= u(x,t) - h(Au_x + Bu_y) \\ &\quad + \tfrac{1}{2}h^2[(A^2u_x + ABu_y)_x + (ABu_x + B^2u_y)_y] + \mathcal{O}(h^3) \\ &= u(x,t) - h(Au_x + Bu_y) + \tfrac{1}{2}h^2 \nabla \cdot [D_{\mathcal{LW}} \nabla u] + \mathcal{O}(h^3) \end{aligned}$$

with

$$D_{\mathcal{LW}} = \begin{pmatrix} A^2 & AB \\ AB & B^2. \end{pmatrix}$$

□

Hence the Lax-Wendroff dissipation model is an excellent device for smooth regions since it guarantees dissipation parallel to characteristic curves (Note that this is only true using an anisotropic discretization which can be found in [16]). So we weight our dissipation tensor with $A^2 + B^2$ and obtain

$$D_{\mathcal{C}} = D_{\mathcal{LW}}.$$

4.2. Construction of the diffusion tensor in non smooth regions

Having constructed the diffusion tensor for the smooth regions of the data we look at domains where the entropy inequality is violated. This corresponds to entropy production. Now we compute the discrete gradient of the cell entropy production:

$$\nabla e^+_{i,j} = \tilde{e_1} := \begin{pmatrix} \dfrac{e^+_{i+1,j} - e^+_{i-1,j}}{2\Delta x} \\ \dfrac{e^+_{i,j+1} - e^+_{i,j-1}}{2\Delta y} \end{pmatrix}.$$

Based on this the normed eigenvectors of the dissipation tensor are computed as parallel and perpendicular directions to the gradient of the entropy production, i.e.

$$e_1 := \frac{1}{|\tilde{e_1}|}\tilde{e_1}, \quad e_1^\perp := \frac{1}{|\tilde{e_1}|}\tilde{e_1}^\perp \tag{9}$$

Now we are able to denote the matrix of eigenvectors $V := (e_1, e_2) := (e_1, e_1^\perp)$ and get the diffusion tensor

$$D_{\mathcal{E}} = V\Lambda V^{-1}.$$

The question is how to choose the eigenvalues which control the amount of dissipation with their size. We choose them according to (4) as

$$\begin{aligned} \lambda_1 &= (A^2 + B^2)(1 - \exp\left(\frac{-C_m}{(0.2 \cdot e^+_{i,j}/e^{\max})^6}\right)) \\ \lambda_2 &= (A^2 + B^2). \end{aligned} \tag{10}$$

Hence we use the same amount of dissipation as in $D_{\mathcal{C}}$. In areas where the entropy gradient grows, the diffusion across this gradient tends to zero. Note that there is a scaling with $e^{\max}$ which is the maximal entropy production in the whole domain, i.e. $e^{\max} := \max_{i,j} e^+_{i,j}$.

4.3. The resulting nonlinear filter

Having constructed the diffusion tensors $D_{\mathcal{C}} = D_{\mathcal{L}W}$ and $D_{\mathcal{E}}$ we are able to design a numerical scheme with the following features.

- In smooth region dissipation parallel to the characteristics .
- Near discontinuities dissipation parallel to the important features.
- Avoid smearing and enhance the shock .
- Remove oscillations in the vicinity of shocks.

Near shocks the dissipation matrix for the smooth part reduces, so the governing dissipation occurs parallel to the shock. This enables us to add both dissipation tensors, i.e.

$$D = D_{\mathcal{C}} + D_{\mathcal{E}}.$$

The explicit nonnegative discretization is based on avoiding negative diffusion coefficients and to take care of the diagonal directions which can guarantee this (see [16, 2, 4]. Having designed this nonlinear filter the resulting scheme reads as

$$\begin{aligned} w^0_{i,j} &:= \mathcal{C}(\Delta t)U^n_{i,j} \\ U^{n+1}_{i,j} &= w^0_{i,j} - \Delta t\Delta x\Delta y\nabla\cdot\left(D(w^0_{i,j})\nabla w^0_{i,j}\right) \end{aligned} \tag{11}$$

5. Numerical results

In this section we present numerical results of the new scheme (11). We consider the scalar conservation law (1) with fluxes $f(u) = u^2/2, \quad g(u) = u$ and entropy $\eta = u^2/2$. The test case we use is a initial-boundary value problem with

$$u(x,y,0) = \begin{cases} 1.5 & ; \quad x = 0 \\ -2.5x + 1.5 & ; \quad y = 0 \\ -1.0 & ; \quad x = 1 \end{cases}$$

at the boundary and $u(x,0) = 0$ at the interior with 50×50 points. We determine the boundary condition on the upper side of the unit square through simple extrapolation, which results in a steady solution as shown in Figure 1. On the right we see the numerical solution for an unfiltered Lax-Wendroff scheme. Succeeding

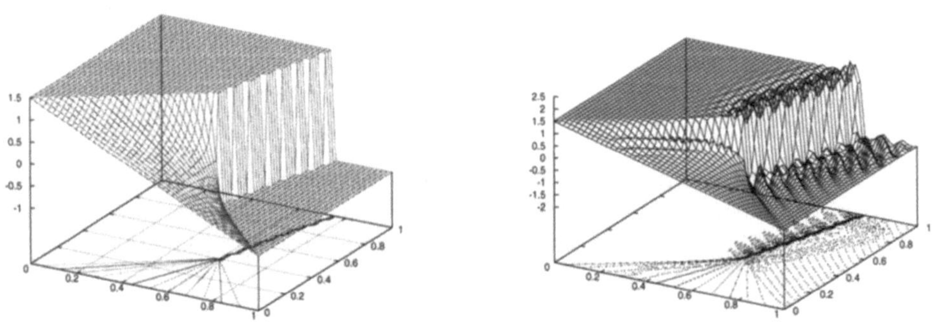

FIGURE 1. Exact solution of the test problem (left) and the solution of the unfiltered Lax-Wendroff Scheme (right)

we present the development of the initial value problem computed with the filter algorithm under CFL condition of 0.35.

6. Conclusions

We have presented a numerical algorithm for two-dimensional scalar conservation laws based on the discrete entropy production as an indicator for different diffusion areas. Using this information we have constructed a nonlinear diffusion filter to remove the spurious oscillations occurring from second order methods. A similar approach was used for the Euler equations [3] but the results are not quite satisfying. So a deeper inside into the steering of the dissipation is needed. This is one of the major tasks for the future work.

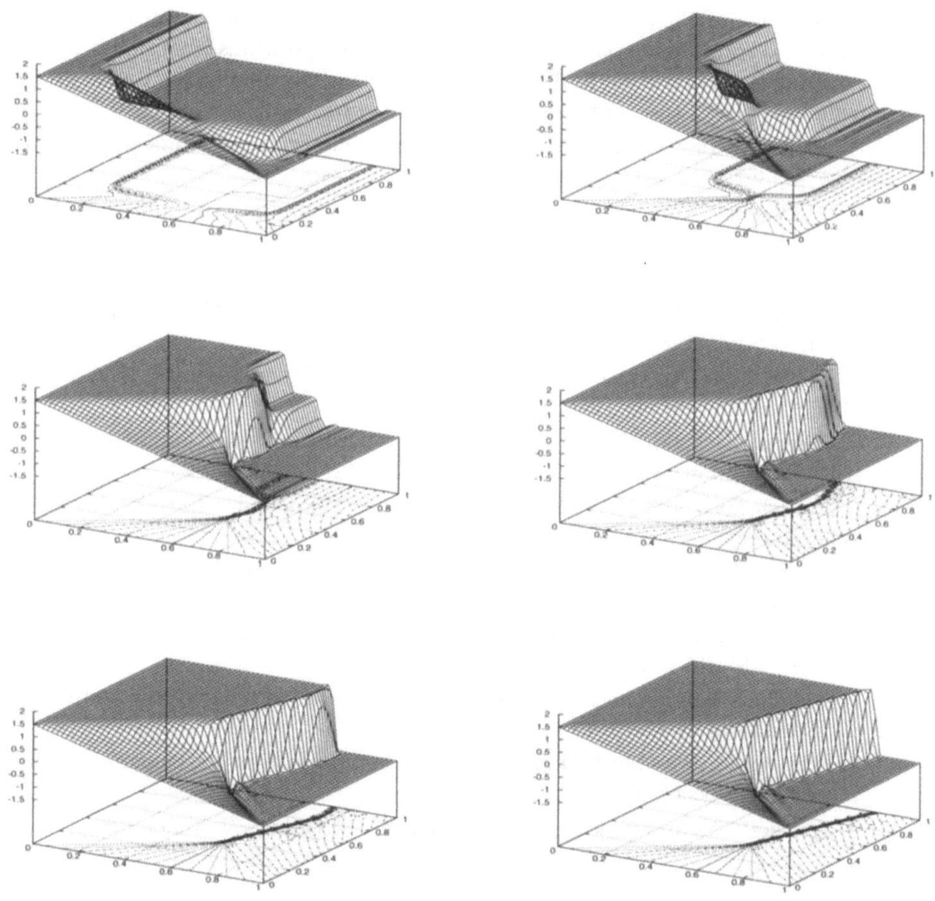

FIGURE 2. Solution after 50,100,150,200,250 and 1000 time steps

References

[1] E. Godlewski and P.-A. Raviart, "Hyperbolic Systems of Conservation laws", Ellipses 1991.

[2] Th. Grahs, A. Meister and Th. Sonar, "Image processing for numerical approximation of conservation laws: nonlinear anisotropic artificial diffusion", Reihe F, Computational Fluid Dynamics and Data Analysis 8, Hamburger Beiträge zur Angewandten Mathematik, 1998 (accepted for publication in SIAM J. Sci. Comput.).

[3] Th. Grahs, A. Meister and Th. Sonar, "Nonlinear anisotropic artificial dissipation - Characteristic filters for computation of the Euler equations", in Finite Volumes for Complex Applications II, Eds.: R. Vilsmeier, F. Benkhaldoun, D. Hänel, HERMES Science Publication 1999

[4] Th. Grahs and Th. Sonar, "Entropy controlled artificial anisotropic diffusion for the numerical solution of conservation laws based on algorithms from image processing" accepted for publication in J.Vis.Comm. *special issue on PDE in image processing, computer vision and computer graphics*, to appear.

[5] A. Harten, "High resolution schemes for hyperbolic conservation laws", J. Comp. Phy. 49, 357-393, 1983.

[6] R.J. LeVeque, "Numerical Methods for Conservation Laws", Birkhäuser Verlag, 1992.

[7] A. Majda and S. Osher, "Numerical viscosity and the entropy condition", Comm. Pure Appl. Math., 32, 797-838, 1979.

[8] M.L. Merriam, "Smoothing and the second law", Comp. Meth. App. Mech, Eng. 64(1), 177-193, 1987.

[9] M.L. Merriam, "An entropy-based approach to nonlinear stability", NASA Technical Memorandum 101086, 1989.

[10] P. Perona, J. Malik, "Scale space and edge detection using anisotropic diffusion", IEEE Trans. Pattern Anal. Mach. Intell., Vol. 12, 629-639, 1990.

[11] R.D. Richtmyer and K.W. Morton, "Difference Methods for Initial-value Problems" Wiley-Interscience, 1967.

[12] Th. Sonar, "Entropy production in second-order three-point schemes", Num. Math. 62, 371-390, 1992.

[13] E. Tadmor, "Numerical viscosity and the entropy condition for conservative difference schemes", Math. Comp. 43, 369-381, 1984.

[14] E. Tadmor, "The numerical viscosity of entropy stable schemes for systems of conservation laws I", Math. Comp. 49, 91-103, 1987.

[15] Shokin, Yu.I., "The Method of Differential Approximation", Springer Series in Computational Physics, 1983

[16] J. Weickert, "Anisotropic Diffusion in Image Processing", Teubner 1998.

T. Grahs
Technische Universität Braunschweig,
Pockelsstr. 14, D-38106 Braunschweig, Germany
`t.grahs@tu-bs.de`

T. Sonar
Technische Universität Braunschweig,
Pockelsstr. 14, D-38106 Braunschweig Germany
`t.sonar@tu-bs.de`

Author Index

MIX
Papier aus verantwortungsvollen Quellen
Paper from responsible sources
FSC® C105338

If you have any concerns about our products,
you can contact us on
ProductSafety@springernature.com

In case Publisher is established outside the EU,
the EU authorized representative is:
Springer Nature Customer Service Center GmbH
Europaplatz 3, 69115 Heidelberg, Germany

Printed by Libri Plureos GmbH
in Hamburg, Germany